BARRON'S

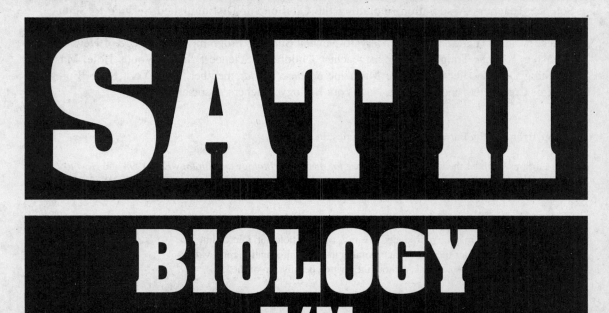

HOW TO PREPARE FOR THE

SAT II

BIOLOGY
E/M

13TH EDITION

Including:

Modern Biology in Review

Maurice Bleifeld
Principal Emeritus
Martin Van Buren High School
Queens Village, New York

BARRON'S

About the Author

Maurice Bleifeld is principal emeritus of Martin Van Buren High School, Queens Village, N.Y. He was also formerly principal of Benjamin Franklin High School, New York City, and of Samuel Huntington Junior High School, Jamaica, N.Y.; chairman of the Department of Biological Sciences at Newtown High School, Elmhurst, N.Y.; and teacher of biology at the Bronx High School of Science, New York City. He is the recipient of several awards from the National Science Teachers Association, including the Citation for Distinguished Service to Science Education. Among his other former positions and activities are the following: Lecturer on Science Education, New York University; Harvard-Newton Summer Program; City College of New York Intensive Teacher Training Program; teacher, Children's Science School, Woods Hole, Mass.; Teaching Guide writer, Scholastic Magazine *Science World;* member, New York State Biology Regents Committee; and president, New York Biology Teachers Association.

© Copyright 2002 by Barron's Educational Series, Inc.

Prior editions © under the title *How to Prepare for the SAT II Biology and Biology E/M* 1998 and *How to Prepare for the College Board Achievement Test Biology* 1994, 1991, 1987, 1984, 1981, 1978, 1973, 1972, 1970, 1969, 1963 by Barron's Educational Series, Inc.

All rights reserved.
No part of this book may be reproduced in any form, by photostat, microfilm, xerography, or any other means, or incorporated into any information retrieval system, electronic or mechanical, without the written permission of the copyright owner.

All inquiries should be addressed to:
Barron's Educational Series, Inc.
250 Wireless Boulevard
Hauppauge, New York 11788
http://www.barronseduc.com

Library of Congress Catalog Card No. 2001056708
International Standard Book No. 0-7641-1788-2

Library of Congress Cataloging-in-Publication Data

Bleifeld, Maurice.
How to prepare for the SAT II, biology E/M / Maurice Bleifeld.—13 ed.
p. cm.
Rev. ed. of: How to prepare for the SAT II, biology and biology E/M. 12th ed. c1998.
ISBN 0-7641-1788-2
1. Biology—Examinations, questions, etc. 2. College entrance achievement tests—United States—Study guides. I. Title: SAT II, biology E/M. II. Title: SAT two, biology E/M. III. Title: SAT 2, biology E/M. IV. Bleifeld, Maurice. How to prepare for the SAT II, biology and biology E/M. V. Title.
QH316 .B55 2002
570'.76—dc21

2001056708

PRINTED IN THE UNITED STATES OF AMERICA

14 13 12 11 10 9 8

Contents

Preface to the Thirteenth Edition

This latest revision, the 13th edition, prepares students for the College Board SAT II: Biology E/M Subject Test. The test offers a choice between questions on Biology E (Ecology) and Biology M (Molecular Biology).

This book is organized into four parts.

Part I consists of two chapters that offer general information about the test, advice on when to take the test, the content of the test, and test score Conversion Table. Also included are test-taking strategies, instructions on how to record answers, and sample questions with hints for answering seven different types of multiple-choice questions.

Part II contains a Mini-Diagnostic Test, constructed specifically for this book to help the student diagnose his or her strengths and weaknesses. It is similar to SAT II: Biology E/M Subject Test in content, type of questions, and degree of difficulty. An Answer Key and explanations of all the answers are provided. A revised chart for self-evaluation identifies specific weaknesses. A special section analyzes the student's overall performance on the Mini-Diagnostic Test.

Part III—Modern Biology in Review—provides a comprehensive review of the subject. It has been revised and updated to include the latest information in all areas of biology, such as: Completion of the Human Genome Project, and the Plant Genome Project; other genome progress; stem cells; new evidence of global warming at the North Pole; cloning limitation; ability of brain cells to reproduce; reducing atmospheric carbon dioxide; Archaea ancestor; the drug Ecstacy; advances in the fight against cancer and Alzheimer's disease; new diabetic treatment; and anthrax problems.

Throughout the book, key definitions, charts, tables, illustrations, and problems present the essential concepts in clear fashion. Each section is followed by a multiple-choice review quiz, with an Answer Key and a set of explanations for all answers.

Part IV—Four Practice Tests—contains tests similar to the College Board test in content, type of questions, and degree of difficulty. Each practice test, which was developed specifically for this book, includes challenging questions based on the five official subject areas covered by the test: (1) Cellular and Molecular Biology; (2) Ecology; (3) Classical Genetics; (4) Organismal Biology; (5) Evolution and Diversity. In addition to these general core questions, there are also specific questions on Biology E (Ecology) and Biology M (Molecular Biology). Each test is followed by an Answer Key and explanations of all the answers. A Self-Evaluation Chart after each test provides the equivalent College Board score and further opportunity for learning.

This revision takes into account the current New York State Biology Syllabus, and deals comprehensively with each of its units. Consequently, the book may be used to good advantage by students studying this course and by those who are preparing for the New York State Regents Examination in Biology.

Because of its comprehensive treatment of modern concepts in biology, the book will also be useful to those preparing for College Board tests, scholarship examinations, standardized tests, college biology courses, teaching examinations, and other challenging tests in biology. Despite the technical and advanced nature of some of the material, the terminology and vocabulary have been kept as simple as possible.

Grateful acknowledgment is made to the students, teachers, and scientists who were kind enough to offer a number of constructive comments and suggestions which have been incorporated in this edition

A special debt of gratitude is due my wife, Belle K. Bleifeld, for her valuable help in typing the manuscript. This book is dedicated to the generation of Max, Annie, and Sam Kramer, Dylan and Sydney Diamond, and Spencer Bleifeld, who will be using it on their road to higher learning.

Maurice Bleifeld

TAKING SAT II: BIOLOGY E/M SUBJECT TEST

PART

ABOUT SAT II: BIOLOGY E/M SUBJECT TEST

CHAPTER

1

General Information

A high school student preparing to apply for college admission may be asked to take one or more college entrance examinations. Among these is SAT II: Biology E/M Subject Test. The College Board describes it and provides sample questions in a booklet entitled *Taking the SAT II: Subject Tests*. You can obtain a copy at your high school guidance office, or you can write to College Board SAT Program, P.O. Box 6200, Princeton, NJ 08541-6200.

Should You Take SAT II: Biology E/M Subject Test?

The answer to the above question depends on which colleges you are considering applying to. Some of them specify certain tests you are required to take. Others give you a choice. The best way to find out about the requirements of the various colleges regarding the SAT II tests is to look in their catalogs.

If you have a choice of SAT II: Subject Tests and if you have made good progress in your biology course, you may want to take the test in biology. Another reason for choosing the biology test is that, if a college requires an SAT II: Subject Test in one of the sciences, you may believe that your best chance of getting a good score is in biology.

Although the college you are interested in may not require Subject Tests, it may consider the Biology E/M Test useful as a way of learning more about your academic background.

When you take the Biology E/M Test, you can choose either the Biology E Test, which emphasizes ecology, or the Biology M Test, which emphasizes molecular biology. Both tests start with a common core of 60 questions, which is followed by a collection of 20 questions in the specialized section of either Biology E or Biology M. Thus you answer a total of 80 questions when you take the test. You have one hour in which to answer them.

When Should You Take the Test?

The College Board recommends that you take the SAT II: Biology E/M Test right after you complete your biology course, while the material is still fresh in your mind. The Biology E/M Test is offered six times a year: in October, November, December, January, May, and June.

Be sure to find out when the colleges you are interested in require that the test be taken. They usually specify that you take the test no later than December or January of your senior year. If you are interested in an Early Decision program, you should take the test no later than June of your junior year. On the other hand, if your selected colleges use the SAT II: Subject Test only for guidance purposes in placement decisions, the catalogs may state that you can take the test as late as May or June of your senior year.

You may, if you choose, take one, two, or three SAT II: Subject Tests at one sitting. Each test lasts one hour and consists of a varying number of multiple-choice questions, except for SAT II: Writing. The College Board recommends that students who earn high scores on SAT II: Subject Tests be encouraged to take Advanced Placement courses.

How Do You Register for the Test?

Copies of the *Registration Bulletin for the SAT Program* are sent to high schools each year. You should be able to pick up a copy of the Registration Form at the guidance office. You can also write College Board SAT Program, P.O. Box 6200, Princeton, NJ 08541-6200, or call, Monday to Friday, (609) 771-7600, 8:30 a.m. to 9:30 p.m. (Eastern time).

The *Bulletin* also contains information about having your score reported to the colleges of your choice. Be sure to send in your Registration Form at least five weeks before the test date; otherwise you will have to pay a late fee.

You can also register online at www.collegeboard. com or you can *re*register by telephone (800) SAT-SCORE if you previously registered for tests while in high school. When you register, you must indicate which Subject Tests you are planning to take.

Content of the Biology E/M Test

You have one hour in which to answer 80 multiple-choice questions. The questions are based on a one-year course in high school biology that is on a suitable level for college preparation. Since the content of such a course differs throughout the country, you may find some questions on topics that are unfamiliar to you. But don't let this worry you. It is not expected that a student will know all the answers on any one test. It may be of some comfort to realize that the other students taking the test undoubtedly find it just as challenging.

The following table describes the scope of the test:

Biology E/M List of Topics Covered	Approximate Percentage of Test
BASIC CORE:	
Cellular and Molecular Biology	
Cell structure and function, mitosis, photosynthesis, cellular respiration, enzymes, molecular genetics, biological chemistry	12

Ecology	
Energy flow, cycles, populations, communities, ecosystems, biomes	12
Classical Genetics	
Meiosis, Mendelian, inheritance patterns	10
Organismal Biology	
Structure, function, and development of plants and animals, animal behavior	30
Evolution and Diversity	
Origin of life, evidence of evolution, natural selection, patterns of evolution, classification	11
PLUS:	
Ecology/Evolution Section (Biology E Test)	25
OR	
Molecular/Evolution Section (Biology M Test)	25

The test is different from most comprehensive examinations taken by high school students in that it tests not only knowledge of the subject matter but also the ability to apply that knowledge and to reason with it. These skills are described below:

Biology E/M List of Skills Specifications	Approximate Percentage of Test
Knowledge of Fundamental Concepts: knowledge of factual information and terminology	30
Application: understanding concepts; evaluating information; applying knowledge; solving problems using mathematics	35
Interpretation: using data to form conclusions; identifying unstated relationships	35

Helpful Background: The Biology E/M Test assumes that you have also taken a one-year course in algebra which included such concepts as ratios and direct and inverse proportions. In addition, it is assumed that you have had experience with laboratory procedures, enabling you to develop reasoning and problem-solving skills. Some questions may ask you to interpret or draw conclusions from graphs, tables, and other experimental data. Incidentally, because mathematical calculations are limited to simple arithmetic, you are not allowed to use a calculator. Also, the metric system is referred to in these tests.

Deciding on Biology E or Biology M: You should take the test for which you feel best prepared. If you feel more prepared for answering questions that deal with biological communities, cycles, populations, and energy flow, you may want to choose Biology E. If you feel more prepared for answering questions that deal with cellular structure, photosynthesis, molecular genetics, and cellular processes, you may want to choose Biology M. So, if you take Biology E, use only ovals 1 to 80 on your answer sheet. But, if you take Biology M, use only ovals 1 to 60 (Basic Core) and 81 to 100 on your Answer Sheet. On the line under the words Subject Test print Biology-E or Biology-M.

On the day you appear for the test, you can indicate your choice of E or M by marking the appropriate grid on your answer sheet. Because there is a common core of 60 questions for both the Biology E and the Biology M, you are not allowed to take both tests on the same date. If you do, your test results will be cancelled. However, if you want to take both Biology E and Biology M, you can do so on two different dates.

How the Test Is Scored

For each correct answer, you are given one point. For each wrong answer, you lose a fraction of a point. If you omit a question, it is not counted. If you mark more than one answer to a question, that question also is not counted. After all your points are totaled, your paper is given a Raw Score. This Raw Score is then converted to an equivalent College Board Score, which can range from a high of 800 to a low of 200. This is the score that is sent to the colleges you have listed, as well as to your own high school.

When you take the four Practice Tests in this book, use the Self-Evaluation Chart at the end of each of these tests to determine your Raw Score. Then locate your equivalent College Board Score on the Conversion Table on page 6. Simply look for your Raw Score and find the corresponding College Board Score in the column to the right of it. For example, a Raw Score of 65 is equivalent to a College Board Score of 740. This table may differ slightly from year to year, depending on the content of the particular SAT II: Subject Test.

After the Test

You can expect to receive the score of your tests in the mail about three weeks after the test date. The score is also mailed to your high school and to colleges and scholarship programs you listed on your registration form. If you would like your score sooner, use the *Score By Phone* service. The Registration Bulletin has a description of this service, and available dates when scores may be obtained.

How This Book Can Help You Prepare for the Test

Now that you have some general information about the test, you can get ready for the test itself by following these steps:

- Read Chapter 2 on test-taking strategies.
- Try to answer the sample questions.
- Study the hints for answering the seven different types of questions.
- Take the Mini-Diagnostic Test in Part II.
- Score your results.
- Fill out the Self-Evaluation Chart to pinpoint your weaknesses and plan your review.
- Review any weak areas of biology by referring to Part III, Modern Biology in Review.
- Take the four Practice Tests in Part IV to increase your ability to answer the different types of questions and to give you important clues to your strengths and weaknesses.
- Score your results and fill out the Self-Evaluation Chart in each case.

WATCH YOUR SCORE IMPROVE!

TYPICAL CONVERSION TABLE TO OBTAIN COLLEGE BOARD SCORES FOR BIOLOGY E/M TESTS

Raw Score	College Board Scaled Score	Raw Score	College Board Scaled Score	Raw Score	College Board Scaled Score
80	800	45	610	10	390
79	800	44	610	9	380
78	800	43	600	8	370
77	800	42	590	7	360
76	800	41	590	6	350
75	800	40	580	5	340
74	780	39	580	4	340
73	780	38	570	3	330
72	780	37	570	2	330
71	770	36	560	1	320
70	770	35	550	0	310
69	760	34	550	−1	310
68	760	33	540	−2	300
67	750	32	540	−3	300
66	750	31	530	−4	290
65	740	30	520	−5	280
64	740	29	520	−6	280
63	730	28	510	−7	270
62	720	27	510	−8	270
61	710	26	500	−9	260
60	710	25	490	−10	260
59	700	24	490	−11	250
58	700	23	480	−12	250
57	690	22	480	−13	240
56	690	21	470	−14	240
55	680	20	460	−15	230
54	680	19	450	−16	230
53	670	18	430	−17	220
52	660	17	420	−18	220
51	660	16	410	−19	210
50	650	15	410	−20	210
49	650	14	410	−21	200
48	640	13	400	−22	200
47	630	12	400	−23	200
46	620	11	390	−24	200

TEST-TAKING STRATEGIES

CHAPTER

2

General Hints

Here's the first rule to follow in preparing for SAT II: Biology E/M. Don't cram. Don't wait until the last few days to start preparing for the test. The more time you give yourself, the more familiar you will become with the subject matter and the types of questions, and the better prepared you will feel.

What to Bring to the Test

Bring a watch to help you keep track of the time. Don't depend on the proctor to announce the time. Also, take along three or four sharpened No. 2 pencils with erasers.

Do not bring an electronic calculator or any notes, study aids, or books.

Making the Best Use of Your Time

Since you have exactly one hour, you do not want to run out of time. Do not spend too much time on questions that give you difficulty. If you do, you may not have time left for the questions you can answer. Start by going through the entire test, answering all the questions you are reasonably sure about. Write on the test booklet, as you go along, and circle the numbers of the difficult questions. Then, in the remaining time, locate them quickly and work them out.

Should You Guess?

There is a difference between making a "wild guess" and making an "educated guess." You may be up against a tough question about which you know nothing, and are not able to eliminate any of the choices—not even one or two. The only way you could answer such a question would be to guess haphazardly. But since there is a penalty for wrong answers—you lose a fraction of a point for each—it is better not to answer such a question at all. Experience has shown that many students who do well may omit some of the questions.

On the other hand, if you can eliminate one or two of the choices as being definitely wrong, you are on the way to making an "educated guess." At least, you have some knowledge about the question. Cross out the wrong choices in your test book. It is generally to your advantage to pick an answer from the remaining choices. Sometimes you may not be sure of this answer and may want to change it later on. However, studies have shown that it is best to leave the original answer; your first guess is more likely to be correct than a revised one.

Feel free to mark up the test booklet to keep track of questions you want to come back to, but do not place any unnecessary marks on the answer sheet.

Marking Your Answer Sheet

Be careful to mark only one answer to a question. Avoid extra marks on the answer sheet. Completely blacken the space. Make sure you answer the question in the appropriate spot.

Notice how the following question is to be answered:

1. Animal cells do not contain

   (A) nucleus
   (B) nuclear membrane
   (C) cell membrane
   (D) chloroplasts
   (E) cytoplasm

1. Ⓐ Ⓑ Ⓒ ● Ⓔ
   Correct Method

If you answered the question in any of the following wrong ways, or any other incorrect way, you would not receive credit for it. Your answer might be correct, but the scoring machine would not be able to record it.

1. Ⓐ Ⓑ Ⓒ ⓧ Ⓔ   1. Ⓐ Ⓑ Ⓒ ✓ Ⓔ
   Wrong Method   **Wrong Method**

1. Ⓐ Ⓑ Ⓒ ⊘ Ⓔ
   Wrong Method

The Test Directions

You can save time during the test by being familiar, ahead of time, with the directions for answering questions. Study the directions for the following sample questions.

Sample Questions

Try to answer the following sample questions to get an idea of some of the different types of questions that appear on the test. Read the directions carefully; they are like those you will find on the test.

Directions: Each of the questions or incomplete statements below is followed by five suggested answers or completions. Select the best one in each case.

1. If gastric juice is tested with a pH meter, the pH is most likely to be about

   (A) 2
   (B) 6
   (C) 7
   (D) 8
   (E) 14

2. If $\underline{X}$ = chromosome with genes for color blindness and X = chromosome with genes for normal vision, which of the following will produce a color-blind son?

   (A) $\underline{X}$Y × XX
   (B) XY × $\underline{X}$X
   (C) $\underline{X}$Y × XX
   (D) X$\underline{Y}$ × XX
   (E) XY × X$\underline{X}$

3. In this pedigree of a family, brown eyes are indicated as ◯, blue eyes as ●. The eye color of Charles is not given. From this chart, it can be determined that

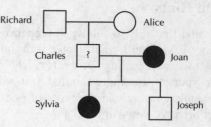

   (A) Richard and Alice are both homozygous for brown eyes
   (B) Charles is homozygous for brown eyes
   (C) Charles is probably heterozygous for brown eyes
   (D) Charles is homozygous for blue eyes
   (E) Sylvia's eyes could not be brown, since her mother is blue-eyed

4. In which of the following parts of a cell is RNA found?

   I. Nucleus
   II. Cytoplasm
   III. Ribosome

   (A) I only
   (B) II only
   (C) I and III only
   (D) II and III only
   (E) I, II, and III

Questions 5–9 refer to the following diagrams depicting the stages of cleavage.

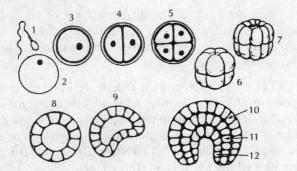

5. The first cell to contain the diploid number of chromosomes is

(A) 2
(B) 3
(C) 4
(D) 6
(E) 9

6. Which layer differentiates into the nervous system?

(A) 6
(B) 7
(C) 8
(D) 9
(E) 10

7. Of the following, which is a gamete containing stored food?

(A) 1
(B) 2
(C) 3
(D) 8
(E) 9

8. A female gamete containing the haploid number of chromosomes is

(A) 2
(B) 3
(C) 4
(D) 5
(E) 8

9. Which is the first cell to undergo mitotic division?

(A) 3
(B) 4
(C) 5
(D) 6
(E) 8

Questions 10–12

In a demonstration of trial-and-error learning, a pupil was given a puzzle consisting of four pieces of cardboard. She was told to arrange them in the form of a letter L. At a signal from another pupil who acted as timekeeper, she worked at the puzzle until the pieces fitted together, as shown in the accompanying diagram.

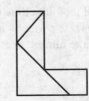

As soon as she finished, the time was recorded. The pieces were then scrambled, and the same pupil attempted to put them together again. This was done for a total of 10 times. Graph *A* was then constructed:

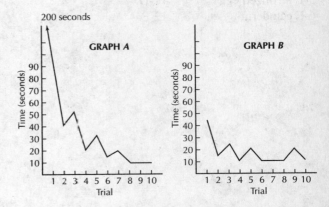

10. On Graph *A*, the greatest improvement is shown in the

(A) first trial
(B) second trial
(C) fourth trial
(D) sixth trial
(E) tenth trial

11. The straight line on Graph *A* after the eighth trial indicates all of the following EXCEPT that the pupil

(A) was putting the pieces together as quickly as possible
(B) could not improve any further
(C) had learned the puzzle perfectly
(D) gave up and made no more efforts to learn the puzzle
(E) made no more errors in solving it

12. The timekeeper, who had been watching the pupil, then attempted, again for a total of 10 times, to put the puzzle together, with the results shown in Graph *B*. Which of the following statements is true about the timekeeper's performance?

(A) He did the puzzle perfectly the first time.
(B) He learned the puzzle in less time than the first student.
(C) His fifth trial took less time than his fourth.
(D) He did the puzzle perfectly after the sixth trial.
(E) He did not time himself accurately.

Answer Key

1-A	7-B ovum
2-E	8-A ovum
3-C	9-A fertilized egg
4-E	10-B
5-E fertilized egg	11-D
6-E ectoderm	12-B

Hints for Answering Different Types of Multiple-Choice Questions

There are a number of different types of multiple-choice questions on SAT II: Biology Subject Test. Let's examine a variety of these and analyze the best way to answer each of them.

Type 1. From your own experience in taking tests, you are probably familiar with the common type of multiple-choice question that requires you to select the best of the choices. It is presented as a question or an incomplete statement. You should also recall that animal cells do not contain chloroplasts.

Type 2. You would apply your knowledge of pH being in the acid range. Gastric juice, which is formed in the stomach, gives an acid response such as 2 to a pH meter because it contains hydrochloric acid.

Type 3. Color blindness is a sex-linked characteristic whose recessive gene is carried on the sex chromosomes, which are XX for female and XY for male. Since the Y chromosome carries few genes, only one gene in its chromosome is sufficient to result in color blindness, whereas a female would require a gene on the two X chromosomes to produce the condition of color blindness. In tracing the inheritance of a characteristic, you would use the typical Punnett square to obtain the possible results:

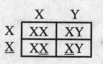

Possible results:
Male: 50% color blind
        50% normal
Female: all normal
        50% carrier

Type 4. You are asked to study the pedigree chart eye color in a family. The eye color for Charles is not given. However, he obviously is heterozygous for brown eyes and contains a recessive gene for blue eyes; his wife, Joan, is blue-eyed. Their daughter, Sylvia, who is blue-eyed, received a gene for blue eyes from each parent. Since Charles has a recessive gene for blue eyes, either one or both of his parents, Richard and Alice, are heterozygous for brown eyes. Sylvia might have had brown eyes if she had received a gene for brown eyes from Charles.

Type 5. You are given a list of three items designated by Roman numerals. This is followed by five choices, each relating to the Roman numerals. See sample question 4. The simplest way to answer this question is to analyze each by the items labeled by the Roman numerals, one at a time. On the sample test, mark each of the items True or False in terms of the question that is asked. Then proceed to the choices and pick them out one at a time, until you have arrived at the correct one.

In sample question 4 the first item of the question is true. Messenger RNA is formed in the nucleus. It then leaves through the nuclear membrane and goes through the cytoplasm to the ribosomes where it functions in protein synthesis. Thus, RNA is found in all three locations.

Type 6. See questions 5–9, which refer to a number of diagrams depicting a simplified view of the stages of cleavage. Each step should be viewed carefully. Stages 1 and 2 represent the sperm and ovum (egg). Each of them has the haploid number of chromosomes. When the sperm unites with the ovum, it forms the fertilized egg, containing the diploid number of chromosomes. See question 5. When cleavage starts, it forms a three-layered gastrula. The outermost layer, the ectoderm, differentiates into the nervous system. See question 6.

At the beginning of the process, the ovum contains stored food. See question 7. The ovum, or female gamete, contains the haploid number of chromosomes. See question 8. After the fertilized egg divides to initiate the stages of cleavage, it divides mitotically, resulting in the same diploid number of chromosomes in each of the cells. See question 9.

Type 7. See questions 10–12. These questions are based on an experimental procedure in which graphs are prepared from the results of trial-and-error learning. Graph A shows the length of time required to solve the puzzle at the end of the first and second trials, especially. See sample question 10, in which the time decreased in the trial from 200 seconds to 40 seconds, an improvement of 160 seconds.

Sample question 11 uses a qualifying word such as EXCEPT (or LESS, LEAST, or NOT). Next to each of the suggested answers, write True or False. In each case, except for choice D, your answer would be True. In sample question 12, the timekeeper apparently made use of another ability other than trial-and-error learning; she must have made use of thinking, or comparing, or other observational skills in order to learn the puzzle in less time than the first student.

The Scientific Method

Some types of questions expect you to be familiar with the scientific method. This is the process by which scientists conduct their research and reach their conclusions. Although individual scientists may work with a variety of different living things, they use, in general, the following similar procedures:

1. They start with a problem. (For example: How is the growth of paramecia affected by the different colors of the spectrum?)

2. They read about the subject in science journals and books in order to broaden their background and to learn what other scientists have discovered about the problem.

3. They form a hypothesis or suggested answer to the problem. (For example: Paramecia grow best in red and blue light.)

4. They conduct an experiment to test the hypothesis. (For example: They set up cultures of paramecia under cellophane envelopes of different colors. At weekly intervals, they examine a drop of each culture with a microscope and estimate the number of paramecia on each slide.)

5. They include a control. This represents the normal procedure of an experiment without the variable conditions that are being tested. (For example: A culture of paramecia is grown in ordinary light under colorless cellophane.)

6. They keep accurate records of their procedures, materials, and observations. They may include photographs, drawings, graphs, and tables to demonstrate and summarize their results.

7. They state their conclusion. It is based on the evidence gathered in the experiment, and provides an answer to the hypothesis. They usually repeat the experiment to confirm their observations.

THE MINI-DIAGNOSTIC TEST BIOLOGY E/M

PART

Answer Sheet: Mini-Diagnostic Test E/M

Core Questions E/M

1. Ⓐ Ⓑ Ⓒ Ⓓ Ⓔ
2. Ⓐ Ⓑ Ⓒ Ⓓ Ⓔ
3. Ⓐ Ⓑ Ⓒ Ⓓ Ⓔ
4. Ⓐ Ⓑ Ⓒ Ⓓ Ⓔ
5. Ⓐ Ⓑ Ⓒ Ⓓ Ⓔ
6. Ⓐ Ⓑ Ⓒ Ⓓ Ⓔ
7. Ⓐ Ⓑ Ⓒ Ⓓ Ⓔ
8. Ⓐ Ⓑ Ⓒ Ⓓ Ⓔ
9. Ⓐ Ⓑ Ⓒ Ⓓ Ⓔ
10. Ⓐ Ⓑ Ⓒ Ⓓ Ⓔ
11. Ⓐ Ⓑ Ⓒ Ⓓ Ⓔ
12. Ⓐ Ⓑ Ⓒ Ⓓ Ⓔ
13. Ⓐ Ⓑ Ⓒ Ⓓ Ⓔ
14. Ⓐ Ⓑ Ⓒ Ⓓ Ⓔ
15. Ⓐ Ⓑ Ⓒ Ⓓ Ⓔ
16. Ⓐ Ⓑ Ⓒ Ⓓ Ⓔ
17. Ⓐ Ⓑ Ⓒ Ⓓ Ⓔ
18. Ⓐ Ⓑ Ⓒ Ⓓ Ⓔ
19. Ⓐ Ⓑ Ⓒ Ⓓ Ⓔ
20. Ⓐ Ⓑ Ⓒ Ⓓ Ⓔ
21. Ⓐ Ⓑ Ⓒ Ⓓ Ⓔ
22. Ⓐ Ⓑ Ⓒ Ⓓ Ⓔ
23. Ⓐ Ⓑ Ⓒ Ⓓ Ⓔ
24. Ⓐ Ⓑ Ⓒ Ⓓ Ⓔ
25. Ⓐ Ⓑ Ⓒ Ⓓ Ⓔ
26. Ⓐ Ⓑ Ⓒ Ⓓ Ⓔ
27. Ⓐ Ⓑ Ⓒ Ⓓ Ⓔ
28. Ⓐ Ⓑ Ⓒ Ⓓ Ⓔ
29. Ⓐ Ⓑ Ⓒ Ⓓ Ⓔ
30. Ⓐ Ⓑ Ⓒ Ⓓ Ⓔ

E Choice

31. Ⓐ Ⓑ Ⓒ Ⓓ Ⓔ
32. Ⓐ Ⓑ Ⓒ Ⓓ Ⓔ
33. Ⓐ Ⓑ Ⓒ Ⓓ Ⓔ
34. Ⓐ Ⓑ Ⓒ Ⓓ Ⓔ
35. Ⓐ Ⓑ Ⓒ Ⓓ Ⓔ
36. Ⓐ Ⓑ Ⓒ Ⓓ Ⓔ
37. Ⓐ Ⓑ Ⓒ Ⓓ Ⓔ
38. Ⓐ Ⓑ Ⓒ Ⓓ Ⓔ
39. Ⓐ Ⓑ Ⓒ Ⓓ Ⓔ
40. Ⓐ Ⓑ Ⓒ Ⓓ Ⓔ

M Choice

41. Ⓐ Ⓑ Ⓒ Ⓓ Ⓔ
42. Ⓐ Ⓑ Ⓒ Ⓓ Ⓔ
43. Ⓐ Ⓑ Ⓒ Ⓓ Ⓔ
44. Ⓐ Ⓑ Ⓒ Ⓓ Ⓔ
45. Ⓐ Ⓑ Ⓒ Ⓓ Ⓔ
46. Ⓐ Ⓑ Ⓒ Ⓓ Ⓔ
47. Ⓐ Ⓑ Ⓒ Ⓓ Ⓔ
48. Ⓐ Ⓑ Ⓒ Ⓓ Ⓔ
49. Ⓐ Ⓑ Ⓒ Ⓓ Ⓔ
50. Ⓐ Ⓑ Ⓒ Ⓓ Ⓔ

THE MINI-DIAGNOSTIC TEST BIOLOGY E/M

CHAPTER

3

This Mini-Diagnostic Test contains a sample of questions that gives you a preview of the core questions and the E/M questions in the SAT II: Biology E/M test. Although you would not answer both E and M questions at the time of the examination, they are presented here for practice purposes. Normally, you would answer only the E choice or the M choice on the actual test, depending on your preference.

The Mini-Diagnostic Test is a shortened version of the questions on a typical SAT II: Biology E/M Test. Set aside 30 minutes for it.

As you take the test, imagine yourself in the actual examination room. Don't allow any interruptions or distractions. Keep the radio and TV off during this time. Also do not take any telephone calls.

Remove the Answer Sheet from the book and mark your answers on it. Ready? Go!

Mini-Diagnostic Test E/M

Part A (Core Questions 1–30)

Directions: Each of the questions or incomplete statements below is followed by five suggested answers or completions. Choose the one that is best in each case and then blacken the corresponding space on the answer sheet.

1. The Hardy-Weinberg principle applies to

   (A) the cell theory
   (B) the theory of spontaneous generation
   (C) experimental embryology
   (D) population genetics
   (E) the germ theory of disease

2. The technique of DNA fingerprinting has been used to identify people in crimes, accidents, and family relationships. In order to analyze the DNA of members of the same family, it is necessary to examine samples of their tissue for

   (A) similarities in the nucleotide uracil
   (B) similarities in deoxyribose sugar
   (C) similarity in recombinant DNA
   (D) similarities in nucleotide base pairs
   (E) RNA differences

3. A chemical produced in an axon is

   (A) auxin
   (B) acetylcholine
   (C) adenine
   (D) autosome
   (E) antigen

4. Which statement describes a feedback mechanism involving the thyroid gland?

   (A) The production of estrogen stimulates the formation of gametes for sexual reproduction.
   (B) The level of oxygen in the blood is related to the heartbeat.
   (C) The production of urine allows for excretion of cell waste.
   (D) The amount of carbon dioxide in the blood increases with exercise.
   (E) The hypothalamus reduces its supply of thyroid-stimulating hormones to the pituitary.

5. When a pregnant woman ingests alcohol and nicotine, the embryo is put at risk because these toxins can

   (A) interfere with ovulation
   (B) transfer to the embryo since the mother's blood normally mixes with the embryo's blood in the placenta
   (C) diffuse from the mother's blood into the embryo's blood within the placenta
   (D) enter the uterus through the mother's navel
   (E) diffuse into the mother's milk on which the young baby is nourished

6. All of the following veins carry deoxygenated blood in the human body EXCEPT the

   (A) superior vena cava
   (B) inferior vena cava
   (C) pulmonary vein
   (D) renal vein
   (E) hepatic vein

7. Which of the following would be most likely to occur in an ecosystem?

   (A) As the number of prey decreases, the number of predators increases.
   (B) As the number of predators increases, the number of prey increases.
   (C) As the number of prey increases, the number of predators increases.
   (D) As the number of prey increases, the number of predators decreases.
   (E) As the number of predators decreases, the number of prey decreases.

8. The transpiration rate of a tree would most probably be increased by

   (A) increase in temperature only
   (B) increase in both humidity and temperature

   (C) increase in humidity and decrease in temperature
   (D) increase in both humidity and air movement
   (E) increase in both temperature and air movement

9. Astronauts are not likely to find active heterotroph aggregates in any future explorations of the moon because

   (A) the force of gravity on the moon is too weak
   (B) the force of gravity on the moon is too strong
   (C) there are no X rays, ultraviolet radiation, or cosmic rays on the surface of the moon
   (D) there is no atmosphere or water on the moon
   (E) it is now suspected that there are moonquakes on the surface

10. The scientific name of the dog is *Canis familiaris* and that of the coyote is *Canis latrans*. This indicates that both the dog and the coyote are members of the

    (A) same genus but different species
    (B) same species but different genera
    (C) same genus but different classes
    (D) same species but different classes
    (E) same class but different genera

11. An overweight condition may be caused by any of the following factors EXCEPT

    (A) overeating
    (B) lack of exercise
    (C) eating too many high-calorie foods
    (D) lack of iron in the diet
    (E) thyroid gland disturbance

12. In which of the following organisms does aerobic respiration take place?

    I. Mammals
    II. Higher green plants
    III. Protozoa

    (A) I only
    (B) II only
    (C) I and III only
    (D) II and III only
    (E) I, II, and III

13. The structure of a lipid contains all of the following EXCEPT

    (A) a carboxyl group
    (B) a CH_2O basic structure
    (C) a glycerol molecule

(D) a fatty acid molecule

(E) an OH group

14. The main function of a nephron is to

(A) break down red blood cells to form nitrogenous wastes

(B) form urea from the waste products of protein metabolism

(C) regulate the chemical composition of the blood

(D) absorb digested food from the contents of the small intestine

(E) collect urine after it flows through the urethra

15. Which of the following represents the normal sequence of stages in an embryo's development?

(A) Gastrula formation—blastula formation— mesoderm formation—cleavage—muscle formation

(B) Cleavage—blastula formation—muscle formation—gastrula formation—mesoderm formation

(C) Cleavage—blastula formation—mesoderm formation—gastrula formation—muscle formation

(D) Cleavage—blastula formation—gastrula formation—mesoderm formation—muscle formation

(E) Blastula formation—gastrula formation— mesoderm formation—cleavage—muscle formation

16. RNA is found in

(A) the nucleus only

(B) the cytoplasm only

(C) both the nucleus and the cytoplasm

(D) proteins

(E) amino acids

17. The humerus, the bone in the upper arm of a human, is directly connected to other bones in the arm by

(A) cartilage   (D) ligaments

(B) tendons   (E) adipose tissue

(C) extensors

18.

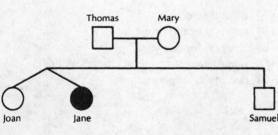

In the pedigree chart of a family, brown eyes are indicated as ○ and blue eyes as ●. Joan and Jane are twins. From this chart, it can be determined that:

(A) Thomas and Mary are homozygous for brown eyes

(B) Joan and Jane are identical twins

(C) Jane is heterozygous for blue eyes

(D) Jane is homozygous for blue eyes

(E) Joan and Samuel are homozygous for brown eyes

19. A plant breeder obtained 252 red, 235 white, and 503 pink flowers, when he crossed pink snap-dragons. From this information, which of the following conclusions can be drawn?

(A) The phenotype of the hybrid is the same as its genotype.

(B) The phenotype of the pure type is different from its genotype.

(C) The phenotype of the hybrid is different from its genotype.

(D) The offspring show the phenotype of the dominant gene.

(E) All of the above statements are false.

20. Because urea is a nitrogen compound, it cannot be derived from the metabolism of

(A) peptides   (D) proteins

(B) polypeptides   (E) amino acids

(C) polysaccharides

21. In which of the following organisms does transpiration take place?

   I. Cat

  II. Maple tree

 III. Paramecium

(A) I only   (D) II and III only

(B) II only   (E) I, II, and III

(C) I and III only

22. During the process of respiration, energy is transferred from glucose molecules to molecules of

(A) ACTH   (D) ATP

(B) DNA   (E) BCG

(C) RNA

23. In the nitrogen cycle, nitrogen compounds are broken down by decomposers to release

(A) ammonia   (D) vinegar

(B) carbon dioxide   (E) urea

(C) oxygen

24. Rh blood disease can occur when a Rh-negative mother

   (A) develops erythroblastosis
   (B) fails to form antibodies
   (C) has to have her blood replaced
   (D) has an Rh-negative baby
   (E) has an Rh-positive baby

25. A vacant lot may have organisms such as timothy grass, dandelions, ants, soil bacteria, earthworms, ailanthus trees, and mice living in it. Together, all of these organisms comprise a

   (A) population   (D) community
   (B) biome   (E) climax
   (C) biosphere

26. The relationship described in a food chain may also be expressed as a food pyramid. The major difference is that in the food pyramid, as depicted in the accompanying diagram, there is an indication of both the relative numbers of individuals and the amounts of energy involved at each level.

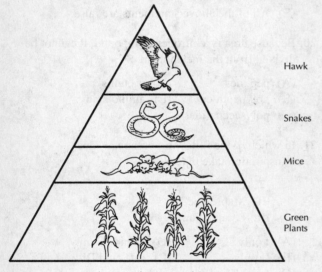

Food Pyramid

Which of the following best summarizes this relationship?

   (A) When the top of the pyramid is reached, the number of individuals decreases but the amount of energy increases.
   (B) When the top of the pyramid is reached, the number of individuals increases, and the amount of energy remains the same as at the other levels.

   (C) At the bottom level of the pyramid, both the number of individuals and the amount of energy involved are greatest.
   (D) At the bottom level of the pyramid, both the number of individuals and the amount of energy involved are lowest.
   (E) At all the levels, the number of individuals and the amount of energy are the same.

27. A study was made of the daily activities of the average farmer's wife, weighing 62 kilograms, in New York State. The following table shows a day's activities, the time spent on each, and the output of energy in calories per kilogram per hour.

Activity	Hours	Output of Energy (cal/kg/h)
Sleeping	8	1.0
Sitting quietly	2	0.4
Eating	1 1/2	0.4
Laundering	1 1/2	1.3
Sweeping	1	1.4
Dishwashing	2	1.0
Driving car	2	0.9
Going upstairs	1	14.7
Going downstairs	1	4.1
Walking	4	2.0

Which of the following represents the total energy requirements of the farmer's wife, in calories, for such a day?

   (A) 2,000–2,200
   (B) 2,200–2,400
   (C) 2,400–2,600
   (D) 2,600–2,800
   (E) 2,800–3,000

28. In 1889, a biologist named August Weissmann attempted to see if he could produce a strain of mice without tails. He surgically removed the tails of mice and permitted them to reproduce. The result was that all of the offspring had long tails. He repeated the procedure for 22 generations of mice but always observed the same result. His research helped disprove the theory of

   (A) natural selection
   (B) survival of the fittest
   (C) struggle for existence
   (D) inheritance of acquired characteristics
   (E) continuity of germplasm

29. A color-blind man marries a normal woman who is heterozygous for color vision. What are the chances of their two sons being color-blind?

   (A) 0%   (D) 75%
   (B) 25%   (E) 100%
   (C) 50%

30. The number of different gene combinations possible in the gametes of a trihybrid pea plant, *TtYySs*, is

   (A) 2   (D) 8
   (B) 4   (E) 10
   (C) 6

Part B—BIOLOGY E (Questions 31–40)

Directions: Each group of questions below concerns a laboratory or experimental situation. In each case, first study the description of the situation. Then select the one best answer to each question following it and blacken the corresponding space on the answer sheet.

Questions 31–33

A technique has been perfected by which the tip of a root can be cut off and placed in a sterile nutrient solution. Here, under controlled conditions of temperature, it will grow in length and produce secondary roots. A research scientist made daily measurements of a tomato root tip growing at 33°–35°C this way. The accompanying graph shows the results.

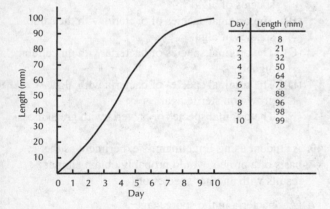

31. The root tip showed no growth

   (A) at the end of the 1st day
   (B) at the end of the 3rd day
   (C) at the end of the 8th day
   (D) at the end of the 10th day
   (E) at no time shown on the graph

32. The smallest increment in growth was achieved

   (A) during the first 3 days
   (B) during the last 3 days
   (C) during the middle 3 days
   (D) on the 3rd day
   (E) on the 8th day

33. The graph would have appeared as a straight line if

   (A) the growth after the 8th day had been at the same rate as on the preceding day
   (B) the root tip had increased in length at the same rate daily
   (C) the temperature had been maintained at a constant 37°C
   (D) the growth increment of the first 2 days had been identical
   (E) the nutrient solution had been re-enforced with root growth hormone

Questions 34–36

A young scientist studied the response of plant lice, or aphids, to various wavelengths of light. These small insects are parthenogenetic, and give rise to living young. They may be either wingless or winged, depending on various environmental conditions, such as temperature, humidity, length of day, intensity of light, abundance of food, and quality or color of light. In this experiment, the aphids were grown on nasturtium plants that were covered with light filters of red, yellow, blue, or gray (the control). The appearance of winged aphids is shown in Graph I, and the reproduction rate in Graph II.

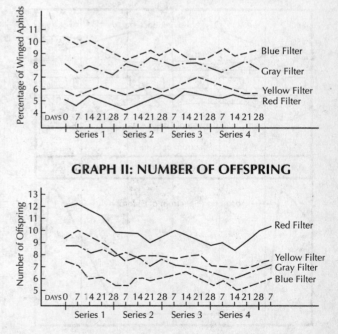

34. Which of the following occurred under the red filter?

(A) There were the fewest winged aphids and the highest reproduction rate.
(B) There were the fewest winged aphids and the lowest reproduction rate.
(C) There were the most winged aphids and the highest reproduction rate.
(D) There were the fewest wingless aphids and the highest reproduction rate.
(E) There were the most wingless aphids and the lowest reproduction rate.

35. The production of wingless aphids was stimulated to the greatest extent by both

(A) blue and gray light
(B) gray and yellow light
(C) yellow and red light
(D) red and blue light
(E) yellow and blue light

36. The highest reproduction rate was stimulated by both

(A) blue and gray light
(B) gray and yellow light
(C) yellow and red light
(D) red and blue light
(E) yellow and blue light

Questions 37–40

The accompanying diagram shows the results of Englemann's experiment, in which he found clusters of bacteria gathered in different amounts along a green alga filament which was being illuminated with the spectrum of light from a prism.

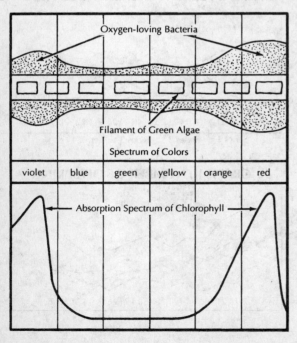

37. Which of the following would Englemann have noted?

(A) The bacteria concentrated where oxygen production by photosynthesis was the greatest in violet and blue light.
(B) The bacteria concentrated where oxygen production by photosynthesis was the greatest in violet and red light.
(C) The bacteria concentrated where oxygen production by photosynthesis was the greatest in green light.
(D) The filament of green algae prevented the bacteria from absorbing light in the green zone.
(E) The bacteria produced the greatest amounts of oxygen in the red zone.

38. If Engelmann had substituted direct light for the spectrum light from a prism, he probably would have observed

(A) identical results
(B) a greater concentration of bacteria throughout the length of the alga filament
(C) a lower concentration of bacteria throughout the length of the alga filament
(D) equal concentrations of bacteria throughout the length of the alga filament
(E) a greater concentration of bacteria at the ends of the alga filament

39. Which one of the following could Engelmann have used as a control in his experiment?

(A) The original species of bacteria with a different species of alga
(B) The original species of bacteria with the same species of alga
(C) The original species of bacteria with unicellular algae
(D) The original species of bacteria with algae grown on sterile agar
(E) The original species of bacteria with green moss

40. A student using Engelmann's experiment as the basis of a project would probably obtain similar results with another species of

(A) bacteria in the spore stage
(B) anaerobic bacteria
(C) aerobic bacteria
(D) bacteria that had been heated at 100°C for 20 minutes
(E) bacteria that had been kept at a temperature of 0°C for 20 minutes

Part C—BIOLOGY M (Questions 41–50)

41. The structural formulas of certain organic molecules are shown below:

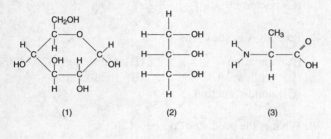

(1)   (2)   (3)

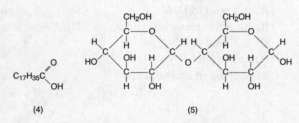

(4)   (5)

Which structural formulas represent carbohydrate molecules?

(A) 1 and 5   (D) 4 and 3
(B) 2 and 3   (E) 4 and 5
(C) 4 and 2

Questions 42–45

In comparing the effectiveness of enzymes at various pH concentrations, the accompanying graph was prepared to show the maximum rate of activity for maltase.

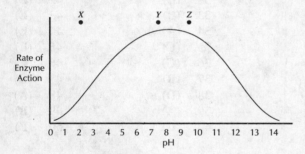

It was also determined that pepsin is most active at pH 1.5–2.2, and trypsin at pH 7.9–9.0.

42. The graph shows that

(A) point X represents pepsin and point Y represents trypsin
(B) point X represents maltase and point Y represents pepsin
(C) point X represents trypsin and point Y represents pepsin

(D) point X represents pepsin and point Z represents maltase
(E) point X represents pepsin and point Y represents maltase

43. The pH concentrations at which maltase is least effective are

(A) 1 and 7   (D) 2.2 and 9.0
(B) 1.5 and 7.9   (E) 1 and 14
(C) 7 and 14

44. On the basis of the graph, the temperature at which maltase is most active

(A) is the same as the temperature at which pepsin is most active.
(B) is different from the temperature at which pepsin is most active
(C) is the same as the temperature at which trypsin is most active
(D) is different from the temperature at which trypsin is most active
(E) cannot be determined

45. The nutrient substrate acted on by maltase is most likely

(A) glycerol   (D) amino acid
(B) carbohydrate   (E) lipid
(C) protein

Questions 46–47

In an experiment, a few drops each of concentrated $MnSO_4$ (manganese sulfate) and NaOH (sodium hydroxide) added to a solution containing O_2 (oxygen) will cause a brown precipitate. The greater the O_2 content, the deeper the brown color of the precipitate. Three tubes, marked *A*, *B*, and *C* respectively, were filled with water. A sprig of *Elodea* was placed in tubes *A* and *B*. Tubes *A* and *C* were placed in the sunlight. Tube *B* was placed in the dark. After three hours the *Elodea* was removed and three drops of each testing substance were added to all three tubes. The precipitate of darkest color was found in tube *A*.

46. Which of the following is most clearly illustrated by this experiment?

(A) Only green plants carry on photosynthesis.
(B) Green plants carry on respiration.
(C) Green plants give off O_2 during photosynthesis.
(D) Sunlight alters the O_2 content of the water.
(E) Green plants use up O_2 in the dark.

47. For an additional control, the experimenter should have

(A) used another type of water plant
(B) used another test for oxygen
(C) used bromthymol blue
(D) set up a tube in the dark with no plant
(E) added oxygen to the three drops of the testing substance

Questions 48–50

48–50. Base your answers on the diagram below, which contains arrows representing different processes occurring in a cell.

48. Which processes take place in the nucleus?

(A) A and B   (D) C and E
(B) B and C   (E) D and E
(C) C and D

49. Process A is known as

(A) mutation   (D) translation
(B) replication   (E) transduction
(C) nondisfunction

50. What is the product of process C?

(A) a strand of DNA
(B) two complementary strands of DNA
(C) a strand of RNA
(D) messenger RNA
(E) a chain of amino acids

Answer Key: Mini-Diagnostic Test

1. **(D)**	11. **(D)**	21. **(B)**	31. **(E)**	41. **(A)**
2. **(D)**	12. **(E)**	22. **(D)**	32. **(B)**	42. **(E)**
3. **(B)**	13. **(B)**	23. **(A)**	33. **(B)**	43. **(E)**
4. **(E)**	14. **(C)**	24. **(E)**	34. **(A)**	44. **(E)**
5. **(C)**	15. **(D)**	25. **(D)**	35. **(C)**	45. **(B)**
6. **(C)**	16. **(C)**	26. **(C)**	36. **(C)**	46. **(C)**
7. **(C)**	17. **(D)**	27. **(D)**	37. **(B)**	47. **(D)**
8. **(E)**	18. **(D)**	28. **(D)**	38. **(D)**	48. **(A)**
9. **(D)**	19. **(A)**	29. **(C)**	39. **(A)**	49. **(B)**
10. **(A)**	20. **(C)**	30. **(D)**	40. **(C)**	50. **(E)**

Analyzing Your Performance on the Mini-Diagnostic Test

Now that you have taken the Mini-Diagnostic Test, try to make it a learning experience by reviewing and analyzing your performance on it.

1. *Did you discover weaknesses in your knowledge of biology?*
   Since this was your first tryout on an SAT II: Biology E/M test, it may have opened your eyes to topics of biology that you need to review to a greater extent. You can correct your weaknesses by referring to the next section of this book—Part III, Modern Biology in Review.

2. *Did you have difficulty in answering certain types of questions?*
   In the many biology tests you have undoubtedly taken in class up to now, it is likely that you never had experience with some of the types of questions on this test. They look for more than mere recall of information. Some of these "new" types of questions test your ability to apply your knowledge to unfamiliar or practical situations. Others require that you interpret the information presented and arrive at appropriate conclusions. It might be a good idea to review Chapter 2 of this book, particularly the sections entitled "Sample Questions" and "Hints for Answering Different Types of Multiple-Choice Questions."

3. *Are you satisfied with your test score?*
   Since this type of test is a new experience for you, you may feel that you were not fully prepared for either the content or the types of questions. However, there are ways of improving your scores. First, follow the suggestions offered above. Then, when you have finished your "training" period, take the four Practice Tests in Part IV of this book. After completing each, use the Self-Evaluation Chart, and see whether your scores improve.

4. *Did you have enough time to answer all the questions on the test?*
   If not, you should keep the time element in mind when you take the four Practice Tests in Part IV. Review the suggestions in the third section of Chapter 2, entitled "Making the Best Use of Your Time." Remember: you can write on the test booklet and circle the numbers of the difficult questions, so you can come back to them after answering the questions you are sure about. But do not write anything on the Answer Sheet. It should have no extra marks.

5. *Did you guess haphazardly at difficult questions?*
   Keep in mind that there is a penalty for wrong answers; you lose a fraction of a point for each. Would you have been better off not answering the difficult questions at all? Keep in mind that there are two kinds of difficult questions: those you know nothing about, and those about which you have some knowledge. Review the fourth section of Chapter 2, which analyzes such questions and offers answers to the query, "Should You Guess?"

6. *Were you familiar with the directions for each group of questions?*
   The directions on the Mini-Diagnostic Test are similar to those actually used on SAT II: Biology E/M. If you spent too much time figuring out the directions when you took the Mini-Diagnostic Test, you had that much less time to answer the questions themselves. Be sure to become familiar with the directions in advance so that you will be ready for the four Practice Tests. Once you know what to expect, the directions will not seem complicated.

Self-Evaluation Chart: Your Road to More Knowledge and Improved Scores

Your Score

Using the Answer Key at the end of the test, place a ✔ next to each correct answer and an ✘ next to each incorrect answer.

A) Number of correct (✔) answers _____

B) Number of incorrect (✘) answers _____

C) *Score* (A – B) _____

Improving Your Score

1. In column *A* below, list the numbers of the questions that you did not answer correctly.

2. Turn to the Answers Explained section, and for each question number listed in column *A* write in column *B* the key word or phrase that best summarizes the main topic or point of the answer.

3. Look up the topic in the Index and review the material.

4. Go back to the test and try to answer again each of the questions you answered incorrectly the first time. Write your new answers in column *C*.

5. Compare the answers in column *C* with the Answer Key.

6. Calculate your revised Score:
Number of correct answers (*A* above + number of column *C* correct answers) _____

A. Incorrectly answered questions	B. Main point(s) of the answer	C. Answers to questions in column A
_____	_____	_____
_____	_____	_____
_____	_____	_____
_____	_____	_____
_____	_____	_____
_____	_____	_____
_____	_____	_____
_____	_____	_____
_____	_____	_____
_____	_____	_____
_____	_____	_____
_____	_____	_____
_____	_____	_____
_____	_____	_____
_____	_____	_____

Answers Explained: Mini-Diagnostic Test

1. (D) The Hardy-Weinberg principle states that the gene pool of a population remains constant from generation to generation if the population is large, there are random matings, and no new factors are introduced, such as mutations or migration.

2. (D) DNA segments of different people have different sets of nucleotides. People who are related would have many similar sets of nucleotides. When samples of blood, semen, hair cells, or other tissues containing nuclei are analyzed, the DNA sequences can be analyzed by a special procedure which includes a technique called gel electrophoresis.

3. (B) At the end of an axon, the end brushes (terminal branches) secrete chemicals known as neurotransmitters. One type of neurotransmitter is acetylcholine. It travels across a synapse to stimulate the dendrites of an adjacent neuron. In this way, a nerve impulse passes from one neuron to another.

4. (E) If the concentration of thyroxin rises in the blood, the hypothalamus reduces the supply of its thyroid-releasing hormone to the pituitary. This, in turn, inhibits the production of the thyroid-stimulating hormone by the pituitary. The result is a decrease in the secretion by the thyroid gland. This type of regulation is referred to as a negative feedback mechanism. It helps to maintain homeostasis throughout the body. In the case of the thyroxin feedback system, the rate of metabolism is kept at a constant level.

5. (C) In the uterus, the embryo receives its nourishment from the mother through the placenta. This consists of a set of thick membranes composed of tissues of both the parent and embryo lying in intimate contact with each other. These membranes are well supplied with many capillaries connected to the separate blood systems of both individuals. As blood flows through the two different sets of capillaries, food and oxygen diffuse from the mother's bloodstream into the embryo's bloodstream. If alcohol and nicotine are ingested by the mother, they too will enter the embryo's bloodstream.

6. (C) All of the veins carry deoxygenated blood (blood lacking in oxygen) except the pulmonary vein, which carries blood from the lungs to the left atrium of the heart. The pulmonary vein is the only vein with oxygenated blood.

7. (C) The number of predators depends on their ability to catch prey. If more prey are available, the predators will be more successful and their numbers will increase.

8. (E) With an increase in temperature there is an increase in the activities of the plant, including transpiration. Air movement removes moisture from the immediate vicinity of the stomates, making it possible for more water vapor of transpiration to evaporate.

9. (D) According to the heterotroph hypothesis, heterotroph aggregates are supposed to have formed in the ancient seas present on the earth's surface during its early history, but not now.

10. (A) *Canis* is the genus name for both animals, but each has a different species name, either *familiaris* or *latrans*.

11. (D) Iron is part of hemoglobin, the red protein found in red blood cells. If iron is lacking, not enough red blood cells are formed and anemia results. Foods rich in iron are liver, beef, raisins, and spinach. An overweight condition would not be due to a lack of iron in the diet. The other factors may be responsible for fat to be stored by the body. If the thyroid gland is not producing enough thyroxin, the person may become overweight.

12. (E) Aerobic respiration takes place in the presence of oxygen. It releases the chemical bond energy of food into energy that can be used by all three types of organisms listed.

13. (B) Like carbohydrates, lipids contain carbon, hydrogen, and oxygen. However, they differ in that the hydrogen and oxygen in carbohydrates are present in the same ratio as in water, 2:1. CH_2O is therefore a basic structure of carbohydrates, not lipids.

14. (C) The nephron is the excretory unit in the kidney of vertebrates. At one end, it contains the glomerulus, a collection of capillaries within Bowman's capsule. A plasmalike liquid containing water, urea, salts, fatty acids, and glucose diffuses into the capsule. As it moves into the long tubule of the nephron, water, minerals, and the digestive end-products are reabsorbed into the capillaries by active transport. The remaining liquid collects as urine and passes out of each kidney through a long tube, the ureter, into the urinary bladder. Urine leaves the bladder at intervals through the urethra.

15. (D) After fertilization, the fertilized egg undergoes a series of cell divisions known as cleavage. In many animals, the embryo next takes the appearance of a hollow ball of cells, one cell thick, known as the blastula. One part of the blastula then grows inward, forming a second layer. The outer layer is called the ectoderm; the inner, the endoderm. A third or middle layer then develops between these two; it is known as the mesoderm. From this point on, the embryo elongates as the three layers of cells begin to differentiate into different tissues and organs. The mesoderm layer differentiates into muscles, bone, the circulatory system, kidneys, and the reproductive organs.

16. (C) RNA is formed in the nucleus from the pattern, or template, of DNA. It then passes into the cytoplasm.

17. (D) Ligaments are examples of white fibrous connective tissue which connects bones together forming joints. The humerus is connected to the shoulder joint at the upper end and the elbow joint at the lower end by several ligaments. Tendons are also made of white fibrous tissue which connect muscles to bones.

18. (D) Blue eyes are a recessive trait; that is, two genes are needed for the trait to appear. Thomas and Mary are heterozygous for brown eyes, as is seen from the appearance of blue-eyed Jane. Joan and Jane are fraternal twins, since they do not have the same eye color. Joan and Samuel could be heterozygous for brown eyes, since their parents also produced a blue-eyed child.

19. (A) In incomplete dominance, the hybrid shows a blending of the colors of both parents, and therefore is seen to have one gene from each. The phenotype is the pink color of the hybrid. The genotype is the genetic makeup of the pink hybrid. This generation shows a 1:2:1 ratio.

20. (C) Urea is a nitrogenous waste resulting from the breakdown, or deaminization, of amino acids in the liver. Polysaccharides are large carbohydrate molecules consisting of many repeating glucose molecules bonded together; the only elements they contain are carbon, hydrogen and oxygen.

21. (B) Transpiration refers to the evaporation of water vapor from the leaves of a plant, such as the maple tree.

22. (D) During cellular respiration, energy is released from glucose by a series of complex steps. When glucose is broken down, some of its hydrogen is removed, releasing energy that is transferred to ADP with the addition of a high-energy phosphate molecule to convert the ADP to ATP.

23. (A) In one part of the nitrogen cycle, bacteria of decay in the soil decompose animal and plant remains that contain nitrogen compounds into ammonia.

24. (E) When an Rh-positive fetus develops in an Rh-negative mother, its red blood cells may be destroyed by antibodies from the mother.

25. (D) The population of all the plants, animals, and other organisms living together in a given environment comprise a community.

26. (C) Green plants are the basis of the food chain. They are also the base of the food pyramid, where they occur in the largest number. Green plants serve as the source of energy for the other organisms involved. There is a decrease in the amount of energy as it is passed along from the producers to the consumers.

27. (D) The output of energy for each activity is computed as follows:

Activity	Hours	Output per Kilogram	Output for 62 Kg
Sleeping	8	1.0 calories × 62	496.0 calories
Sitting quietly	2	0.4 calories × 62	49.6 calories
Eating	1 1/2	0.4 calories × 62	37.2 calories
Laundering	1 1/2	1.3 calories × 62	120.9 calories
Etc.			
		Total	2,718.7 calories

28. (D) According to Lamarck's explanation of evolution, if an animal uses an organ to adjust to its environment, that organ will become well developed; if it is not used, the organ will remain small and undeveloped. He believed that such acquired characteristics are inherited and passed on to future generations. Weissmann's experiment showed that this is not so; the mice were still born with normal tails.

29. (C) The possible results are arrived at in this way:

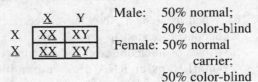

Male: 50% normal;
        50% color-blind
Female: 50% normal
        carrier;
        50% color-blind

30. (D) When the gametes are being formed, the alleles separate out during reduction-division to form the following possible types of gene combinations: *TYS, TYs, TyS, Tys, tYS, tYs, tyS, tys.*

31. (E) Growth occurred at all times.

32. (B) The curve began to level off during the last 3 days, indicating a slower rate of growth. This was due either to an accumulation of waste products, or to a decrease in the amount of available nutrients, or to both.

33. (B) At a constant rate, there would have been a constant increase in length, leading to a straight-line graph.

34. (A) In Graph I, the lowest percentage of winged aphids occurred under the red filter. In Graph II, the highest number of offspring occurred under the red filter.

35. (C) Under the yellow and red filters, the percentage of wingless aphids was highest, being in inverse proportion to the percentage of winged aphids.

36. (C) As shown in Graph II, the greatest number of offspring were produced under the red and yellow filters.

37. (B) Chlorophyll absorbs certain wavelengths of light more effectively than others. Most absorption takes place at the blue-violet and orange-red ends of the spectrum. Here the rate of photosynthesis is highest, and the most oxygen is liberated.

38. (D) With direct light, there is no spectrum of colors, as is the case with a prism. Photosynthesis would therefore occur equally throughout the length of the alga filament, where equal amounts of O_2 would be liberated. Bacteria would not concentrate at any particular part of the filament.

39. (A) To observe that bacteria concentrate where oxygen production is the greatest, Engelmann could simply have substituted another species of alga filament. If the results were the same, he would conclude that bacteria concentrate at the spectrum colors where photosynthesis is most active.

40. (C) Aerobic bacteria grow best in the presence of oxygen. Anaerobic bacteria cannot grow when oxygen is present.

41. (A) Number (1) represents a glucose molecule, consisting of atoms of carbon, hydrogen, and oxygen. Hydrogen and oxygen are present in the same ratio as in water, 2:1. Another carbohydrate is depicted in (5) and it represents the structural formula of the disaccharide maltose. Glucose molecules may be linked together to form maltose by a process called dehydration synthesis. This refers to the building up of complex molecules from simpler molecules with the release of water. Number (2) is a lipid molecule of glycerol which, like carbohydrates, contains carbon, hydrogen, and oxygen. However, there is proportionately much less oxygen in relation to hydrogen. Number (3) is an amino acid, which has a carbon atom bonded with a hydrogen atom, an amino group (NH_2), a carboxyl acid group (COOH), and a side group, which is variable. In this case, the side group is CH_3, a methyl group. The molecule is a simple amino acid, alanine. Number (4) represents a fatty acid, oleic acid, present in the oil olein. Like other lipids, it has much less oxygen than hydrogen.

42. (E) The maximum activity of pepsin (X) is shown on the graph at pH 1.5–2.2. The maximum activity of maltase (Y) is shown to take place at a pH of about 7.

43. (E) At pH concentrations of 1 and 14, the graph shows a decline in maltase activity to the zero level.

44. (E) The graph does not show a record of temperature activity.

45. (B) The enzyme maltase acts on maltose, which is a carbohydrate.

46. (C) The dark brown color in tube A indicates that more O_2 was formed in that tube. We may assume that photosynthesis carried on by the *Elodea* sprig was responsible for the O_2. There was little color in tube C because there was no *Elodea* sprig to produce O_2. The *Elodea* in tube B could not photosynthesize because of lack of light.

47. (D) The experimenter could see that there was less O_2 in the tube with *Elodea* placed in the dark (tube B) than in the tube with *Elodea* placed in the light (tube A). He could not, however, compare the O_2 content in a tube with *Elodea* that was placed in the dark, with the O_2 content of a tube without *Elodea,* also in the dark.

48. (A) DNA forms an exact copy of itself (A) in the nucleus. Here, it separates into two parts and adds free nucleotides to bond adenine with thymine and cytosine with guanine to complete the new structure. It also serves as a template or pattern for the synthesis of messenger RNA (B). The newly formed RNA molecule, which is a "reverse copy" of the DNA that produced it, separates from the DNA strand and moves through a pore in the nuclear membrane to enter the cytoplasm.

49. (B) The process by which DNA produces an identical copy is known as replication. It starts when the relatively weak hydrogen bonds holding the nitrogen bases together break and the double spiral ladder separates into two half-ladders. Free nucleotides move into position to form a bond with matching nucleotides of the half-ladder. Adenine bonds with thymine, and cytosine with guanine. Gradually a new upright portion is added to the half-ladder to complete two sections of a new spiral ladder. As a result, identical DNA molecules have been formed.

50. (E) When messenger RNA leaves the nucleus, it becomes located in a ribosome. It carries a triplet of nucleotides called a codon. Now a small transfer RNA picks up amino acid molecules in the cytoplasm that line up with messenger RNA. Each transfer RNA has a group of three nucleotides called an anticodon, and fits the appropriate codon of messenger RNA. Thus, if transfer RNA has an anticodon containing the sequence UGC, it will fit in with the messenger codon ACG. The arrangement of the nucleotides on messenger RNA dictates the order in which the amino acids are lined up and bonded together into polypeptide chains.

MODERN BIOLOGY IN REVIEW

PART

III

HOW LIVING THINGS ARE CONSTRUCTED

CHAPTER

4

4.1 THE CELL, LIFE ACTIVITIES, AND SIMPLE ORGANISMS

Cells

How Cells Are Studied

The average school microscope usually has an *eyepiece* (ocular) magnification of 10; the *low-power objective* also has a magnification of 10. Objects viewed under low power will therefore be magnified 10 × 10, or 100 times. The *high-power objective* generally has a magnification of 43; therefore objects studied under high power will be magnified 430 times (10 × 43). Not only is the image of an object enlarged by the lenses of the microscope, but also it is reversed and inverted. After first focusing on the image with the *coarse adjustment,* a sharper view can be obtained with the *fine adjustment.* The depth of the object can also be studied by careful use of the fine adjustment under high power. It is important for the *diaphragm* under the stage to be adjusted properly in order to admit the best amount of light for the clearest view. An even, north light is best; direct sunlight is to be avoided because of its glare and uneven qualities.

With the best equipment employing oil immersion lenses, the *compound microscope* is capable of magnifying up to about 1,800 times. The *electron microscope* permits magnifications of over 100,000

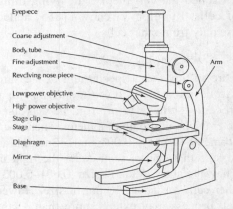

Average School Microscope

times. A stream of electrons is focused on the object by means of a magnetic field, and the image is formed either on a fluorescent screen or on a photographic film. The specimen is kept in a vacuum chamber because any air in the tube would interfere with the flow of electrons. Since living cells cannot exist in a vacuum, it is not possible to study them with an electron microscope. However, viruses, heretofore invisible, have been photographed, as have specific cell structures.

Additional knowledge about the structure of cells is being obtained through the use of modern techniques involving the *phase microscope* for examining living cell structures without the use of stains which may distort or kill them; *time-lapse cinematography,* for observing the reactions of living cells treated with beams of X rays, ultraviolet light, or streams of atomic nuclei; *acoustic microscope,* which utilizes very short wavelengths of sound and has already been used to scan such living tissue as human chromosomes, blood cells, and embryonic chicken cells; *differential interference-contrast microscope,* which uses a television camera to view in greater detail the way microfilaments carry on transportation in living cells; *laser beams* for studying the organelles of the cell; and the *ultracentrifuge* for whirling cells in a tube and separating them into their parts.

Early Discoveries about Cells

The first discoveries of cells were made by such early microscopists as Anton van Leeuwenhoek (1632–1723), who made his own simple lens microscopes, and saw bacteria and protozoa; and Robert Hooke (1635–1703), who had several lenses in his *compound* microscope, and who first named cells, after studying the tiny boxlike structures in cork. Robert Brown (1831) first named the nucleus in plant cells. In 1838–1839, Matthias Schleiden and Theodor Schwann stated the cell theory: all living things are made of cells. Rudolph Virchow (1855) added that cells arise only from other cells.

Cell Size

Cells are so tiny that a special unit of measurement is used to describe them. It is the *micrometer* (μm). (Until recently, the term micron, μ, was used.) A micrometer is one millionth of a meter. There are 1,000 in 1 millimeter (1 μm = 0.001 mm). Bacteria are the smallest cells that can be seen under the light microscope. They average 1–3 μm in size (0.001–0.003 mm). Most other cells are about 10 μm in size. However, a nerve cell process may extend more than a meter in length. And the yolk of a bird's egg, which is a single cell, may measure several centimeters across.

Types of Cells

There are two large groups of cells:

1. *Eukaryotes.* These cells contain a nucleus and several types of organelles. Such cells are found in all living things, except for bacteria and blue-green algae.

2. *Prokaryotes.* These cells do not contain a nucleus or organelles; however, they do have small ribosomes. These cells are only found in simple organisms, such as bacteria and blue-green algae.

Cell Structure

Until quite recently, the structure of plant and animal cells was considered to be relatively simple, as shown in the following diagrams:

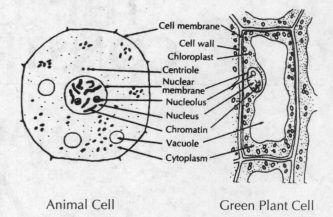

Animal Cell   Green Plant Cell

With the aid of the electron microscope, many details of the structure of the cell are being revealed, as shown in the modern version of a generalized cell in the next column.

Protoplasm

A cell is a tiny unit of living material called protoplasm. Protoplasm is a very complex substance composed chiefly of the elements carbon, oxygen, hydrogen, and nitrogen, with smaller amounts of other elements such as phosphorus, sulfur, calcium, and iron. It also contains trace elements, such as copper, zinc, cobalt, and fluorine, which are present in relatively minute amounts. It contains molecules of water (from 65 to 90 percent of the makeup of cells), protein, carbohydrates, fats, and minerals. Its content varies somewhat in the various living things. Even the protoplasm of the same organism may differ in specific parts—for example, the protoplasm of skin cells is different from that of muscle cells.

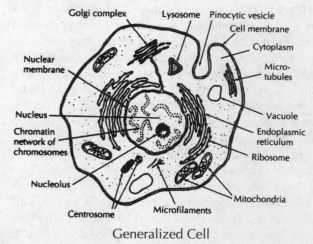

Generalized Cell

The cell is divided into a nucleus, which contains nucleoplasm, and cytoplasm, all of the protoplasm outside the nucleus. Throughout the cell specialized areas called *organelles* perform certain functions.

1. **Nucleus.** The nucleus is a small, spherical, dense body. It controls the activities of the cell, cell reproduction, and heredity. It is made of a *chromatin* network of *chromosomes,* which contain the *genes,* the hereditary units. The genes consist of molecules of DNA (deoxyribonucleic acid). There is usually at least one *nucleolus,* a tiny, dotlike structure inside the nucleus. It also contains RNA (ribonucleic acid) which helps direct the synthesis of proteins. The nucleus is surrounded by the *nuclear membrane*, which controls the transport of materials into and out of the nucleus through tiny openings called *nuclear pores*.

2. **Cytoplasm.** This granular, thick, grayish liquid occupies most of the cell. It appears to be somewhat like the white of a raw egg. It carries on most of the other activities of the cell. Many organelles are contained in the cytoplasm:

* *Endoplasmic reticulum (ER)*—a network of channels, or tubes, that extends throughout the cytoplasm. Its membranes connect with the nuclear membrane and the cell membrane. It is thought to function in the transport of materials throughout the cell. There are two types of endoplasmic reticulum: *rough* and *smooth. Rough ER* appears rough because of the ribosomes attached to it. *Smooth ER* does not have any ribosomes. Rough ER is present mainly in cells that make abundant amounts of proteins for use outside of the cells, e.g., pancreas cells that produce enzymes, which are protein in structure, that flow into the small intestine for digestion. Smooth ER is present in large amounts in cells that specialize in the synthesis of lipids, e.g., gland cells that produce steroid hormones. Smooth ER also helps build the lipids present in the cell membrane.
* *Ribosomes*—tiny granules that are either distributed along the endoplasmic reticulum or are free in the cytoplasm. They contain RNA and function as sites for protein synthesis.
* *Mitochondria*—rodlike structures with inner, folded surfaces called *cristae.* They contain enzymes associated with cellular respiration, leading to the release of energy from food. Most of the ATP molecules are formed here, serving as centers of energy storage. Mitochondria are called the "powerhouses" of the cell. They contain their own supply of DNA.
* *Lysosomes*—saclike organelles that contain digestive enzymes that break down large organic molecules, and worn-out organelles within the cell.

* *Vacuoles*—reservoirs for water and dissolved materials. They are especially large in plant cells.
* *Golgi complex*—a series of organelles that concentrate protein molecules that are produced in the ribosomes and are passed along the endoplasmic reticulum to the Golgi complex. The proteins are enclosed in vesicles or membranes that migrate to the cell membrane and then are released outside the cell; in plants they assemble and secrete carbohydrates for cell wall formation. This organelle is also sometimes called the Golgi body or Golgi apparatus.
* *Centrosome*—present in animal cells, outside the nucleus. It contains a pair of *centrioles,* which are composed of a bundle of small filaments, and are active during nuclear division.
* *Microtubules*—thin, tubular organelles that help support the structure of the cell. They also make up other cell structures such as cilia, flagella, centrioles, and spindle fibers.
* *Microfilaments*—minute, widely distributed filaments involved in movement by pseudopodia in cells such as ameba and white blood cells, and in the tips of growing nerve cells. They appear to consist of very thin strands of the proteins *actin* and *myosin,* which have also been found to be involved in the contraction of skeletal muscle. It has been determined that food and wastes are transported within a cell through the tiny microfilaments.

3. **Cell membrane.** The cell, or plasma, membrane is the outer living layer of cytoplasm. It has a complex structure consisting of a double lipid layer in which large proteins float. According to S. J. Singer and G. L. Nicolson, who proposed the "fluid-mosaic model" of the cell membrane in 1972, the proteins are partially embedded in the two lipid layers, resembling "floating icebergs in a lipid sea."

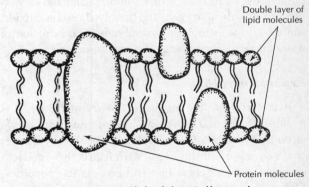

Double layer of lipid molecules

Protein molecules

Fluic-Mosaic Model of the Cell Membrane

The cell membrane is semipermeable and controls the passage of dissolved substances in and out of the cell.

Many small particles are able to pass through it. Larger molecules, such as proteins and starch, cannot pass through unless they have been digested and reduced in size and complexity. The movement of materials across the cell membrane into or out of the cell is called *transport*. In general there are two types of transport: passive and active. In *passive transport* the cell does not contribute energy for the movement of molecules across the membrane. Molecules move from an area of greater concentration to an area of lesser concentration by *diffusion*. In contrast, in *active transport,* the cell uses the energy of ATP to move molecules across the membranes. In such cases, carrier proteins contained in the cell membrane may serve as carriers in transporting molecules across. Such movement may be from a region of low concentration to a region of higher concentration. This is against the concentration gradient. *Pinocytosis* is one form of active transport.

Diffusion Through a Cell Membrane

Diffusion can be observed in the following demonstration:

1. Fill a cellophane bag (or some dialyzing tubing) with a light starch suspension made by boiling some cornstarch in water and allowing it to cool.
2. Seal the bag tightly with a rubber band, and place it in a jar of iodine solution. The effect of iodine on starch may be reviewed at this point by placing a drop of iodine directly on a small amount of starch suspension; a blue-black color results.
3. After a few minutes, observe that the starch contents of the cellophane bag have turned a blue-black color. The iodine solution still retains its original color.

From this demonstration, it can be seen that iodine, which is in solution (is soluble), diffused through the membrane into the bag to stain the starch. The starch, which is not in solution (is *insoluble*), did not pass through the membrane. Other soluble materials, such as glucose, dissolved oxygen, and dissolved carbon dioxide, can also diffuse through a membrane. The diffusion of water through a membrane is known as *osmosis*.

Explanation of Diffusion

By referring to the *molecular theory* of matter, it is possible to explain the passage of these materials through a membrane. According to this theory, matter is made up of molecules, which are the smallest particles of a substance that still possess its properties. Molecules are in a constant state of movement, and tend to *diffuse*, or move from one place to another, especially in a gas or a liquid. Thus, you can smell ammonia from a distant bottle of ammonia water,

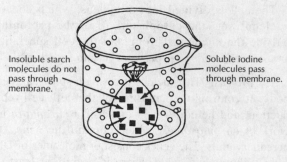

Insoluble starch molecules do not pass through membrane.

Soluble iodine molecules pass through membrane.

Demonstration: Diffusion

because the ammonia molecules diffused through the air. Also, if a lump of sugar is added to a cup of tea, after a while without stirring you can taste the sugar in the first sip of tea, because the sugar molecules have diffused through the tea.

In the demonstration described above, the iodine solution contained a high concentration of iodine molecules; on the inside of the membrane, there were no iodine molecules. As the molecules of iodine moved about constantly, many of them passed through the membrane into the cellophane bag and stained some of the starch. As more and more iodine molecules continued to pass through the membrane, the entire starch suspension took on the blue-black color. The insoluble starch molecules, on the other hand, could not pass through the membrane, and so could not react with the iodine solution in the jar, which therefore retained its original light brown color.

In other words, during diffusion through a membrane, there is a flow of dissolved molecules from a region of high concentration (the iodine solution) to a region of low concentration (the inside of the cellophane bag).

Pinocytosis and Phagocytosis

A *pinocytic vesicle* may form as a pocket of the cell membrane to engulf liquids or relatively small particles. These pockets are brought into the cell as vacuoles, which then release their contents into the cytoplasm. This process is known as pinocytosis.

Some one-celled organisms, such as the ameba, engulf solid materials, including bacteria. White blood cells in our own bodies engulf bacteria in a similar way. This process is called *phagocytosis*, from the Greek word *phage*, which means "to eat." By contrast, pinocytosis deals with the intake of liquids or tiny particles.

Plant Cells

In addition to the nucleus, cytoplasm, and cell membrane, which are present in all living things, plant cells have certain other structures. If the cells of an onion skin are stained with a dye such as Lugol's iodine or methylene blue, the parts of the protoplasm can be readily seen.

The following nonliving structures of plant cells can also be observed:

1. A *cell wall* containing cellulose, surrounding the cell.

2. One or more clear vacuoles distributed throughout the cytoplasm and containing *cell sap,* which is largely water with dissolved minerals.

3. *Chloroplasts,* small, oval bodies containing *chlorophyll,* which is necessary for food-making (photosynthesis), in all green plant cells.

Life Activities

Living things are able to carry on the following basic life functions:

Ingestion: Food is taken in.

Digestion: Food is broken down into simpler, soluble form, with the aid of enzymes.

Secretion: Useful substances, such as enzymes, are formed.

Absorption: Dissolved materials are passed through the cell membrane, into and out of the cell.

Respiration: Energy is released from food.

Excretion: Waste products are passed out of the cell through the cell membrane.

Transport: Materials are circulated throughout the organism.

Regulation: Stability of the organism's chemical makeup is maintained under a constantly changing internal and external environment (homeostasis).

Synthesis: Complex molecules are formed by chemical processes from simple compounds.

Assimilation: Nonliving materials in food are changed into more protoplasm, resulting in growth and repair.

Reproduction: More living things are produced.

Irritability: Stimuli produce responses.

Movement: Most living things have the ability to change position. This function is largely restricted to animals, although a few plants, such as the sensitive plant (mimosa) and the insectivorous plants (Venus-flytrap, sundew), have moving parts. Plant growth may also result in a change of position.

Living things perform the basic life functions with the help of specific structures or organs. In addition some of them have other abilities, such as bioluminescence, the production of light (firefly; some bacteria, protozoa, and fungi; deep-sea fish), and bioelectricity, the production of electric current (electric eel).

Simple Organisms

Some simple organisms serve to illustrate other structures and activities common among living things. Among these organisms are the ameba, paramecium, euglena, and spirogyra.

Ameba

One of the simplest of the protozoa is the one-celled ameba, which may be found in pond water. Although it consists of only one cell containing a tiny drop of streaming protoplasm, it is able to carry on all the life functions. It has no specific shape. As it crawls along, it extends projections called *pseudopods* (false feet), and flows into them. Ingestion is accomplished when the ameba comes into contact with a food particle, which the pseudopods surround and engulf. Once inside the cell, the food particle becomes located within a *food vacuole.* Here, digestion occurs with the aid of enzymes which are secreted by the surrounding protoplasm into the food vacuole. The process is referred to as *phagocytosis.* As the food is digested and becomes soluble, it diffuses out of the food vacuole into the protoplasm.

The contractile vacuole contracts to eliminate excess water. After all the food in the food vacuole has been digested, the indigestible remains are eliminated in a simple manner—the cell membrane opens momentarily, and these wastes are left behind as the ameba flows along.

When the ameba has grown to its maximum size, the nucleus divides in two. Half the cytoplasm collects around each nucleus, and the cell splits in two. This simple method of reproduction is called binary fission.

The protoplasm of this simple cell also shows the property of irritability, or sensitivity to stimuli about it. It will stop moving in the direction of a grain of salt when it comes into contact with it, will reverse the flow of its protoplasm, and move away from it. A beam of strong light at one end of the ameba will cause it to flow in the opposite direction.

If conditions become unfavorable, such as drying of the pond, the ameba forms a hard, protective wall around itself, called a *cyst,* and its protoplasm becomes dormant. Under suitable conditions, the cyst breaks open and the ameba emerges.

Paramecium

Another microscopic, one-celled inhabitant of pond water is the well-known paramecium. It is a slipper-shaped organism that has a somewhat higher level of organization than the ameba for the performance of its life functions. It has a permanent shape maintained by

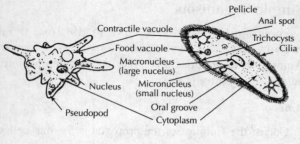

Ameba   Paramecium

the presence of a flexible exterior *pellicle.* The outside of the cell is covered with tiny, hairlike extensions of protoplasm called *cilia.* These beat rapidly and move the paramecium through the water. There is an opening on one side called the *oral groove,* or mouth. It is also lined with cilia, which create a current of water that leads inward to the *gullet,* carrying with it tiny particles of food. At the end of the gullet, food vacuoles are formed, which are carried slowly within the paramecium by the streaming protoplasm. As in the ameba, this is where digestion takes place. After the food within a food vacuole has been digested and absorbed, the vacuole will move close to a weak spot in the cell membrane called the *anal spot,* which will open suddenly and the indigestible solid material will be eliminated.

There are two contractile vacuoles, one at each end, surrounded by radiating canals, that alternately contract and expand to eliminate excess water. Since the water is being removed into an environment of fresh water, it occurs against a concentration gradient or level. Active transport is used to eliminate the water, in addition to the pumping action of the contractile vacuoles. The energy for active transport is derived from ATP molecules in the cell. The maintenance of the water balance is an example of *homeostasis.*

There are two nuclei, a large *macronucleus* and a small *micronucleus.* During fission, both of them divide. In addition to binary fission, the paramecium carries on a type of sexual reproduction called *conjugation,* in which two cells lie close to each other and exchange parts of their micronuclei.

Trichocysts lining the inside of the cell membrane discharge into long, fine threads under certain conditions. The reason may be to defend itself or to anchor itself while feeding on bacteria.

Euglena

The one-celled euglena, found in ponds, is considered to be on the borderline between animals and plants, having characteristics of both. It moves by means of a whiplike *flagellum,* at the base of which is a contractile vacuole. Nearby is an *eye spot* that is sensitive to light. It contains chloroplasts and makes its own food by photosynthesis.

Spirogyra

In the spirogyra, one of the green algae that live in ponds, the cells are arranged lengthwise in a long thread or filament. Each cell has a spiral chloroplast; in some species there may be two or more chloroplasts in each cell. The spirogyra manufactures its own food by carrying on photosynthesis. Starch is stored in structures on the chloroplast called *pyrenoids.* The individual cells reproduce by binary fission. At times, two filaments lying close to each other will send out projections from the cells, which touch and form a bridge; the protoplasm of each cell in one of the filaments streams across into the cells of the adjoining filament and combines with it. Hard-walled *zygospores* result, which eventually form new filaments. This is another example of the form of sexual reproduction called *conjugation.*

Viruses

Viruses represent an enigma to biologists. Are they living or nonliving? These ultramicroscopic materials are best known because of diseases they cause in living things (AIDS, smallpox, polio, rabies, tobacco mosaic disease, etc.). They spread and invade living tissue. Once there, they reproduce. They thus show properties of protoplasm. On the other hand, they have been purified from infected tissue, and have appeared as dry protein crystals. In this condition, they are as nonliving as crystals of salt. However, once dissolved and placed in contact with living tissue, they become active and reproduce. Perhaps viruses are on the borderline between the living and the nonliving. Investigations have shown that a virus generally contains a core of nucleic acid consisting of DNA or RNA, and a covering made of protein material.

COMPARISON OF AMEBA AND PARAMECIUM

Characteristics	Ameba	Paramecium
Shape	No specific shape	Slipper-shaped
Locomotion	By pseudopods—slow	By cilia—rapid
Ingestion	Engulfs food anywhere along cell membrane	Food enters through oral groove
Digestion	In food vacuoles	In food vacuoles
Elimination of solid wastes	Through temporary opening anywhere along cell membrane	Through definite anal spot
Excretion of wastes	By diffusion through cell membrane	By diffusion through cell membrane
Elimination of excess water	One contractile vacuole	Two contractile vacuoles having canals
Behavior	Moves toward favorable stimulus, away from unfavorable stimulus; forms cysts during unfavorable conditions	Same; also expels trichocysts as protection or for anchoring
Reproduction	Binary fission	Binary fission; conjugation
Nucleus	One	Two (macronucleus and micronucleus)

Section Review

Select the correct choice to complete each of the following statements:

1. The average microscope having a high-power objective marked 43X and an ocular marked 10X gives a magnification of   (A) 33X   (B) 43X   (C) 53X   (D) 100X   (E) 430X

2. In order to admit the proper amount of light, the part of the microscope that should be adjusted is the   (A) coarse adjustment   (B) fine adjustment   (C) diaphragm   (D) eyepiece   (E) stage

3. All of the following scientists studied cells *except*   (A) Hooke   (B) van Leeuwenhoek   (C) Schleiden   (D) Schwann   (E) Linnaeus

4. The nucleus contains all of the following structures *except*   (A) mitochondria   (B) chromatin   (C) genes   (D) nucleolus   (E) nuclear membrane

5. The most abundant substance in protoplasm is   (A) protein   (B) fat   (C) carbohydrate   (D) water   (E) minerals

6. The conversion of nonliving material into living protoplasm is known as   (A) assimilation   (B) respiration   (C) reproduction   (D) absorption   (E) digestion

7. A cell obtains energy during the process of   (A) ingestion   (B) respiration   (C) irritability   (D) excretion   (E) secretion

8. Viruses resemble living things because they   (A) circulate   (B) move   (C) reproduce   (D) are crystalline   (E) are able to respond to stimuli in the environment

9. Animal cells do not possess   (A) a cell membrane   (B) a cell wall   (C) cytoplasm   (D) a nucleus   (E) a nuclear membrane

10. An ameba moves by means of   (A) cilia   (B) flagella   (C) pseudopods   (D) pseudonyms   (E) microscopic hairs

11. Both the ameba and the paramecium possess a (an)   (A) contractile vacuole   (B) anal spot   (C) oral groove   (D) trichocysts   (E) gullet

12. Digestion in one-celled animals takes place in the (A) cyst (B) contractile vacuole (C) food vacuole (D) pellicle (E) gullet

13. Paramecium may reproduce by (A) binary fission only (B) conjugation only (C) both binary fission and conjugation (D) budding only (E) binary fission and budding

14. Dissolved gases pass in and out of paramecium through the (A) nuclear membrane (B) cell membrane (C) food vacuole (D) micronucleus (E) macronucleus

15. Plant and animal cells are alike in possessing (A) chlorophyll (B) chloroplast (C) cell wall (D) cellulose (E) cell membrane

16. The euglena is different from the ameba and paramecium in possessing (A) cytoplasm (B) a nucleus (C) a cell membrane (D) chloroplasts (E) a contractile vacuole

17. The spiral structure in spirogyra is the (A) spireme (B) chloroplast (C) chromatin (D) nucleus (E) filament

18. The only one of the following structures found in a spirogyra is (A) a flagellum (B) an eyespot (C) a pyrenoid (D) a food vacuole (E) cilia

19. ATP is a chemical that is essential for (A) digestion (B) appetite (C) absorption (D) oxidation (E) assimilation

20. Enzymes are useful (A) only during digestion (B) only in respiration (C) in both digestion and respiration (D) only during ingestion (E) in both digestion and ingestion

21. All of the following are organelles *except* the (A) endoplasmic reticulum (B) mitochondria (C) ribosome (D) Golgi complex (E) ultracentrifuge

22. The life activity concerned with the taking in of food is known as (A) ingestion (B) digestion (C) secretion (D) excretion (E) assimilation

23. The life activity dealing with the stability of the organism's chemical makeup under its constantly changing environment is (A) respiration (B) irritability (C) reproduction (D) regulation (E) photosynthesis

24. DNA is found in the cell's (A) vacuole (B) nucleolus (C) nucleus (D) ribosomes (E) cell membrane

25. All of the following have recently given us greater knowledge about the cell *except* (A) the electron microscope (B) the phase microscope (C) laser beams (D) time-lapse cinematography (E) the single-lens microscope

Answer Key

1-E	6-A	11-A	16-D	21-E
2-C	7-B	12-C	17-B	22-A
3-E	8-C	13-C	18-C	23-D
4-A	9-B	14-B	19-D	24-C
5-D	10-C	15-E	20-C	25-E

Answers Explained

1. (E) The magnification is obtained by multiplying the magnifying power of the high-power objective (43) by that of the ocular (10). In this example $43 \times 10 = 430$ times, or 430X.

2. (C) The diaphragm controls the amount of light. A larger opening of the diaphragm will allow more light to reach the object on a slide. If the light is too bright, a smaller opening will reduce it and allow more details to be observed.

3. (E) Carolus Linnaeus (1707–1778) devised the present system of classifying living things according to genus and species. He did not study cells.

4. (A) Mitochondria are organelles located in the cytoplasm of the cell that are active in releasing energy from food.

5. (D) Water makes up from 65% to 90% of the protoplasm in various living things.

6. (A) Living things are able to grow and reproduce by changing the nonliving material in food into the living material of the protoplasm; this process is known as assimilation.

7. (B) During the process of cellular respiration, the chemical bond energy of food is released into energy needed for life activities.

8. (C) Viruses are ultramicroscopic units that can reproduce only when they invade living tissue.

9. (B) The cell wall is made largely of cellulose and is present only in plant cells.

10. (C) An ameba moves by extending projections, called pseudopods (false feet), and flowing into them.

11. (A) Both the paramecium and the ameba possess contractile vacuoles. The contractile vacuole fills with excess water and then contracts, expelling it from the cell.

12. (C) After food has been taken into the cell, it becomes located in a food vacuole, where it is digested by enzymes.

13. (C) A paramecium divides asexually into two cells by the simple process of binary fission. During conjugation, which is a form of sexual reproduction, two paramecia lie next to each other and exchange parts of their micronuclei. Afterwards, they separate and reproduce asexually by binary fission.

14. (B) The cell membrane is semipermeable and controls the passage of dissolved materials into and out of the cell.

15. (E) The cell membrane is the outer layer of the cytoplasm in all types of living cells.

16. (D) Although a euglena is able to move through the water like a protozoan, it can also carry on photosynthesis in its chloroplasts.

17. (B) Spirogyra takes its name from its green, spiral chloroplast that winds along the length of its cell.

18. (C) In a spirogyra, starch is stored in special tiny structures in the chloroplast called pyrenoids.

19. (D) ATP is found largely in the mitochondria, where it serves an important role in the oxidation of food during the process of respiration.

20. (C) Enzymes help digest food during digestion. They also play an essential role in cellular respiration.

21. (E) The ultracentrifuge is a device that whirls cells around in a tube at very high speeds, separating them into layers of their organelles.

22. (A) Ingestion, one of the essential life functions of living things, is the taking in of food.

23. (D) Although a living organism's internal and external environment is constantly changing, the life process of regulation helps keep the organism's chemical makeup stable.

24. (C) DNA, deoxyribonucleic acid, is located in the chromosomes, which are found in the nucleus.

25. (E) The single-lens microscope was used by the early microscopist Antony van Leeuwenhoek (1632–1723).

4.2 MOLECULAR BIOLOGY

Although only microscopic in size, the cell is a veritable chemical factory in which there is a constant interplay of biochemical processes and reactions. *Biochemistry* is the science that deals with the chemical compounds and processes in living things.

Basic Chemistry

Elements

All matter is composed of one or more elements. An element is the simplest form of a substance that cannot be broken down any further by ordinary chemical means, for example, hydrogen, oxygen, carbon, nitrogen. Ninety-two elements occur naturally in nature. A number of additional elements, such as neptunium, plutonium, and curium, have been produced artificially, bringing the total number to more than 109.

Compounds

Elements are combined to form compounds in a definite proportion by weight. A compound has different properties from the elements that compose it. Thus, water, H_2O, is a compound composed of two parts of hydrogen and one part of oxygen. Table salt, NaCl, is composed of sodium combined with chlorine.

Molecules

The smallest part of a substance that still has the properties of the substance, and is capable of stable independent existence, is called a molecule. Water can be broken down into smaller and smaller droplets until the smallest unit is reached, consisting of a molecule, H_2O. If it is broken down still further, it releases hydrogen and oxygen, two different substances.

Atoms

Elements are made of invisible building blocks called atoms. Each atom has a central nucleus surrounded by a definite number of moving, negatively charged *electrons.* The electrons move so rapidly that they are thought of as a cloud surrounding the nucleus. Although the nucleus is only five-thousandth part of the atom in size, it is so dense that it represents nearly all of the atom's weight. The nucleus contains two kinds of particles, positively charged *protons,* and *neutrons,* which have no charge. It has also been found to contain additional particles such as positrons, antiprotons, mesons, neutrinos, and hyperons. Recent research has shown that smaller particles, *quarks,* combine to form protons and neutrons. Electrons are part of a class of particles named *leptons.*

For each proton in the nucleus, there is one electron flying around it, thus making the atom electrically balanced. Uranium, which for some time was the heaviest known element, has 92 protons and 92 planetary electrons. The number of protons makes one element different from another. If one proton could be removed from an atom of oxygen, the result would be an atom of nitrogen.

Atoms of different elements differ in their number of electrons, protons, and neutrons. The simplest atom is that of hydrogen, the lightest element. It contains only 1 proton in its nucleus, and has 1 electron on the outside. Its atomic weight is 1. An atom of helium has 2 protons and 2 neutrons in its nucleus, giving it an atomic weight of 4. A carbon atom has 6 protons and 6 neutrons in its nucleus for an atomic weight of 12. The table below summarizes these facts for a number of elements.

Isotopes

Different forms of the same element, whose nuclei contain the same number of protons but have different numbers of neutrons, are called isotopes. Thus uranium-238 has 92 protons and 146 neutrons, for a total atomic weight of 238. Uranium-235 also has 92 protons, but only 143 neutrons, giving a total atomic weight of 235. There are three isotopes of hydrogen: the common type of hydrogen, with one proton and no neutrons; heavy hydrogen, or deuterium, with 1 proton and 1 neutron; and a heavier hydrogen, or tritium, with 1 proton and 2 neutrons. Some isotopes occur naturally, while others have been made artificially. Some isotopes are stable and remain the same at all times; others are radioactive and are constantly undergoing change, either to another isotope or to another element (e.g., the carbon-14 isotope pictured below).

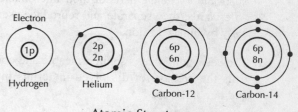

Atomic Structure

| Element | Symbol | Nucleus | | Atomic Weight |
		Protons	Neutrons	
Hydrogen	H	1	0	1
Helium	He	2	2	4
Carbon	C	6	6	12
Nitrogen	N	7	7	14
Oxygen	O	8	8	16
Aluminum	Al	13	14	27
Potassium	K	19	20	39
Iron	Fe	26	30	56
Cobalt	Co	27	32	59
Iodine	I	53	74	127
Lead	Pb	82	126	208
Uranium	U	92	146	238

Chemical Reactions

In a diagram of atomic structure, the electrons are shown arranged in rings or shells around the nucleus. Each shell holds a certain number of electrons. The innermost shell, designated as *K*, has 2; the next shell, *L*, can hold 8; the third shell, *M*, also 8; etc. When an atom has a shell that does not have a complete set of electrons, it is said to be structurally unbalanced, and will tend to interact with other atoms. The chemical activity of an atom, then, arises from the number of electrons in its outer shell. Hydrogen has 1 electron in its *K* shell; it can hold 2 electrons. The numbers of electrons present in the outermost shells of some other atoms (and the numbers of electrons the shells can hold) are as follows: sodium—1 (8); carbon—4 (8); oxygen—6 (8); chlorine—7 (8).

Each of these orbits represents a different energy level. The electrons in the inner shell are bound tightly to the nucleus. Those in the outermost shell are not held as tightly, and may react with the outermost shell of other atoms to form compounds. When chemical reactions normally occur, electrons are involved, not the nucleus. However, the nucleus may be changed under certain conditions involving atomic reactions.

Chemical Bonds

Atoms are combined to form molecules. The forces of attraction between them are referred to as chemical bonds. When bonds are made or broken, chemical reactions occur, involving energy changes.

A chemical bond may be made when two atoms share electrons. Thus, the hydrogen atom has 1 electron. Since its shell can hold 2 electrons, 2 atoms of hydrogen can share their electrons to form a molecule of hydrogen gas (H_2). In forming a molecule of water, H_2O, 2 hydrogen atoms share electrons with an atom of oxygen, which has 6 electrons in its outermost shell (8 is the stable number of electrons in its outermost *L* shell). Similarly, 4 atoms of hydrogen can share their electrons with an atom of carbon, which has only 4 electrons in its outermost shell, to form methane, CH_4. Such bonds formed when atoms share electrons are called *covalent bonds*.

In another type of bond, the *ionic bond*, electrons are transferred from one atom to another. This is the case in table salt or sodium chloride, NaCl. Sodium has only 1 electron in its outermost shell, which can hold 8 electrons; chlorine has 7 electrons in its outermost shell, and needs 1 more electron to complete this shell. When sodium and chlorine are united, an electron passes from the sodium atom to the chlorine atom. The chlorine atom is referred to as an electron acceptor.

When the sodium atom loses an electron, it loses its neutral status. Up to this time, it had equal number of electrons and protons. Now, after losing an electron, it has an extra positive charge, and becomes an ion of sodium, Na^+. Similarly, the chlorine atom gains an electron and has an extra negative charge, becoming a chloride ion, Cl^-. An *ion* is thus an electrically charged atom with either a positive or negative charge.

Ionization

In ionization, the bonds of a substance are broken, and it separates into its ions. Sodium chloride ionizes when it is dissolved in water, to form ions of Na^+ and Cl^- in solution. Water also ionizes, with some of the H_2O molecules breaking down into hydrogen ions, H^+, and hydroxide ions, OH^-.

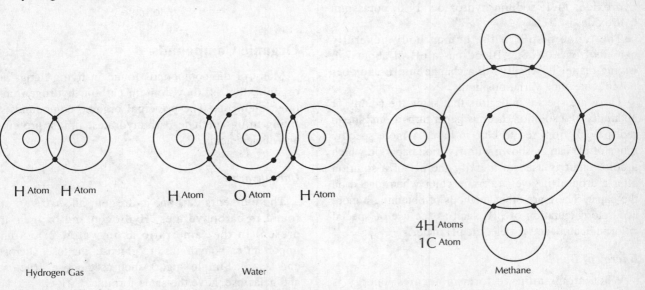

H Atom   H Atom

Hydrogen Gas

H Atom   O Atom   H Atom

Water

4H Atoms
1C Atom

Methane

Sharing of Electrons in Covalent Bonds

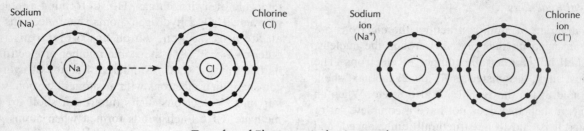

Transfer of Electrons in Ionic Bonds

pH Scale

The number of hydrogen ions in a solution is the basis of pH. (Strictly speaking, this unit is the negative logarithm of the mass of hydrogen ions in a liter of solution. Thus, if 1 liter of a certain solution contains 10^{-7} gram H^+ ions, the pH = -7. The negative log is then equal to 7.)

By various methods, the relative numbers of free H^+ and OH^- ions in solution can be determined. The pH scale extends from 0 to 14. A pH of 7 is neutral. Water, which has equal numbers of H^+ and OH^- ions, has a pH of 7, and is neutral. Below the pH of 7, the number of H^+ ions increases and the solution becomes more and more *acid*. The lower the pH number, the stronger the acid; the nearer the pH number approaches 7, the weaker the acid. Some common pH values for acids: boric acid, 5.2; acetic acid, 2.9; hydrochloric acid, 1.1.

Similarly, above pH 7, the relative number of H^+ ions decreases, and the concentration of HO^- ions increases, to make the solution more *base,* or alkaline. The greater the number above 7, the stronger the base; the nearer the pH number approaches 7, the weaker the base. Some common pH values for bases: ammonium hydroxide, 11.1; sodium hydroxide, 14.0; potassium hydroxide, 14.0.

The range of pH in the human body generally extends from 3 to 8.5. Blood has a pH of about 7.4. Living matter is sensitive to a change in pH, and does not tolerate wide variations in it.

There are special indicators that show the acidity or alkalinity of a solution. Litmus paper turns from blue to red in acid; from red to blue in base. A more specific range of pH can be shown with Hydrion paper, in which a color chart is used to identify the pH of a solution after a drop of it gives a certain color when placed on the paper. There are other methods of obtaining a more accurate calculation of pH, such as the use of special color indicators and the electric pH meter.

Energy of Bonds

When atoms unite to form molecules, energy is involved in the establishment of their bonds. Activation energy may be of several types, such as heat, light, or electricity. There are varying amounts of bond energy in different compounds. During chemical reactions, such bonds are created or broken.

Structural Formulas

The chemical bond between the atoms of hydrogen that form a molecule of hydrogen gas, H_2, can be shown as H—H. In the case of oxygen gas, O_2, the bond can be shown as O=O; here, two pairs of electrons are shared by the two atoms. Water, H_2O, can be designated as H—O—H, where oxygen shares 2 electrons with the 2 hydrogen atoms. In the gas methane, CH_4, a carbon atom shares electrons with 4 hydrogen atoms; this can be represented as follows:

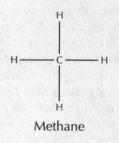

Methane

Organic Compounds

Many of the compounds found in living things are organic, that is, they contain carbon, hydrogen, and other elements. The principal organic compounds of living things include carbohydrates, lipids, proteins, and nucleic acids.

Carbohydrates

The elements carbon, hydrogen, and oxygen are found in carbohydrates. Hydrogen and oxygen are present in the same ratio as in water, 2:1. Some important carbohydrates are glucose, maltose, sucrose, and starch. Simple sugars such as glucose, fructose, and galactose, have the same formula, $C_6H_{12}O_6$, and are called monosaccharides. Although they contain the same atoms, their characteristics are different because of the different arrangements within their molecules. Thus, their structural formulas differ to a slight degree. Glucose is the source of energy for most organisms. Its structural formula is as follows:

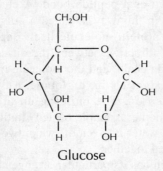

Glucose

Glucose molecules may be linked together to form the disaccharide maltose, by a process called *dehydration synthesis*. This refers to the building up of complex molecules from simpler molecules, with the release of water. Other disaccharides are sucrose, which is ordinary table sugar, and lactose or milk sugar. Their formula is $C_{12}H_{22}O_{11}$. The structural formula of maltose is as follows:

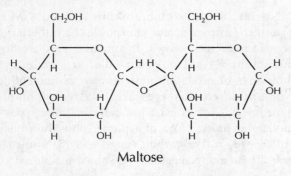

Maltose

Hundreds of glucose units bonded together make up more complex carbohydrates, called polysaccharides. Such a collection of many similar, repeating units to form a large molecule is referred to as a *polymer*. Examples are starch, cellulose, and glycogen. Starch is a storage form of sugar. Glycogen is known as animal starch. Cellulose is the supporting material found in the cell walls of plant cells.

Lipids

Lipids include fats, oils, and waxes. Like carbohydrates, lipids also contain carbon, hydrogen, and oxygen. However, there is proportionately much less oxygen in relation to hydrogen. Lipid molecules are relatively small and are not arranged as polymers. Typically, a lipid molecule consists of a glycerol molecule bonded to 3 fatty acid molecules. A glycerol molecule can be represented as follows:

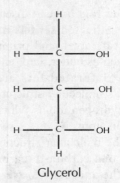

Glycerol

A fatty acid molecule consists of a long chain of carbon and hydrogen (hydrocarbon) atoms with a carboxyl (—COOH) group. The number of carbon atoms may vary from 4 to 24, although most of the common fatty acid molecules contain 16–18 carbon atoms. Here is the general structural formula of a fatty acid:

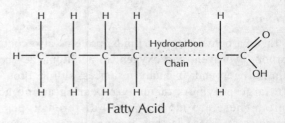

Fatty Acid

When the 3 fatty acid molecules are linked to a glycerol molecule, a lipid molecule is formed. Each of the glycerol's 3 OH groups becomes attached to 1 of the 3 fatty acid molecules, with the splitting off of 3 water molecules at the sites of linkage. This is another example of dehydration synthesis. The structural formula of a lipid can be represented as follows:

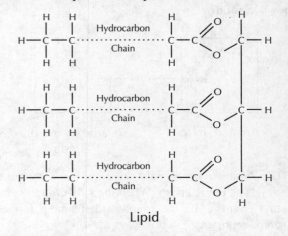

Lipid

Fats and oils have the same molecular structure, although fats are solid at room temperature and oils are liquid. A common fatty acid is stearic acid, $C_{17}H_{35}COOH$. When combined with glycerol, it forms stearin, a fat. Other fatty acids are palmitic acid ($C_{15}H_{31}COOH$), present in the fat palmitin, and oleic acid ($C_{17}H_{33}COOH$), in the oil olein.

Because of their role in heart disease, *saturated fats* in the diet have attracted serious medical attention. These types of fats are found primarily in foods of animal origin, such as butter, lard, and whole milk and milk products. Saturated fats are solid at room temperature. In its molecular structure, saturated fat is saturated with hydrogen atoms which are attached to each of the carbon atoms.

By contrast, the molecular structure of *unsaturated fat* has at least one carbon-to-carbon double bond, meaning that the molecules are not completely saturated with hydrogen atoms. Two hydrogen atoms are missing for every such double bond. Unsaturated fats are liquid at room temperature. Dietary sources of these types of fat include plant oils such as corn oil, olive oil, and sunflower oil, and margarine and fish oils.

Proteins

Like carbohydrates and lipids, proteins also contain carbon, hydrogen, and oxygen; in addition, they contain nitrogen and, in many instances, sulfur. Proteins are large polymers of many repeating amino acid units. There are more than 20 different types of amino acids. These can be combined in various ways to form many types of proteins.

The general structure of an amino acid shows a carbon atom bonded in four places with (1) a hydrogen atom, (2) an amino group (NH_2), (3) a carboxyl acid group (COOH), and (4) a side group (R), which is variable and is the basis for the variety of different amino acids. The structure of an amino acid can be shown as follows:

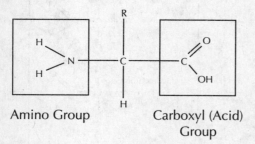

Amino Group   Carboxyl (Acid)
                          Group

The simplest amino acid, glycine, has a hydrogen atom (H) as the R group. When the side group, R, is a methyl group (CH_3), the amino acid alanine if formed; it can be represented in this way:

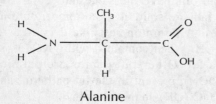

Alanine

Proteins are built up by the bonding of many amino acids. There may be more than 3,000 amino acids in a protein. In the bonding of amino acids, a C—N bond forms between the carboxyl group of one amino acid and the amino group of the next amino acid. The C—N bond is known as a peptide bond. A chain of amino acids bonded in this way is called a *polypeptide*. The shape of the protein molecule itself depends on the nature of the attraction between the different parts of the polypeptide chain, and the molecule may be in the form of a straight chain, a globule, a twisted chain (helix), and so on. The building up of the protein molecule is another example of dehydration synthesis; a molecule of water is given off when two amino acids become bonded together.

Each protein has its particular arrangement of amino acids. This gives the protein it properties. Since the structure of proteins is so complex, it has been difficult to analyze them. In 1954, the structural arrangement of the amino acids in a protein, insulin, was determined for the first time, by English biochemist Frederick Sanger.

Nucleic Acids

Nucleic acids, which contain carbon, oxygen, hydrogen, nitrogen, and phosphorus, are the largest organic molecules known. In structure, they are high-molecular-weight polymers that are made up of thousands of repeating units known as nucleotides. Nucleotides themselves are relatively complex molecules. They consist basically of three types of molecular units: (1) a phosphate (phosphoric acid, H_3PO_4), (2) a five-carbon sugar, and (3) a nitrogen base. Their arrangement may be shown as follows:

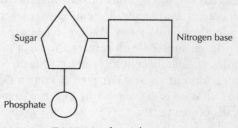

DNA Nucleotide

The sugar is either ribose or deoxyribose. Deoxyribose, as the name indicates, has one oxygen less than ribose. The nitrogen base may be a purine, either adenine or guanine; or it may be a pyrimidine, either cytosine, thymine, or uracil. These bases are joined by weak hydrogen bonds. The significant nucleic acids are DNA (deoxyribonucleic acid) and RNA (ribonucleic acid). DNA plays a key role in the determination of heredity. RNA is important in the synthesis of protein. Further details about the nucleic acids will be considered in Section 10.4.

Enzymes

Living matter is constantly in a state of dynamic chemical activity. The reactions that take place are made possible by organic catalysts called enzymes. Each cell contains over a thousand enzymes. Enzymes help provide energy for the cell, assist in the building of new cell structures, digest food, and play a part in almost every activity that takes place.

Enzymes are considered organic catalysts, because they affect the rate of a chemical reaction without being changed. They can be used over and over again. They are protein in nature, and specific in their action. They often work together with coenzymes in reactions. Coenzymes have smaller molecules than enzymes and are not protein. They are active only with enzymes. Such B-complex vitamins as thiamin, riboflavin, and niacin have been found to act as coenzymes in cellular respiration, leading to energy release.

How Enzymes Function

Enzyme molecules are huge compared to the molecules with which they interact. Apparently, only a small portion of the enzyme functions when it is active. The localized region is called the active site of the enzyme. The action of enzymes has been explained on the basis of the "lock and key" model. This analogy arises from the fact that a particular enzyme interacts only with a single type of substrate molecule. The substrate is the substance the enzyme acts on. The name of an enzyme usually has the ending -ase, added to the stem of the word, which is taken from the substrate. Here are three examples:

Enzyme	Substrate
Maltase	Maltose
Lipase	Lipids
Protease	Protein

The association between enzyme and substrate is thought to be a close physical one, but does not lead to the formation of bonds between them. It is sometimes referred to as an enzyme-substrate complex. Enzyme action takes place while the enzyme-substrate is formed.

The following illustration shows how an enzyme may cause a complex molecule, designated as *A-B,* to separate into two smaller molecules, *A* and *B*.

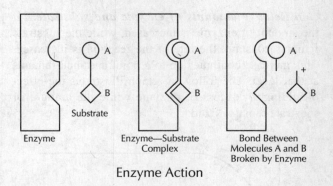

Enzyme Action

Factors Affecting Enzyme Action

The rate of enzyme action is not fixed, but varies with such conditions as pH, temperature, and relative amounts of enzyme and substrate.

1. *pH.* Some enzymes act at a pH in the vicinity of 7; for example, the enzyme maltase. If the solution became acid or base, this enzyme's action would be slowed down. The following graph illustrates the effect of varying the pH:

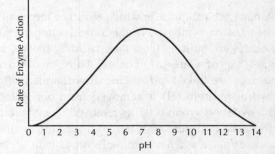

However, many enzymes function best at other pH values. Pepsin, found in the stomach, has its best activity at a pH of 1.5–2.2. Trypsin, in the small intestine, acts best at pH 7.9–9.0.

2. *Temperature.* Most enzymes function best at body temperature, 37°C. As the temperature is lowered, their rate of activity decreases. As the temperature is raised, their activity increases, until a maximum is reached at about 40°C. Beyond this point, the shape of the enzyme molecule becomes distorted, and enzyme deactivation occurs. The effects of varying temperatures on the action of many enzymes are shown in the following graph:

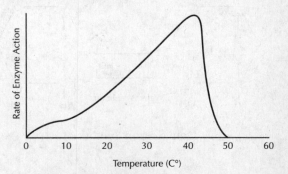

3. ***Relative amounts of enzyme and substrate.*** If the amount of enzyme is increased, while the substrate remains constant, the rate of the reaction is increased. This increase continues up to a point and then remains at that level. The following graph illustrates the effect of adding an excess of enzyme while the amount of substrate remains fixed:

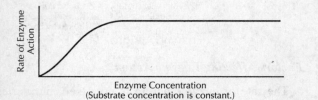

Enzyme Concentration
(Substrate concentration is constant.)

Similarly, if the amount of substrate is increased while the concentration of enzyme remains the same, the rate of the reaction will increase. This will continue up to the point where every available enzyme molecule is actively involved in the reaction.

Dehydration Synthesis

Among the reactions in which enzymes are involved is dehydration synthesis. In this process, large organic molecules are built up from their building blocks, with the release of water molecules. Thus, two glucose molecules are linked to form the disaccharide maltose. A hydrogen atom (H) is removed from one glucose, and a hydroxyl group (OH) is removed from the other, to form the molecule of water. Dehydration synthesis also takes place when a lipid molecule is built up from fatty acids and glycerol; also, when a protein molecule is being synthesized from amino acids.

Hydrolysis

In hydrolysis, the reverse process takes place. Large molecules are broken down to their building blocks, with the addition of water. The enzyme maltase, for example, acts on maltose and breaks the —O— bond between the two glucose subunits. A hydrogen atom (H) from the water molecule combines with the O to form the OH part of the glucose molecule. The OH from the water molecule is added to make the other glucose molecule complete.

The reversible reactions, involving enzymes in dehydration synthesis and hydrolysis, are shown in the illustrations below.

Cellular Respiration

Energy of the Cell

Energy is needed for such activities as the contraction of a muscle cell, conduction by a nerve cell, synthesis of organic compounds by a cell, movement by a paramecium, active transport by a root cell, and division of a cell. The energy to carry on these activities is derived from food, just as a candle flame burns fuel. However, the similarity ends there. In a fire, the fuel is burned all at once. In the cell, the organic bonds of food are broken down in a series of many small steps involving dozens of enzymes, which are located in the mitochondria. Each enzyme is specific in performing only one step of the entire process. Glucose is the chief source of energy, but proteins and lipids may also be used. The release of the chemical-bond energy of food into energy that can be used for the life activities of the cell is known as *respiration.*

Every cell contains molecules of the nucleotide ATP (adenosine triphosphate), which is the actual source of energy. Each ATP molecule stores energy in its phosphate bonds. When one of these bonds is broken, energy is released to the cell for its various activities and ATP is converted to ADP (adenosine diphosphate), giving off a phosphate group, P. It is this change of ATP into ADP which supplies the cell with energy:

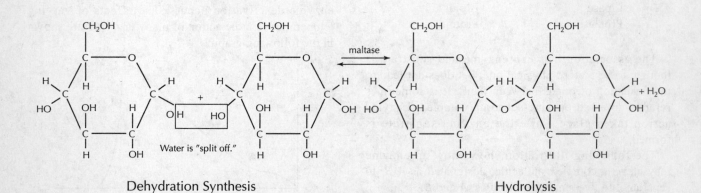

Dehydration Synthesis   Hydrolysis

$$ATP \underset{\longleftarrow}{\xrightarrow{enzymes}} ADP + P + energy$$

Without respiration taking place, the ATP supply of the cell would soon be used up. To produce more ATP, high-energy phosphate bonds must be added to ADP. The energy for changing ADP into ATP comes from the chemical bonds of glucose. In short, the breakdown of glucose leads to the storage of energy in ATP. The breakdown of ATP provides the cell with energy. When the whole process takes place in the presence of oxygen, it is referred to as aerobic respiration; without oxygen, it is anaerobic respiration.

Aerobic Respiration

There are three phases in the process: (1) glycolysis, (2) the Krebs cycle, (3) the electron transport chain.

1. *Glycolysis.* This takes place in the cytoplasm of the cell. Through a series of steps, glucose, which is a six-carbon compound ($C_6H_{12}O_6$), is converted to 2 three-carbon molecules of pyruvic acid ($C_3H_4O_3$). Hydrogen is removed; this is oxidation. Energy is released in this stage and is stored in 4 molecules of ATP. However, in order for some of these processes to take place, the energy from 2 ATP was used up. As a result, the total gain of energy during glycolysis is 2 ATP molecules. In this stage, oxygen has not yet been involved and it is considered to be an anaerobic phase.

2. *The Krebs (or citric acid) cycle.* The next phase of aerobic respiration is named after Sir Hans Krebs, who received the Nobel Prize (1953) for his research.

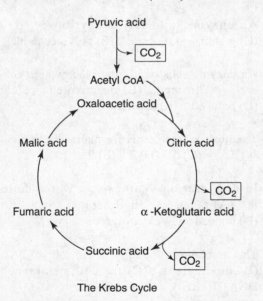

The Krebs Cycle

The remaining steps of respiration take place within the mitochondria of the cell and involve many enzymes and coenzymes. The pyruvic acid molecules are decarboxylated (CO_2 removed) in a series of reactions. Some of the intermediate compounds formed are oxaloacetic acid, citric acid, succinic acid, and ketaglutaric acid. Hydrogen is also removed during this cycle. Two additional ATP are formed.

3. *The electron transport chain.* During this phase, the hydrogen released previously is passed on to a series of compounds, which are referred to as hydrogen acceptors: NAD (nicotinamide-adenine dinucleotide), riboflavin coenzyme, and cytochrome. The electrons of the hydrogen atoms are passed along to these acceptors to release energy in small units. This energy is stored in 32 ATP molecules along the way. The final hydrogen acceptor is oxygen and the combination forms water.

The detailed steps may be summarized as follows:

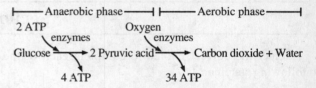

During the succession of steps involved in the release of energy from 1 molecule of glucose, it can be seen that the energy is stored in a total of 36 ATP molecules, as follows: anaerobic phase (glycolysis), 2 ATP; aerobic phase (Krebs cycle), 2 ATP + (electron transport chain) 32 ATP = 36 ATP. Aerobic respiration of glucose may be summarized in the following equation:

$$C_6H_{12}O_6 + 6O_2 \xrightarrow[enzymes]{numerous} 6CO_2 + 6H_2O + 36 ATP$$

Anaerobic Respiration

Most cells use oxygen and carry on aerobic respiration. However, in some cases, as in yeast, certain bacteria, and fungi, and in the work of muscle cells, oxygen is not used and anaerobic respiration occurs. This is also referred to as fermentation. As in aerobic respiration, glycosis takes place, resulting in the formation of pyruvic acid. The net gain of energy is 2 ATP molecules. Following this, either lactic acid or ethyl alcohol and carbon dioxide are formed, without any additional yield of energy. A great deal of energy is still contained in the lactic acid or alcohol molecules, where the C—C and C—H bonds remain intact. Thus, anaerobic respiration is not as efficient a process as aerobic respiration in converting glucose to energy.

Lactic acid fermentaion may be summarized in the following equation:

$$Glucose \xrightarrow{enzymes} 2 Lactic acid + 2 ATP$$

Lactic acid is produced in muscles that are very active. Under such conditions, anaerobic respiration takes place. The accumulation of lactic acid results in muscle fatigue.

Alcohol fermentation is the basis of the brewing, baking, and wine industries. It may be summarized as follows:

$$\text{Glucose} \xrightarrow{\text{enzymes}} \text{2 ethyl alcohol} + 2CO_2 + 2\text{ ATP}$$

COMPARISON OF AEROBIC AND ANAEROBIC RESPIRATION

Characteristic	Aerobic	Anaerobic
Net yield of energy	36 ATP	2 ATP
Products	$CO_2 + H_2O$	Lactic acid or CO_2 + alcohol
Use of oxygen	Yes	No

Section Review

Select the correct choice to complete each of the following statements:

1. Uranium-238 is different from uranium-235 in having (A) more neutrons (B) more protons (C) more electrons (D) fewer neutrons (E) fewer protons

2. All of the following are elements *except* (A) hydrogen (B) oxygen (C) carbon (D) plutonium (E) water

3. The *K* and *L* shells are full when they hold the following numbers of electrons, respectively: (A) 2 and 2 (B) 2 and 4 (C) 2 and 6 (D) 2 and 7 (E) 2 and 8

4. Chlorine has 7 electrons in its outer shell. The number of electrons it needs to complete the shell is (A) 1 (B) 2 (C) 3 (D) 4 (E) 5

5. In a covalent bond, electrons are (A) shared (B) split (C) lost (D) added (E) transferred

6. In an ionic bond, electrons are (A) hydrolyzed (B) transferred (C) fused (D) neutralized (E) shared

7. The strongest acid has a pH of (A) 1 (B) 5 (C) 7 (D) 9 (E) 11

8. A substance composed of a large molecule that is made of many smaller, similar, repeating units is called (A) glycerol (B) a polymer (C) a fatty acid (D) a monosaccharide (E) a disaccharide

9. A lipid molecule is composed of glycerol and fatty acid molecules in a ratio of (A) 1:1 (B) 1:2 (C) 1:3 (D) 1:4 (E) 1:5

10. An amino acid molecule contains all of the following *except* (A) a hydrogen atom (B) phosphate (C) an NH_2 group (D) a COOH group (E) an R group

11. A combination of a phosphate, a ribose, and a nitrogen base is characteristic of a (A) carbohydrate (B) lipid (C) protein (D) nucleotide (E) fatty acid

12. An enzyme is a (A) deoxyribose (B) lipid (C) protein (D) ribose (E) polysaccharide

13. An enzyme-substrate is the place where enzymes are (A) formed (B) deactivated (C) active (D) reduced (E) diluted

14. The most favorable pH for maltase action is (A) 1 (B) 3 (C) 5 (D) 7 (E) 9

15. In dehydration synthesis, a water molecule is (A) taken in (B) liberated (C) absorbed (D) bonded (E) electrolyzed

16. The storehouse of energy in the cell is (A) ATP (B) the nucleus (C) the cell membrane (D) DNA (E) RNA

17. The release of the chemical-bond energy of food is called (A) digestion (B) irritability (C) respiration (D) ingestion (E) diffusion

18. The net numbers of ATP molecules formed in aerobic and anaerobic respiration, respectively, are (A) 30 and 2   (B) 32 and 2   (C) 34 and 2 (D) 35 and 2   (E) 36 and 2

19. As the result of glycolysis, the three-carbon molecule formed is   (A) DPN   (B) oxalacetate (C) citric acid   (D) succinic acid   (E) pyruvic acid

20. During fermentation, the final products formed are CO_2 and either lactic acid or (A) water (B) glucose (C) alcohol (D) riboflavin coenzyme (E) cytochrome

Answer Key

1-A	5-A	9-C	13-C	17-C
2-E	6-B	10-B	14-D	18-E
3-E	7-A	11-D	15-B	19-E
4-A	8-B	12-C	16-A	20-C

Answers Explained

1. (A) Both types of uranium are isotopes. Uranium-238 has a greater atomic weight because it has 146 neutrons in its nucleus, whereas uranium-235 has only 143 neutrons.

2. (E) Water is a compound composed of 2 elements: hydrogen and oxygen. One molecule of water contains 2 atoms of hydrogen and 1 atom of oxygen.

3. (E) The electrons of an atom are arranged in shells around the nucleus, each shell holding a certain number of electrons. The innermost shell, designated as *K*, can hold 2 electrons; the next shell, *L*, can hold 8 electrons.

4. (A) The outermost shell of chlorine can hold 8 electrons. If it already has 7, it needs 1 more to complete the shell.

5. (A) Covalent bonds are formed when atoms share electrons. For example, in a molecule of H_2O, 2 hydrogen atoms share electrons with 1 atom of oxygen.

6. (B) In an ionic bond, electrons are transferred from one atom to another. Thus, in table salt, NaCl, sodium (Na) has only 1 electron in its outermost shell, which is passed on to chlorine, (Cl), which has 7 electrons in its outermost shell and needs 1 more electron to complete this shell.

7. (A) The lower the pH number, the stronger the acid is. At low pH, the number of H+ ions increases and the solution becomes more acid.

8. (B) A polymer is made up of many smaller, similar, repeating units. Complex carbohydrates, called polysaccharides, are polymers, composed of many similar, repeating units (saccharides). Examples: starch, cellulose, glycogen.

9. (C) A lipid molecule consists of a glycerol molecule bonded to three fatty acid molecules.

10. (B) Typically, an amino acid molecule contains a carbon (C) atom bonded in four places with: (1) a hydrogen (H) atom; (2) an amino group (NH2); (3) a carboxyl acid group (COOH); and (4) a side group (R), which varies.

11. (D) A nucleotide is made up of three types of molecular units: (1) a phosphate; (2) ribose or deoxyribose, a five-carbon sugar; and (3) a nitrogen base, either a purine or a pyrimidine.

12. (C) An enzyme is a protein that acts as an organic catalyst in chemical reactions.

13. (C) The substrate is the substance acted on by an enzyme. For example, maltose is the substrate acted on by the enzyme maltase.

14. (D) The enzyme maltase functions best at the neutral pH of 7. If the solution becomes acid or base, enzyme action is slowed down.

15. (B) During dehydration synthesis, large organic molecules are built up from their building blocks, with the release of water. The word "dehydration" means the withdrawal, or liberation, of water.

16. (A) The ATP molecule, adenosine triphosphate, stores energy in its phosphate bonds. When one of these bonds is broken, energy is released and ATP gives off a phosphate group, becoming ADP (adenosine diphosphate).

17. (C) The release of the chemical-bond energy of food into energy that can be used for the life activities of the cell is known as respiration.

18. (E) Aerobic respiration takes place in the presence of oxygen, and results in a net yield of 36 ATP molecules from each molecule of glucose. Anaerobic respiration takes place in the absence of oxygen and results in a net yield of 2 ATP.

19. (E) In the first stage of aerobic respiration, glycolysis, the six-carbon compound glucose ($C_6H_{12}O_6$) is broken down into 2 three carbon molecules of pyruvic acid ($C_3H_4O_3$); hydrogen is removed.

20. (C) During fermentation, which is also referred to as anaerobic respiration, oxygen is not used, and there is a net gain of only 2 ATP molecules. Either lactic acid or alcohol and carbon dioxide are formed as waste products.

4.3 TISSUES, ORGANS, AND SYSTEMS

Single-celled organisms such as the ameba, paramecium, and euglena carry on the basic life activities through special structures contained within the cell. Higher organisms consisting of millions of cells carry on similar activities by means of tissues, organs, and systems that are specialized to perform the specific activities. They have a differentiation, or division of labor, of their parts and functions in which all the different activities are coordinated for the mutual benefit of the individual cells and the total organism.

Animal Tissues

A tissue is a group of similar cells performing the same function. Mucous membrane cells from the tissue lining the inside of the cheek can readily be examined by scraping it lightly with a toothpick and depositing the material gently on a slide in a drop of stain. The arrangement of the cells next to each other can be seen, as well as the five- or six-sided shape of each cell. The typical structures of protoplasm are noticeable, that is, nucleus, cytoplasm, and cell membrane. By special techniques, tissues from other parts of the body can also be examined. The study of tissues is known as *histology*. Tissues are classified as epithelial, muscle, nerve, connective and supporting, blood, and reproductive.

Epithelial Tissue

This tissue consists of a continuous layer of cells covering the body surfaces and lining the cavities within the body. It is found on the outer layer of skin; in the lining of the digestive tract, the respiratory system, and the blood vessels; and in the glands. In these various parts of the body, it may have one or more of the following functions: protection, absorption, secretion, and sensation. Epithelial cells may also differ in appearance, some being flat, some boxlike or cuboidal, and others tall and rectangular (columnar).

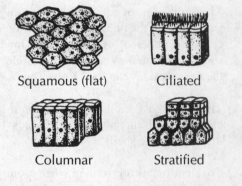

Squamous (flat)   Ciliated

Columnar   Stratified

There are several types of epithelial tissue:

1. *Epidermis of the skin*. The outer layers of the skin protect the underlying cells from injury, from bacteria, and from drying out. There are layers of these cells that are constantly being worn away and replaced from underneath.

2. *Flat (squamous) epithelium.* These cells are broad and flat, like tiles on a floor. They form a protective membrane that lines the inside of the mouth (mucous membrane) and the esophagus. They have

special cells that secrete *mucus* to lubricate the passages and to trap dust and bacteria. They also line the smallest blood vessels (capillaries), where they allow dissolved materials to pass through them (absorption).

3. *Ciliated epithelium.* The epithelial cells lining the nasal cavities and the trachea (windpipe) have cilia that are constantly in motion, beating foreign particles upward and outward.

4. *Columnar epithelium.* These cells line the small intestines, where they absorb digested food. They contain special cells called goblet cells, which secrete mucus, a sticky fluid that lubricates the interior of the intestine.

5. *Glandular epithelium.* The complex glands of the digestive system (e.g., salivary glands), the endocrine glands, which secrete hormones (e.g., thyroxin), and the tear glands are specialized to secrete special liquids used by the body.

6. *Sensory epithelium.* Sensations from the outside of the body are received by certain other epithelial cells which are specialized to receive stimuli. The olfactory cells lining the inside of the nose help us to smell. The cells on the retina of the eye permit us to see.

Muscle Tissue

This tissue has the ability to contract and produce movement. There are three types of muscle tissue:

1. *Voluntary or striated muscles.* These contract according to our will. Muscles of our arms and legs are examples of this type of muscle: they are usually attached to the bones of the skeleton, and so may also be referred to as skeletal muscles.

Striated Muscle   Heart Muscle   Smooth Muscle

Under the microscope, a small piece of beef that has been teased apart and stained shows the striated structure and many nuclei. Striated muscles consist of *fibrils*, which have a striped or striated appearance, with alternating dark and light bands. The fibrils consist of two types of proteins, called *actin* and *myosin*. According to current hypothesis, contraction is produced by the sliding of thin filaments of actin between thick filaments of myosin. This results in skeletal movement.

2. *Smooth muscle.* This type is involuntary, and is found in the walls of the alimentary canal and the blood vessels. We are rarely aware of the activities of these muscles. Food is moved along and churned in the stomach and intestines by the wavelike contractions and expansions of the smooth muscles in the walls of these organs; this muscular activity is known as *peristalsis*. The relaxation of the muscles of the blood vessels in the face enlarges their diameter and increases the amount of blood they contain; such a sudden increased supply of blood in the skin results in blushing.

3. *Cardiac or heart muscle.* This type is also involuntary, and is found only in the heart. Its cells have a striated appearance.

Nerve Tissue

Nerve cells are specialized for transmitting messages or impulses through the body. They appear different from other cells in having branched projections that are in close contact with each other. A *neuron* (nerve cell) has a cell body containing the nucleus and most of the cytoplasm of the cell. Extending out from the cell body are branched threads, called *dendrites*. Many neurons also have a very long extension, or *axon*, which may be up to several feet long, and is covered with a fatty myelin sheath. Bundles of axons make up a nerve.

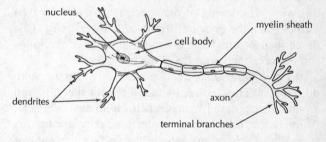

Motor Neuron

There are three types of neurons: (1) sensory neuron—brings impulses from the sense organs (eye, ear, skin, etc.) to the brain or spinal cord; (2) motor neuron—carries impulse from the brain or spinal cord to a muscle to make it move, or to a gland, causing it to secrete; (3) interneuron—connecting neuron between other neurons.

Connective or Supporting Tissue

This tissue consists of several types of cells quite different from each other, whose function is either to support the body or to connect its parts. The cells secrete a nonliving substance, or *matrix*, around them.

1. *Bone.* "Solid bone" is not as solid as it seems. It consists of cells separated from each other by non-living intercellular deposits of calcium and phosphorus, which are secreted by the cells themselves. These cells are connected with each other by fine, rootlike projections of cytoplasm. Blood vessels supply the cells with food and oxygen and remove the wastes. Bone tissue makes up the skeleton, which supports the body and protects vital organs such as the brain and lungs. The bones of the skeleton also serve as levers for body movement, and as anchor sites for muscle action.

2. *Cartilage.* The cells of cartilage are contained in a nonliving matrix, which is smooth, firm, and flexible, but not as hard as bone. Cartilage can be felt in the ears, nose, and windpipe. It also occurs in the joints at the ends of bones, where it allows smooth action. When cartilage breaks down, *arthritis* results, with pain and swelling at the joints.

3. *White fibrous connective tissue.* This consists of a very tough matrix of white fibers containing the living cells. It is found in *tendons,* which connect muscles to bone, and *ligaments,* which connect bones together, forming joints, such as the knee and elbow.

4. *Yellow elastic connective tissue.* This contains fibers that stretch when necessary, and then return to their normal size. The arteries expand as each beat of the heart sends spurts of blood through them and then relax before the next beat or "pulse." The elastic fibers in the walls of the arteries permit them to expand in this way. "Hardening of the arteries" (*arteriosclerosis*) results when the elastic fibers lose their elasticity. This tissue is also found in the bronchial tubes of the lungs and between the vertebrae of the backbone.

5. *Fat (adipose) tissue.* The cells of fat tissue store oil and fat. This tissue, located beneath the skin and around the heart and kidneys, helps to retain body heat. Animals that hibernate in winter gradually use up this stored fat. Overweight people have a large amount of adipose tissue.

Blood Tissue

The cells of this tissue are carried in a liquid called *plasma.* Plasma is a straw-colored fluid consisting of water and dissolved proteins, salts, nutrients, antibodies, hormones, and wastes. There are three kinds of cells in it:

1. *Red blood corpuscles*—round disc cells that lose their nuclei after being manufactured in the marrow of the bones. They contain hemoglobin, which unites freely with oxygen. Their function is to bring oxygen to all the cells of the body. Blood appears red because of the great number of these red corpuscles in the plasma.

2. *White blood corpuscles*—ameba-like cells that move through the circulatory system and the body tissues. They engulf and destroy bacteria that may have entered the body. They are also called *phagocytes.*

3. *Platelets*—very small groups of cells that play a role in blood clotting.

Reproductive Tissue

Special tissues of the body are set aside for the formation of the reproductive cells. *Egg* cells (female cells) are produced in the *ovaries,* and *sperm* cells (male cells) are formed in the *testes.* Sperm cells are extremely small, and have a tail of cytoplasm for movement. The nucleus is contained in the head end. Egg cells are much larger and are round. Both types of sex cells are formed by a special type of nuclear division in which each receives half the amount of chromatin material normally found in cells. When these two types of cells unite during *fertilization,* the fertilized egg then has the full amount of chromatin material.

Organs

Tissues are grouped together to form *organs.* For example, the stomach is an organ that contains epithelial tissue, covering its outer and inner surfaces; smooth muscle tissue, by which it carries on peristalsis; nerve tissue, which controls its muscular and glandular activity; connective tissue, which holds it together; and blood tissue, which supplies its cells with food and oxygen and takes away wastes. All of these tissues work together as the stomach performs its function of digestion. Similarly, other organs such as the heart, kidney, and brain are also made of tissues working together.

Systems

Besides the stomach, there are other organs concerned with digestion, such as the mouth, esophagus, small intestine, large intestine, liver, and pancreas. All together, they make up the digestive *system.* There are seven other systems composed of groups of organs working together to perform their particular functions: circulatory system, respiratory system, excretory system, nervous system, reproductive system, skeletal system, and muscular system.

Organisms

In many-celled living things, cells are grouped together to form tissues; tissues work together to form organs; organs are organized into systems. All of the systems working together comprise an organism. An organism may also be one-celled, such as an ameba.

SUMMARY OF ANIMAL TISSUES

Tissue	Types	Location	Functions
Epithelial (covering tissue)	1. Epidermis	Outer layer of skin	Protection
	2. Flat (squamous)	Lines digestive and respiratory systems; also blood vessels	Protection; lubrication; absorption
	3. Ciliated	Lines trachea and nasal cavities	Protection by sweeping out bacteria and dust
	4. Columnar	Lines small intestines	Absorption; lubrication
	5. Glandular	(1) Digestive system; (2) endocrine glands	Secretion of (1) enzymes; (2) hormones
	6. Sensory	Sense organs: (1) nose; (2) retina of eye	Receive sensations: (1) smell; (2) sight
Muscles	1. Striated	Attached to bones	Voluntary movement
	2. Smooth	(1) Digestive organs; (2) blood vessels	Involuntary: (1) peristalsis; (2) control of diameter of blood vessels
	3. Cardiac	Heart	Heart beat
Nerve	1. Sensory	Connects sense organs with brain or spinal cord	Transmit impulses from sense organs to brain or spinal cord
	2. Motor	Connects brain or spinal cord with muscles or glands	Make muscles contract or glands secrete
	3. Interneuron	Brain, spinal cord	Make connections between neurons
Connective or Supporting	1. Bone	Makes up the skeleton	Support body; protect organs
	2. Cartilage	Ear, nose, trachea, ends of bones	Support; frictionless movement of joints
	3. White fibrous	(1) Tendons; (2) ligaments	Connect (1) muscle to bone; (2) bone to bone
	4. Yellow elastic	Walls of blood vessels; between vertebrae	Elasticity
	5. Fat (adipose)	Under skin; around internal organs	Insulation against heat loss; stores fat
Blood	1. Red blood corpuscles	Plasma	Carry oxygen
	2. White blood corpuscles	Plasma	Destroy bacteria
	3. Platelets	Plasma	Cause clotting of blood
Reproductive	1. Male reproductive tissue	Testes	Produce sperm
	2. Female reproductive tissue	Ovaries	Produce eggs

Plant Tissues

Outer Covering

The epidermis is the outside layer of cells in a plant.

1. *Leaf.* In a leaf the epidermis is waterproof, containing a waxy substance called *cutin.* It covers the upper and lower sides of the leaf. In it are numerous tiny openings, *stomates,* on either or both surfaces. Stomates permit the passage of oxygen and carbon dioxide in and out of the leaf, and the liberation of water.

2. *Stem.* The epidermis of a stem has thickened outer cell walls. It prevents the loss of water, and protects the underlying tissues from injury. On woody stems, the outside layer consists of a *bark.* Stems have openings called *lenticels* for the exchange of gases.

3. *Roots.* The epidermis of a root is soft and thin, permitting the absorption of water and minerals. *Root hairs* occur as single-celled outgrowths of root epidermis. They absorb water and minerals and also help anchor the plant in the ground.

Conducting Tissue

Special tissue located in ducts called *fibrovascular bundles* conducts materials throughout the plant. The ducts start in the roots, and continue up the stem into the leaves. Here, they branch out into the midrib and the veins. They are composed of two kinds of conducting tissue, xylem and phloem.

1. *Xylem*—consists of long, thick-walled, dead, woody cells. They are specialized for conducting water up from the roots to the stem and leaves. They also serve to support the plant.

2. *Phloem*—consists of thin-walled sieve tube cells that have no nucleus and that are located close to the xylem ducts. Attached to each sieve tube cell is a *companion cell* containing a nucleus, which helps direct its activities. The sieve tubes are arranged longitudinally, end to end, with tiny holes in their adjacent walls. These openings permit the passage of food from cell to cell. It is through the phloem tissue that manufactured food is transported from the leaves to other parts of the plant.

In the stem of dicots, the fibrovascular bundles are arranged in the form of a concentric ring. In the stem of monocots, these bundles are scattered about.

Growing Tissue

The growing tip of a stem contains actively dividing cells known as *meristem* cells. At the other end of the plant, the tips of roots are similar in having a growing region also composed of meristem cells.

Another part of the plant that consists of actively dividing cells is the *cambium.* This important growing region is found in the fibrovascular bundles, between the phloem and the xylem. As cambium cells divide, they form more cambium, as well as more xylem and phloem, enlarging the stem and causing the trunk of a tree to grow in diameter. Trees in the Temperate Zone show such growth only during the spring and summer of each year. A cross section of a tree shows the presence of *annual rings,* which are formed during these growing periods. Annual rings consist of large cells of xylem, formed in the spring, and smaller, woody cells in the summer. The age of a tree can be determined by counting these rings. They also give a clue as to the climate conditions that prevailed, with wide rings indicating favorable seasons.

Food-making Tissue

Between the upper and lower epidermis of a leaf, there are cells containing chloroplasts arranged in two regions, a *palisade* layer and a *spongy* layer. The palisade cells are elongated and are arranged in column-like fashion under the upper epidermis. Here, they receive sunlight and carry on their chief function, food-making or photosynthesis. The spongy layer is located below the palisade cells. These cells are more irregular, with large air spaces among them that are adjacent to the stomates. These cells are thus well suited for the exchange of gases with the surrounding atmosphere, and for the passage of carbon dioxide to the palisade layer above them. Cells of the spongy layer have fewer chloroplasts and so are not as important from the viewpoint of photosynthesis as the palisade cells. The green cells of young stems also carry on photosynthesis.

Supporting Tissue

Plants are generally strengthened and supported in four ways.

1. The fibrovascular bundles have woody xylem cells that give support to the stem. These long, pointed dead cells with greatly thickened walls make up a considerable proportion of the wood of trees.

2. There are also groups of living cells with greatly thickened walls that lend support to various parts of a plant, as in the leaf, just above and below the midrib; similar cells occur in stems, next to the epidermis.

3. In addition, there are scattered thick-walled, dead cells in stems, located just outside of the phloem, which help support the plant.

4. In parts of the leaf, especially between the midrib and the epidermis, there are large, thin-walled cells that strengthen the leaf by their turgidity; they are stretched

full of water and are quite rigid in the same way that a hose full of water is rigid. When there is a loss of water from these cells, the leaf becomes softer and *wilted*. Similar groups of cells perform the same function of support by turgidity in succulent stems and in the young stems of woody plants.

Storage Tissue

Large, thin-walled cells in stems, known as *pith*, store food. They may be either located in the center of the stem or scattered through the stem as rays of cells (medullary rays). These pith rays also serve to conduct food and water within the stem. Roots may become thickened and also serve to store food, as in the case of the sweet potato, carrot, and radish.

Reproductive Tissue

The reproductive organs of a plant are located in the flower. Male nuclei (sperm nuclei) are formed in the *pollen* grains, produced by the stamens. Female nuclei (egg nuclei) are formed in the *ovules* of flowers, located within the ovary. When pollination occurs, pollen is deposited on the stigma; a pollen tube develops from each pollen grain, and grows down into an ovule, carrying the sperm nuclei within itself. The sperm and egg nuclei fuse (*fertilization*), and the ovule develops into a seed. Part of the seed contains the embryo of a new plant; stored food for the embryo occupies most of the remaining part of the seed.

SUMMARY OF PLANT TISSUES

Tissue	Types	Location	Functions
Covering	1. Epidermis	Leaf, stem, root	Protect; prevent moisture loss;
	2. Bark	Woody stem	form root hairs on roots
Conducting	1. Xylem	Fibrovascular bundles in leaf, stem, root	Conduct water up from root into stem, leaves
	2. Phloem	Fibrovascular bundles in leaf, stem, root	Conduct food from leaves to rest of plant
Growing	1. Meristem	Tips of stems, roots	Growth; differentiation
	2. Cambium	Fibrovascular bundles, between xylem and phloem	Growth in thickness; formation of new xylem, phloem
Food-making	1. Palisade cells	Upper layer of leaf	Photosynthesis
	2. Spongy layer	Lower layer of leaf	Some photosynthesis; help circulate gases for palisade layer
	3. Green cells	Stems	Photosynthesis
Supporting	1. Wood	Xylem	Mechanical support
	2. Thick-walled living cells	Leaf, stem (under epidermis)	Mechanical support
	3. Thick-walled dead cells	Stem, outside of phloem	Mechanical support
	4. Thin-walled, large cells	Leaf, stem	Support by turgidity
Storage	1. Pith cells	Stems	Store food
	2. Root tissue	Enlarged roots	Store food
Reproductive	1. Pollen	Stamen of flower	Produce sperm nuclei
	2. Ovule	Ovary of flower	Produce egg nuclei

Section Review

Select the correct choice to complete each of the following statements:

1. A group of similar cells performing the same function is known as a(n)   (A) organ   (B) system   (C) tissue   (D) layer   (E) membrane

2. Epithelial tissue may have all of the following functions *except*   (A) protection   (B) absorption   (C) contraction   (D) secretion   (E) sensation

3. The chief tissue of which glands are composed is   (A) epithelial   (B) muscle   (C) nerve   (D) connective   (E) supporting

4. The muscles attached to the bones are   (A) voluntary and smooth   (B) involuntary and smooth   (C) voluntary and striated   (D) involuntary and striated   (E) smooth and striated

5. An axon is part of a(n)   (A) muscle cell   (B) nerve cell   (C) epithelial cell   (D) bone cell   (E) cartilage cell

6. A tissue containing deposits of intercellular material is   (A) nerve   (B) muscle   (C) epithelial   (D) striated   (E) connective

7. Of the following, the one that contains the largest number of different types of cells is the   (A) mucous membrane   (B) cardiac muscle   (C) nerve   (D) small intestine   (E) blood

8. A tissue specialized for contraction is   (A) nerve   (B) muscle   (C) adipose   (D) blood   (E) reproductive

9. Of the following, the one that includes all the others is   (A) blood   (B) connective   (C) epithelial   (D) kidney   (E) nerve

10. The correct order of arrangement, in terms of increasing complexity, of the following: 1-organ, 2-cell, 3-tissue, 4-organism, 5-system, is   (A) 2–1–3–4–5   (B) 2–3–4–1–5   (C) 2–3–1–4–5   (D) 2–3–4–5–1   (E) 2–3–1–5–4

11. Stomates are largely found in a   (A) leaf   (B) stem   (C) root   (D) root hair   (E) leaf hair

12. Structures that conduct materials throughout a plant are called   (A) lenticels   (B) meristem   (C) fibrovascular bundles   (D) cambium   (E) epidermis

13. Food-making cells in plants are located in the   (A) xylem   (B) phloem   (C) pith   (D) palisade layer   (E) ducts

14. The growing region of a plant stem is called the   (A) bark   (B) meristem   (C) wood   (D) spongy layer   (E) phloem

15. Cambium is located next to the   (A) stomates   (B) guard cells   (C) epidermis   (D) root hairs   (E) xylem and phloem

Answer Key

1-C	4-C	7-D	10-E	13-D
2-C	5-B	8-B	11-A	14-B
3-A	6-E	9-D	12-C	15-E

Answers Explained

1. (C) A tissue is a group of similar cells performing the same function. Examples: epithelial, muscle, connective, blood, reproductive tissues.

2. (C) Contraction is a function of muscle tissue.

3. (A) Glandular epithelium is found in the glands of the digestive and endocrine systems and in the tear glands.

4. (C) The muscles attached to the bones are considered voluntary because they are under our will; they are also called striated, because under the microscope their fibrils have a striped appearance, with alternating dark and light bands.

5. (B) An axon is a very long extension of a nerve cell, or neuron. Although microscopic, it may be several feet long.

6. (E) Bone and cartilage are examples of connective tissue in which there is considerable firm intercellular material between the cells.

7. (D) The small intestine is an organ that contains many types of tissues, such as epithelial, smooth muscle, nerve, connective, and blood.

8. (B) Muscle tissue has the ability to contract. There are three types of muscle: (1) voluntary, or striated, muscle, which is usually attached to the bones of the skeleton; (2) smooth (not striated), involuntary muscle, found in the alimentary canal and the blood vessels; and (3) cardiac muscle, striated but involuntary, found in the heart.

9. (D) The kidney is an organ in which blood, connective, epithelial, and nerve tissue are found.

10. (E) Cells (2) are organized into tissues (3), which make up organs (1). A system (5), such as the digestive or respiratory system, consists of various organs. All the systems working together comprise an organism (4).

11. (A) Stomates are tiny openings on the surface of a leaf through which oxygen, carbon dioxide, and water vapor pass.

12. (C) Fibrovascular bundles conduct materials in plants. They contain two kinds of conducting tissue: xylem, which conducts water up from the roots to the stem and leaves; and phloem, which conducts manufactured food from the leaves to other parts of the plant.

13. (D) Palisade cells, elongated cells in a leaf that are arranged in columnlike fashion, contain chloroplasts, in which photosynthesis, the manufacture of food in the presence of light, takes place.

14. (B) The meristem contains actively growing cells and is contained in the growing tip of a stem. Its cells give rise to the different types of tissue in the stem.

15. (E) Cambium contains actively growing cells and is found in the fibrovascular bundles, between the xylem and the phloem. As its cells divide, they form more cambium, xylem, and phloem, adding to the growth in width of the stem.

GREEN PLANTS— BASIS OF ALL LIFE

CHAPTER

5

5.1   HOW PLANTS MAKE FOOD

In the fall of 1966, using radioactive carbon and phosphorus on chloroplasts that had been separated from freshly ground spinach leaves, Dr. James A. Bassham and Dr. R. G. Jensen duplicated *photosynthesis*, the food-making process of green plants, outside of a cell, and studied its successive stages. Thus, another step was taken in our understanding of photosynthesis.

Autotrophs are organisms that can make their own food from inorganic raw materials. A few types of bacteria carry on *chemosynthesis*, by which they produce organic molecules from inorganic compounds of sulfur, iron, and nitrogen. Recent exploration of the blackness in the deep-sea floor of the Pacific has revealed flourishing communities of giant tube worms, mussels, clams, and crabs clustered around areas where hot, mineral-rich waters bubble up from the sea floor. These creatures feed on chemosynthetic bacteria that make their own food by combining the sulfur and oxygen in the mineral-rich hot water.

However, practically all other food is made by green plants during *photosynthesis*, and green plants serve as the basis of practically all life on earth. Some animals eat plants as food. Some animals eat other animals, which in turn obtained their food from plants. However, all animals depend, either directly or indirectly, on autotrophs (green plants or chemosynthetic bacteria).

Photosynthesis—Making Carbohydrates

The food-making process of green plants depends on the presence of chlorophyll, light, water, and carbon dioxide. It takes place inside plant leaves, involves several complex steps, and produces oxygen in addition to carbohydrates.

Chlorophyll and Chloroplasts

Green plants contain *chlorophyll*, which is necessary for photosynthesis. It is formed only in the presence of sunlight. This explains why plants that are germinated from seeds in the dark are white. When analyzed, chlorophyll is found to be a mixture of pigments, at least two green ones, chlorophyll a and chlorophyll b, and the yellow, orange, or red pigments carotene and xanthophyll, which are hidden by the green color. The latter colors show up especially in the fall, when leaves change to yellow, orange, and red colors before dropping from the trees.

Chlorophyll contains the elements carbon, hydrogen, oxygen, nitrogen, and magnesium. In order for chlorophyll to be formed, minute quantities of magnesium and iron must be available to the plant. Chlorophyll is similar to hemoglobin in chemical structure. In chlorophyll, a magnesium atom is surrounded by four nitrogen atoms in the same way that iron in hemoglobin is surrounded by four nitrogen atoms.

Chlorophyll is found in tiny oval bodies called *chloroplasts*. They have been found to reproduce when the cell divides. Under the electron microscope, each chloroplast is seen to be composed of disc-shaped units called *grana* that are arranged in stacks. Each of these grana has layers, or lamellae, of membranes containing the chlorophyll pigments as well as protein and fatty material. The grana are located within a granular dense solution called the *stroma*, which has other lamellae extending through it.

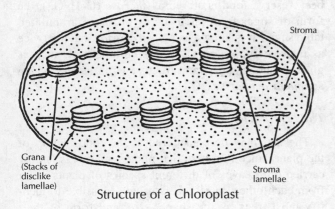

Structure of a Chloroplast

The value of chlorophyll in photosynthesis can be shown by the following demonstration. Choose a plant such as the silver-leaf geranium, in which there is a white fringe around the outer border of each leaf, or a coleus, whose variegated leaves have colored areas that are not green. No chlorophyll is present in these parts of the leaf, and therefore no food should be formed there.

1. Extract the chlorophyll from the leaf by boiling the leaf gently in alcohol. If the leaf is plunged into hot water before this is done, the waxy epidermal covering is broken down and the chlorophyll can be dissolved out more readily.

2. When the alcohol has turned green, remove the leaf and rinse it with water. It is now completely white.

3. Spread the leaf out in a flat dish, and cover it with iodine solution. Iodine reacts with starch and turns it a blue-black color.

4. Observe that the parts of the leaf that contained chlorophyll, and stored starch as the result of photosynthesis, are now deeply stained. The parts of the leaf that did not contain chlorophyll, and therefore did not store starch, merely have the light brown color of the iodine solution.

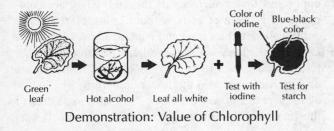

Demonstration: Value of Chlorophyll

The Basic Process of Photosynthesis

Photosynthesis is the manufacture of food (carbohydrates) from carbon dioxide and water in the presence of light. It takes place in the cells of green plants containing chlorophyll. During this process, the raw materials, carbon dioxide and water, are combined by the chlorophyll to form sucrose and starch. Oxygen is given off as a by-product. This is the source of most of the O_2 in the air around us.

Much information about the various steps in photosynthesis has been obtained through the use of isotopes of oxygen and carbon, and the application of refined techniques of chromatography. With the aid of heavy oxygen, oxygen-18, it has been shown that the oxygen that is given off comes from the water. Radioactive carbon, carbon-14, has been traced as a "tagged" molecule through the complex stages of carbon fixation with the aid of instruments such as the Geiger counter. In chromatography, various compounds are separated out at different levels as they are absorbed on a length of filter paper.

Three Sets of Reactions

Photosynthesis takes place in three major steps: (1) absorption of light energy; (2) a light photochemical reaction in which light energy is converted into chemical energy and in which oxygen is liberated; (3) Calvin cycle in which chemical energy is used to convert carbon dioxide into organic compounds.

1. *Absorption of light energy.* Light comes in units or packets called photons. When a photon strikes chlorophyll, its energy is transferred to an electron in chlorophyll, raising it to a higher or an excited level.

2. *Light (photochemical) reaction.* The light reactions take place in the grana of the chloroplasts where high-energy electrons are released to a protein compound named ferrodoxin. Ferrodoxin releases the electrons to form ATP (adenosine triphosphate) and to change the coenzyme NADP (nicotinamide adenine dinucleotide phosphate) to NADPH. The hydrogen atoms for the latter come from water molecules which are split. Oxygen atoms are given off as O_2 gas. The splitting of water into hydrogen and oxygen is known as *photolysis*.

3. *Calvin cycle.* The energy stored in ATP and NADPH is used to convert CO_2 into carbohydrates during the Calvin cycle. Formerly, this was known as the dark phase of photosynthesis. But current belief is that it takes place in the light. Sufficient CO_2 would not be available in the dark when stomates are closed. Also, the essential enzyme *rubisco* is activated by light. Rubisco is the most abundant protein in leaves.

A series of intermediate compounds is formed from the carbon dioxide molecules and hydrogen atoms. A number of enzymes necessary for the reactions are located in the stroma of the chloroplasts. Two of the intermediate compounds are PGA (phosphoglycerate), formed by CO_2, water, and RDPL (ribose diphosphate). PGAL (phosphoglyceraldehyde) is formed when hydrogen is added to PGA from the hydrogen carrier NADPH. Some of the PGAL is rearranged in a series of steps to form sucrose and starch, the end products of photosynthesis. Formerly, glucose was commonly considered to be the end product.

The compound PGAL has several values in addition to forming sucrose and starch. Some of it is recycled to continue carbon fixation, with the formation of various intermediate products; it may also be used in the synthesis of proteins and lipids; it may serve as fuel to provide new energy for carbon fixation.

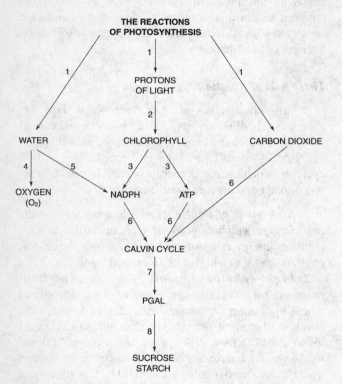

Key to steps in photosynthesis:

1. Essential components of photosynthesis are water, carbon dioxide, and light.

2. Photon of light activates chlorophyll.

3. High energy electrons pass through electron acceptors; ATP and NADPH are formed.

4. Photolysis: splitting of water forms hydrogen atoms and oxygen gas.

5. Hydrogen atoms from water go to form NADPH.

6. NADPH, ATP, and carbon dioxide enter the Calvin Cycle.

7. Series of intermediate compounds are formed resulting in PGAL.

8. PGAL is rearranged to form sucrose and starch.

When photosynthesis is active and sugar is being formed faster than it can be used by the leaf, it is converted into starch for storage. Starch can be seen under the microscope as small grains. At night, starch is changed back into sugar and is conducted away from the leaf. In some plants, such as sugar cane and the beet, reserve food is stored as *sucrose* ($C_{12}H_{22}O_{11}$), the form of sugar familiar to us as ordinary granulated table sugar. It is known as a *disaccharide*, having twice as many carbon atoms as the simple sugars, such as glucose (grape sugar) and fructose (fruit sugar), which are *monosaccharides*, $C_6H_{12}O_6$.

The Leaf Factory

The leaf may be considered as the food factory of the plant, since it is here that photosynthesis is largely carried on. Leaves of different species of plants differ from each other; in fact, it is usually possible to identify most trees by the shape, size, and form of their leaves. In general, most leaves have an expanded broad portion, the *blade*, and a thin stalk, the *petiole,* by which the blade is attached to the stem. The veins, which contain the fibrovascular bundles, are arranged as a network in leaves of dicotyledonous plants (oak, maple, elm), but are parallel to each other in the leaves of monocotyledonous plants (grasses, wheat, corn).

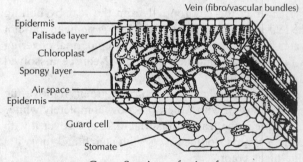

Cross Section of a Leaf

A microscopic study of the cross section of a leaf reveals several different layers of cells and tissues which are adapted for photosynthesis:

1. The **upper** and **lower epidermis** make up the upper and lower surfaces of the leaf. They are composed of a layer of cells without chloroplasts whose outer walls are thickened and contain a waxy substance called *cutin.* They are thus well adapted to prevent the loss of water. Numerous tiny openings, *stomates*, are usually found only on the lower epidermis. The leaves of some plants have hundreds of stomates in an area of the leaf only 1/4 inch square. Each stomate is sur-

rounded by two kidney-shaped cells called *guard cells*, which contain chloroplasts. These guard cells move in such a way as to open or close the stomates. The stomates allow an exchange of oxygen and carbon dioxide between the moist surface of the cells lining the air spaces on the inside of the leaf and the external atmosphere; they also permit the passage of water vapor from the inside of the leaf to the outside.

2. The *palisade layer* is present below the upper epidermis as a row of columnlike, long cells, each containing many chloroplasts. Being near the top of the leaf, they are in a favorable position to receive much sunlight; their chief function is to carry on photosynthesis. Between the palisade cells are small air spaces that are connected with larger air spaces in the lower part of the leaf, which in turn are connected with the stomates. These air spaces make it possible for carbon dioxide to diffuse to the palisades cells and oxygen to diffuse out.

3. The *spongy layer*, below the palisade layer, consists of irregular cells surrounded by large air spaces that are connected with the stomates. There is a free circulation of air around them and with the air spaces between the palisade cells. The spongy cells contain fewer chloroplasts, and are not in as favorable a position to carry on photosynthesis as the palisade cells.

4. The *veins* run through the spongy layer and contain the *fibrovascular bundles*, which help support the leaf, and distribute materials through the xylem and phloem. Water is brought up from the roots through the stem into the petiole and through the blade by means of the *xylem*. Water diffuses from cell to cell and is combined with the carbon dioxide to form glucose. The glucose diffuses out through the cells to the *phloem*, which carries it out of the blade, through the petiole, into the stem and roots.

The Need for Light

Chlorophyll is active in making carbohydrates only if light is present as the source of energy. In nature, sunlight is used, but artificial light can be substituted. When there is no light, photosynthesis stops. Chlorophyll absorbs certain wavelengths of light energy more effectively than others. Most absorption takes place at the blue-violet and orange-red ends of the spectrum. Here the rate of photosynthesis is highest. Green light is reflected by chlorophyll and is least effective in photosynthesis.

The following demonstration can be conducted to show the necessity of light:

1. Cover the upper and lower epidermis of part of a leaf on a geranium plant with two thin, round pieces of cork. Place the plant in sunlight. The covered part of the leaf is now in the dark and cannot make carbohydrates. The rest of the leaf is in the light and can make carbohydrates.

2. Extract the chlorophyll as was done in the preceding demonstration, and cover the leaf with iodine solution.

3. Observe that the leaf now shows two areas of color: the part that was exposed to the light, and was able to store starch, is deeply stained by the iodine. The covered part, which did not make food, has only the light brown color of the iodine.

The Need for Carbon Dioxide

The air around us normally contains about 0.03 percent of carbon dioxide. This gas is constantly being excreted by living things as they carry on respiration. Volcanoes, fires, automobiles, and industrial plants also add carbon dioxide to the air. Green plants use up some of this carbon dioxide as they carry on photosynthesis. The following demonstration shows the necessity of carbon dioxide for food-making:

1. Place a geranium plant under a bell jar (*A*) in which there is a container of dry potassium hydroxide. Seal the edges of the bell jar with Vaseline to make it airtight. The potassium hydroxide absorbs carbon dioxide, so none should be present under the bell jar. Set up another bell jar (*B*) with a plant in the same way, but substitute sand for the potassium hydroxide. This is the control.

2. After a day in sunlight, remove a leaf from each plant, extract the chlorophyll, and cover with iodine solution.

3. Observe that leaf *A* has only the light brown color of the iodine, showing that no starch was formed because of the absence of carbon dioxide. Leaf *B* is stained blue-black, showing the presence of starch resulting from the availability of carbon dioxide for photosynthesis.

The Need for Water

In addition to carbon dioxide, another raw material, water, is needed for photosynthesis. When water is split during photolysis, hydrogen and oxygen atoms are released. The hydrogen atoms are used in synthesizing sucrose and starch. Oxygen is given off as a by-product.

The Giving Off of Oxygen

Photosynthesis is an important source of the oxygen in the air. An aquarium is stocked with green water plants to furnish fish with oxygen; on the other hand, the fish give off carbon dioxide, which is used by the plant in photosynthesis. This *oxygen-carbon dioxide cycle* shows the interrelationship that exists between animals and green plants. The giving off of oxygen by green plants during photosynthesis can be demonstrated as follows:

1. Place a quantity of the green water plant *Elodea* under a funnel in a wide-mouthed jar full of water. Cover the stem of the funnel with an inverted test tube filled with water. Keep in sunlight.

2. In a short time, observe bubbles of gas streaming from the plants up the funnel into the test tube. After some time, the test tube will be filled with the gas, all of the water having been displaced.

3. Insert a glowing splint into the test tube. It will glow brighter or even burst into flame, because of the presence of pure oxygen in the test tube.

Formation of Other Types of Food

Some of the sucrose produced during photosynthesis is used for energy; some is stored as starch; and some is converted into other types of food materials, including fats and oils, proteins, and vitamins.

Fats and Oils

The natural fats and oils are composed of carbon, hydrogen, and oxygen, with oxygen occurring in very small amounts. Thus, the formula of one fat, olein of olive oil, is $C_{57}H_{104}O_6$. The oxidation of fats produces large amounts of energy. Fats and oils are liberally stored in the seeds of the coconut, olive, and peanut, and are extracted for their commercial value. The formation of fats from carbohydrates is known as *fat synthesis*.

Proteins

PGAL is combined with such elements as nitrogen, sulfur, and phosphorus to form basic compounds known as *amino acids*. From these, complex proteins are built up. The formula of the protein in corn, zein, is $C_{736}H_{1161}N_{184}O_{208}S_3$. Proteins are the active constituents of protoplasm and are used for forming new protoplasm, by the process of assimilation, for growth and repair. The production of proteins from carbohydrates is known as *protein synthesis*. The minerals used in this process are transported in the water up from the roots through the stem in the xylem ducts.

Vitamins

Plants are excellent sources of vitamins. Vitamin A is made from carotene, one of the pigments of chlorophyll. It naturally follows that green, leafy vegetables are good sources of this vitamin. The yellow vegetables, such as sweet potato and carrot, are also good sources of vitamin A. The vitamin B complex is found in the seeds of many plants, including the grains and peas. Vitamin C is abundant in fresh vegetables and fruits.

Section Review

Select the correct choice to complete each of the following statements:

1. Practically all living things depend for their food on the activity of   (A) chromatin   (B) chlorotone   (C) chlorophyll   (D) colchicine   (E) cholesterol

2. For the process of photosynthesis, green plants require   (A) carbon monoxide and water   (B) carbon dioxide and carbon monoxide   (C) carbon dioxide and water   (D) oxygen and water   (E) oxygen and carbon dioxide

3. A by-product of photosynthesis is   (A) carbon dioxide (B) water   (C) carbon monoxide (D) oxygen   (E) nitrogen

4. The substance produced as the direct result of photosynthesis is   (A) sucrose   (B) protein   (C) fat   (D) iodine   (E) vitamin A

5. All of the following are carbohydrates *except*   (A) sucrose   (B) starch   (C) fructose   (D) fruit sugar   (E) carotene

6. The passage of gases into and out of a leaf is controlled by the   (A) petiole   (B) guard cells   (C) palisade layer   (D) spongy layer   (E) veins

7. Chlorophyll can be extracted from a leaf by   (A) soaking it in water   (B) boiling it in alcohol   (C) electrolysis   (D) adding iodine to it   (E) adding glucose to it

8. When a leaf that was kept in the dark is treated with iodine after its chlorophyll was extracted, it will turn   (A) blue-black   (B) brick-red   (C) green   (D) the color of the iodine   (E) white

9. The food made during photosynthesis may be used for all of the following purposes *except* (A) protein synthesis   (B) fat synthesis   (C) storage   (D) energy   (E) transpiration

10. In a balanced aquarium, the fish supply the green plants with   (A) oxygen   (B) water   (C) carbon dioxide   (D) glucose   (E) starch

11. An autotroph makes food from   (A) organic molecules   (B) inorganic molecules   (C) glucose   (D) fats   (E) proteins

12. The source of oxygen in photosynthesis is   (A) CO_2 (B) glucose   (C) ATP   (D) H_2O (E) TPN

13. Chloroplasts are made of   (A) grana   (B) stomata   (C) cutin   (D) guard cells   (E) xylem

14. All of the following are intermediate products of carbon fixation *except*   (A) PGA   (B) DNA   (C) PGAL   (D) RDP   (E) fructose

15. The part of the spectrum in which photosynthesis is most active is   (A) green-yellow   (B) yellow-orange   (C) blue-violet   (D) blue-green   (E) green-orange

Answer Key

1-C	4-A	7-B	10-C	13-A
2-C	5-E	8-D	11-B	14-B
3-D	6-B	9-E	12-D	15-C

Answers Explained

1. (C) Chlorophyll is the green compound in plants, where food is manufactured, in the presence of light, out of carbon dioxide and water. This process is called photosynthesis.

2. (C) In order for green plants to carry on photosynthesis in the presence of light, they need two raw materials: carbon dioxide and water.

3. (D) Oxygen is a by-product of photosynthesis. Photosynthesis is, in fact, the source of most of the oxygen in the air around us. (Green plants are placed in an aquarium so that the fish can use the oxygen given off during photosynthesis.)

4. (A) When a green plant carries on photosynthesis, it combines carbon dioxide and water into the final sucrose and starch.

5. (E) The first four choices are carbohydrates because they contain hydrogen and oxygen in the same ratio as in water, 2:1. Carotene is one of the pigments in chlorophyll that make vitamin A.

6. (B) The guard cells surround the stomates (openings) in the epidermis of a leaf. They move to open or close the stomates. In this way, they control the passage of carbon dioxide and oxygen into and out of the leaf; they also control the passage of water vapor out of the leaf.

7. (B) When a leaf is boiled gently in alcohol, the chlorophyll of the leaf becomes dissolved in the alcohol, turning it green. If the leaf is taken out of the alcohol and rinsed in water, it is seen to be completely white.

8. (D) The leaf will merely have the color of the iodine solution because it does not contain any starch. It needs light to carry on photosynthesis and form starch.

9. (E) Some of the glucose made during photosynthesis is used for energy; some is stored as starch; some is converted into proteins and fats. "Transpiration" refers to the evaporation of water from leaves.

10. (C) When fish carry on respiration, they give off carbon dioxide as a waste. Green plants take in carbon dioxide and use it during photosynthesis.

11. (B) An autotroph is an organism that can make its own food from inorganic molecules. Most autotrophs are green plants that can make food from carbon dioxide and water. A few types of bacteria can use such inorganic molecules as sulfur, iron, and nitrogen to produce food in the form of organic molecules.

12. (D) Experiments using heavy oxygen, oxygen-18, revealed that the oxygen given off during photosynthesis comes from the water that is a raw material for the process. The splitting of water into hydrogen and oxygen by light energy is known as photolysis.

13. (A) Under the electron microscope, a chloroplast is seen to consist of disc-shaped units called grana. Each of the grana has layers of membranes (lamellae) that contain chlorophyll as well as protein and fatty material. The grana

are contained within a granular, dense solution called the stroma.

14. (B) DNA is involved in determining the heredity of an organism and is not an intermediate product of carbon fixation. Carbon fixation occurs in the dark reaction of photosynthesis, during which carbon dioxide molecules and hydrogen atoms form intermediate products that will eventually lead to the formation of glucose.

15. (C) Chlorophyll absorbs the blue-violet light of the spectrum most effectively, resulting in a high rate of photosynthesis.

5.2   HOW PLANTS LIVE

In addition to carbon dioxide and water, which are used in photosynthesis, plants need other materials for their activities, including oxygen and minerals. They also use water for a variety of other purposes.

Respiration

Aerobic Respiration

Green plants, like animals, carry on respiration; that is, they take in oxygen, and combine it with food to release energy for their various activities, such as food-making and growth. In the process, carbon dioxide and water are excreted as wastes. Thus, respiration is the reverse process of photosynthesis. As a matter of fact, during the day, both processes go on side by side, in different parts of the green cell. In photosynthesis, an excess amount of oxygen is produced by the chloroplasts. Some of it is used directly by the protoplasm of the cell for oxidizing food, while the rest of it diffuses out and is given off through the stomates. At the same time, while respiration is being carried on by the protoplasm, carbon dioxide and water are given off. These waste products are used up almost at once by the chloroplasts, which are carrying on photosynthesis. Thus two different parts of the cell carry on contrasting activities—the protoplasm takes in oxygen and gives off carbon dioxide and water, to release energy, during respiration; while the chloroplasts in the same cell take in carbon dioxide and water and give off oxygen, storing energy, during photosynthesis.

At *night,* however, when photosynthesis does not go on, only respiration is carried on; oxygen is taken in from the air through the stomates, and carbon dioxide and water are given off.

A demonstration to show that plants give off carbon dioxide during respiration may be performed as follows:

1. Place a geranium plant under a bell jar (A) containing a beaker of fresh limewater. Seal the edges with Vaseline to make it airtight.
2. Set up another bell jar (B) in the same way, but without a plant. Keep both setups in the dark overnight.
3. The next day, observe the limewater in both bell jars. In bell jar A, it will be cloudy because of the effect of the carbon dioxide given off by the plant. In bell jar B, it will be clear, since there was no carbon dioxide.

During respiration, some of the energy is released as very small amounts of *heat.* It is possible to measure this by performing the following demonstration with bean seeds that are germinating and actively carrying on respiration:

1. Fill a thermos bottle (A) that is lined with moist absorbent paper with bean seeds that have begun to germinate. Seal with a one-holed rubber stopper containing a thermometer whose bulb is present inside the bottle. Make a record of the thermometer reading.
2. As a control, set up another thermos bottle (B) in the same way, but without any seeds. Make a record of the thermometer reading.
3. After 24 hours, compare the thermometer readings. In thermos bottle A, there will be a rise, indicating the liberation of heat by the seedlings. Thermos bottle B will show no change.

Anaerobic Respiration

Some simple organisms, such as yeast and certain bacteria, carry on anaerobic respiration, in which oxygen is not needed. They produce as wastes carbon dioxide and ethyl alcohol. The process, known as *alcoholic fermentation,* has commercial value in the brewing, baking, and wine-making industries. Other types of bacteria and certain molds carry on *lactic acid fermentation,* in which lactic acid is produced. The amount of energy released during anaerobic respiration is relatively small compared to that released in aerobic respiration, since most of the C—C and C—H bonds in the alcohol or the lactic acid have not been broken and still retain their chemical energy.

Transport

Food, water, and minerals are moved throughout the plant, from the roots to the leaves, through xylem and phloem tissues, as well as from cell to cell by active transport and diffusion. The movement of these materials is referred to as *translocation.*

COMPARISON OF PHOTOSYNTHESIS AND RESPIRATION

Characteristics	Photosynthesis	Respiration
When activity occurs	Day only	Day and night
What is taken in	CO_2 and H_2O	O_2
What is given off	O_2	CO_2 and H_2O
Where activity takes place	In green cells only	In all cells
What happens to glucose	Built up	Broken down
What happens to energy	Used and stored up	Liberated

The Roots

Minerals and water are absorbed from the soil by the roots of the plant. Roots also serve to anchor the plant to the soil. A study of a root shows several regions. The *tip* consists of a growing point made of rapidly dividing cells. There is a *root cap* over this growing tip, which protects it as it pushes through the soil. Just back of the growing tip is a region that elongates considerably and is the most rapidly growing part of the root.

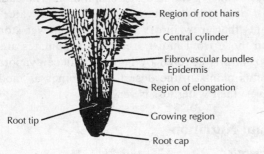

Structure of Root Tip

Farther back, the cells of the epidermis produce projections called *root hairs*. This part of the root appears fuzzy to the naked eye because of the large number of root hairs. Most of the water and minerals absorbed by a root are taken in by these root hairs. A root-hair cell is a single cell with a large vacuole, in which the nucleus is usually found in the root-hair portion. The formation of root hairs increases the absorbing surface of the root; it also allows for closer contact with the film of water around the soil particles. The contact with these soil particles is so close that the root hairs adhere to them. This results in two effects: the root binds the soil together, and the root is firmly anchored in the soil.

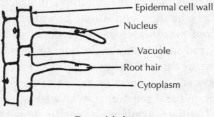

Root Hairs

There are several kinds of roots, which vary in structure and function.

1. *Taproot.* This is the long, strong primary root of a plant such as the dandelion and the oak tree.
2. *Fibrous roots.* These form a network of roots just below the stem in, for example, grasses.
3. *Fleshy root.* This is the large root that stores food, as in the carrot and radish.
4. *Brace roots.* In some plants such as corn and the mangrove tree, large roots grow down from the main stem into the ground, and brace or support the stem.
5. *Aerial roots.* Vines such as ivy attach themselves to walls or trees by aerial roots.

Diffusion of Soil Water

Water in the ground containing minerals is known as *soil water.* This soil water enters the root hair by active transport and diffusion. It diffuses through the cell membrane into the root-hair cell because (1) there is a higher concentration of minerals in it than there is inside the cell, and (2) the water outside is purer, and therefore in higher concentration than the water of the cell sap in the vacuole. After having entered the root hair, soil water then diffuses into the adjoining cells inside the root, where there is a lower concentration of these molecules. As this process continues from cell to cell, the water and minerals then diffuse into the xylem of the fibrovascular bundle and are conducted up the stem to the leaves.

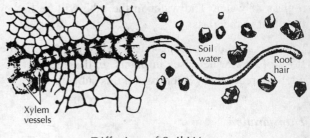

Diffusion of Soil Water

A cell membrane does not permit all materials to pass through it. Only certain types of dissolved substances can pass through it. Cell membranes are therefore said to be *semipermeable.* In certain cells of the plant, the cell membrane will permit more of one substance to accumulate than in others: thus, in some roots (carrots), stored food accumulates as it enters the root cells; also, in leaf cells, the element magnesium will pass through the cell membranes to form more chlorophyll.

A demonstration to show the rise of dissolved substances in a stem to the leaves may be performed as follows:

1. Place a celery stalk containing leaves in a jar of red ink.
2. After a day, observe that the leaves are stained red, showing that the ink has risen up the stem.
3. To see the conducting tubes of xylem, hold the stalk against a bright light. The tubes are stained red and can be seen through the light-colored stalk, running the length of the stalk. To see them more clearly, cut away, with a razor blade, some of the tissue over them.
4. Cut a thin cross section of the stalk. You will see that the fibers are stained red, and surrounded by unstained cells.

The rise of water in a tree several hundred feet high may be explained as resulting from several factors:

1. *Cohesion.* There is a long, fine continuous column of water in the ducts. As the water at the top end evaporates, the water column is pulled upward, with more water entering from the roots. The force holding the water particles together is known as cohesion. It gives the column of water considerable tensile strength. The resulting upward pull is known as *transpiration pull.*
2. *Root pressure.* If the stems of some plants are cut off near the root, water can be seen coming out of the cut end of the stumps. The pressure in the root that causes the rise of water may be due to the turgidity of the root cells. This causes water to diffuse inwardly from the epidermal region of the root, and forces water up through the xylem.
3. *Capillarity.* Water rises in fine capillary tubes. This can be shown by placing a piece of glass tubing in a glass of water. The water in the tube will be higher than that in the glass.

These explanations probably tell only part of the story of how water rises in plants. Full agreement has not been reached by scientists on this subject.

Transpiration

After water is brought up to the leaves, it is passed through the veins to the spongy layer, and then evaporates into the air spaces and diffuses out through the stomates. This evaporation of water from the leaves is known as *transpiration.*

By keeping the cells of the spongy layer around the air spaces of the leaf constantly moist, transpiration makes the absorption of carbon dioxide for photosynthesis possible. An average tree may transpire 50 gallons of water a day. Transpiration may be demonstrated in the following way:

1. Cover the pot and soil of a potted geranium plant completely with a rubber sheet. Place it under a bell jar (*A*) whose edges are sealed with Vaseline.
2. In the same way, cover completely another pot that does not contain a plant, and place it under another bell jar (*B*). This is the control.
3. After several hours, observe that the inside of bell jar *A* is covered with moisture resulting from transpiration. Bell jar *B,* however, will have no moisture in it.
4. If desired, weigh the potted plant at intervals of 3 hours, and keep a record of the weight of the water lost.

If water is lost in transpiration faster than it is replaced by the roots, the leaves will lose their turgidity and will wilt. During such periods, the stomates close, reducing the rate of transpiration. The cutinized wall of the epidermis also helps retard water loss. Plants that grow in dry regions have fewer stomates than those from moist areas; their stomates are also sunk in pits covered by hairs to reduce evaporation. Cactus plants in the desert have spines as modified leaves; their water loss is very small.

Plant Nutrition

Minerals

Minerals enter a plant when there is a greater concentration of them outside than inside the plant. There are 14 elements that are regarded as essential for green plants. The 3 in carbohydrates—carbon, hydrogen, and oxygen—are obtained from carbon dioxide and water during photosynthesis. They make up most of the plant. The other 11 are obtained from the soil—nitrogen, phosphorus, potassium, calcium, magnesium, iron, sulfur, manganese, boron, zinc, and copper. The first 4 of these are needed in larger amounts. Nitrogen, sulfur, and phosphorus are required in protein synthesis. Iron and magnesium are necessary for the production of chlorophyll: although iron is not part of the chlorophyll molecule, chlorophyll cannot be formed without it; magnesium is part of the chlorophyll structure. For a completely healthy condition, however, plants also need about 30 other elements, such as chlorine, aluminum, and molybdenum, in minute amounts; these are known as *trace elements.*

Farmers and gardeners make use of their knowledge about the mineral requirements of plants by adding to the soil fertilizers containing specific amounts of such elements as nitrogen, phosphorus, and potassium. Another way of increasing the amount of nitrogen compounds in the soil is to grow legume plants, which bear nodules on their roots containing *nitrogen-fixing bacteria*; these bacteria fix nitrogen into nitrates. One interesting method of growing plants is to do without soil and instead to use a liquid solution of the required minerals. This method is known as *hydroponics*.

A demonstration to show the absorption of phosphorus by a potted geranium plant may be performed as follows:

1. Water the soil with radioactive sodium phosphate containing the radioactive isotope of phosphorus ^{32}P.

2. At 5-minute intervals, hold a Geiger counter close to the lower and then the upper leaves of the plant. Record the time required for the ^{32}P to enter the lower leaves and finally all the leaves of the plant.

Radioactive isotopes that can be traced through a living organism by a Geiger counter are known as *tagged atoms*. Radioactive carbon in carbon dioxide has added to our knowledge of photosynthesis.

Methods of Plant Nutrition

There are various ways by which plants obtain their food:

1. *Independent plants* are green plants that make their own food and so may be considered to be autotrophs. They are the basis of food for heterotrophs such as saprophytes and parasites.

2. *Saprophytes* do not contain chlorophyll and obtain their nourishment from organic matter; examples are various fungi such as mushrooms, molds, and bacteria. Mushrooms live on soil containing the decayed remains of plants. Molds may grow on bread, fruits, and leather. Bacteria of decay in the soil help to break down plant and animal remains, thereby enriching the soil.

3. *Parasites* attach themselves to living things for their nourishment. Wheat rust is a fungus that lives on the wheat plant, destroying it. Part of the life span of this parasite is spent on another host, the barberry plant. Another fungus causes athlete's foot in humans. Some bacteria are the cause of serious human diseases.

4. *Symbiotic organisms* live together in an intimate relationship. The lichen is made up of a fungus and algae cells. The algae make food by photosynthesis, while the fungus supplies water and minerals. Legume plants (peas, clover) have nodules, or swellings, on their roots which contain nitrogen-fixing bacteria. The bacteria live on the plant material and fix nitrogen, converting it into nitrates, which the plant uses.

5. *Insectivorous plants* have leaves that trap insects and use them as food. All of these plants also carry on photosynthesis. In some cases the trapped insects are digested by enzymes secreted by the leaves. In other cases, they are decomposed by bacteria; the soluble nitrogenous materials are then absorbed by the plants:

- *Pitcher plant* has its leaves modified in the form of an enclosed pitcher, in which water collects. Small insects are drowned in the water, being unable to climb out because there are hairs on the inner walls of the pitcher that point downward. The insects are decomposed by bacteria, and the products of decay are absorbed by the plant.
- *Sundew* has small leaves covered with glands borne at the ends of many thin stalks. Insects become stuck on the sticky fluid given off by these glands. The thin stalks bend over and enclose the insect, which is then digested by enzymes secreted by the glands.
- *Venus's-flytrap* has paired leaves, each with three bristles. If an insect alights on either leaf and touches the bristles, the leaf closes, trapping it. Enzymes are secreted from glands on the leaves, digesting the insect. After the digested material is absorbed, the leaf opens and the insect remains fall out.
- *Bladderwort*, a plant that lives in freshwater ponds, has many tiny bladders the size of a pinhead along its stem. A trapdoor entrance to each bladder prevents tiny water animals from escaping, once they have touched the little bristles around it; they are swept into the bladder by the sudden closing of the trapdoor. Enzymes then digest them.

Digestion of Food

After the green plant has made food, it stores it as starch, fat, or proteins. These stored nutrients are insoluble and remain in their area of storage until they are either to be used by the plant, or to be transported from one part of the plant to another. Then they must be made soluble. This change of nutrients from insoluble to soluble form is known as *digestion*. Certain compounds called *enzymes* bring about these changes: they are examples of catalysts, substances that accelerate chemical changes without being used up in the process. Since this digestion takes place within the cells, it is called *intracellular* digestion.

The enzyme *diastase* (also known as amylase) converts starch, which is insoluble, into glucose, which is soluble. Starch grains that are stored in the leaves during photosynthesis are broken down at night into glucose which is then transported to other parts of the plant. A demonstration to show digestion in plants may be performed with bean seeds as follows:

1. Germinate six bean seeds. Starch may be shown to be present by testing some bean seeds with iodine; the blue-black color indicates the presence of starch. Glucose may be shown to be absent by heating some

crushed beans in Benedict's solution; the Benedict's solution remains blue.

2. In a few days when the bean seeds have started to germinate and are beginning to produce roots, crush them and heat in Benedict's solution. The change of the liquid to an orange color now indicates the presence of glucose.

Without digestion taking place, the growing embryo would be unable to make use of the insoluble stored starch. Glucose, however, is soluble and can move from the cotyledons to the various parts of the young plant when it is needed. Other enzymes produced by plants are *lipase*, which digests fat into fatty acids and glycerin, and *bromelin,* which splits proteins into amino acids.

Extracellular Digestion

Heterotrophic organisms such as fungi and bacteria do not have chlorophyll, and cannot make their own food. These organisms exhibit plantlike characteristics and so are included here. They secrete enzymes that digest food outside of them, changing it into simpler form, which is then absorbed. Because such digestion takes place outside the organism's cells, it is known as *extracellular* digestion.

Flowering and Growth

Photoperiodism

The production of flowers by a plant depends on the changing daily cycle of light and darkness. This was first demonstrated by Wrightman Garner and Harry Allard in 1918 in experiments with a tobacco mutant called Maryland Mammouth. This plant normally does not flower before the onset of frost in the fall. In early summer, these two plant physiologists reduced the hours of daylight for the plant by carrying potted plants into a dark room at 6 p.m. each evening, where the plants remained until 8 a.m. the next morning. Before the end of July, the experimental plants flowered, while the control group living under normal long July days did not flower.

Plants can be classified according to their photoperiodism into three main groups:

1. *Short-day plants*—flower when the hours of daylight are less than 12–13. Examples: aster, chrysanthemum, soybean.
2. *Long-day plants*—flower when the day length is more than about 12 hours. Examples: hollyhock, black-eyed Susan, iris, sweet clover.
3. *Day neutrals*—flower over a wide range of periods. Examples: tomato, rose, nasturtium.

It has been demonstrated that the length of darkness is more important than the length of daylight.

Plant-Growth Substances

In recent years scientists have discovered the existence of certain substances that have pronounced effects on the growth of plants.

Gibberellin The group of compounds known as *gibberellin* was discovered in 1926 by a Japanese scientist on Formosa named E. Kurosawa, who was investigating a fungus disease of rice plants that caused them to grow very tall and spindly. He grew the fungus on a liquid culture medium, and was surprised to discover that this liquid would produce the symptoms of the disease if he rubbed it on healthy rice plants. The substance in the liquid was isolated and named gibberellin after the fungus that produced it, *Gibberella fujikuroi*. It was not until 1950 that gibberellin came to the attention of the western world. Research since then has shown it to have the following unusual effects on plants:

1. It causes corn, wheat, and many other plants to grow very rapidly, showing an increase in height that is three to five times the normal in a short period of time.
2. It makes dwarf plants that by heredity should always be stunted, such as dwarf pea or dwarf corn, grow to the size of normal plants.
3. Seeds that are soaked overnight in it germinate ahead of time.
4. It makes biennial plants such as the foxglove and carrot, which flower in the second year of their life cycle, burst into bloom in only one year.
5. Tomatoes and cucumbers develop from flowers that are not pollinated if the flower buds are sprayed with it.
6. It makes garden and house plants, such as geraniums and petunias, bloom ahead of time, and have larger flowers.

Auxin The growing tips of leaves and stems produce growth-promoting substances or hormones called auxins. They cause the cell walls of young cells to elongate. Auxins also play a part in the tropisms of plants, making them grow toward the light. Besides these two effects, auxins and other growth-promoting substances also produce the following results: they induce the formation of roots on cuttings; they speed up the germination of seeds; they increase the growth of weeds to such an extent that they interfere with their life functions and cause them to die; they prevent the premature dropping of apples from trees; they produce seedless tomatoes.

Cytokinins (kinetin) Coconut milk was originally found to contain factors that stimulated cell division in plant tissue cultures. These cytokinins have also been found in corn kernels and other plant sources. They have been used together with auxins to cause the differentiation of roots, stems, leaves, and buds from small sections of plant tissue growing in tissue culture. They also have been found to break seed dormancy and to promote leaf expansion.

Ethylene This is a gas that causes fruit to ripen by stimulating the change of starch into sugar, and also produces the change in color of the fruit.

Phytochrome (chromophore) The bluish pigment in plants named phytochrome seems to have an effect on photoperiodism. It is affected by light near the limit of vision in the far-red end of the spectrum.

Florigen This is believed to be a hormone that brings about flowering in plants. Although it has not yet been isolated in pure form, there is much experimental evidence that it exists.

Dormin Some growth factors inhibit or prevent growth and development. Dormin keeps buds, leaves, tubers, seeds, and fruits from developing. Its effects have been found to be reversed by auxin, gibberellin, and cytokinin.

Section Review

Select the correct choice to complete each of the following statements:

1. Green plants carry on respiration (A) only during the day (B) only at night (C) during both day and night (D) only when photosynthesis is going on (E) only when photosynthesis is not going on

2. During respiration, green plants take in (A) water (B) carbon dioxide (C) oxygen (D) carbon dioxide and water (E) oxygen and water

3. In a green plant, respiration takes place (A) in all the cells (B) only in the green cells (C) only in the cells without chlorophyll (D) only in the guard cells (E) only in the stomates

4. A root may perform all of the following functions *except* (A) absorption of water (B) absorption of minerals (C) anchorage (D) storage (E) photosynthesis

5. All of the following materials can diffuse through a membrane *except* (A) soil water (B) dissolved minerals (C) dissolved iodine (D) starch (E) red ink

6. A cell membrane (A) stores food (B) is semipermeable (E) is not permeable (D) is soluble (E) keeps molecules out of a cell

7. The evaporation of water from a leaf is known as (A) conduction (B) circulation (C) translocation (D) transpiration (E) transformation

8. Mineral salts enter a plant largely through its (A) root hairs (B) root cap (C) root tip (D) stomates (E) guard cells

9. All of the following factors may be involved in transpiration *except* (A) evaporation (B) cohesion (C) root pressure (D) condensation (E) capillarity

10. Two elements needed for the production of chlorophyll are (A) phosphorus and sulfur (B) chlorine and copper (C) magnesium and iron (D) calcium and phosphorus (E) manganese and boron

11. Green plants need nitrates for making (A) carbohydrates (B) sucrose (C) starch (D) fats (E) proteins

12. A method of growing plants on a liquid solution of minerals is known as (A) hydrotropism (B) hydroponics (C) herpetology (D) symbiosis (E) commensalism

13. In the dark, a green plant excretes (A) hydrogen (B) carbon dioxide (C) oxygen (D) nitrogen (E) none of the above.

14. A fibrous type of root is present in (A) grasses (B) carrot (C) radish (D) dandelion (E) oak tree

15. All of the following are used for digestion by plants *except* (A) enzymes (B) diastase (C) nodules (D) lipase (E) bromelin

16. Organisms that live together in a relationship that benefits both of them illustrate (A) saprophytism (B) mutualism (C) parasitism (D) predation (E) fermentation

17. A mushroom cannot manufacture its own food because it lacks (A) minerals (B) cytoplasm (C) cellulose (D) chlorophyll (E) root hairs

18. An example of a saprophyte is (A) clover (B) bread mold (C) wheat rust (D) athlete's foot fungus (E) legume

19. Pitcher plant, sundew, and Venus's-flytrap are alike in that they (A) are saprophytes (B) are symbiotic (C) trap insects (D) lack chlorophyll (E) do not carry on respiration

20. A plant growth substance produced by a fungus is   (A) auxin   (B) bromelin   (C) a trace element   (D) gibberellin   (E) molybdenum

Answer Key

1-C	5-D	9-D	13-B	17-D
2-C	6-B	10-C	14-A	18-B
3-A	7-D	11-E	15-C	19-C
4-E	8-A	12-B	16-B	20-D

Answers Explained

1. **(C)** Green plants carry on respiration at all times as they release energy for their activities, such as food-making and growth. They take in oxygen and combine it with food to release energy. They give off the wastes, carbon dioxide and water.

2. **(C)** During respiration, green plants take in oxygen and give off carbon dioxide and water as wastes.

3. **(A)** All the cells of a green plant carry on respiration to obtain energy from food.

4. **(E)** A root does not contain chlorophyll and so is unable to carry on photosynthesis.

5. **(D)** Starch is insoluble, and its molecules cannot diffuse through a membrane.

6. **(B)** The cell membrane does not permit all materials to pass through it. Because only certain types of dissolved substances can do this, the membrane is said to be semipermeable.

7. **(D)** During transpiration, water evaporates out of the leaves through their stomates.

8. **(A)** Mineral salts are dissolved in soil water, which enters a plant through its root hairs. A root hair is a single cell which absorbs the soil water by active transport and diffusion. After soil water enters a root hair, it diffuses into the adjoining cells inside the root.

9. **(D)** In a plant the rise of water from the roots takes place through several factors, including cohesion, the force holding water particles together; root pressure, the pressure that forces water up through the xylem; capillarity, which is the rise of water in fine tubes; and evaporation, the process through which water is given off from the leaf through the stomates.

10. **(C)** Magnesium is part of the chlorophyll molecule. Iron is not part of chlorophyll, but without it, chlorophyll cannot be formed.

11. **(E)** Nitrogen is one of the elements in the structure of proteins. When nitrates are taken in by the green plant, the nitrogen in them is added to glucose to form basic compounds known as amino acids; from these, complex proteins are built up.

12. **(B)** Hydroponics is a method of growing plants on a liquid solution containing the required minerals, instead of using soil.

13. **(B)** In the dark, when photosynthesis is not going on, plants take in oxygen and give off carbon dioxide and water during the process of respiration. In the light, respiration is still going on, but now so is photosynthesis; the waste products of respiration—carbon dioxide and water—are used up almost at once by the chloroplasts, which are carrying on photosynthesis.

14. **(A)** A fibrous root, such as occurs in grasses, is a network of roots formed in the soil just below the stem.

15. **(C)** Nodules are small swellings on the roots of legume plants (i.e., clover, beans), which contain nitrogen-fixing bacteria. Nodules are not used for digestion.

16. **(B)** In mutualism, organisms live together in a beneficial relationship. Nitrogen-fixing bacteria live on the nodules of legume plants, where they receive nourishment. The bacteria fix nitrogen, converting it to nitrates, which the plant uses.

17. **(D)** A mushroom does not contain chlorophyll and so is not able to make its own food. It obtains nourishment from the decayed remains of plants in the soil.

18. **(B)** A saprophyte does not contain chlorophyll and obtains its nourishment from organic matter. Bread mold grows on bread.

19. **(C)** Pitcher plant, sundew, and Venus's-flytrap are insectivorous plants, obtaining their nourishment from the bodies of insects that they trap.

20. **(D)** Gibberellin is produced by a fungus that grows on rice plants. When isolated and purified, it causes other plants to grow very rapidly.

HOW WE LIVE

CHAPTER
6

6.1 VALUE OF GOOD NUTRITION

"Man does not live by bread alone." Every food we eat contains one or more substances, known as *nutrients*, that are useful to the body. There are six such groups of nutrients: carbohydrates, lipids, proteins, minerals, water, and vitamins. Some foods, such as sugar, consist of only one type of nutrient; most, of several types. For good health, the body must receive a sufficient supply of them all. Nutrients are needed for several purposes: (1) to provide energy; (2) to allow for growth; (3) to repair body parts; (4) to maintain good bodily health; (5) to supply the requirements of specific parts of the body (e.g., teeth, bones, blood, and thyroid gland); (6) to synthesize compounds needed for the maintenance of the body such as enzymes, hormones, neurotransmitters, other complex molecules. Plant foods are also useful in providing "roughage," which contributes indigestible material such as cellulose and stimulates the large intestine to function properly in excreting solid wastes.

Food and Energy

Carbohydrates and fats are chiefly used for energy, although proteins may also serve this purpose. One of the results of oxidation of food in warm-blooded animals is the maintenance of a constant body temperature—in humans, it is 37°C; in some birds, it may vary from 37.7°C to 44.4°C. Oxidation in animals is basically similar to that in plants. Certain respiratory enzymes in each cell are used to break the food down, with the release of energy and the formation of the waste products, carbon dioxide and water.

Nature of Oxidation

Research has revealed the true nature of this oxidation. The food nutrients are not used directly for energy. Instead, the cells use a chemical compound called *adenosine triphosphate* (ATP), which is derived from the food. When a cell needs energy, part of ATP is broken away, thereby liberating the needed energy. More ATP is then built up from the nutrients by a complex process involving numerous enzymes. The energy is used by the cells to carry on their activities and to grow.

We use energy all the time. Even when we are asleep, the heart is beating, we are breathing, and digestion is going on. The minimum amount of energy required to keep the body alive is known as *basal metabolism*. It can be determined by measuring the amount of oxygen consumed and the amount of carbon dioxide given off when the body is at rest. It is usually measured in the morning before a person has had any food. *Metabolism* is the sum total of all of the body's activities. It is possible to measure a person's rate of metabolism while he or she is engaged in various activities. A person who is reading a book has a lower rate of metabolism than one shoveling snow or swimming.

Calories

The energy value of food is measured in *calories*. A calorie is the amount of heat needed to raise the temperature of one gram of water one degree Celsius. To determine the number of calories in a portion of food, the food is placed in a *calorimeter*, a metal container surrounded by a water jacket, and is burned in pure oxygen. The heat that is released causes the temperature of the water to rise. This increase is measured by a thermometer, and the number of calories is determined. In this way, it has been found that fat has twice the calorie value of carbohydrates or protein. There are tables showing the calorie values of different foods: for example, a slice of white bread has 60 calories; an average-size egg, 100 calories; a slice of pie, 300 calories; a portion of cabbage, 40 calories.

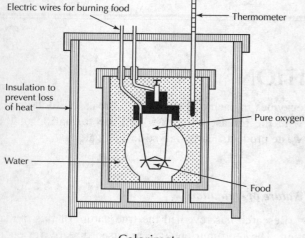

Electric wires for burning food

Thermometer

Insulation to prevent loss of heat

Pure oxygen

Water

Food

Calorimeter

The calorie requirements of people vary with their age, sex, weight, and activity, and with the climate. Basal metabolism is higher for babies and children than for grownups; for boys than for girls; for heavy than for average-weight people; in a cold climate than in a warm one. An average-size man may have a basal metabolism of 1,800 calories. A growing boy may need up to 4,000 calories for his daily activities; a clerk, only about 2,500 calories; a ditch digger, at least 5,000 calories. When more food is eaten than can be used by the body, it is stored as fat under the skin and elsewhere in the body. Some of this fat is useful in insulating the body and preventing heat loss. Part of it serves as a reserve source of energy. The rest of it is excess and results in an overweight condition, being carried around as an added burden to the body's work. In middle-aged people it may cause a strain on the heart, since extra blood must be pumped to the added fat tissues. Overweight is usually a sign of overeating. It may also result from eating too many high-calorie foods, from lack of exercise, or from a glandular disturbance, especially of the thyroid gland. An overweight person should consult a doctor as to the proper steps to be taken to modify his or her diet and to reduce the calorie intake.

Types of Nutrients

Carbohydrates

The carbohydrate nutrients are used primarily as a source of energy. They should make up about 50 percent of the daily diet. Complex carbohydrates such as starch and cellulose are found in whole-grain bread and cereals, fruits, and vegetables. They supply not only energy, but also roughage, vitamins, and minerals. Before starch can be utilized by the cells of the body, it must be digested into glucose. Excess sugar is converted for storage either into fat or into a form of animal starch known as *glycogen*. Glycogen is stored in the liver and the muscles. When required, glycogen is changed back again into glucose and used for energy. Refined carbohydrates found in cake, soda, and candy offer little more than a high sugar content. They are often a main source of excess calories and overweight; they also contribute to tooth decay.

Lipids

Like the carbohydrates, lipids consist of the elements carbon, hydrogen, and oxygen. They have twice the energy value of carbohydrates and proteins. Excess fats are stored under the skin and in the internal body cavity. Fats serve as an energy reserve system. They are an important part of the cell membrane structure. They also are important in the transport of fat-soluble vitamins such as A, D, E, and K. Nutrition experts consider fats as "empty" calories because they add few other nutrients. They should make up no more than 30–35 percent of the daily calorie intake. Foods containing large amounts of fats are butter, mayonnaise, lard, bacon, ham, peanuts, and sweet cream.

Proteins

Like carbohydrates and fats, proteins may be used for energy. However, since they contain nitrogen, sulfur, and sometimes phosphorus, in addition to carbon, hydrogen, and oxygen, proteins are useful for making new protoplasm and for repairing the body tissues (assimilation). Proteins are complex compounds composed of various *amino acids*. Some of these are more useful to the body than others. Among the most valuable amino acids are lysine and tryptophane. Growing children need more proteins than adults.

Severe protein malnutrition in very young children probably results in retarded brain development; it may thus be responsible for low educational achievement. Proteins are not stored in the body and so should be part of the daily diet. On the other hand, a continued excess of proteins may be harmful to the kidneys, which have the function of removing nitrogenous wastes from the body. Foods rich in proteins are meat, fish, eggs, bread, beans, and milk.

Minerals

Minerals in small amounts are important for maintaining the health of various parts of the body. Some of the most important mineral compounds are those containing calcium, phosphorus, sodium, potassium, and chlorine.

1. *Calcium* and *phosphorus* are needed for building bones and teeth. Calcium is also essential for the proper clotting of blood. These two elements are found in milk and other dairy products and in leafy vegetables.

2. *Sodium* and *potassium* help maintain the proper water balance in the body fluids. Sodium is found in table salt, and both minerals occur in vegetables.

3. *Chlorine* is part of hydrochloric acid, made in the digestive fluid of the stomach, gastric juice. It is found in table salt and in vegetables.

Trace Elements

Trace elements, or micronutrients, are required by the body in tiny amounts to maintain normal health and growth. However, they can be deadly poisons if consumed in large doses. In addition to such trace elements as iron, copper, iodine, and fluorine, others also known to be needed for good health are zinc, chromium, cobalt, and manganese.

1. *Iron* is part of hemoglobin, the red protein found in red blood cells. If iron is lacking, not enough red blood cells are formed and *anemia* results. Foods rich in iron are liver, beef, raisins, and spinach.

2. *Copper* is needed for the formation of red blood cells. It is also an essential ingredient of several respiratory enzymes. Good dietary sources are nuts, cocoa powder, beef, liver, raisins, and spinach.

3. *Iodine* is needed for a healthy condition of the thyroid gland, an endocrine gland located in the neck. If iodine is lacking, the gland enlarges, producing a condition known as simple *goiter*. Sea foods contain large amounts of this element.

4. *Fluorine* seems to be needed for proper tooth formation. In a number of cities and towns, small traces of it are being added to the water supply, as the result of scientific studies like those conducted in Newburgh

and Kingston, N.Y. In the former city, fluorine was added to the drinking water, and the teeth of the children were observed over a 10-year period. They were found to show less tooth decay than in the comparable city of Kingston, which was used as a control, and which did not receive fluorine.

5. *Zinc* is a constituent of nearly 100 enzymes involved in major metabolic processes. Deficiency causes loss of appetite and poor growth. A good source is a varied diet including animal proteins, as well as eggs, seafood, milk, and whole-grain products.

6. *Chromium* helps maintain normal metabolism of glucose. Good sources are meat, cheese, dried beans, and brewer's yeast.

7. *Cobalt* is an essential part of vitamin B_{12}. Beef is a good source.

8. *Manganese* is needed for normal bone structure, reproduction, and proper functioning of the central nervous system. It also makes up several enzyme systems. It is found in nuts, whole grains, vegetables, and fruit.

Some other elements whose importance, in minute quantities, for human beings has not been fully established are molybdenum, selenium, nickel, tin, vanadium, and silicon.

Water

Water makes up most of protoplasm. About 70 percent of the body consists of water. It serves to carry dissolved nutrients, oxygen, and wastes for diffusion into and out of cells. Most of blood consists of water, which makes it a circulating medium of exchange. By evaporating from the skin, it helps regulate the body temperaure. The body needs about eight glasses of water every day.

Vitamins

Eighty years ago, vitamins, originally named "vitamines" by the Polish scientist Casimir Funk in 1911, were exciting, mysterious substances in food that were just beginning to be recognized as essential for health. Today their value is known and accepted. Minute quantities of vitamins are needed to help keep the body in vigorous health and to prevent the *deficiency diseases*. Vitamins also serve as important coenzymes.

Vitamin A. This vitamin is formed in the intestinal wall from the yellow pigment *carotene,* found in yellow vegetables such as corn, carrots, and sweet potatoes, and in green leafy vegetables such as spinach and lettuce. It is also obtained from liver, butter, and eggs. Vitamin A is needed for growth. It is also used by the body to make *visual purple,* a part of the retina required for vision, especially in dim light. A deficiency of the vitamin results in night blindness, in which a person has difficulty in seeing at night. A healthy, moist condition of the eyes also depends on vitamin A; in its absence a dry, scaly condition, *xerophthalmia,* results. It also keeps the epithelial tissue of the respiratory system healthy, reducing susceptibility to colds. This vitamin is stored in various parts of the body, including the liver, and the fatty layer under the skin.

Vitamin B complex. Vitamin B is known to consist of a number of vitamins, many of which have been isolated in crystalline form.

1. *Thiamin*, or *vitamin B₁*, is used by the body to keep the nerves in healthy condition; it stimulates a good appetite and is one of the substances needed in cell respiration. For many years, natives of Oriental countries suffered from *beri-beri*, a disease characterized by weakness and inflamed nerves. It was discovered by Christiaan Eijkman and others that this condition was brought on by a diet consisting largely of polished rice; when rice is polished, the husks containing thiamin are discarded, leading to the disease. Thiamin is found in whole grains, pork, liver, yeast, and fresh green vegetables.

2. *Riboflavin*, or *vitamin B₂*, is one of the essential substances needed in oxidation and energy release in all cells. A deficiency of it leads to weakness and sores around the mouth. It is found in such foods as milk, liver, green vegetables, and yeast.

3. *Niacin* is also important in cell respiration. It is required in maintaining the good health of the skin and nervous system. Without it, a severe disease, *pellagra,* results. This was common in underprivileged parts of the South until Goldberger discovered that the basic diet of corn, molasses, and pork was deficient in the necessary vitamin, which at first was called vitamin G. Foods containing niacin, such as liver, lean meats, yeast, and peanut butter, prevent the disease.

4. *Vitamin B₁₂* is found in liver, kidney, meats, and milk. It is needed by the bone marrow to form red blood cells. Without it, a person develops pernicious anemia. It is also found in the mitochondria of cells, where it plays a role in cell metabolism.

5. *Folic acid*, which has been primarily used to treat anemia, has recently been found to be important in preventing heart disease, cancer, and neural tube defects in newborn infants (spina bifida). Folic acid reduces the amount of homocysteine in the blood. Homocysteine can cause heart attacks and strokes. Folic acid is found in green leafy vegetables, fruits, and dried beans.

6. *Other B complex vitamins*, whose functions in human beings are not entirely understood, include *biotin*, needed for the growth of bacteria; *pantothenic*

acid, needed for cellular respiration and for general body health; and *pyridoxine* (vitamin B₆), needed for cellular respiration and good skin condition.

Vitamin C, ascorbic acid, helps in the formation of connective tissue. It also keeps the walls of the capillaries in good condition. It is found in citrus fruits (orange, lemon, etc.) and in other fresh fruits and vegetables. For many years its absence in the diet of sailors was the cause of *scurvy*, until, during the eighteenth century, the British navy began to include limes in its menu. This disease leads to hemorrhages (bleeding), swollen gums, loose teeth, and general weakness.

Vitamin D, calciferol, is also known as the sunshine vitamin. It is formed from dehydrocholesterol present in the skin, upon exposure to the direct rays of the sun. It is also formed in milk, bread, and yeast when they are irradiated with ultraviolet rays: in this process, the substance *ergosterol* is transformed into a form of vitamin D. It occurs naturally in cod-liver oil and other fish oils, egg yolk, and butter. Vitamin D is needed by the body in order for it to assimilate calcium and phosphorus in the proper formation of bones and teeth. Children lacking the vitamin have bad teeth and soft leg bones which become bowed, resulting in *rickets*.

Vitamin E, tocopherol, seems to be needed for reproduction in rats. It also plays a part in the health of the muscles. Its value for humans has not been definitely established. It is found in vegetables and whole grains.

Vitamin K, naphthaquinone, plays a part in the clotting of blood. It seems to be necessary for the formation of prothrombin, one of the substances involved in the clotting process. Vitamin K is made in the liver from material present in such foods as green leafy vegetables and liver. It is also formed by bacteria in the intestines and absorbed.

Conserving the Nutrients in Food

Some of our foods are *enriched* by having vitamins and minerals added to them. When flour is being refined, valuable B complex vitamins and minerals are lost. In the baking of bread, fixed amounts of thiamin, riboflavin, niacin, iron, and calcium are added to the dough. Milk is usually enriched by the addition of vitamin D. Cereals, spaghetti, and other foods are also enriched, as can be determined by reading the label. Improper cooking may result in a loss of valuable vitamins and minerals. Thus, the water in which vegetables are cooked will contain a large amount of these dissolved nutrients; it should not be thrown away. Since vitamin C is oxidized and destroyed at high temperatures, vegetables should not be overcooked. Milk loses practically all of its vitamin C content during the pasteurization process; for this reason, young babies are given orange juice or ascorbic acid at an early age. Fruits and vegetables that were picked long before they are eaten lose part of their vitamin content. Quick-frozen vegetables retain a high percentage of their vitamins. Canned fruit juices keep their vitamin C value because they are heated in the absence of air.

Improving the American Diet—The Food Guide Pyramid

In 1992, the U.S. Department of Agriculture issued a new set of guidelines, known as the Food Guide Pyramid, for a person's daily diet. The large base of the pyramid advises that bread, cereals, rice, and pasta should make up most of the daily food choices; 6 to 11 servings should be eaten from this group every day. As one moves up the pyramid, specific servings of the different food groups are recommended, as can be seen in the accompanying illustration. At the top, the small peak of the pyramid suggests that fats, oils, and sweets be eaten only sparingly.

Revised recommendations for an increase in the recommended daily allowance of calcium were issued in 1997 by the Institute of Medicine. Because the average adult was found to be consuming only 500–700 mg of calcium per day, the new requirements were increased to 1000 mg per day, the equivalent of two cups of milk. Other good dietary sources are cheese, sardines, and vegetables. Calcium helps to increase bone density and to prevent osteoporosis caused by the weakening of bones from calcium loss. For people older than 50 years of age, the new daily recommendation is 1200 mg of calcium; for growing teenagers, it is 1300 mg.

We can build a vigorous, disease-resistant condition by eating foods containing the proper amounts of vitamins, minerals, water, proteins, fats, and carbohydrates daily.

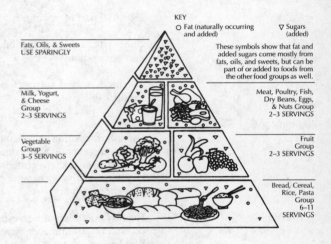

Food Guide Pyramid
SOURCE: U.S. Department of Agriculture/U.S. Department of Health and Human Services

SUMMARY OF THE NUTRIENTS

Nutrient	Uses	Good Sources
Carbohydrates	Energy	Potatoes, bread, spaghetti
Lipids (fats and oils)	Energy; storage; insulation	Butter, mayonnaise, bacon
Proteins	Energy; growth; repair	Meat, fish, eggs, beans, milk
Mineral salts of:		
Calcium	Formation of bones and teeth; clotting of blood	Milk and dairy products, vegetables
Phosphorus	Bone and tooth formation	Whole-wheat bread, cheese, meat
Potassium	Maintains water balance in body fluids	Vegetables
Sodium	Maintains water balance in body fluids	Vegetables, table salt
Chlorine	Formation of hydrochloric acid in stomach	Table salt, vegetables
Trace elements:		
Iron	Formation of hemoglobin for red blood cells	Liver, beef, spinach, raisins

SUMMARY OF THE NUTRIENTS *(Continued)*

Nutrient	Uses	Good Sources
Copper	Red blood cell formation; respiratory enzymes	Nuts, raisins, beef, liver, spinach
Iodine	Proper activity of thyroid gland	Seafood
Fluorine	Tooth formation; prevents tooth decay	Dissolved in drinking water
Zinc	Proper enzyme activity	Animal proteins, eggs, seafood, milk
Chromium	Normal glucose metabolism	Meat, cheese, dried beans, brewer's yeast
Cobalt	Essential part of vitamin B_{12}	Beef
Manganese	Bone structure, reproduction, central nervous system	Nuts, whole grains, vegetables, fruits
Water	Carries dissolved nutrients, oxygen, wastes, for diffusion	Liquid and solid foods
Vitamins: A	Growth; formation of visual purple in retina; prevents night blindness, xerophthalmia; keeps epithelial tissues healthy; deactivates oxygen radicals for cancer protection	Yellow and green vegetables, liver, butter, eggs
B-complex Thiamin (B_1)	Healthy condition of nerves; good appetite; cellular respiration coenzyme; prevents beri-beri	Whole grains, liver, pork, fresh green vegetables
Riboflavin (B_2)	Cellular respiration coenzyme; prevents sores around mouth	Milk, liver, green vegetables; yeast
Niacin	Cellular respiration coenzyme; prevents pellagra; healthy condition of nerves and skin	Liver, lean meats, yeast, peanut butter
B_{12}	Prevents pernicious anemia; cell metabolism	Liver, kidney, meats, milk
Folic acid	Prevents heart disease, cancer, and neural tube defects; treatment of anemia	Green leafy vegetables, fruits, dried beans
C (Ascorbic acid)	Formation of connective tissue; formation of capillary walls; prevents scurvy; deactivates oxygen radicals for cancer protection	Citrus and other fruits, fresh vegetables
D (Calciferol)	Assimilation of calcium and phosphorus in tooth and bone development; prevents rickets	Cod liver oil, other fish oils, made in skin in sunlight, irradiated foods
E (Tocopherol)	Reproduction in rats; healthy muscles; function in humans uncertain; deactivates oxygen radicals for cancer protection	Vegetables, whole grains
K (Naphthaquinone)	Clotting of blood; formation of prothrombin	Liver, green leafy vegetables, intestinal bacteria

Section Review

Select the correct choice to complete each of the following statements:

1. A food rich in carbohydrates is   (A) bacon (B) bread   (C) lard   (D) butter   (E) ham

2. Of the following, the function that includes all the others is   (A) digestion   (B) excretion (C) metabolism   (D) irritability   (E) respiration

3. ATP is used directly in the process of   (A) growth (B) repair   (C) roughage   (D) respiration (E) reproduction

4. A calorimeter is used in testing food for its value of   (A) carbohydrate   (B) fat   (C) protein (D) vitamin   (E) energy

5. An overweight condition may be caused by any of the following factors *except*   (A) overeating (B) eating too many high-calorie foods   (C) lack of exercise   (D) lack of iron in the diet (E) thyroid gland disturbance

6. A nutrient used to build protoplasm is   (A) protein (B) fat   (C) starch   (D) sugar   (E) iodine

7. All of the following are rich in proteins *except* (A) potatoes   (B) meat   (C) fish   (D) eggs (E) milk

8. Two elements needed for building bones and teeth are   (A) iron and calcium   (B) calcium and phosphorus   (C) phosphorus and iron   (D) iron and sodium   (E) sodium and potassium

9. Scientific evidence indicates that an element needed to prevent tooth decay is   (A) chlorine (B) iodine   (C) bromine   (D) fluorine   (E) iron

10. All of the following are true of vitamin A *except* (A) it prevents night-blindness   (B) it is formed from carotene   (C) it is used in making visual purple   (D) it is used in making hemoglobin   (E) it is found in green, leafy vegetables

11. A vitamin used in the formation of red blood cells is   (A) vitamin B_1 (B) vitamin B_2 (C) vitamin B_{12} (D) calciferol   (E) ascorbic acid

12. All of the following are members of the vitamin B complex *except*   (A) thiamin   (B) gibberellin (C) riboflavin   (D) niacin   (E) biotin

13. A vitamin formed in the skin upon exposure to ultraviolet rays is vitamin   (A) A   (B) B_2 (C) C (D) D   (E) E

14. The proper clotting of blood depends on an adequate supply of vitamin   (A) K   (B) B_1 (C) B_6 (D) C   (E) A

15. If equal quantities of the following are burned, the highest calorie value will be found in (A) fat   (B) protein   (C) glucose   (D) starch (E) minerals

Answer Key

1-B	4-E	7-A	10-D	13-D
2-C	5-D	8-B	11-C	14-A
2-D	6-A	9-D	12-B	15-A

Answers Explained

1. (B) Bread is rich in the complex carbohydrate starch. Bacon, lard, butter, and ham contain large amounts of fat.

2. (C) Metabolism is the total of all the body's activities.

3. (D) During the process of respiration, energy is released from the ATP molecules, which were derived from food. Part of the ATP molecule is broken away to form ADP, while energy is liberated.

4. (E) The calorimeter is a device in which a fixed amount of food is burned. The heat that is released causes the temperature to rise, giving a measurement of the energy value of the food in calories.

5. (D) Lack of iron in the diet may result in anemia; it does not result in an overweight condition. (Foods rich in iron are liver, beef, raisins, and spinach.)

6. (A) Proteins are used for making new protoplasm and for repairing body tissue. The process of making the living material of protoplasm from the nonliving material of proteins is known as assimilation.

7. (A) Potatoes contain starch and are a good source of carbohydrates, not proteins.

8. (B) Calcium and phosphorus are used in the building and maintenance of bones and teeth. They are found in milk and dairy products and in leafy vegetables.

9. (D) Small amounts of fluorine in the drinking water have been shown to prevent tooth decay.

10. (D) Iron is part of hemoglobin, the red protein found in red blood cells. If iron is lacking, not enough hemoglobin is formed and one type of anemia may result.

11. (C) Vitamin B$_{12}$ is needed by the bone marrow to form red blood cells. This vitamin is found in liver, kidney, other meats, and milk.

12. (B) Gibberellin is a substance produced by a fungus that grows on rice plants. When purified, it can promote very rapid growth in many plants.

13. (D) Vitamin D, known as the sunshine vitamin, is formed in the skin upon exposure to the ultraviolet rays of the sun.

14. (A) Vitamin K is necessary for the formation of prothrombin, one of the substances involved in the clotting of blood. It is made in the liver from substances in certain foods such as green leafy vegetables and liver. It is also produced by bacteria that grow in the intestines.

15. (A) When a calorimeter is used to measure calorie values, fat is shown to have twice the calorie content of carbohydrates or proteins. There are no calories in minerals.

6.2   INGESTION AND DIGESTION

Ingestion refers to the activities of animals whereby they take in food. Before food can be absorbed into the cells of the body, it must be made soluble, or digested. During *digestion* the complex molecules of food are reduced into smaller ones capable of passing through the membranes of the body. Digestion may occur within the cell (intracellular digestion), as in protozoa; or outside the cell, in a special cavity, or tube, as in most animals (extracellular digestion), after which the soluble molecules are absorbed into the cells.

The Digestive System in Higher Animals

Food is passed along through the body in the tube called the *alimentary canal*. This consists of the mouth, esophagus or gullet, stomach, small intestine, large intestine, and rectum. The rectum leads to the outside through the anus. There are digestive glands leading into the alimentary canal from the mouth, stomach, and small intestine. Two large glands, the liver and the pancreas, are outside the food tube, and send their digestive fluids into it through ducts. The sum total of the alimentary canal and the digestive glands comprises the *digestive system*.

Mouth

In the brief time that food is in the mouth, two main things happen to it: (1) It is chewed and broken down to smaller pieces by the teeth, aided by the tongue. This is known as *mechanical digestion*. (2) It is attacked by *enzymes* in the salivia, a digestive liquid; this is *chemical digestion*.

Teeth

Human beings and most other mammals develop two sets of teeth during a lifetime. The first, or milk teeth, number 20. In an adult, there are 32 teeth arranged in a semicircle, half on the upper jaw and half on the lower jaw. These include 8 front, chisel-like teeth, the *incisors,* which cut the food; 4 *canines*, or eye teeth, which are pointed and which help bite the food; 8 *bicuspids*, or premolars, for grinding; and 12 *molars,* also for grinding. The last set of molars, the wisdom teeth, usually appears late in adolescence. A tooth consists of several parts.

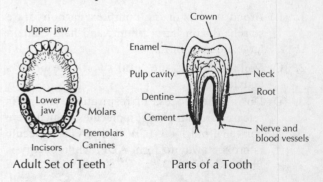

Adult Set of Teeth   Parts of a Tooth

1. The *crown*, the exposed part, is covered with hard *enamel*. Enamel is the hardest substance in the body. When it is chipped or decayed by bacteria, cavities *(caries)* develop which reach into inner parts of the tooth, causing pain and sometimes infections. It is then that the dentist installs a filling to seal the cavity.

2. The *root*, the part of the tooth below the level of the gum, fits into sockets in the jawbone. It is covered with a layer of *cement*.

3. The *dentine* makes up the main body of the tooth. It has a very strong bony structure and is beneath the enamel.

4. The *pulp cavity*, a chamber in the center of the dentine, contains pulp, consisting of blood vessels, connective tissue, and nerves. From this section, food and oxygen are brought into the tooth. The nerves spread into the dentine.

5. The *gum* is the living tissue that holds the teeth in place. Brushing the teeth not only removes particles of food that may decay and start cavities, but also massages the gums and helps to keep them in a firm, healthy condition, which is one of the requirements for a set of good teeth.

Salivary Digestion

While the food is being chewed into smaller particles, it is being moistened by saliva, a digestive liquid produced by three pairs of salivary glands. These glands are located under the tongue, in the cheek near the ear, and in the lower jaw. They empty into the mouth through *ducts*. Saliva contains the enzyme *pytalin,* which digests starch into a sugar, *maltose.* A demonstration to show the chemical digestion of starch by saliva may be performed in the following way:

1. To some boiled 1 percent starch suspension in a test tube *(A)*, add a drop or two of iodine. After this has turned blue-black, add about an inch of saliva. Warm by holding the tube in your hand. Set up a control test tube *(B)* in the same way, but without saliva.

2. After a few minutes, observe that the blue-black color in test tube *A* is fading, a sign that the amount of starch is being decreased. There is no change in the control *(B)*.

3. When the liquid in test tube *A* has become clear, heat a small amount of it in a test tube of Benedict's solution. It will become orange, showing the presence of sugar.

4. If the liquid in tube *B* is heated with Benedict's solution, there will be no change. Also, if saliva is tested with Benedict's solution, it will be shown not to have any sugar, either.

The conclusion is reached, therefore, that the only way that sugar appeared was from the digestion of starch by saliva.

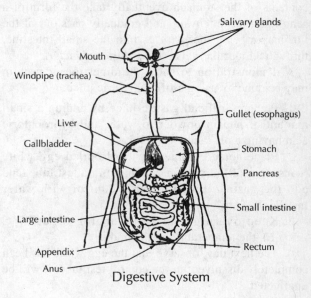

Digestive System

Esophagus

After food has been chewed, lubricated by saliva, and partially digested by ptyalin, a small mass of it (bolus) is pushed into the throat (pharynx), where it is swallowed. A flap of tissue, the *epiglottis,* closes over the adjacent windpipe and prevents the food from going into the wrong tube. The esophagus is lubricated by mucus, which is secreted by glands in its lining. *Peristalsis,* the alternate wavelike contractions and expansions of the involuntary muscles in the wall of the esophagus, moves the food along into the stomach.

Stomach

Food remains in the stomach from 3 to 4½ hours. During this time, it is constantly being churned and mixed by peristaltic action with gastric juice, the secretion of the stomach glands (mechanical digestion). For a brief period, saliva continues to act on starch; this ceases when the contents of the stomach become acid. Gastric juice contains hydrochloric acid, water, and the enzymes *pepsin* and *rennin* and a tiny amount of *lipase*. Pepsin partially digests proteins into *proteoses* and *peptones*, which are smaller combinations of amino acids and are intermediate products of digestion. Rennin curdles the protein of milk, casein, and prepares it for digestion by pepsin. Complete digestion of proteins will take place in the small intestine. Pepsin is active only in an acid medium, so that advertisements for patent medicines that imply that "acid stomach" is undesirable are misleading. Hydrochloric acid is further useful in dissolving minerals for absorption, and in killing bacteria that may have been swallowed. The very small amounts of the enzyme lipase digest only a tiny amount of fat into fatty acids and glycerol. The contents of the stomach eventually take the form of a semiliquid paste, *chyme,* and gradually pass out of the far, narrow end, the *pylorus,* into the small intestine, through a muscular valve, the *pyloric sphincter.*

A demonstration to show the digestion of protein may be done by using artificial gastric juice:

1. Prepare artificial gastric juice by adding a small amount of pepsin powder to water and hydrochloric acid having a pH (acidity) of 1.5.

2. Place some chopped-up hard-boiled egg white (consisting of the protein, albumen) in a test tube, and add the gastric juice. Set up a control with water instead of gastric juice.

3. Keep overnight in an incubator set at 37°C (98.6°F), the temperature of the body.

4. The next day, observe that the egg white has been completely dissolved; in the control test tube, it will be unaffected.

Small Intestine

Digestion of food is completed in the small intestine, a narrow tube over 20 feet long. Mechanical digestion continues as peristalic action moves the food along, and constantly mixes it with the digestive liquids. There are three such liquids, secreted by (1) numerous intestinal glands in the wall of the small intestine, (2) the pancreas, (3) the liver.

Intestinal juice contains the following enzymes:

Erepsin—digests proteins into amino acids.

Maltase—digests maltose into glucose.

Lactase—digests lactose (milk sugar) into glucose and galactose.

Sucrase—digests sucrose (cane sugar) into glucose and fructose.

Small quantities of *lipase*—digest fats into fatty acids and glycerol.

Pancreatic juice enters the upper end of the small intestine from the pancreas by means of the pancreatic duct. It contains three enzymes:

Trypsin—digests peptones and proteoses into amino acids.

Amylopsin—changes starch into maltose.

Lipase—changes fats into fatty acids and glycerol.

Liver

This organ produces bile, which is stored in the *gallbladder.* When food is being digested in the small intestine, the gallbladder contracts, and sends bile into it by means of the bile duct. Bile does not contain any enzymes. Instead, it prepares fat for digestion by *emulsifying* it, that is, breaking it into very small particles. This is really a type of mechanical digestion, and per-

mits lipase to digest the small fat globules more readily. The liver is the largest organ in the body. Beside forming bile, it also has a number of other functions. It changes excess glucose to glycogen and stores it; it destroys old, worn-out red blood cells; it converts amino acids into glucose and a nitrogenous waste called urea, which is removed from the blood in the kidneys; it produces prothrombin, one of the substances needed in the clotting of blood; it prepares fat for use by the body; and it serves as a reservoir for storing blood.

Absorption of Digested Food

As the result of digestion in the mouth, stomach, and small intestine, the nutrients are converted into their *end products*: carbohydrates into glucose; lipids into fatty acids and glycerol; proteins into amino acids. These soluble end products are absorbed through the walls of the small intestine. Here, there are numerous microscopic, fingerlike projections called *villi,* which increase the absorptive surface of the small intestine. Each villus is covered by a single-celled layer of epithelial cells. Inside, there are a network of *capillaries* and a central *lacteal,* a lymph vessel. The nutrients diffuse into the capillaries and become dissolved in the plasma of the blood. The capillaries of the villi lead into the *portal vein,* which goes into the liver. Here, most of the rich supply of glucose is removed and changed into glycogen, or animal starch, for storage. Some glycogen is also stored in the striated muscles. The blood that leaves the liver circulates the nutrients throughout the body. Several hours after a meal, when the glucose level in the blood has fallen, glycogen is changed back into glucose and is transferred out of the liver into the blood. The fatty acids and glycerol do not enter the blood directly, but diffuse into the lacteals. These lead into larger lymph vessels, and finally into the *thoracic duct,* which empties into the blood stream in the neck region.

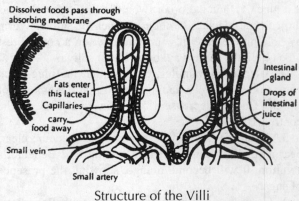

Structure of the Villi

Large Intestine (Colon)

After food has been digested and absorbed in the small intestine, the indigestible remains pass into the large intestine. A short distance from this point, there is a small outpocketing of the large intestine, about the length of the little finger, called the *appendix*. It has no apparent use in humans, although in herbivorous animals it serves for digestion. Like the esophagus, the large intestine has no digestive functions. The mucous cells in its lining secrete mucus, which provides lubrication for the solid matter being moved along by peristalsis. Water is absorbed during this time and is used over again by the body. The solid wastes, consisting of indigestible food, especially the cellulose remains of plant food, intestinal secretions, and bacteria, are known as *feces,* and are eliminated from the lower end of the large intestine (*rectum*) through the *anus*.

Digestive System Disorders

The digestive system in humans is affected by a number of disorders, including the following:

1. *Ulcer:* An open sore in the lining of the alimentary canal first discovered to be caused by a bacterium, *Heliobacter pylori*, in the early 1980s. Until then, most doctors attributed ulcers to nervous stress.

2. *Constipation:* A condition in which it is difficult for the large intestine to be emptied of its wastes. As water is reabsorbed from the undigested food wastes, the feces become hard.

3. *Diarrhea:* Rapid removal of the watery wastes in the large intestine, brought on by increased peristalsis.

4. *Appendicitis:* An inflammation of the appendix, characterized by pain in the lower right abdominal area and fever. Removal by surgery may be necessary to prevent the appendix from rupturing.

5. *Gallstones:* Hardened deposits of cholesterol that block the duct of the gallbladder, causing pain and discomfort.

Ingestion and Digestion in Lower Animals

All animals require food for energy, growth, and repair. Their survival depends on their ability to obtain food—in fact, their lives may be said to be dedicated to this purpose.

Protozoa

Although protozoa are classified by many biologists as protists, they are included here because they show animal-like characteristics. Many of them take in food through a special structure, the *oral groove,* as in the case of the paramecium; or they may *engulf* their food, as is done by the ameba. Other protozoa have specific modifications for food getting. Food is digested in the *food* vacuoles. Digestion is intracellular.

Hydra

This commonly studied member of the coelenterate phylum shows the adaptation of a simple multicellular animal for ingestion and digestion. Food is obtained with the aid of *tentacles* that surround the mouth. These have numerous stinging cells (*nematocysts*) that paralyze small animal prey. The tentacles then deposit this food into the mouth leading to the central body cavity. Digestion takes place with the aid of enzymes secreted by special gland cells located in the endoderm layer (extracellular digestion). The food is circulated by means of currents created by the beating of long flagella attached to some of the cells. Many of the broken-down food particles are engulfed by other cells lining the cavity, and digestion is completed within their food vacuoles (intracellular digestion). The indigestible remains left in the central cavity are eliminated through the mouth.

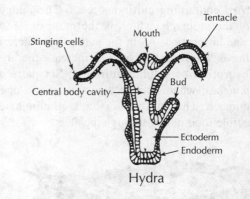

Hydra

Earthworm

The earthworm has a digestive tube with a mouth opening at one end, and an anus at the other. It swallows soil containing organic plant and animal remains, which passes from the mouth through a muscular *pharynx* into the *esophagus,* and then into an enlarged *crop,* where it is collected. Behind the crop, the food is thoroughly ground up in the *gizzard*, which has thick, muscular walls, with the help of small stones. Then the food passes into the intestine, where it is digested by enzymes secreted by gland cells of the lining epithelium. Soluble food is absorbed into the blood vessels of the intestinal wall. The lining of the intestine is folded, providing a greater area for absorption. The solid wastes are eliminated through the anus. Earthworms spend most of their time swallowing soil below the surface and depositing it on the surface as waste known as *castings*. Charles Darwin first pointed out the immense value of the lowly earthworm to agriculture because of its method of improving the soil as it burrows through it, aerating it, grinding it, enriching it, and bringing it to the surface in castings.

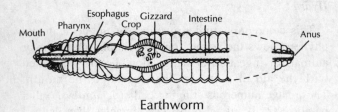

Earthworm

Insects

Some insects, such as the butterfly and the moth, have sucking mouth parts. The grasshopper has chewing mouth parts, consisting of toothed horny *mandibles* that work from side to side and *maxillae* that help to hold and cut the food; the other mouth parts consist of the *labrum* (the upper lip), and the *labium* (the lower lip). The digestive tract has three parts: foregut, midgut, and hindgut. The foregut consists of the mouth, into which the salivary glands secrete saliva; the esophagus leading to the *crop*, where food is stored; and the muscular *gizzard* containing chitinous teeth. The midgut consists of the stomach, which is the main organ of digestion and absorption. As in the earthworm, there is an infolding of the digestive tube to increase the area for the absorption of soluble nutrients. Six pairs of stomach *pouches*, which secrete digestive juices, open into the stomach. The intestine, or hindgut, eliminates the indigestible food matter through the anus.

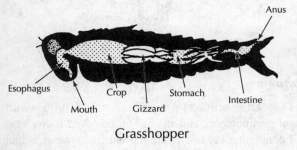

Grasshopper

Fish

The mouth of a fish contains small teeth that point backward and function mostly in holding large masses of food before they are swallowed whole. Tiny particles of food are strained by long projections of the gills called *gill rakers*, as the particles are carried in the

water passing over the gills. After food is swallowed, it enters the stomach, where digestion is started by strong digestive juices. Just behind the stomach, where it joins the intestine, there are digestive glands, or *caeca*, which send secretions into the intestines. The *liver* forms bile, which is stored in the *gallbladder* until it enters the intestine through a duct. Indigestible wastes are excreted through the anus.

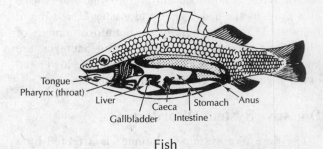

Fish

Frog

The *tongue* of a frog is well adapted for catching insects and worms. It is soft and sticky and is attached at the front end of the lower jaw. It can flip out quite a distance to trap its food. There are small toothed structures in the mouth for holding the food. The *esophagus* leads to the *stomach*, where digestion starts. It continues in the *small intestine* with the aid of secretions from the *pancreas*, and bile from the *gallbladder* and the *liver*. After soluble food has been absorbed, the indigestible remains are passed into the large intestine. This opens into the *cloaca*, which also collects wastes from the bladder and leads out of the body through the anus.

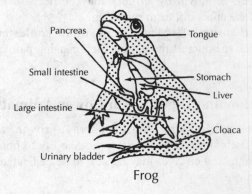

Frog

SUMMARY OF HUMAN DIGESTION

Nutrient	Where Digested	Enzyme	Products
Starch	1. Mouth (starts, then continues in stomach for short time)	Ptyalin	Maltose
	2. Small intestine	Amylopsin (pancreatic juice)	Glucose
Maltose	Small intestine	Maltase (intestinal juice)	Glucose

SUMMARY OF HUMAN DIGESTION *(Continued)*

Nutrient	Where Digested	Enzyme	Products
Lactose	Small intestine	Lactase (intestinal juice)	Glucose, galactose
Sucrose	Small intestine	Sucrase (intestinal juice)	Glucose, fructose
Lipids	Small intestine	Lipase (pancreatic juice)	Fatty acid and glycerol
	Small intestine	Bile (not an enzyme)	Emulsifies fats
Protein	Stomach	Pepsin	Proteoses, peptones
		Rennin	Curdles casein for pepsin digestion
	Small intestine	Erepsin (intestinal juice)	Amino acids
	Small intestine	Trypsin (pancreatic juice)	Amino acids

Section Review

Select the correct choice to complete each of the following statements:

1. A part of the digestive system that is not in contact with food is the   (A) small intestine   (B) stomach   (C) liver   (D) large intestine   (E) trachea

2. Digestion of food starts in the   (A) mouth   (B) esophagus   (C) stomach   (D) pancreas   (E) gullet

3. All of the following are parts of a tooth *except*   (A) enamel   (B) dentine   (C) root   (D) caries   (E) pulp

4. A secretion that digests both carbohydrates and proteins is   (A) ptyalin   (B) saliva   (C) pepsin   (D) gastric juice   (E) pancreatic juice

5. The end product of carbohydrate digestion is   (A) sucrose   (B) glucose   (C) cellulose   (D) peptone   (E) trypsin

6. Pepsin is active only in the presence of   (A) rennin   (B) erepsin   (C) amylopsin   (D) hydrochloric acid   (E) lipase

7. Villi are present in the   (A) stomach   (B) small intestine   (C) large intestine   (D) appendix   (E) colon

8. Large amounts of carbohydrate are stored in the   (A) liver   (B) stomach   (C) gallbladder   (D) adipose tissue   (E) portal vein

9. Bile is useful in the digestion of   (A) starch   (B) sucrose   (C) fat   (D) protein   (E) amino acids

10. Fatty acids are absorbed by the   (A) capillaries   (B) lacteals   (C) pylorus   (D) thoracic duct   (E) colon

11. Food is moved along the alimentary tract by the contractions known as   (A) epiglottis   (B) circulation   (C) secretion   (D) proteoses   (E) peristalsis

12. Castings are produced by the   (A) hydra   (B) paramecium   (C) crayfish   (D) grasshopper   (E) earthworm

13. The food tube in humans is known as the   (A) digestive system   (B) trachea   (C) alimentary canal   (D) crop   (E) gizzard

14. Greater surface for food absorption is provided in the earthworm by   (A) food vacuoles   (B) infolding of the intestinal wall   (C) the pharynx   (D) stomach pouches   (E) nematocysts

15. Intracellular digestion takes place in both the   (A) protozoan and earthworm   (B) protozoan and insects   (C) protozoan and fish   (D) protozoan and frog   (E) protozoan and hydra

Answer Key

1-C	4-E	7-B	10-B	13-C
2-A	5-B	8-A	11-E	14-B
3-D	6-D	9-C	12-E	15-E

Answers Explained

1. (C) Food is passed along the alimentary canal, where it is digested. The liver is a part of the digestive system outside of the alimentary canal; it sends bile into the alimentary canal to assist in digestion. The trachea, or windpipe, is part of the respiratory system.

2. (A) Both mechanical digestion (by which food is broken down to small size by the teeth) and chemical digestion (by which food is acted on by enzymes in the saliva) start in the mouth.

3. (D) Caries is the condition of tooth decay caused by bacteria in the mouth.

4. (E) Pancreatic juice, which enters the small intestine from the pancreas, contains the enzymes amylopsin, which digests starch (a carbohydrate), and trypsin, which digests proteins.

5. (B) As a result of digestion, nutrients are converted into their end products; carbohydrates are digested into glucose.

6. (D) Pepsin is an enzyme present in gastric juice in the stomach. Gastric juice also contains hydrochloric acid. Without the presence of acid, pepsin becomes inactive and will not act on proteins.

7. (B) Villi are numerous, fingerlike projections on the walls of the small intestine. They increase the surface area and absorb the nutrients, which have been digested into soluble end products.

8. (A) When digestion of carbohydrates is completed in the small intestine, the end product—glucose—is carried by the portal vein to the liver. There it is changed to glycogen, or animal starch, for storage.

9. (C) Bile does not contain any enzymes. It emulsifies fat, breaking it into very small particles. This is a type of mechanical digestion, which permits the enzyme lipase, in the small intestine, to digest fat more readily.

10. (B) Lacteals are small lymph vessels located in the villi of the small intestine that absorb digested fatty acids.

11. (E) Peristalsis is the alternate wavelike contractions and expansions of the involuntary muscles in the walls of the alimentary canal that move food along.

12. (E) Earthworms feed by swallowing soil below the surface, which contains organic plant and animal remains. They then digest the nutrients and deposit waste known as castings on the surface.

13. (C) The alimentary canal is the tube through which food passes. It consists of the mouth, esophagus, stomach, small intestine, large intestine, and rectum.

14. (B) The lining of the intestinal wall of the earthworm is folded, providing a greater surface for the absorption of digested food.

15. (E) Intracellular digestion occurs within the cell. In the protozoan, it takes place within the food vacuoles. In the hydra, it takes place in food vacuoles located within cells lining the central body cavity.

6.3 CIRCULATION

After food is digested, it is distributed throughout the body by means of the circulatory system. Circulation is the transport of materials throughout organisms, and within cells. Each of the billions of cells in a human being depends for its existence on the continuous delivery of food, oxygen, and other materials to it, and the removal of wastes from it.

The Human Circulatory System

The circulatory system of the body is arranged in a closed system of blood vessels, in which the heart serves as a pump that keeps the blood moving. This was first demonstrated by William Harvey (1578–1657) in the early part of the seventeenth century.

Heart

This four-chambered, muscular organ, about the size of a man's fist, is located approximately in the center of the chest, pointing to the left. It has two smaller upper chambers, the *atria* (auricles), which receive the blood, and two thick-walled ventricles below them, which pump the blood away from the heart. Between the atria and ventricles are valves, which prevent the

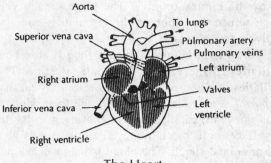

The Heart

blood from flowing backward when the ventricles contract. Located in the right atrium is an area of specialized muscle tissue called the *pacemaker*, where the pulsations of the heart appear to originate, and from which they spread like a wave to the rest of the heart.

A doctor listening to the heartbeat with a stethoscope may hear a "murmur" if there is a defect in the *valves*, when some of the blood flows back into the atrium. The right and left sides of the heart are completely separated from each other by a thick wall. The cells of the heart itself are nourished by the coronary arteries; if any of them become blocked by a clot, as in *coronary thrombosis*, the heart cells are deprived of their food and oxygen, and death may result. The right and left sides of the heart really function as two pumps, since they pump blood to different parts of the body: the right to the lungs (pulmonary circulation), and the left to the rest of the body (systemic circulation).

The rhythmic pumping of the heart takes place in two parts: the period when the ventricles are contracting is called the *systole*, and the phase immediately afterward, when they are relaxing, is called the *diastole*. When a doctor measures a person's blood pressure with an instrument called a sphygmomanometer, he or she records both the systolic and the diastolic pressure. Thus, the average blood pressure for a young woman might be 120/80. In a baby, it might be 90/15; in an old man, 150/90. In high blood pressure, or *hypertension*, the pressure is higher than the average. Its cause is not well understood; it may arise from nervous tension, a kidney condition, obesity, or smoking.

In 1984, it was announced that the heart is more than just a muscular pump that circulates the blood. It apparently also produces chemicals that influence blood pressure and volume. These chemicals stimulate the kidneys to excrete salt and water and also make the blood vessels relax. It is believed that the substances come from the atria (auricles) chambers of the heart; for that reason, they have been named *auriculin*. Auriculin is thought to be a hormone. Its chemical makeup has been shown to consist of peptides, or strings of amino acids. Since auriculin is produced in the atria, which receive blood from the body's circulation, it may be formed in response to changes in blood pressure and volume. The gene responsible for making auriculin has been cloned; this may make possible the large-scale production of auriculin in the near future.

Arteries

These large blood vessels carry blood *away from* the heart. They are thick-walled, containing much elastic fiber tissue. There are also layers of involuntary muscle that control the size of the opening of the arteries. In older people, the walls of the arteries sometimes lose their elasticity, resulting in a condition known as *arteriosclerosis*, or hardening of the arteries. Each time the heart contracts, it forces blood into the arteries, making them stretch. Following a contraction, the heart rests momentarily and then the arteries relax. This alternate pulsation of the arteries can be felt as the *pulse* beat in parts of the body where the arteries lie close to the surface, that is, the wrist, the temple, and the neck. The average pulse rate varies with age, sex, activity, and other factors. It may vary from 60 to 85 per minute in different people. Exercise will cause it to increase. The largest artery of the body is the *aorta*, which leads away from the left ventricle to most of the body. All arteries except the pulmonary artery carry oxygenated blood.

Veins

These blood vessels carry blood *to* the heart. They have thinner walls than the arteries, with less elastic fiber and muscular tissue. Blood passes through them in a steady flow, as compared with the spurting action of the arteries. Blood in the veins is kept moving toward the heart by both the pressure behind it, and the movements of the voluntary muscles. Each time these muscles move they squeeze the veins, forcing the blood forward. *Valves* along the length of the veins prevent the blood from flowing backward. All veins carry deoxygenated blood (blood lacking in oxygen), with the exception of the pulmonary vein. Some of the veins lie close to the surface and can be seen through the skin as having a bluish color; deoxygenated blood actually has a dark red color. The largest vein is the *vena cava*, leading into the right atrium.

Capillaries

Capillaries connect the smallest arteries (arterioles) and the smallest veins (venules). Capillaries are microscopic, with walls only one cell thick. They are located everywhere in the body, close to the cells. When the skin is pricked with a needle, a drop of blood will appear, coming from broken capillary walls. Food, oxygen, and other materials diffuse out of the blood through the capillary walls, into the intercellular fluid (ICF) surrounding the cells, and then into the cells. By the reverse process, wastes leave the cells, diffuse into the ICF, and then enter the blood through the capillary walls.

The Circulatory Route

Blood enters the *right atrium* of the heart through the (1) *superior (upper) vena cava,* carrying blood from the head and upper part of the body, and (2) *inferior (lower) vena cava,* carrying blood from the rest of the body. When the right atrium contracts, blood is forced into the *right ventricle.* It in turn contracts, and sends blood to the *pulmonary artery* leading to the *lungs.* Here, the blood enters capillaries surrounding the air sacs. Oxygen is absorbed through the capillary walls; carbon dioxide and water vapor are excreted. The oxygenated bright red blood then collects in the *pulmonary veins,* which lead back to the heart, this time into the *left atrium.* It contracts and forces the blood into the *left ventricle.* This chamber has the thickest walls of the heart; it pumps blood into the *aorta,* which carries blood to the entire body except the lungs. The aorta divides into arteries that go into the head, the arms, the legs, the heart itself (coronary arteries), and various other parts of the body. The arteries branch into *arterioles,* which in turn lead to the *capillaries.* Oxygen and food enter the cells via the lymph; carbon dioxide and other wastes diffuse into the capillaries. These combine to form *venules,* which lead to the veins. The blood is now largely deoxygenated and dark red in color; it is carried once more into the heart through either *vena cava.* This circuit of blood takes less than 20 seconds, and includes:

1. ***Pulmonary circulation.*** Blood from the right side of the heart is pumped through the pulmonary artery to the lungs, where it absorbs oxygen and excretes carbon dioxide and water vapor, and then back to the left side of the heart through the pulmonary veins. Deoxygenated blood is refreshed, or oxygenated, in this part of the circulatory route. The pulmonary artery is the only artery containing deoxygenated blood. The pulmonary veins are the only veins with oxygenated blood.

2. ***Systemic circulation.*** In the "body" circulation, blood from the left ventricle goes through the aorta to the upper and lower parts of the body and then back to the right atrium. Arteries deliver blood containing food and oxygen to the internal organs. Veins leaving these organs collect the blood, now containing carbon dioxide and nitrogenous wastes. The arteries entering the kidneys carry excretions of the cells, which are filtered

out by the kidney tubules. The large kidney veins, which return the blood to the circulatory route, are relatively free of wastes.

3. ***Portal circulation.*** Digested food is absorbed in capillaries of the villi in the small intestine. These capillaries lead out of the small intestine by means of the *portal vein,* which then enters the liver. Here, the blood goes into capillaries, which bring it to the liver cells. Excess glucose is removed and stored as glycogen, or animal starch. The blood is then carried out of the liver by means of the *hepatic vein* to the inferior vena cava. The portal circulation is essentially a part of the systemic circulation.

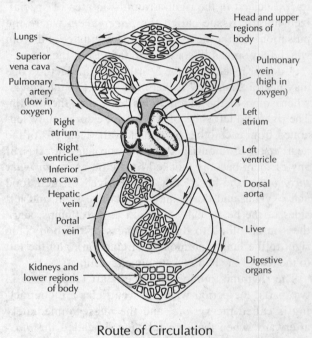

Route of Circulation

Lymph and Lymph Circulation

All the cells of the body are surrounded by, and bathed by, the watery intercellular fluid, which originates in the blood plasma. It slowly oozes out through the capillary walls into the surrounding spaces. It consists of plasma without its large protein molecules, and is largely water, with small amounts of dissolved nutrients that will be used by the cells, wastes given off by them, and secretions. It also contains white blood cells that pass through the capillary walls. It serves as the intermediate fluid.

It collects in tiny lymph vessels called *lymphatics,* which, like the capillaries, are present in every part of the body. The fluid is called *lymph* when it is absorbed into the lymph vessels. It moves slowly along, under the gentle pressure of the fluid behind it, and as a result of the body's movements. Valves along the lymphatics keep lymph moving in a forward direction. The lymphatics join into larger vessels and include the *lacteals* of the small intestine, which carry digested *fats.* They finally form two large lymphatics, one from the upper part of the body and one from the lower. The latter is the *thoracic* duct, the largest lymph vessel. It joins the circulatory system in the neck region, returning lymph back into the bloodstream. In various parts of the body, such as the armpits and the groin, there are small *lymph glands,* or *nodes,* containing masses of lymphatic tissue. They manufacture certain white blood cells and also serve to filter germs out of the body. During an infection, the lymph glands may become swollen.

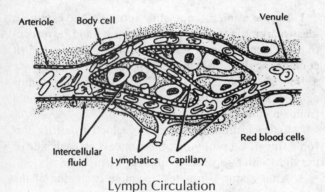

Lymph Circulation

The Blood

There are almost 6 quarts of blood in an average-sized person. Blood consists of a liquid, *plasma,* in which two types of cells are carried, the *red blood corpuscles,* and the *white blood corpuscles.* Also present are the *platelets,* which are cell fragments.

Plasma

The liquid part of the blood is strawcolored, consisting of 90 percent water, and containing dissolved nutrients (glucose, amino acids, fatty acids, salts, vitamins), wastes, antibodies, hormones, respiratory enzymes, and fibrinogen. When a person donates a pint of blood at a blood bank, the plasma is separated, dried, and later dissolved in distilled water, for use in operations and emergencies. Recently, specific parts of plasma have been fractionated (separated out), such as gamma globulin, which contains antibodies against measles and other diseases, and serum albumin, a protein used in transfusions to restore the blood volume.

Red Blood Corpuscles

Almost all the cells in the blood are red blood cells. Their function is to transport oxygen. The red blood cells, or *erythrocytes*, are so small that 1 cubic millimeter of blood contains about 5,000,000 of them, and so numerous that blood appears red—without them, blood is pale yellowish in color. They are disclike cells

that are concave on both sides. They contain the protein *hemoglobin*, which gives them their color. When blood passes through the lungs, the oxygen of the air combines with the hemoglobin of the red blood corpuscles to form *oxyhemoglobin*. As these cells circulate through the blood, oxygen diffuses from them into the cells of the body. New research is pointing to the importance of *nitric oxide* (NO) in hemoglobin's role of transporting oxygen. Nitric oxide appears to be a messenger molecule that is also involved in controlling blood pressure and regulating intestinal movements and the functioning of the immune system.

Red blood corpuscles are manufactured in the marrow of the bones. They lose their nuclei when they enter the bloodstream, and have recently been shown to remain active for about 4 months. Worn-out red cells are removed from the blood by the liver and the spleen. The element iron is needed for the formation of hemoglobin. If it is lacking in the diet, *anemia* results.

White Blood Corpuscles

Approximately one out of every 500 blood cells is a white blood cell. The white blood cells, or *leucocytes,* move like amebae and engulf bacteria that have entered the body, as well as other foreign matter. For this reason, they were named *phagocytes* (eating cells) by their discoverer, Elie Metchnikoff (1845–1916). The white blood cells form pseudopodia and can crawl through the capillary walls into the tissues. They thus help protect the body against disease germs. Their action is known as phagocytosis. *Pus* is a thick, yellowish liquid that often results when there is an infection; it consists of bacteria, white blood corpuscles, both dead and alive, and broken-down tissue. The white blood cells are formed in the marrow of the bones and in the lymph glands. There are several types of such cells. During a case of *appendicitis* or other infection, there is a striking increase in these cells above the average number normally found in 1 cubic millimeter (about 7,000). In another disease, *leukemia,* which is believed to be a form of cancer, enormous numbers of white blood corpuscles are produced.

Platelets

Platelets are smaller than red corpuscles. They are activated at the occurrence of a wound. When activated, platelets break down and give off an enzyme called *thromboplastin.* This causes the substance *prothrombin* to change into *thrombin.* Thrombin, in turn, acts on *fibrinogen,* one of the dissolved proteins in plasma, making it insoluble, and turning it into threads of *fibrin.* The red blood cells become entangled in this network, and a clot forms. This whole process takes

SUMMARY OF THE BLOOD GROUPS

Blood Type	Agglutinogens in Red Blood Cells	Agglutinins in Plasma	Donor to	Recipient from
A	A	Anti B	A, AB	A, O
B	B	Anti A	B, AB	B, O
AB (Universal recipient)	A and B	None	AB	A, B, AB, O
O (Universal donor)	None	Anti A and Anti B	A, B, AB, O	O

from 3 to 6 minutes. Fresh blood that is allowed to stand in a jar will slowly separate into two parts—an upper clot, and a clear, yellowish liquid below called *serum.* Serum is like plasma except that it has no fibrinogen. Some people inherit a disease known as *hemophilia,* in which the blood does not clot. This is produced by a hereditary factor, or gene, and is not curable. In other cases, there may be poor clotting of a cut or wound, if a person is lacking in vitamin K or the element calcium. At times, older people may suffer from the appearance of a clot in the circulatory system; if it occurs in the coronary arteries that bring blood to the cells of the heart, *coronary thrombosis* results, and the heart may be damaged, or death may result. Doctors inject *heparin*, a substance extracted from the liver and lungs, or *dicumarol*, which is obtained from clover, to reduce or eliminate such clots.

Blood Types

Although all blood contains plasma and the two types of blood cells mentioned above, it is chemically different in various people. At the beginning of the century, Dr. Karl Landsteiner discovered why blood transfusions from one person to another sometimes result in death. Some red blood cells contain substances called *agglutinogens,* which react with antibodies known as *agglutinins* in the plasma of other people. On the basis of these differences, he classified blood into four types, now referred to as A, B, AB, and O. The agglutinogens of Type A are different from those of Type B blood. Therefore, if blood from one is mixed with the other, the blood will agglutinate or clump, because of the action of the agglutinins in the opposite type of blood. Type AB has both agglutinogens A and B, and Type O has neither. Therefore, it can be seen that Type AB blood can receive blood from A, B, AB, or O; it is known as the "universal recipient." Since Type O blood has no agglutinogens, it may be transfused to any of the other three types, and is known as the "universal donor." A demonstration to show the blood type of a person may be performed as follows:

1. Divide a microscope slide into two halves with a glass marking pencil. Mark the halves I and II, respectively. Place a drop of serum from Type A on side I (it contains agglutinins against Type B); on side II, place a drop of serum from Type B (it contains agglutinins against Type A). (Blood typing serum is obtainable from a medical supply company.)

2. To each drop, add a drop of blood from the person's finger. Carefully mix the contents on each side of the slide with a separate toothpick.

3. After about 3 minutes, examine each side of the slide with the low power of the microscope.

If the red blood cells on side I have gathered together in clumps, but not those on side II, the blood is Type B.

If the cells on side I have not clumped, but those on side II have clumped, the blood is Type A.

If both sides I and II have clumped, the blood is Type AB.

If neither side has clumped, the blood is Type O.

Other types of blood have also been described, including the M, N, Hr, and P factors, and the Rh factor.

Rh Factor

About 85 percent of white Americans are Rh *positive*; that is, their red blood cells contain the Rh antigen, or factor. The rest are Rh *negative,* indicating that their red blood cells do not have the Rh antigen. These blood types are inherited in definitive ways. If both parents are Rh positive, there is normally no difficulty in the development of their child. However, if the father is Rh positive and the mother is Rh negative, the developing fetus is usually Rh positive. The fetus sensitizes the mother to produce antibodies against its red blood cells, destroying them (*erythroblastosis*). This situation usually occurs after the birth of a first baby, when a large supply of such antibodies has been formed in the mother's bloodstream.

Doctors have learned how to save the life of such a

baby; they drain out all of its blood as soon as it is born and replace it with a fresh transfusion. The Rh factor of expectant mothers is classified ahead of time, so that emergency transfusions can be ready, if necessary.

In the 1960s, a vaccine was developed that has conquered Rh blood disease. The vaccine contains passive Rh antibodies obtained from the blood of Rh-negative women who have had Rh-positive babies. Injected into an Rh-negative mother, the vaccine prevents her from becoming actively sensitized to the Rh factor in the red blood cells of the newborn baby. As a result, she does not produce her own antibodies, and the red blood cells of her next baby will not be destroyed.

Circulation in Lower Animals

Food and oxygen are needed by all living cells. In single-celled protists such as paramecia, the circulating protoplasm distributes these needed materials throughout the cell by its cytoplasmic streaming, known as *cyclosis,* and by diffusion.

Hydra

The hydra has cells with long flagella lining its central cavity. The beating of these flagella creates currents which circulate the food throughout the cavity, where it diffuses into the cells, and where the larger particles are engulfed by special cells.

Earthworm

The earthworm has a closed transport system. It has a long, large blood vessel on its upper (dorsal) side, just above the digestive system, and two long blood vessels below it. They are connected by side branches. Blood moves forward by the contractions of the dorsal vessel. Five enlarged side vessels around the esophagus (aortic arches) serve as hearts, and pump the blood back through the lower, or ventral, blood vessel. Valves in the hearts and the dorsal vessel prevent the blood from flowing backward. Muscular movements of the body also keep the blood moving along. Digested food enters the capillaries in the intestine. Oxygen is combined with hemoglobin, which is dissolved in the blood fluid rather than being contained in blood cells. Capillaries branch out to the various parts of the earthworm carrying food and oxygen to the cells, and removing the wastes.

Insects

In insects, the *heart* is a tubular structure that is located in the abdomen and pumps blood forward through the *aorta.* From this blood vessel, the blood flows through spaces among the head tissues and passes back through the thorax and the abdomen, bathing the various structures in them. It absorbs digested food from the stomach and reenters the heart from the rear part of the aorta. Circulation is thus through an *open system,* rather than a closed one of specific blood vessels; there are no capillaries or veins. Oxygen is not carried in the circulatory system to any great extent because of the insects's branching system of air tubes.

Fish

A fish has a two-chambered tubular heart consisting of an *atrium* and a *ventricle.* It is located in the front part of the body, below the gills. Blood from the body enters the *atrium,* which contracts and pumps the blood into the thick-walled *ventricle.* It in turn pumps the blood into the *ventral aorta,* which sends the blood to the gills, where it branches into capillaries that absorb oxygen from the water and give off carbon dioxide. Then the capillaries combine to form the *dorsal aorta,* which runs along the back and branches into arteries going throughout the body. The veins carry blood back to the heart. Digested food is absorbed as the blood passes through the intestine. The red blood cells are oval and contain a nucleus.

Frog

The heart of a frog has three chambers, consisting of two *atria* and one *ventricle.* Blood from the body enters the *right atrium,* from which it is pumped into the *ventricle.* Most of the blood is pumped out into the *pulmonary artery* and then to the lungs. Here, the blood flows in capillaries and absorbs oxygen and excretes carbon dioxide. The *pulmonary vein* carries blood back to the *left atrium.* It is then pumped into the ventricle, from which it is sent through arteries to the various parts of the body. There is thus a mixing, to some extent, of deoxygenated and oxygenated blood in the ventricle. As the blood circulates, food is absorbed in the small intestine. The blood consists of plasma, oval red blood cells containing hemoglobin and having a nucleus, and white blood cells. There is a lymph circulation, in which four contracting lymphatics, or lymph hearts, pump lymph and eventually send it into the veins.

Section Review

Select the best choice to complete each of the following statements:

1. In human circulation, a part of the heart that receives blood from the rest of the body is the (A) left ventricle (B) right ventricle (C) right atrium (D) valves (E) aorta

2. The cells of the heart are nourished by the (A) carotid arteries (B) coronary arteries (C) jugular vein (D) portal vein (E) pulmonary arteries

3. All of the following statements about arteries are true *except* (A) they are thick-walled (B) they pulsate (C) they contain much elastic fiber tissue (D) they carry blood away from the heart (E) they contain valves.

4. All of the following veins carry deoxygenated blood *except* the (A) superior vena cava (B) inferior vena cava (C) pulmonary vein (D) renal vein (E) hepatic vein

5. In the portal circulation, blood flows through the (A) liver (B) lungs (C) skin (D) kidneys (E) entire body

6. The liquid that bathes all the cells of the body is called (A) fibrinogen (B) lymph (C) plasma (D) blood (E) fibrin

7. The thoracic duct is part of the (A) skeletal system (B) digestive system (C) excretory system (D) lymphatic system (E) enzyme system

8. All of the following are found in plasma *except* (A) fibrinogen (B) antibodies (C) hormones (D) starch (E) amino acids

9. Oxygen is carried in the (A) phagocytes (B) platelets (C) white blood cells (D) red blood cells (E) gamma globulin

10. Iron is needed for the (A) formation of white blood cells (B) formation of red blood cells (C) formation of platelets (D) prevention of goiter (E) prevention of diabetes

11. The circulation of blood was first demonstrated by (A) Metchnikoff (B) Pasteur (C) Cohn (D) van Leeuwenhoek (E) Harvey

12. Red blood cells are formed in the (A) kidneys (B) skeletal muscles (C) marrow of the bones (D) cartilage (E) heart

13. All of the following statements about white blood cells are true *except* (A) they are formed in lymph glands (B) they are formed in bone marrow (C) they move like paramecia (D) they destroy bacteria (E) they have a nucleus

14. The correct sequence of steps in the clotting of blood (1-platelet; 2-fibrin; 3-fibrinogen; 4-thrombin; 5-prothrombin) is (A) 1–3–2–5–4 (B) 1–2–3–5–4 (C) 1–2–3–4–5 (D) 1–5–4–3–2 (E) 1–5–4–2–3

15. An inherited disease in which blood does not clot is (A) thrombosis (B) atherosclerosis (C) hemophilia (D) heparin (E) anemia

16. A person with Type A blood may safely receive a transfusion of (A) Type AB (B) Type A and Type AB (C) Type AB and Type O (D) Type A and Type O (E) none of these

17. Worn-out red blood cells are removed from the blood by the (A) lungs (B) spleen (C) small intestine (D) large intestine (E) pancreas

18. Most people (A) are Rh positive (B) are Rh negative (C) have anemia (D) have erythroblastosis (E) have hemophilia

19. Circulation of blood takes place in an open system in the (A) fish (B) frog (C) paramecium (D) grasshopper (E) toad

20. Serum is like plasma except that it has no (A) water (B) fibrinogen (C) hemoglobin (D) hemophilia (E) red blood cells

Answer Key

1-C	5-A	9-D	13-C	17-B
2-B	6-B	10-B	14-D	18-A
3-E	7-D	11-E	15-C	19-D
4-C	8-D	12-C	16-D	20-B

Answers Explained

1. (C) The right atrium receives blood from the upper part of the body through the superior vena cava and from the lower part of the body through the inferior vena cava.

2. (B) The cells of the heart receive nourishment through the blood coming from the coronary arteries around it.

3. (E) Valves are found in veins, not in arteries. Valves prevent the blood from flowing backward.

4. (C) All veins carry deoxygenated blood (blood lacking in oxygen), except the pulmonary vein, which carries blood from the lungs to the left side of the heart. The blood becomes oxygenated in the lungs.

5. (A) Blood carrying digested food from the small intestine enters the portal vein and is carried to the liver.

6. (B) All the cells of the body are surrounded by, and bathed by, a watery intercellular fluid, called lymph, which slowly oozes out of the blood plasma through the capillary walls. Lymph consists of plasma without its large protein molecules; it is largely water, containing dissolved nutrients, wastes given off by the cells, secretions, and white blood cells.

7. (D) The lymphatic system consists of lymph vessels in which lymph is carried. The lymph vessels join to form larger ones, the largest of which is the thoracic duct. It empties into a large vein in the neck region, thereby returning lymph to the circulatory system.

8. (D) Starch is an insoluble nutrient. After it has been digested, its end product, glucose, is absorbed into the capillaries of the villi and is dissolved in plasma.

9. (D) Red blood cells contain the protein hemoglobin. When blood passes through the lungs, the oxygen of the air combines with hemoglobin to form oxyhemoglobin. As the red blood cells circulate through the body, oxygen diffuses from them into the cells.

10. (B) Iron is needed to form hemoglobin, a protein in red blood cells. If iron is lacking in the diet, not enough hemoglobin-containing red blood cells are produced, and the result can be a type of anemia.

11. (E) William Harvey (1578–1657) first demonstrated that blood is pumped by the heart to circulate through the body.

12. (C) Red blood cells are manufactured in the marrow of the bones. In humans, red blood cells lose their nuclei when they enter the bloodstream.

13. (C) White blood cells move like amebae and form pseudopods, or false feet, that engulf bacteria that may have entered the body.

14. (D) When a wound occurs, the platelets give off an enzyme that causes prothrombin to change into thrombin. Thrombin acts on fibrinogen, a dissolved protein in plasma, making it insoluble and turning it into threads of fibrin. Red blood cells become entangled in this network, and a clot forms.

15. (C) In people with hemophilia, the blood does not clot normally. This condition is genetic and is not curable.

16. (D) A person with Type A blood can safely receive a transfusion from another person with Type A blood, which has the same agglutinogens, and from a person with Type O blood, who is called a universal donor, because there are no agglutinogens in Type O blood and it may be safely injected into a person with any of the blood types.

17. (B) Red blood cells remain active for about 4 months. Worn-out red blood cells are removed from the body by the spleen and the liver.

18. (A) About 85% of white Americans are Rh positive, having red blood cells with the Rh factor.

19. (D) In insects, such as the grasshopper, blood is pumped through spaces in the head tissues and passes back through the thorax and the abdomen, bathing the structures in them. Circulation is thus through an open system, without capillaries or blood vessels.

20. (B) If fresh blood is allowed to stand in a jar, it will separate into an upper clot and a clear, yellowish liquid called serum. Serum is like plasma, but without fibrinogen, which was used up in the clotting process.

6.4 RESPIRATION

After food has been delivered to the cells, it must be combined with oxygen in order to release energy. When this occurs, carbon dioxide and water are given off as wastes. During this process of *respiration,* numerous respiratory enzymes participate in the series of reactions by which energy is produced in every cell of the body.

The Human Respiratory System

The respiratory system includes the structures through which oxygen comes into the body to reach the bloodstream, and through which carbon dioxide and water vapor leave. Air first enters through the two *nostrils* in the nose, where large hairs hold back some of the dust. Farther along, the *nasal passages* are lined with cilia, which beat bacteria and dust outward. The sticky mucus secreted by the epithelial cells also traps bacteria and dust. Thus, the air is filtered as it enters the body. It is also moistened and warmed as it passes over the thin lining layer of cells, and through the sinuses. The *sinuses* are cavities in the head connected with the nasal passages by narrow openings. During a cold, their membranes may become infected and swollen, blocking the openings and causing the troublesome condition known as *sinusitis,* or "sinus trouble."

Air continues into the *pharynx,* or throat, at the back of the mouth. It is obvious that breathing by mouth does not permit the air to be cleansed, moistened, or warmed, as occurs in nasal breathing. The *tonsils,* and the *adenoids,* lymph masses at the entrance to the throat and at the back of the nasal cavity, often have to be removed when they become diseased. The pharynx also has two small openings leading to each ear through the narrow *Eustachian tubes.*

The *larynx,* or voice box, is located just below the opening of the trachea. It contains the vocal cords, and sometimes protrudes in the neck as the "Adam's apple."

The *trachea,* or windpipe, next receives the air. The upper end is covered with a flap of tissue, the *epiglottis,* which is open during breathing. However, when food is being swallowed into the nearby opening of the esophagus, the epiglottis closes and keeps the food from entering the trachea. The wall of the trachea contains rings of cartilage that keep the passageway open. The inner surface is lined with ciliated epithelial cells, which beat dust and bacteria upward.

The trachea divides into two *bronchi,* tubes that carry air into the lungs. Each bronchus branches into smaller and smaller *bronchial tubes,* or bronchioles,

which spread through the lungs. *Bronchitis* is a condition in which the membranes of the bronchial tubes become inflamed.

At the end of each of the millions of microscopic bronchial tubes there is an *alveolus,* or air sac. The wall of the air sac is made up of one layer of cells surrounded by a network of capillaries. As the blood flows by in these capillaries, oxygen diffuses from the air through the walls of the air sacs and combines with hemoglobin in the red corpuscles to form oxyhemoglobin. Carbon dioxide and water vapor diffuse out of the blood into the air sacs, and are eliminated when the air is carried out.

The route of air through the respiratory system may be summarized as follows: (1) nostrils, (2) nasal cavity and sinuses, (3) pharynx, (4) larynx, (5) trachea, (6) bronchi, (7) bronchial tubes in the lungs, (8) air sacs (alveoli).

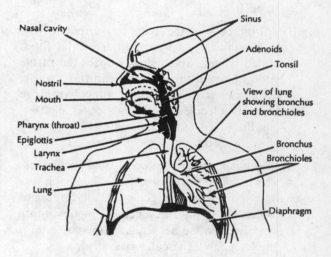

Structure of Respiratory System

Breathing

Inhaled and Exhaled Air

When air is *inhaled,* it contains about 79 percent nitrogen, 20.9 percent oxygen, and 0.04 percent carbon dioxide. After it has reached the air sacs, and diffusion has occurred, it is *exhaled* with 79 percent nitrogen, 16.3 percent oxygen and 4.5 percent carbon dioxide. The increased content of carbon dioxide may be demonstrated as follows:

1. Blow through a straw into a test tube of clear limewater *(A).* As a control, use an air pump to bubble air through another test tube of clear limewater *(B).*

2. In a short time, observe that test tube *A* of limewater has become cloudy because of the accumulation of carbon dioxide. Test tube *B*, however, will remain clear since it is unaffected by the minute amount of carbon dioxide present in the air.

The presence of water vapor in exhaled air is easily demonstrated by breathing on a cold window pane, and observing the collection of moisture that results.

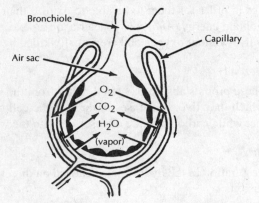

Gas Exchange in an Air Sac (Alveolus)

Breathing Mechanism

The lungs are elastic, spongy organs located in the *thorax,* or chest cavity, which is a closed chamber. The outside of the lungs is covered with a moist, smooth membrane, the *pleura,* which also lines the thorax. A condition known as *pleurisy* results when the pleura becomes infected. The *diaphragm,* the muscular floor of the thorax, consists of voluntary striated muscle, and separates the thorax from the abdominal cavity below. The walls of the thorax contain the *ribs* in a cagelike arrangement; most of them are attached to the breastbone in front, and all of them are attached to the spinal column in the back. Layers of voluntary muscle are attached to the ribs.

The act of breathing involves the movement of both the diaphragm and the ribs, to enlarge or decrease the size of the chest cavity. During this process, the lungs are either filled with air or are made to give up their air. They are thus passive, and cannot expand or contract by themselves. Specifically, breathing takes place as follows:

1. *Inspiration,* or inhaling. The diaphragm, which is curved upward when it is relaxed, contracts and flattens out. The rib muscles also contract, causing the ribs to be raised and spread apart. Both of these actions enlarge the chest cavity and reduce the internal pressure on the lungs. Air from the outside, where the atmospheric pressure is higher, is then forced into the lungs, and they inflate. This brings air into the air sacs, where the exchange of gases takes place.

2. *Expiration* or exhaling. The diaphragm relaxes, curving upward. The rib muscles also relax, lowering the position of the ribs. The chest cavity is thus decreased in size, compressing the air in the lungs and forcing it out.

Normal breathing takes place about 15–18 times a minute. Exercise or emotional stress can cause it to increase. When breathing stops altogether, because of drowning, gas suffocation, or electric shock, *artificial respiration* must be applied in order to continue the supply of oxygen to the cells. The most common method is CPR (cardiopulmonary resuscitation). This includes rhythmic hand pressure on the chest to stimulate the heart, and mouth-to-mouth breathing. The rescuer blows into the mouth of the victim about 15 times a minute, thus inflating the lungs and enlarging the chest cavity. The exhaled air of the rescuer contains enough oxygen to supply the cells of the victim and revive him.

Control of Breathing

Although breathing is a voluntary activity involving voluntary muscles, it is automatically controlled by a part of the brain, the medulla. It therefore goes on without our having to think about it. The rate of breathing is controlled by the amount of carbon dioxide in the blood. Carbon dioxide originates in the cells as they perform their metabolism. It then diffuses into the bloodstream, where a small amount combines with the hemoglobin of the red blood cells and is exhaled in the lungs. Most of the carbon dioxide reacts with water in the blood plasma to form carbonic acid (H_2CO_3), which is then converted by the enzyme carbonic anhydrase into bicarbonate ions (HCO_3^-) and hydrogen ions (H^+). This process prevents the blood from becoming too acidic. In the lungs, the bicarbonate ions are reconverted to carbonic dioxide which is exhaled along with water vapor.

These reactions, which are reversible, may be expressed as follows:

$$CO_2 + H_2O \rightleftharpoons H_2CO_3 \text{ (carbonic acid)}$$
$$H_2CO_3 \rightleftharpoons H^+ + HCO_3^- \text{ (bicarbonate ion)}$$

As the blood circulates, the increased amount of carbon dioxide in it will stimulate the cells of the medulla to send impulses to the chest muscles and diaphragm, causing them to contract and expand more rapidly; the breathing rate then increases. This is what happens when we are exercising and the cells are giving off large amounts of carbon dioxide. When we are resting, less carbon dioxide accumulates, and the cells of the medulla are not stimulated as much, so that the breathing rate is lower. This is one example of a feedback mechanism that is important in the maintenance of homeostasis.

Lung Disorders

1. *Lung cancer.* This is the most common cause of cancer death in the United States. The chief cause is smoking. As the rate of smoking by women has increased in recent years, so has the incidence of lung cancer among them.

2. *Asthma.* The elastic tissue in the bronchial tubes may contract, diminishing their diameter and reducing the amount of air passage. This happens in asthma and other allergies, requiring the person to take antihistamines or adrenalin to relax the elastic tissue.

3. *Emphysema.* This is a reduction in lung capacity. The lungs have lost their elasticity and the alveoli walls have broken down, usually because of smoking.

4. *Infectious diseases.* Germs that cause such serious diseases as tuberculosis, influenza, and pneumonia may enter the lungs with the air that is breathed in. These germs are spread by airborne droplets via coughs, sneezes, speech, and proximity. This subject will be discussed in greater detail in Chapter 8.

Effects of High Altitudes

The coming of the air age has presented us with the problem of breathing at high altitudes. At sea level, the weight of the air produces a pressure of 15 pounds per square inch. Higher up, there is less pressure, and less oxygen in a given volume of air. Therefore, it is necessary to "pressurize" the cabins of airplanes flying above 3,000 meters, and to add oxygen to the air being breathed. Without this additional oxygen, a condition called *anoxia* results, which is characterized by a loss of coordination, vision, hearing, and judgment, and finally results in unconsciousness.

Another problem of reduced pressure at high altitudes concerns the nitrogen that is normally dissolved in the body fluids. This nitrogen has the same concentration as in the air, about 78 percent. As a plane climbs to higher altitudes, where the air pressure is low, there is less pressure on the nitrogen, and it begins to come out of solution and leave the body. If the ascent is too fast, bubbles of nitrogen form in the tissues and block small blood vessels, producing a very painful case of the "bends." This is similar to the condition affecting deep-sea divers who ascend too rapidly from underwater regions of high pressure to sea-level pressure. Besides the use of pressurized cabins, high-altitude fliers may also wear pressurized suits. These and other problems must be overcome before astronauts can explore outer space in rockets.

Mountain climbers who ascend the tallest mountains must be equipped with oxygen tanks if they are to survive and supply their cells with sufficient oxygen. It has been found that natives of Peru who live high up in the Andes Mountains have a much higher red blood count than the average person at sea level. This greater supply of hemoglobin enables these Peruvians to absorb enough oxygen from the thin air to maintain the life of their cells.

Respiration in Lower Animals

Oxygen must be obtained by practically all living things in order for them to be able to oxidize their food and obtain energy. *Anaerobic* bacteria are an exception; they live without oxygen and are actually killed by it.

Protozoa

These protists absorb oxygen directly from the water in which they live; it diffuses in through the cell membrane, while carbon dioxide diffuses out into the water.

Hydra

Respiration in the hydra is also carried on directly by diffusion.

Earthworm

The earthworm has a thin, moist skin through which oxygen diffuses into the blood. The hemoglobin dissolved in the blood then carries oxygen throughout the body. Carbon dioxide diffuses out of the body as the blood passes through the skin.

Insects

Insects such as the grasshopper have small openings, called *spiracles,* on the side of the segments of the abdomen. These lead to a system of branched air tubes, *tracheae,* spreading throughout the body to all the cells. Air is pumped in and out of the tracheae by the muscular contractions and expansions of the abdomen. In this way, oxygen is brought into, and carbon dioxide is expelled from, the body.

Fish

Water is constantly entering a fish's mouth, and passing out through openings on both sides of the head. *Gills*, located in gill chambers on each side, have numerous thin-walled, threadlike *gill filaments,* which are well supplied with capillaries. As the water passes over the gill filaments, oxygen diffuses through the thin membrane into the blood, while carbon dioxide leaves the blood and diffuses out into the water. The oxygenated blood is circulated through the fish by means of the arteries. In order for a fish to survive in an aquarium, it must be supplied with oxygen, either by stocking the aquarium with green plants that will give off oxygen during photosynthesis, or by pumping air into the water.

Section Review

Select the correct choice to complete each of the following statements:

1. Energy is obtained from food in the process of (A) assimilation (B) digestion (C) excretion (D) respiration (E) storage

2. Ciliated epithelial cells are located in the (A) esophagus (B) trachea (C) villi (D) aorta (E) cardiac muscle

3. The correct order of structures (1-nasal cavity; 2-bronchi; 3-larynx; 4-air sacs; 5-trachea) through which air passes is (A) 1–5–3–2–4 (B) 1–5–3–4–2 (C) 1–3–4–5–2 (D) 1–3–5–4–2 (E) 1–3–5–2–4

4. Oxyhemoglobin is formed in the (A) lungs (B) spleen (C) right auricle (D) left auricle (E) bone marrow

5. The percentages of nitrogen, oxygen, and carbon dioxide in exhaled air are about (A) 79–20–0.04 (B) 79–16–4 (C) 20–79–0.04 (D) 16–79–4 (E) 79–4–16

6. The passageway of the trachea is kept open by rings of (A) striated muscle (B) adenoids (C) cartilage (D) mucus (E) sinuses

7. During inhalation (A) the diaphragm flattens out, and the ribs are raised (B) the diaphragm is raised, and the ribs are lowered (C) the diaphragm flattens out, and the ribs are lowered (D) the diaphragm is raised, and the ribs are raised (E) the diaphragm is raised, and the ribs are stationary

8. The rate of breathing is controlled by the (A) lungs (B) bronchi (C) air sacs (D) medulla (E) diaphragm

9. The breathing rate is increased by an increase in the content of (A) oxygen (B) nitrogen (C) water vapor (D) carbon monoxide (E) carbon dioxide

10. Lack of oxygen at high altitudes produces (A) "bends" (B) anoxia (C) asthma (D) artificial respiration (E) antihistamines

Answer Key

| 1-D | 3-E | 5-B | 7-A | 9-E |
| 2-B | 4-A | 6-C | 8-D | 10-B |

Answers Explained

1. (D) During respiration, the chemical-bond energy of food is released and stored in ATP molecules. The breakdown of ATP into ADP occurs when a phosphate group is given off and energy is released for the life activities of the cell.

2. (B) The inner surface of the trachea is lined with ciliated epithelial cells that sweep bacteria and dust upward.

3. (E) As air passes through the respiratory system, it goes through the nose and then on to the nasal cavity, the pharynx, the larynx, the trachea, the bronchi, the bronchial tubes, and finally the air sacs in the lungs.

4. (A) As blood passes through the lungs, the hemoglobin in the red blood cells absorbs the oxygen in the air sacs to form oxyhemoglobin.

5. (B) Exhaled air contains less oxygen and more carbon dioxide than inhaled air. Exhaled air has 16.3% oxygen, 4.5% carbon dioxide and 79% nitrogen. By comparison, inhaled air has 20.9% oxygen, 0.04% carbon dioxide, and 79% nitrogen.

6. (C) Rings of cartilage provide support and firmness to the trachea, so that its passageway is always kept open.

7. (A) During inhalation, the diaphragm contracts and flattens out; at the same time the rib muscles contract, causing the ribs to be raised. Both of these actions enlarge the chest cavity and reduce the internal pressure on the lungs. Air from the outside, where the atmospheric pressure is higher, is forced into the lungs and they inflate.

8. (D) Breathing is automatically controlled by the part of the brain called the medulla. This automatic action takes place without our having to think about it.

9. (E) An increased amount of carbon dioxide in the blood stimulates the cells of the medulla to send impulses to the diaphragm and the rib muscles, causing them to contract and expand more rapidly. The breathing rate then increases.

10. (B) At high altitudes, there is less air pressure and less oxygen in a given volume of air. If a person does not receive sufficient oxygen, the condition called anoxia can result. The symptoms are loss of coordination; impaired vision, hearing, and judgment; and unconsciousness.

6.5   EXCRETION

A review of the activities carried on by the various parts of the body reveals that the cells and tissues are really members of a well-organized community. Each organ or system does its specific job for the good of all. The digestive system prepares food for use by the cells. The circulatory system distributes food, oxygen, and other materials to the cells, and removes carbon dioxide and other wastes. The respiratory system provides for the intake of oxygen and the excretion of carbon dioxide and water vapor. The excretory system completes the job of removing wastes from the body.

The Human Excretory System

The wastes given off by the cells would poison them if they accumulated in the body. The lungs, kidneys, and skin serve as organs of excretion and remove such wastes as carbon dioxide, water, urea, and mineral salts from the blood. The large intestine does not excrete indigestible wastes; it merely eliminates them, since they were never really part of the body.

Lungs

When food is oxidized by the cells, carbon dioxide and water are given off as waste products. They diffuse out of the blood into the alveoli of the lungs, and are carried out of the body in the exhaled air.

Kidneys

The two kidneys are dark red, kidney-bean-shaped organs located in the rear of the abdominal cavity, above the small of the back. A large artery, the renal artery, carries blood containing a high percentage of wastes into each kidney. It branches into smaller blood vessels that become capillaries arranged in a dense network, or *glomerulus*. Each glomerulus is surrounded by a thin-walled cup, or *Bowman's capsule,* which extends into a long *nephron tubule.* The nephron is enclosed in a separate network of capillaries, and it connects with other tubules to form small tubes that lead into the *central cavity* of the kidney. As the blood flows through each glomerulus, a plasmalike liquid containing water, urea, salts, fatty acids, amino acids, and glucose diffuses out into the capsule end of the nephron. As this liquid moves along the long tubule of the nephron, water, minerals, and the digestive end products are reabsorbed into the capillaries by active transport. The remaining liquid collects as *urine* at the far end of the tubule and joins urine from the million other tubules in the central cavity of the kidney. From here, the urine passes out of each kidney through a long tube, the *ureter,* leading into the urinary *bladder.* Urine leaves the bladder at intervals through the tube called the *urethra.*

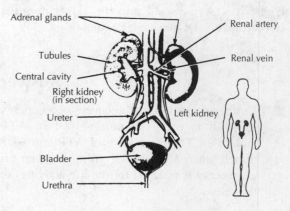

Kidney Excretory System

Urine

Urine contains about 95 percent water. Much of the remainder is the nitrogenous waste—urea—and mineral salts, uric acid, and other materials. Urea is produced as a waste when the liver breaks down amino acids in a process called *deamination.* The amino group ($-NH_2$) is removed from the amino acid molecule through enzyme action and rearranged into ammonia (NH_3). This is then combined with carbon dioxide and, through several reactions, becomes urea.

Under normal conditions, there are predictable amounts of each part of the urine, giving a clue to the healthy functioning of the various cells. If an abnormal condition is suspected, a doctor makes a *urinalysis.* Extra amounts of glucose may point to diabetes; high amounts of protein may indicate a disease of the kidney itself; pus is a symptom of an infection.

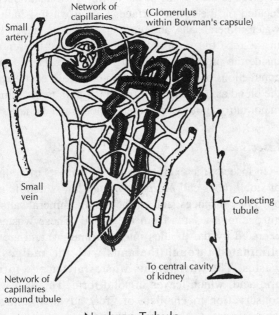

Nephron Tubule

The blood that was flowing through the capillaries around the tubules collects into small veins, and finally leaves each kidney through a large vein, the renal vein. This leads into the inferior vena cava, and carries blood that has been purified of much of its liquid wastes. The kidney thus is a vital organ for removing the waste materials that would otherwise poison the body. It also helps to maintain the correct balance of water and mineral salts in the body.

Excretory system problems The excretory system may be affected by infection, by diseases involving other systems, and by other disorders. Among these are the following:

1. *Kidney diseases.* The health of the kidneys may be adversely affected by infections, heavy-metal pollutants (lead, mercury), and extremely high-protein diets. In patients with severe kidney disease, "artificial kidneys," or dialysis machines, may be used to help keep the bloodstream free of wastes.

2. *Gout.* In this condition, there is impaired uric acid excretion by the kidneys, and uric acid crystals develop, concentrating in the big toe or the joints, causing painful inflammations.

Skin

The skin consists of two main layers of cells.

1. The thin, outer *epidermis* consists of layers of flat squamous epithelial cells. The upper part is made up of dead cells that are constantly being replaced by the cells below them.

2. The underlying, thicker *dermis* contains sweat glands, connective tissue, blood vessels, nerves, nerve endings, oil glands, muscles, and hair follicles.

A sweat gland consists of a coiled section that leads up through a tube to an opening on the skin called a *pore.* A network of capillaries surrounds the coiled part of the sweat gland. As blood flows through these capillaries, the thin wall of the sweat glands absorbs a large amount of water, as well as some salts. This liquid fills the tube, and comes out from the pore as *perspiration,* or sweat. The sweat glands excrete large quantities of water, to maintain the balance of various salts in the body.

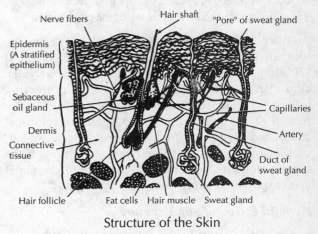

Structure of the Skin

Regulation of body temperature The excretion of perspiration helps to regulate the temperature of the body. As the perspiration evaporates from the surface of the skin, it cools it. The cooling effect of an evaporating liquid may be demonstrated in the following way:

1. Wrap some absorbent cotton around the bulbs of three thermometers. Wet one with a rapidly evaporating liquid, such as ether or alcohol; wet the second with water; leave the third dry, as a control.

2. After a few minutes, compare the readings of the three thermometers. It will be seen that the first will show the lowest temperature, since the rapid evaporation cooled it the most. The second will show a higher temperature than the first, since the water did not evaporate as quickly; however, it will show a lower temperature than the dry thermometer because evaporation was taking place, thereby cooling the thermometer.

When perspiration evaporates on the skin, it cools it, and lowers its temperature. This also lowers the temperature of the blood as it circulates through the skin. During vigorous exercise, when the cells of the body are releasing much heat energy, the blood vessels in the skin increase in diameter, thus carrying more blood to the skin, which may appear flushed. This results in more perspiration being excreted from the sweat glands. As it evaporates, it removes heat from the skin, cooling it. The reverse is true in cool weather. The blood vessels in the skin decrease in diameter, thus carrying less blood, and resulting in less excretion of perspiration. With less perspiration, less evaporation and less cooling occur.

On a sunny, dry day, perspiration evaporates quickly, giving one a sensation of coolness and comfort. On a humid day, when the air contains a high percentage of moisture, there is less evaporation of perspiration, resulting in an uncomfortable, muggy feeling. An electric fan helps to provide a feeling of coolness by removing moist air that was directly in contact with the surface of the skin; this allows more rapid evaporation of the perspiration.

Other functions of the skin Besides (1) excreting water, to maintain the balance of salts in the body; and (2) helping to regulate the temperature of the body, the skin also has four other functions: (3) protection against the entrance of bacteria into the body; (4) protection of the underlying tissues against mechanical injury, and drying up; (5) sensation of outside stimuli, such as touch, heat, cold, and pain, by means of the nerve endings located in it; (6) formation of vitamin D upon exposure to ultraviolet rays.

Harmful Effects of Sun Exposure

Although exposure to the sun's ultraviolet (UV) rays is useful in the formation of vitamin D, too much sunning can harm your skin. Two UV components can be damaging: (1) UVB, the so-called burning rays, harm the surface layer of the skin and may lead to cancer and (2) UVA, the so-called slow tanning rays, penetrate deeper layers of skin and can give it a leathery, wrinkled appearance. Sunscreen lotions, which block UVB radiation from reaching the skin, are recommended for anyone spending time in the sun, especially between 10 A.M. and 2 P.M.

Excretion in Lower Animals

When living cells carry on their activities, they give off wastes such as carbon dioxide, water, and nitrogenous compounds.

Protozoa

These wastes diffuse directly out through the cell membrane. The nitrogenous wastes in protozoa are in the form of ammonia.

Hydra

Excretion also takes place in the hydra by diffusion.

Earthworm

Each segment of the earthworm contains a pair of excretory organs called *nephridia*. These are tiny tubes that lead to the outside of the worm by means of openings located on the lower surface. Liquid wastes, including water, mineral salts, and nitrogenous wastes, are collected in the tube by means of a ciliated, funnel-like entrance inside the preceding segment. Wastes are passed out by contractions of the body. Carbon dioxide is excreted from the blood as it circulates through the thin, moist skin. The major nitrogenous wastes are ammonia and urea.

Insects

In insects, the excretory system consists of a number of small tubes call *Malpighian tubules*, which lie in the open blood spaces, and extract water, mineral salts, and nitrogenous wastes from the blood. These wastes are excreted by the tubules into the hindgut, and are then eliminated through the anus with the indigestible wastes. The nitrogenous wastes are in the form of uric acid, which leaves in solid form. This is a water-conservation mechanism of great advantage in a dry environment. Carbon dioxide diffuses out of the blood into the air tubes (tracheae) and is excreted through the spiracles by the pumping action of the abdominal muscles.

Fish

In fish, two long *kidneys* remove nitrogenous wastes from the blood and send them to the urinary bladder. From here, the wastes are expelled to the outside. Carbon dioxide diffuses out of the blood as it passes through the gill filaments, and is taken up in the water flowing past the gills and out of the openings on the sides of the head.

Frog

The frog's *kidneys* are deep red, flat, oval structures located in the back of the body cavity. They remove nitrogenous compounds from the blood, and send them through *ureters* into the *cloaca,* from which the urine passes into the urinary *bladder.* The cloaca is a common passageway for the liquid wastes of the bladder and the indigestible solid wastes of the large intestine, both of which leave the frog's body at the same time through the *anus*. Carbon dioxide is excreted through the lungs, as well as through the moist skin.

Section Review

Select the correct choice to complete each of the following statements:

1. Two wastes resulting from the oxidation of food are (A) oxygen and water (B) carbon dioxide and water (C) oxygen and carbon dioxide (D) oxygen and ATP (E) carbon dioxide and ATP

2. Urea is removed from the blood by the (A) lungs (B) liver (C) kidneys (D) spleen (E) bladder

3. The microscopic structures in the kidneys that remove wastes from the blood are (A) renal arteries (B) renal veins (C) nephrons (D) ureters (E) urethra

4. The dermis of the skin contains all of the following *except* (A) a layer of dead cells (B) blood vessels (C) nerve endings (D) hair follicles (E) oil glands

5. The part of the skin that removes liquids from the blood is the (A) pores (B) epidermis (C) network of capillaries (D) sweat glands (E) flat squamous layer

6. All of the following are functions of the human skin *except* (A) formation of vitamin D (B) sensation (C) protection against entrance of bacteria (D) excretion of carbon dioxide (E) regulation of body temperature

7. On a humid day, a person is uncomfortable because (A) he or she perspires more (B) there is less evaporation of perspiration (C) perspiration evaporates quickly (D) larger amounts of salt are excreted by the skin (E) the surface of the skin is too dry.

8. Urine leaves the body through the (A) glomerulus (B) epiglottis (C) alveoli (D) ureters (E) urethra

9. The structure through which protozoa excrete wastes is the (A) food vacuole (B) cell membrane (C) trichocyst (D) oral groove (E) cilia

10. Malpighian tubules are used for excretion in the (A) earthworm (B) insect (C) hydra (D) fish (E) frog

Answer Key

1-B	3-C	5-D	7-B	9-B
2-C	4-A	6-D	8-E	10-B

Answers Explained

1. (B) When food is oxidized by the cells, carbon dioxide and water are given off as waste products. They diffuse out of the blood into the alveoli of the lungs and are carried out of the body in the exhaled air.

2. (C) As blood flows through the kidneys, a plasma-like liquid containing water, urea, salts, and nutrients diffuses out of it into the capsule end of the nephrons. Much of the water, salts, and nutrients are reabsorbed in the tubular part of the nephron, leaving urea and water to be excreted by the kidney.

3. (C) The kidney consists of numerous microscopic nephrons that remove wastes from the blood. Each nephron consists of Bowman's capsule, which encloses a network of capillaries called the glomerulus. Bowman's capsule leads to a long tubule which deposits its wastes into a collecting tubule that extends to the central cavity of the kidney.

4. (A) The dermis is the underlying layer of the skin, which contains blood vessels, nerve endings, hair follicles, oil glands, muscle tissue, and sweat glands. The upper part of the outer layer, the epidermis, contains dead cells.

5. (D) The coiled part of a sweat gland is surrounded by blood capillaries. As blood flows by, the thin wall of the sweat gland absorbs a large amount of water and some salts. This liquid fills the tube and comes out of the pore as perspiration.

6. (D) Carbon dioxide is excreted into the air sacs of the lungs and then exhaled out of the body during breathing.

7. (B) When sweat evaporates from the skin, it cools it and lowers its temperature. On a humid day, there is more moisture in the air, allowing for less evaporation of sweat. This results in an uncomfortable feeling.

8. (E) After being formed in the kidneys, urine passes through the ureters and collects in the urinary bladder. At intervals, it leaves the bladder and is excreted through the tube called the urethra.

9. (B) Protozoa give off carbon dioxide, water, and ammonia as they carry on their life activities.

These wastes diffuse directly out through the cell membrane into the surrounding water.

10. (B) The excretory system of insects consists of a number of small Malpighian tubes, which lie in the open blood spaces and extract nitrogenous wastes from the blood. These tubules excrete the wastes into the hindgut, from which they are eliminated through the anus, along with indigestible wastes.

HOW LIVING THINGS ADJUST TO THEIR ENVIRONMENTS

CHAPTER

7

7.1 THE ENDOCRINE SYSTEM

The body carries on many activities, with the help of its various organs and systems. These activities are all coordinated so that they go on at the proper rate, at the proper time, and in the proper amounts. For example, when food leaves the stomach and enters the small intestine, the pancreas and the liver send their secretions to join intestinal juice in digesting it. These digestive organs do not send their secretions until that time. How is this controlled?

Secretin

Certain cells in the lining of the small intestine secrete a chemical substance called *secretin,* when food enters from the stomach. Secretin diffuses directly into the bloodstream and is carried through the body. When it reaches the pancreas, it stimulates the digestive glands there to produce pancreatic juice, which pours into the small intestine through the pancreatic duct. Secretin thus serves as a chemical messenger, or *hormone* (endocrine). A hormone is a chemical produced in one part of the body that is transported to other parts, where it exerts specific effects.

Other Hormones of the Alimentary Canal

When food enters from the stomach, the lining of the small intestine also secretes two other hormones that serve similar functions: one, *cholecystokinin,* stimulates the gallbladder to contract, and send bile into the small intestine; the other, *enterocrinin,* stimulates the digestive glands of the small intestine to secrete intestinal juice. Like secretin, these hormones are circulated in the bloodstream. Thus, the start of digestion in the small intestine is coordinated by three different hormones that are produced at the time when food is arriving from the stomach. The lining of the stomach itself also secretes a hormone, *gastrin,* when food enters it. Gastrin is carried in the bloodstream to the digestive glands in other parts of the stomach, which are then stimulated to produce gastric juices.

Endocrine Glands

Endocrine glands are different from digestive glands, which have a duct leading to the digestive system, in that they are "ductless"; their secretions, the hormones, diffuse directly into the blood and are transported to other parts of the body where they are used for specific purposes. Hormones may be secreted by groups of individual cells, such as those that produce secretin and the other hormones of the alimentary canal, or by entire endocrine glands. These endocrine glands are the thyroid, parathyroid, hypothalamus, pancreas (islands of Langerhans), adrenal, pituitary, and reproductive organs.

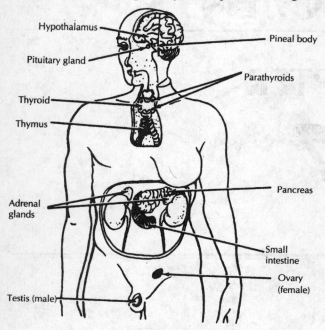

Glands of the Endocrine System

Thyroid Gland

The thyroid gland is located at the base of the neck, below the larynx. It absorbs large amounts of iodine from the blood to make the hormone *thyroxin*. When the diet is deficient in iodine, the thyroid gland enlarges and produces a swelling on the throat. This condition is known as simple or *endemic goiter*. It used to be much more prevalent in the inland areas of the country, where iodine is not available in the food. Along the seacoast, iodine is plentiful in seafood. Iodized salt is now used to prevent goiter.

Thyroxin is brought by the bloodstream to all the cells of the body, where it controls the rate at which oxidation takes place, and energy is released. The *basal metabolism test* is used by doctors to measure the rate of oxidation in a person who is at rest. A rate that is too low indicates an underactive thyroid; a high rate of oxidation is a sign of an overactive thyroid.

Undersecretion of thyroxin (hypothyroidism) (1) In children: Infants whose thyroid glands are not producing enough thyroxin are known as *cretins*. They are underdeveloped, physically and mentally. Their bodies are stunted. However, treatment with thyroxin helps correct the condition. (2) In adults: Grownups whose thyroxin production is below normal develop a condition known as *myxedema*. Their bodies become puffy, and they lose their alertness and even intelligence. They have a slow pulse and a low rate of oxidation. Treatment with thyroxin makes the abnormal condition disappear.

Oversecretion of thyroxin (hyperthyroidism) When the thyroid gland is overactive and is producing an excess of thyroxin, an individual has a higher rate of oxidation. He remains thin because his food is being oxidized rather than being stored. He is restless and sleeps less. His pulse beat is rapid. He becomes irritable. His eyes bulge, and the thyroid gland is usually enlarged. Such a person is said to have *exophthalmic goiter*. Treatment may consist of removing part of the thyroid gland, as a method of reducing the production of thyroxin. Instead, radioactive iodine may be given to the individual; it is absorbed and concentrated in the thyroid gland, where its rays cause a reduction in the production of thyroxin. The chemical *thiouracil* also may be used to decrease the activity of the thyroid cells.

Role of thyroxin in frogs Frog metamorphosis from the tadpole stage depends on thyroxin. Tadpoles have been made to develop into frogs in a very short time by feeding them with thyroxin.

Parathyroid Glands

There are four small parathyroid glands located next to the thyroid gland. They secrete the hormone *parathormone*, which regulates the absorption and use of the element calcium by the body. It is also important in bone formation. If it is lacking, the muscles do not function properly, and the individual may develop involuntary muscle contractions or convulsions, known as *tetany*. This condition can be relieved by injecting either parathormone or calcium salts.

Pancreas

In addition to producing pancreatic juice, the pancreas has collections of cells, called the *islands of Langerhans,* that secrete the hormones insulin and glucagon. *Insulin* serves two purposes in the body: (1) it helps the cells to utilize glucose for energy; and (2) it also causes the liver to convert excess glucose in the blood into glycogen for storage. In these ways, the amount of glucose in the blood is always reduced to a constant level. The hormone *glucagon* seems to counteract insulin. It increases the blood level of glucose by causing the liver and small intestine to release it. In the condition called *diabetes mellitus,* the production of insulin is affected. As a result, the cells cannot use glucose for energy, and the liver does not store it. The person becomes lacking in energy and loses weight. There is an excess of glucose in the blood, and a urinalysis reveals the presence of glucose.

Diabetes cannot be cured, but a diabetic can achieve a normal and useful life span if he or she receives daily injections of insulin. In mild cases, the drug tolbutamide (or orinase) may be taken orally. A diabetic should eat a diet low in carbohydrates.

Recent success: Transplants of insulin-producing cells from the pancreas of dead organ donors are inserted into the patient's liver through the portal vein. The cells disperse through the liver where they begin producing insulin.

Adrenal Glands

The adrenal glands are located on top of the kidneys. They consist of two parts: (1) the *medulla,* or inner section, and (2) the *cortex,* or outer layer.

1. *Medulla.* The medulla produces the hormone *adrenaline.* (The alternative name of this hormone, *epinephrine,* is used in medical contexts.) During periods of great emotional stress, such as anger or fear, a large amount of adrenaline is secreted into the bloodstream. Because of its effects in giving the body extra energy, the adrenal glands have been called the "glands of emergency" or "glands of combat." Adrenaline produces the following effects:

- The liver is stimulated to convert glycogen to glucose.
- The heart beats faster.
- The smooth muscles of the blood vessels contract, causing an increase in blood pressure.
- The breathing rate is increased, and the diameter of the bronchioles enlarges.
- The blood vessels in the skin and the digestive system contract, and digestion is reduced.
- The clotting rate of the blood is speeded up.
- With extra amounts of glucose and oxygen being supplied more rapidly to the cells of the body, including the skeletal muscles and the brain, there is an increase in energy, strength, and alertness.

Under these conditions, a girl will find that she can run as she has never run before. A man will be able to lift very heavy objects that he would normally be unable to budge. Adrenaline is injected directly into the heart by doctors in order to stimulate it during an operation or during a heart attack. Adrenaline is also used, in cases of asthma or other allergies, to enlarge the size of the bronchial tubes and relieve breathing. During a tooth extraction, a dentist may apply adrenaline to contract the size of the blood vessels, thus reducing bleeding and hastening clotting.

2. *Cortex.* At least 30 hormones are secreted by the cortex of the adrenal glands. In general, they seem to be important in maintaining the proper balance of liquids and mineral salts in the body; they also play a role in carbohydrate metabolism. One of them, *cortisone,* plays a role in the healthy condition of the cartilage in the joints between bones. It has been found to be important medically in successfully treating cases of rheumatoid arthritis, and in relieving the painful swellings of the joints. It has also been used in severe cases of asthma and skin conditions. However, its effects are not completely understood at present; it sometimes produces undesirable side effects.

Cortin, another hormone of the cortex, appears to control the use of water and salts by the cells. It also affects blood pressure. When it is lacking, an individual develops *Addison's disease,* in which there are marked weakness, bronze coloration of the skin, and loss of weight. Injections of cortin relieve this condition.

Other hormones of the cortex appear to have an influence on the development of secondary sexual characteristics. The production of the cortex hormones is stimulated by one of the hormones of the pituitary gland, called ACTH.

Pituitary Gland

This gland is about the size of a pea, and is situated at the base of the brain. It has three main parts. Despite its small size, it secretes many hormones that interact with and control the other ductless glands. For this reason, it is referred to as the "master gland." Some of its hormones have the following effects:

1. *ACTH (adrenocorticotropic hormone)* stimulates the cortex of the adrenal gland to produce a number of hormones, including cortisone. In severe cases of rheumatoid arthritis, it has brought about dramatic relief from disability and pain.

2. *Growth hormone* promotes the growth of the body. In rare cases, too much or too little of it may be produced in children. If there is an excess of it, a child continues to grow to extraordinary height (*gigantism*); if there is too little, the child remains small (*dwarfism*). If an adult begins to produce too much growth hormone, the extremities of the body (hands, feet, face) enlarge to produce a condition known as *acromegaly;* this may be treated with X rays.

3. *Prolactin* is a hormone from the anterior (front) part of the pituitary that stimulates the mammary glands to produce milk.

4. Another hormone stimulates the thyroid gland to produce thyroxin.

5. One of the hormones stored in the posterior pituitary, oxytocin, stimulates contraction of the smooth muscles of the uterus during childbirth.

6. Other hormones interact with the reproductive organs—the testes and the ovaries (gonads)—and stimulate them to develop and to produce additional hormones.

Hypothalamus

The hypothalamus, a small part of the brain located close to the pituitary gland, is known to secrete at least nine hormones from its nerve endings to the pituitary. Among these hormones is one (*TRH, thyrotropin-releasing hormone*) that causes the pituitary to release a hormone (TSH, thyrotropin-stimulating hormone) that stimulates the production of thyroxin by the thyroid gland. Another hormone controls the release of gonad-stimulating hormones by the pituitary. A third hormone inhibits the secretion of prolactin from the pituitary. Another hormone, *oxytocin*, is produced in the hypothalamus and stored in the posterior pituitary until its release to act on the uterus during childbirth. As a part of the brain, the hypothalamus serves as a major link between the nervous system and the endocrine system.

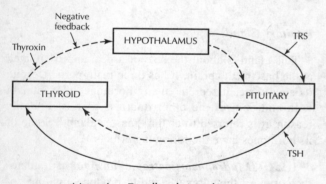

Negative Feedback Mechanism

The activities of the hypothalamus are closely related to the activities of the pituitary gland and of the other endocrine glands that are, in turn, stimulated by it. For example, if the concentration of thyroxin rises in the blood, the hypothalamus reduces the supply of its TRH hormone to the pituitary; this inhibits the production of thyroid-stimulating hormone (TSH) by the pituitary; the result is a decrease in the secretion of thyroxin by the thyroid gland. This type of regulation is referred to as a *negative feedback mechanism*. It helps maintain homeostasis throughout the body. In the case of the thyroxin feedback system, the rate of metabolism is kept at a constant level. A similar negative feedback mechanism affects the production rate of the other hormones.

Reproductive Organs (Gonads)

Various hormones are produced by (1) the male gonads, the *testes,* and (2) the female gonads, the *ovaries.*

1. *Testes* secrete male hormones, including *testosterone*. This hormone stimulates the formation of such male secondary sexual characteristics as appearance of hair on the face, deepening of the voice, increase in height. Other male animals show the effects of this hormone in their own development: the lion has a mane; the male deer has long antlers; the rooster has a comb and can crow; and male songbirds have bright plumage and beautiful voices.

2. *Ovaries* produce female hormones, including *estrogen*. This hormone stimulates the formation of such female secondary sexual characteristics as development of the breasts, retention of a high-pitched voice, and softening of the body features. Other female animals produce this hormone and develop accordingly: the lioness has no mane; the female deer has no antlers; the hen has no comb, and has a clucking voice; female songbirds have dull feathers, providing camouflage in the nest, and lack the ability to sing. Under the influence of the pituitary, the ovaries produce mature egg cells at monthly intervals. A great deal of research is presently being conducted on the value of female sex hormones in treating certain types of cancer.

Thymus Gland

This is a large gland in children and is located in the chest. It decreases in size with age. It is thought to play a part in growth and development, but this has not been confirmed. The thymus is part of the body's immune system, producing *T-cells*, a type of white blood cell, which attack germs.

Pineal Gland

Another little-understood gland is the pineal, which is located in the brain. Its function was clarified when it was found to secrete the hormone *melatonin,* which influences sexual maturity in growing children. It is secreted in relatively large amounts in early childhood, when it appears to restrict the development of the reproductive organs. Relatively little melatonin is secreted once puberty is reached. Melatonin is also thought to be related to the sleep mechanism.

Heart Hormone

In the early 1980s, it was discovered that the atria of the heart secrete a hormone that stimulates the kidneys to excrete more salt and water into the urine. The effect is to reduce the volume of the blood and the blood pressure. The hormone appears to be produced when the blood pressure and the volume of blood entering the atria are too high. Scientists gave the name *atrial natiuretic factor,* or *ANF,* to this hormone, and are conducting experiments to determine its use in people who suffer from high blood pressure.

Cyclic AMP (cAMP)

Cyclic AMP appears to be present in every living cell. Its full name is adenine 3', 5'-monophosphate; it is called cyclic because the atoms of its phosphate group are arranged in a ring. Changes in its level will either increase or decrease the rate at which cells carry out their appropriate functions.

Here is an example of its role: When the adrenal gland is stimulated by fear to secrete adrenaline, the heart begins to beat faster. Adrenaline activates a specific receptor on the surface of the heart muscle cell. The combination of the hormone and the receptor activates the cell enzyme *adenyl cyclase* to convert ATP to cyclic AMP in the cytoplasm. Cyclic AMP, which is often called the *second messenger,* since the hormone was the first messenger, now causes the heart cell to speed up its specialized function of rhythmic contraction.

In the same way, many of the hormones of the body cause a rise or fall in the level of cyclic AMP in specific cells, changing the rate at which they function. In some cases, the specific cells will be caused to secrete other hormones; for example, a hormone produced by the pituitary (*luteinizing hormone,* LH) acts on the ovary, where cyclic AMP initiates the formation of the hormone estrogen.

In other types of cells, a rise in cyclic AMP content changes the rate at which their functions are carried out. Examples are the uptake of calcium by cells that make bone; the breakdown of fat in adipose tissue; the secretion of digestive juices by the stomach lining; in the slime mold, the attraction of the amebalike cells to collect into a sluglike body, which then behaves like a multicellular organism. Dr. Earl W. Sutherland was awarded the Nobel Prize in 1971 for proposing how a hormone's message is transmitted at the cellular level through cyclic AMP.

Coordination of the Body's Activities

Very small amounts of the hormones are needed, yet they play an important part in regulating the various activities of the body. The ductless glands not only stimulate some of these activities, but also interact with each other. Thus, in the oxidation of glucose, thyroxin regulates the rate of oxidation; insulin permits the cells to oxidize glucose, and stimulates the liver to store it as glycogen; adrenaline stimulates the liver to convert glycogen back into glucose, and stimulates the rapid oxidation of glucose; the pituitary hormones control the thyroid and adrenal glands. The development of the reproductive organs is a complex process, involving hormones from the pituitary gland and from the reproductive organs themselves.

Insect Hormones

It has recently been found that the life history of insects is regulated by at least three hormones: (a) a brain hormone, produced when the insect egg hatches, which stimulates the formation of growth and juvenile hormones in the larva; (b) a growth hormone (*ecdysone*), produced in the thorax, which regulates growth and molting of the larva; (c) the juvenile hormone, produced in the head of the larva, which also influences growth, but inhibits metamorphosis. When the juvenile hormone is absent, the larva is induced to develop into the pupa stage. In experiments the juvenile hormone was injected into cecropia moth caterpillars, causing their metamorphosis to be delayed, and resulting in giant adult moths.

Plant Hormones

In animals, it has been shown that a hormone is a substance formed in one part of an organism and transported to another part, where it produces its effects. Plants have also been shown to produce hormones. Growth-promoting hormones called *auxins* are formed in the growing tips of stems. One of them, auxin A, has the formula $C_{18}H_{32}O_5$. Auxins cause the stem cells to elongate, resulting in their growth. It has also been determined that auxins initiate the formation of roots on stems. A number of chemical substances similar to auxins have been found that produce these and other effects. One of them, called *2,4-D* for short, makes broad-leaved weeds grow so rapidly that the metabolism of the plants is upset and they die. Others make unpollinated flowers develop into fruits without seeds; seedless tomatoes have been produced in this way. Auxins are also used to prevent fruits, such as apples, from dropping off a tree too soon. If auxin is distributed evenly on all sides of a stem, a plant grows straight up. If more accumulates on one side, growth proceeds more rapidly there, resulting in a bending toward the other side. This topic will be discussed more fully in Section 7.2 under the heading "Plant Behavior."

Another type of plant hormone, vitamin B, is unusual in that it is also useful to humans as a vitamin. The vitamin B complex consists of plant hormones produced in the leaves of plants and transported down the stem to the roots. Roots have been shown to require the vitamin B complex in order to grow.

SUMMARY OF THE DIGESTIVE AND ENDOCRINE GLANDS

Gland	Location	Hormone	Effects
Lining of alimentary canal	1. Stomach	Gastrin	Stimulates gastric glands to secrete gastric juice
	2. Small intestine	Secretin	Stimulates pancreas to secrete pancreatic juice
	3. Small intestine	Cholecystokinin	Stimulates gallbladder to send bile to small intestine
	4. Small intestine	Enterocrinin	Stimulates flow of intestinal juice
Adrenal	Top of kidneys		
1. Medulla	Inner part of adrenal	Adrenaline (epinephrine)	Stimulates liver to change glycogen to glucose; stimulates heartbeat, breathing rate, blood pressure, clotting rate
2. Cortex	Outer part of adrenal	1. Many hormones	Maintain balance of liquids and mineral salts; carbohydrate metabolism; secondary sexual differences
		2. Cortisone	Useful in treating rheumatoid arthritis, allergies, skin conditions
		3. Cortin	Useful in treating Addison's disease
Islands of Langerhans	Pancreas	1. Insulin	Regulates utilization of glucose by cells, and storage of glucose as glycogen by the liver; treatment for diabetes
		2. Glucagon	Stimulates liver and small intestine to release glucose
Parathyroid	On thyroid gland	Parathormone	Regulates calcium metabolism; lack causes tetany
Pineal	Base of brain	Melatonin	Appears to influence sexual maturity; also affects sleep
Pituitary	Base of brain	1. ACTH	Stimulates production of adrenal cortex hormones, including cortisone
		2. Growth	Promotes growth: deficiency in children causes dwarfism; oversecretion in children causes gigantism; in adults, acromegaly
		3. Prolactin	Stimulates production of milk
		4. Thyroid-stimulating	Stimulates production of thyroxin
		5. Oxytocin	Stimulates contraction of smooth muscles of uterus
		6. Gonad-stimulating	Stimulates development of gonads, and production of their hormones
Hypothalamus	Base of the brain	At least 9 hormones	Stimulate pituitary to secrete hormones to thyroid, gonads, mammary, adrenal glands, and uterus.

SUMMARY OF THE ENDOCRINE GLANDS *(Continued)*

Gland	Location	Hormone	Effects
Reproductive organs (gonads)	1. Testes	Testosterone	Regulates male secondary sexual characteristics
	2. Ovaries	Estrogen	Regulates female secondary sexual characteristics
Thymus	Chest of children	?	(Disappears during growth) Is involved in immunity; produces T-cells.
Thyroid	Base of neck	Thyroxin	Regulates rate of oxidation of glucose. Undersecretion in children causes cretinism; in adults causes myxedema. Oversecretion causes exophthalmic goiter. Lack of iodine causes endemic goiter.

Section Review

Select the correct choice to complete each of the following statements:

1. Hormones are distributed throughout the body by (A) ducts   (B) ductless glands   (C) blood   (D) nerves   (E) connective tissue

2. All of the following hormones are produced in the alimentary canal *except*   (A) secretin   (B) cholecystokinin   (C) enterocrinin   (D) gastrin   (E) insulin

3. The pancreas is stimulated to secrete its digestive enzymes by   (A) the liver   (B) bile   (C) the gallbladder   (D) secretin   (E) insulin

4. The adrenal glands are located above the (A) kidneys   (B) heart   (C) neck   (D) brain   (E) stomach

5. An element required in large amounts by the thyroid gland is   (A) iron   (B) iodine   (C) fluorine   (D) calcium   (E) phosphorus

6. The basal metabolism rate is regulated by the (A) parathyroid   (B) thyroid   (C) spleen   (D) thymus   (E) pineal

7. All of the following conditions are related to the thyroid gland *except*   (A) goiter   (B) cretinism   (C) myxedema   (D) high rate of metabolism   (E) tetany

8. The absorption and use of calcium are regulated by (A) parathormone   (B) adrenaline   (C) thyroxin   (D) thiamin   (E) prolactin

9. A hormone that enables the cells to utilize glucose is   (A) gastrin   (B) insulin   (C) testosterone   (D) cortisone   (E) pitocin

10. The gland known as the "gland of emergency" is the   (A) pituitary   (B) adrenal   (C) thyroid   (D) parathyroid   (E) pancreas

11. All of the following effects are produced by adrenaline *except*   (A) increase in heartbeat   (B) conversion of glucose to glycogen   (C) increase in breathing rate   (D) increase in clotting rate of blood   (E) decrease in rate of digestion

12. A hormone that stimulates the production of cortisone is   (A) adrenaline   (B) cortin   (C) auxin   (D) ATP   (E) ACTH

13. The gland that is referred to as the "master gland" is the (A) thyroid (B) adrenal (C) islands of Langerhans (D) pituitary (E) ovary

14. All of the following are related to the pituitary gland *except* (A) gigantism (B) dwarfism (C) xerophthalmia (D) acromegaly (E) rheumatoid arthritis

15. A hormone that affects secondary sexual characteristics is (A) insulin (B) adrenaline (C) estrogen (D) thyroxin (E) parathormone

16. A plant hormone that stimulates growth is (A) axon (B) auxin (C) adenoid (D) anaphase (E) annelida

17. A giant results from the overproduction of a hormone of the (A) pancreas (B) pituitary (C) parathyroid (D) thyroid (E) lining of esophagus

18. Radioactive iodine is sometimes used to treat an overactive (A) liver (B) pancreas (C) thyroid (D) adrenal cortex (E) adrenal medulla

19. A gland that serves as both a duct and a ductless gland is the (A) pituitary (B) thyroid (C) parathyroid (D) pancreas (E) adrenal

20. During a heart attack, a doctor may inject (A) thyroxin (B) prolactin (C) insulin (D) adrenaline (E) enterocrinin

Answer Key

1-C	5-B	9-B	13-D	17-B
2-E	6-B	10-B	14-C	18-C
3-D	7-E	11-B	15-C	19-D
4-A	8-A	12-E	16-B	20-D

Answers Explained

1. (C) Hormones are formed in ductless (endocrine) glands. They diffuse directly into the blood and are transported by the blood to other parts of the body, where they are used for specific purposes.

2. (E) Insulin is a hormone produced by the islands of Langerhans in the pancreas.

3. (D) Secretin is a hormone secreted by the lining of the small intestine. It is carried through the body by the blood. When it reaches the pancreas, it stimulates the digestive glands there to produce pancreatic juice, which, in turn, enters the small intestine through the pancreatic duct.

4. (A) The adrenal glands are located just above the kidneys, like triangular caps.

5. (B) The thyroid gland absorbs large amounts of iodine from the blood to make the hormone thyroxin. If iodine is deficient in the diet, the thyroid gland enlarges to produce a swelling on the throat called a goiter.

6. (B) The thyroid gland produces the hormone thyroxin, which is circulated by the bloodstream to the cells of the body. Thyroxin controls the rate at which oxidation takes place. The basal metabolism rate (BMR) is the rate of oxidation in a person at rest.

7. (E) Tetany occurs if there is a deficiency in the production of the hormone parathormone by the parathyroid glands. This hormone regulates the use of calcium by the body. Involuntary muscular contractions or convulsions occur if there is a deficiency of parathormone.

8. (A) Parathormone, secreted by four small parathyroid glands located next to the thyroid gland, regulates the use of calcium, which is needed for normal muscle function. Tetany results if there is a deficiency.

9. (B) The hormone insulin, secreted by the pancreas, helps the cells of the body to utilize glucose for energy. If insulin is deficient, diabetes mellitus results, causing a lack of energy, weight loss, and other symptoms. Excess glucose is then excreted and its presence shows up in the urine.

10. (B) The adrenal gland is known as the "gland of emergency" because its hormone, adrenaline, stimulates the body to have extra energy, extra strength, and extra alertness.

11. (B) Under the influence of adrenaline, heart, breathing, and blood-clotting rates are increased and the digestion rate is decreased.

12. (E) ACTH (adrenocorticotropic hormone), secreted by the anterior pituitary, stimulates the cortex of the adrenal gland to secrete a number of

hormones, including cortisone. Cortisone is useful in reducing inflammation.

13. (D) The pituitary gland is known as the "master gland" because it produces many hormones that interact with and control the other ductless glands.

14. (C) Xerophthalmia is a dry, scaly condition of the eyes caused by lack of vitamin A; it is not related to pituitary gland function.

15. (C) Estrogen, a hormone produced in the ovaries, stimulates the formation of female secondary sexual characteristics such as development of the breasts and high-pitched voice.

16. (B) Auxin is a plant hormone produced in the growing tips of stems that stimulates the elongation of stem cells.

17. (B) Gigantism results when there is an oversecretion of growth hormone by the pituitary gland.

18. (C) When the thyroid gland is overactive, the excess of thyroxin causes an increase in the rate of metabolism. Radioactive iodine concentrates in the thyroid gland, causing a reduction in the production of thyroxin.

19. (D) The pancreas secretes pancreatic juice containing digestive enzymes, which pass through the pancreatic duct into the small intestine. The pancreas is also a ductless gland, secreting its hormones, insulin and glucagon, into the bloodstream, which carries these hormones through the body.

20. (D) Adrenaline stimulates the heart to beat faster, and also acts to increase the breathing rate, thus bringing more oxygen to the heart tissue. For these reasons it may be given to a person suffering a heart attack.

7.2 REACTIONS OF PLANTS AND LOWER ANIMALS

One characteristic of protoplasm is its ability to react to the environment, known as *irritability*. Single-celled organisms, as well as multicellular plants and animals, show a *response* to *stimuli* about them. Indeed, their *behavior* may be said to form a basis for their suvival. Without the ability to move away from unfavorable stimuli, or to react positively toward favorable ones, they would soon die out. Higher animals have a well-developed nervous system for receiving stimuli and responding to them. However, plants and lower animals are able to react too.

Plant Behavior

Most plant behavior consists of a turning, or growing, away from or toward a particular stimulus. This type of behavior is known as a *tropism*. When the tropism is in the direction of the stimulus, it is said to be *positive*; when it is away from the stimulus, it is *negative*.

Phototropism

Plants grow toward the light. This is known as *positive phototropism*. The leaves of a geranium plant on a window sill are seen to face the window. If the plant is turned around, the leaves will grow toward the light again. A simple demonstration shows this positive phototropism:

1. Place a flower pot containing a bean seedling that has just begun to germinate in the right-hand corner of a tall box that is entirely enclosed, except for a small opening in the upper left corner. Next to it, as a control, set up another plant in the same way but in a box that has no opening. Water both plants frequently.

2. After a few weeks, observe that the first plant is growing in the direction of the light from the opening, and will grow out through it. The control plant will grow straight up.

Geotropism

Different parts of a plant may respond differently to the same stimulus. Thus the stem and leaves will grow upward, away from gravity; they show *negative geotropism*. On the other hand, the roots will grow downward, toward gravity; they show *positive geotropism*. A demonstration to show this can be performed as follows:

1. Germinate some bean seedlings in moist toweling paper. When the roots are about 2 inches long, arrange three of the seedlings on a strip of moist absorbent cotton, which is then put between two panes of glass. Keep the setup in place with two rubber bands, and stand it upright in a vertical position in a pan containing an inch of water, to keep the seedlings moist. Place it in the dark so that it will not be affected by the light.

2. Arrange the three seedlings in this manner: *(A)* root pointing up and stem down; *(B)* root and stem in a horizontal position; *(C)* root pointing down, with stem up.

3. After 2 or more days, observe the following: in *(A)*, the root has curved around and is growing downward, while the stem has curved around and is growing upward; in *(B)*, the root has begun to grow downward, and the stem to grow upward; in *(C)*, the root is continuing to grow downward and the stem upward.

Hydrotropism

Roots show positive *hydrotropism,* and grow in the direction of water. This tropism sometimes causes complications in streets where roots of poplar trees grow into drain pipes and plug them up. Here is a demonstration to show positive hydrotropism:

1. Germinate a dozen bean seedlings in moist toweling paper. When the roots are about 1 or 2 inches long, place the seedlings with the roots pointing downward, at intervals against the glass of a dry aquarium containing dry sawdust. Put a large sponge against the glass at the far end of the aquarium.

2. Water the sponge liberally so that it is the only source of moisture available to the plants.

3. Within a week, observe that the roots of all the plants are growing toward the end of the aquarium containing the wet sponge.

Thigmotropism

The growing stem of a climbing plant such as a bean or grape shows a positive reaction to contact, as it coils around a supporting stick. This is known as *positive thigmotropism.*

Auxins and Tropisms

The plant hormones known as *auxins* are formed at the growing tip and are transported downward in the stem. They stimulate the cells to elongate, resulting in growth. When auxins are distributed equally on all sides of the stem, the plant grows straight. It has been found that auxins accumulate in higher concentrations on the shaded side of a stem. The increased growth of the cells on this side makes the stem grow toward the light, resulting in positive phototropism.

Our knowledge of auxins also helps explain geotropism. When a young plant is blown over on its side so that it is in a horizontal position, its stem soon turns upward. The reason is that auxins accumulate on the lower side of the stem, under the influence of gravity. They cause the cells there to elongate, resulting in a bending upward. On the other hand, auxins have the reverse effect on root cells. When a root is in a horizontal position, auxins accumulate on the lower surface. Here, they inhibit the growth of the cells. The roots then bend downward, resulting in positive geotropism.

Other Plant Responses

The "sensitive plant" *(mimosa)* has many leaflets on its stems. If a part of it is touched, these leaflets can be seen to fold up along a stem. The petiole of the leaf will droop if the stimulus is a strong one. The heat of a lighted match held near the end of a leaf produces the same results. These movements are thought to result from rapid changes in the turgidity of the cells.

The carnivorous plants also show rather rapid responses. The leaves of *Venus's-flytrap* fold over to trap an insect that has touched any of the three hairs on one of the blades. The *sundew* has numerous slender stalks with sticky glands at their ends, which bend over to trap an insect that is located on the center of the leaf. In the case of the *bladderwort,* a group of sensitive filaments at the entrance to each bladder stimulates the nearby entrance to open suddenly, trapping the small water animal that touched them.

Behavior of Protozoa

Ameba

This simple blob of protoplasm shows the property of irritability when it reacts to a strong beam of light shining on one end; it moves in the opposite direction. If it comes into contact with a particle of food, it reacts positively, and engulfs it. If, on the other hand, it comes into contact with an irritating substance such as a grain of salt, it flows away from it.

Paramecium

A paramecium reponds to a weak acid such as mild vinegar by moving toward it. It moves away from salt and toward sugar. It moves toward a warm region, but away from a hot one. If it bumps into an object, it reverses its movement and then comes forward again at a different angle. If a drop of blue ink is placed on a slide containing paramecia, they are killed by the ink, but first they discharge their long, hairlike trichocysts. It is thought that trichocysts may normally help repel small animals that annoy the paramecium. A demonstration to show its response to a mild electric current may be performed as follows:

1. Place a culture of paramecia in a U-tube. Insert a wire into each arm of the tube, and connect it with the terminals of a dry cell.

2. In a few seconds, observe the paramecia with the naked eye, moving as tiny white specks toward the arm of the tube containing the wire connected to the negative terminal. Soon, practically all of the organisms will be collected in that area.

3. Now reverse the terminals so that the current flows in the opposite direction. The paramecia will migrate to the other arm of the tube, where the negative terminal is located.

The form of behavior exhibited by paramecia and other animals, in which they move in a particular direction in response to a stimulus, is known as *taxis*.

Vorticella

This bell-shaped protozoan is on a slender stalk. The stalk contracts rapidly when another animal touches it.

Euglena

Euglena has a sensitive eyespot which attracts it to light. If a beam of light is directed to one end of a culture dish containing euglena, they will all move toward the light, where they can carry on photosynthesis with the aid of their chloroplasts.

Behavior of Other Lower Organisms

If a hydra is touched, it contracts to a compact mass, little larger than a pinhead. Earthworms move away from the light into a shaded or dark place. Fish in a stream of water align themselves facing into the current; this is an example of positive *rheotaxis*.

Insects are attracted to light, as can often be seen on a summer evening. The positive phototaxis of a moth is well known. Female moths produce sex chemicals called *pheromones* which attract males of the same species for miles around. Apparently, their antennae detect the presence of these chemical compounds in the air. Other pheromones identify members of a species, cause fight reactions, or are used to mark a path toward a source of food.

Section Review

Select the correct choice to complete each of the following statements:

1. The characteristic of protoplasm that enables it to respond to stimuli is known as   (A) absorption   (B) ingestion   (C) digestion   (D) irritability   (E) excretion

2. The response of a plant stem in growing upward in the dark is known as   (A) positive geotropism   (B) negative geotropism   (C) positive phototropism   (D) negative phototropism   (E) negative photosynthesis

3. The response of a plant to contact is known as   (A) chemotropism   (B) geotropism   (C) thigmotropism   (D) rheotropism   (E) hydrotropism

4. The action of the roots of a tree in growing into a drain pipe is an example of   (A) positive hydrotropism   (B) negative hydrotropism   (C) positive geotropism   (D) negative geotropism   (E) negative rheotropism

5. The growth of a leaf toward the light results from   (A) a greater concentration of auxin on the shaded side   (B) a greater concentration of auxin on the lighted side   (C) equal concentrations of auxin on all sides of a stem   (D) a negative phototropism   (E) a positive geotropism

6. All of the following plants show rather rapid movements *except* the   (A) sensitive plant   (B) Venus's-flytrap   (C) bladderwort   (D) sunflower   (E) sundew

7. An organism that shows positive phototaxis is the   (A) earthworm   (B) hydra   (C) vorticella   (D) ameba   (E) moth

8. The growth of a bean vine around a stake is an example of   (A) geotropism   (B) galvanotropism   (C) thigmotropism   (D) chemotropism   (E) phototropism

9. An earthworm will move from a dry area to a moist area; this is an example of   (A) phototaxis   (B) geotaxis   (C) hydrotaxis   (D) rheotaxis   (E) thigmotaxis

10. The upward growth of a stem may result from   (A) positive phototropism and negative geotropism   (B) positive phototropism and positive geotropism   (C) negative phototropism and negative geotropism   (D) negative phototropism and positive geotropism   (E) positive phototropism and negative thigmotropism

Answer Key

1-D	3-C	5-A	7-E	9-C
2-B	4-A	6-D	8-C	10-A

Answers Explained

1. (D) Irritability is the ability of protoplasm to react to the environment by responding to stimuli.

2. (B) Plant stems grow upward away from gravity. This response is an example of negative geotropism. The stems do this in the dark, without their growth being influenced by the factor of light.

3. (C) The growing stem of a climbing plant, such as a bean, shows a positive reaction to contact as it coils around a supporting stick. This reaction is an example of thigmotropism.

4. (A) Roots show positive hydrotropism as they grow in the direction of water.

5. (A) Auxin accumulates in higher concentrations on the shaded side of a stem. The increased growth of the cells on this side makes the stem grow toward the light, resulting in positive phototropism.

6. (D) The sunflower turns very slowly to face the sun at various times of the day.

7. (E) The form of behavior exhibited by animals in response to a stimulus in a particular direction is known as taxis. When a moth is attracted to light, it exhibits positive phototaxis.

8. (C) The response of a plant to touch is called thigmotropism. A bean vine coiling around a stake demonstrates this response.

9. (C) Hydrotaxis is the movement response of a simple animal, such as the earthworm, toward water.

10. (A) In growing upward, a stem may be attracted by light; this is an example of positive phototropism. The stem is also growing away from gravity; this is an example of negative geotropism.

7.3 NERVOUS SYSTEM

The human nervous system is a complex communication system that branches throughout billions of cells, receiving and sending messages, and directing numerous voluntary and involuntary activities. In addition, human beings have the most highly developed brain in the animal kingdom, giving them the unique ability to think, reason, and create. The nervous system is organized into several parts: the central nervous system, consisting of the spinal cord, brain, and nerves; the peripheral nervous system; and the sense organs.

Neurons

The nervous system is composed of neurons, or nerve cells, which are specialized to possess the property of irritability.

Types of Neurons

There are three chief types of neurons:

1. *Sensory (afferent) neuron*—transmits impulses from the various sense organs (eye, ear, skin, etc.) to the brain or spinal cord.

2. *Motor (efferent) neuron*—transmits impulses from the brain or spinal cord to muscles, causing them to contract, or to glands, stimulating them to secrete.

3. **Interneuron (associative) neuron**—makes connections between sensory, motor, and other associative neurons; it is found in the brain and spinal cord.

Structure of a Neuron

Neurons are quite different in appearance from the other cells of the body. A motor neuron has a cell body, or *cyton,* containing the nucleus and most of the cytoplasm. Spreading out from it are many branched, threadlike *dendrites.* On one side there is a very long extension, the *axon,* which may be several feet long, even though it is microscopic. Around the axon is a white, fatty covering, the *myelin sheath,* which insulates it. The axon ends in a terminal branch, which lies next to the dendrites of another neuron, or in a muscle or a gland. Although the other two types of neurons are somewhat different in appearance from a motor neuron, they are basically alike in their structure. Axons of thousands of neurons are arranged in bundles called *nerves,* as they leave the brain or spinal cord. At this point the nerves are quite thick. They become thinner as they branch out to the various parts of the body, very much as the wires in a thick telephone cable spread out in various directions.

Nerve Impulse

Neurons have a high degree of irritability. They transmit messages, or impulses, along their length, and from one neuron to another, at the rate of over 300 feet a second. Their transmission is accompanied by the use of oxygen and the excretion of carbon dioxide. There is a change in the balance of sodium and potassium ions along the length of the neuron, and heat energy is released. In addition to this chemical reaction, there is also an electrical charge, which can be measured. The record of electrical impulses in the brain is called an *electroencephalograph* (EEG); the electrical impulses in the heart can be studied in an *electrocardiogram* (ECG). A type of chemical stimulant—neurotransmitter—is secreted at the point where the end brushes transmit an impulse into a muscle, making it contract. One of these neurotransmitters is called *acetylcholine.* Acetylcholine is also produced along the length of the axon, and is thought to play a part in the transmissions of an impulse.

Synapse

Neurons are not connected with each other, like the wires in an electrical circuit. Instead, the end of an axon lies very close to the dendrites of another neuron, forming a branched network in which the fibers do not touch, but are almost connected. This area is called a *synapse.* The direction of flow of an impulse is always from the axon of one neuron across a synapse into the dendrites of another neuron. A neurotransmitter, acetylcholine, is secreted by the terminal branch during the passage of an impulse across the synapse. When we are first learning a new skill, such as keyboarding, impulses pass with difficulty across the many synapses involved in the new nerve pathways. As the habit becomes established, these impulses cross over with ease, with resulting facility and speed in keyboarding.

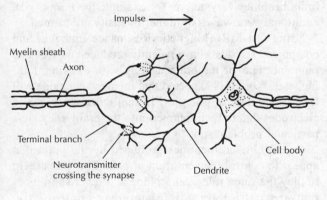

A Synapse

Reflex Arc

One of the pathways through the nervous system may be illustrated by tracing the impulses involved in a simple reflex such as suddenly pulling your finger back when it touches a flame.

1. The sensation is received in the sense organ, the skin, where there are numerous nerve endings.

2. The impulse that is generated there travels along the sensory neuron to the spinal cord.

3. In the spinal cord, the impulse crosses a synapse and enters an interneuron.

4. The impulse passes through this neuron, crosses a synapse, and goes onto the dendrites of a motor neuton.

5. It then travels out of the spinal cord along the axon to the end brush, where it passes on to the muscle fibers of the arm, making them contract to pull the finger away from the flame. The pathway of the impulse involved in this reflex act is known as a *reflex arc.*

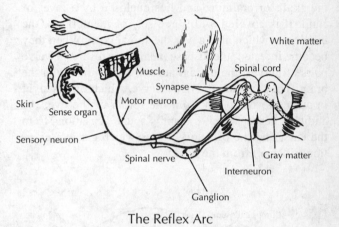

The Reflex Arc

During an actual reflex, many more neurons of all three types, as well as many muscle fibers, are involved. It takes only a moment for the arm to be withdrawn quickly and the finger saved from further contact with the flame. No thought was involved in this reflex act, which was centered in the spinal cord. The individual becomes aware of the incident when the associative neurons of the spinal cord send impulses along each other up to the brain. Many neurons now become involved, as there is thought about what happened, and a sensation of pain. It is interesting to note that, although countless neurons are closely located next to each other, the impulses take just the right path.

The Human Central Nervous System

The central nervous system is made up of the spinal cord and brain, and their nerves.

Spinal Cord

The spinal cord runs down the length of the back from the brain, and is enclosed in the vertebrae, the bones of the backbone. Thirty-one pairs of nerves leave the spinal cord and branch out to the body. One of them, the *sciatic nerve,* which runs into the leg, is the largest in the body and is as thick as a pencil. The center of the spinal cord has a butterfly-shaped mass of *gray matter,* which contains the cytons of the neurons. Surrounding it is the *white matter* in which the axons are located. One function of the spinal cord is to serve as the pathway for impulses from the brain to the various parts of the body and back. Another function, as mentioned in the description of the reflex arc, is to serve as a center of reflex actions.

Brain

The brain of a human being is the largest and most highly developed of all animal brains. It is protected by the skull, or cranium, and lies enclosed by a layer of fluid. It is covered with three sets of membranes, the *meninges,* which also cover the spinal cord. When they become infected, the disease *meningitis* results. Twelve pairs of nerves, the cranial nerves, branch out from the brain. They include the optic nerve, which brings sight impulses to the brain from the retina of the eye; the auditory nerve, which transmits sound sensations from the ears; and the olfactory nerve, which carries sensations of smell from the nose.

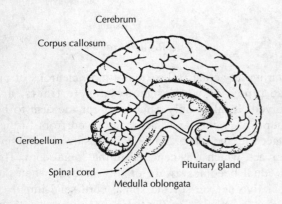

Structure of the Brain

It has recently been discovered that the brain in adults continues to produce new cells, thus overturning the belief that the number of brain cells is fixed at birth and does not change. As a person grows, there is also an increase in the number of dendrites and neuron connections. The brain consists of three main areas: (1) cerebrum, (2) cerebellum, and (3) medulla.

Cerebrum

This is the largest part of the human brain. It has two halves, known as *cerebral hemispheres.* The surface has many folds, or *convolutions,* which increase the volume of the cerebrum. The outer part, the cortex, contains the *gray matter,* which is made up largely of cytons. The *white matter* is within this layer; it contains the nerve fibers, or axons.

The cerebrum is the seat of intelligence and thought. In some way that is not yet understood, it gives us the ability to learn, reason, and remember; it helps us create great music, literature, art, and scientific theories. Various areas of the cerebrum have been mapped. The *sensory* areas are connected by the cranial nerves with the sense organs (eye, ear, tongue, etc.), and interpret the stimuli they receive from these organs. Thus, we "see" with the part of the cerebrum located in the back of the head. A sharp blow there can render a person temporarily or permanently blind, although the eyes themselves may be uninjured. There are also *motor* areas that control the voluntary muscles of the arms, legs, face, and trunk. The nerves to these areas cross over in the brain, so that the left side of the cerebrum sends impulses for physical movement to the right side of the body, and vice versa.

The two halves of the cerebrum are involved in different types of mental activity, as though there are "two minds" present. In most people, the left side deals with learning such as reading, writing, and arithmetic. It has to do with speech and logical thinking. The right side of the brain is the area for creative, artistic, and musical abilities and for spatial reasoning. The two cerebral hemispheres work together. They are joined together by a dense mass of white fibers, the *corpus callosum*. If this is damaged, messages fail to pass between the two hemispheres.

Electrical impulses called *brain waves* can be recorded from the cerebral cortex. In a person who is awake but not thinking hard, continuous waves can be recorded at a frequency of about 10 per second. These are called *alpha* waves. When a particular part of the brain becomes very active, for example the motor area, additional *beta* waves of higher intensity are formed.

Using fast MRI (Magnetic Resonance Imaging) and computers, scientists have recently produced movies that pinpoint areas of the brain dealing with thinking, talking, and so on. It is expected that the new technique will find immediate application in medicine; previously, electrodes had to be implanted into the brain to locate a particular problem spot.

Endorphins are peptides produced by the brain that appear to control the sensation of pain. They thus seem to play the same role as morphine and other substances that are artificially administered to control pain. Endorphins reduce or kill pain by binding to action sites, or receptors, on nerve cells. Their action may also be related to the production of hormones dealing with the emotions. The region of the brain in which endorphins function is not known.

Cerebellum

Below the cerebrum, in the rear of the cranium, is the smaller cerebellum. It coordinates the movement of the voluntary muscles so that they work together. The act of picking up a piece of candy with the fingers and putting it in the mouth requires the coordination of many muscles, all working together. The control of the body's balance is also centered in the cerebellum.

Medulla

The part of the brain located at the top of the spinal cord is the medulla. It is the center for breathing and heartbeat. When it is stimulated by the presence of increased amounts of carbon dioxide in the blood, it sends to the diaphragm and chest muscles impulses that speed up the rate of breathing. The medulla also serves as the area through which impulses travel from the spinal cord into the brain, and back. Certain reflexes, such as blinking, sneezing, swallowing, and the secretion of salivary juice, are centered in the medulla.

The Peripheral Nervous System

The peripheral nervous system consists of two parts: (1) the *somatic nervous system,* which transmits impulses between the voluntary, skeletal muscles and the central nervous system; and (2) the *autonomic nervous system*, which controls internal, involuntary functions.

The autonomic nervous system consists of masses of cytons called *ganglia,* which are located on either side of the backbone. This chain of ganglia is connected by nerves to the spinal cord on the one side and the internal organs on the other. There are also nerves connecting these ganglia with clusters of ganglia, each called a *plexus*, in several parts of the body, which serve as local centers of control. Thus, the *solar plexus* is located near the stomach; others are located at the heart, in the lower part of the abdomen, and in the side of the head.

The involuntary actions controlled by this system of ganglia and plexuses include heartbeat, secretion of digestive juices, size of the opening of the blood vessels, sweating rate, peristalsis, and size of the pupil of the eye. Emotions such as anger, fear, and joy, which stimulate the production of adrenaline, first act on the autonomic nervous system, which in turn sends impulses to the adrenal glands, causing them to secrete. There are two sets of nerves to the inner organs, one stimulating them to action (sympathetic nerves), the other causing them to relax (parasympathetic nerves). Thus, peristalsis of the digestive tract is stimulated by certain nerves of the autonomic system. The end brushes of the motor neurons secrete acetylcholine, which has a stimulating effect on the smooth muscles. There are also nerves whose end brushes secrete a substance like adrenaline, which inhibits and slows down peristalsis. The secretions of the glands, the contractions and relaxations of the heart, the size of the bronchioles in the lungs, the extraction of urine by the kidneys, and the other internal activities are similarly regulated by this balance between the nerves of the autonomic nervous system.

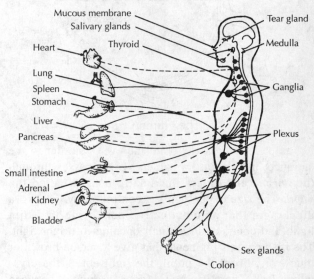

Autonomic Nervous System

The Sense Organs

The sense organs are specialized to receive impulses from the outside. For this reason, they are known as *receptors*. They are sensitive to a specific type of stimulus. The sense organs are (1) the eyes, which are sensitive to light; (2) the ears, which are sensitive to sound; (3) the olfactory nerves in the nose, for smelling; (4) the taste buds in the tongue, for tasting; (5) the receptors for pressure, pain, heat, and cold, in the skin. There are also internal receptors, such as those that give us the sensation of hunger when the stomach is empty, and a sense of balance in the muscles.

The Eye

The eye lies enclosed in a *socket* of the skull, and is protected by the movable *eyelids.* The eyelids close at intervals to keep the eye moistened. Tears are produced by tear glands, and serve not only to moisten the eye but also to keep bacteria out, with the aid of the substance *lysozyme.* The eyeballs are moved by three pairs of muscles. The outside of the eyeball is covered with a tough white covering, the *sclerotic coat.* In the front of the eye, this coat is transparent and is called the *cornea.* Beneath the sclerotic coat is the *choroid* coat, made of black tissue, which absorbs all light rays except those that fall on the retina; this layer also contains a rich supply of blood.

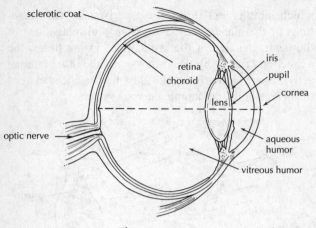

The Human Eye

The path of light rays into the eye is similar to the entrance of light into a camera. Light passes through the cornea and enters the small opening called the *pupil*. The size of the pupil is controlled by the *iris,* a diaphragm that automatically opens wide in a dim light, and contracts to a small opening in bright light. The iris contains pigments that give a person brown or blue eyes. From the pupil, the light enters the crystalline *lens*. There is a liquid, *aqueous humor,* between the lens and the cornea. The lens focuses the light through the large part of the eyeball, which contains the clear liquid, *vitreous humor,* which helps give shape to the eyeball. The light then falls on the sensitive layer, the *retina.*

Here, there are two types of receptors, the rods and the cones. The *rods* provide black and white vision, as well as vision in dim light. They contain the light-sensitive substance *visual purple,* which requires vitamin A for its production. The *cones* are responsible for the perception of colors. A single substance, *retinal,* which comes from vitamin A, activates both rods and cones. It binds to proteins inside thousands of little discs in the rods and cones, providing sensitivity to different wavelengths of light. The *optic nerve* contains fibers from the rods and cones which carry the light impulses to the *cerebrum.* Here, the *visual area* interprets the sensations received from both eyes at the same time; it also corrects the positions of the images received, which are upside down when they fall on the retina.

Color blindness   In this inherited condition there is an inability to distinguish certain colors, such as red or green; instead, they appear gray. The condition is probably due to a defect in the cones of the retina. Details of this type of inheritance are discussed in Section 10.3.

The Ear

The ear has three parts: (1) the external ear, (2) the middle ear, and (3) the inner ear.

The reception of sound waves   Sound waves are received in this way:

1. They enter the external ear and pass through the *ear canal,* at the end of which the *eardrum* is located.

2. The vibrations are transmitted by the eardrum to the parts of the middle ear. Here, there are three small bones, the *hammer, anvil,* and *stirrup.*

3. They pass the vibrations along to the membrane of the inner ear. This chamber contains the *cochlea,* a snail-shaped coiled structure, in which there is a liquid that receives the vibrations.

4. The movement of this liquid stimulates the endings of the *auditory nerve,* which carries the impulse to the *auditory area* of the cerebrum.

Other structures of the ear   The *Eustachian tube* connects the middle ear with the throat. It helps equalize the air pressure on the inside of the eardrum with that of the outside. When we go up rapidly in the elevator of a tall building, we can feel the pressure on the eardrum. By yawning, we are able to open the end of the Eustachian tube in the throat, and to equalize this pressure.

The inner ear contains, in addition to the cochlea, the three *semicircular canals*. These canals are at right angles to each other. They are filled with liquid, and have nerve endings that are connected with the cerebellum. Along with receptors in the skeletal muscles, they give us a sense of balance. Motion sickness in air, on land, or on sea results from a disturbance of equilibrium in this part of the ear. The drug Dramamine helps to overcome the condition.

The Nose

The receptors of the *olfactory nerves* on the inside lining of the nose receive the stimuli of smell. The impulses are then transmitted to the cerebrum. Some animals, such as the dog, rely heavily on the sense of smell for information about their surroundings.

The Tongue

The taste buds, which receive stimuli for the four taste sensations of sweet, sour, bitter, and salty, are located in different parts of the tongue. The senses of taste are related to the sense of smell. When we have a bad head cold, or hold our nose, taste is affected.

The Skin

Different parts of the skin have different receptors for pressure, pain, heat, and cold. The fingertips are more sensitive to touch than the back of the hand, because they contain more pressure receptors. All of the receptors send impulses over sensory nerves to specific areas of the cerebrum, where they are interpreted.

Some Malfunctions of the Nervous System

There are a number of disorders that can affect the nervous system. Among them are the following:

1. *Cerebral palsy.* In this condition, there is difficulty in controlling the voluntary muscles; as a result, the person suffers from involuntary movements. It is caused by injury or infection affecting the brain's motor areas before or during birth.

2. *Meningitis.* Meningitis is an infection of the meninges, the membranes covering the brain and spinal cord. The cause may be a virus or a bacterium.

3. *Strokes.* When a person suffers a *cerebrovascular accident*, or stroke, paralysis often occurs, because of damage to a part of the brain's motor area, from either a blood clot or the bursting of a blood vessel in the brain. A person with high blood pressure is especially liable to have a stroke.

4. *Polio.* This disease, also known as *infantile paralysis* or *poliomyelitis*, is caused by a virus that attacks the nerve cells of the brain or spinal cord that control the muscles. Paralysis results because the muscles stop functioning when the motor neurons that control them are damaged. Immunization has greatly reduced the threat of this disease (see Section 8.3).

The Nervous System in Lower Organisms

Protozoa

Protozoa, which are protists, of course, do not have a nervous system. Their protoplasm possesses the property of irritability, and enables them to respond to stimuli in their environment. Some protozoa, such as the *paramecium,* have been found to have a *neuromotor* mechanism, consisting of fine threads of protoplasm that are connected to the base of the cilia. They are believed to regulate ciliary movement; when they are cut experimentally, certain groups of cilia stop beating.

Hydra

This animal has a *nerve net*, that is, a network of nerve cells extending throughout its body. There is no definite pathway for impulses. A strong stimulus may be spread in all directions, to make the animal contract. Stimuli are received by *sensory* cells, which are sensitive to touch or to dissolved substances in the water. They are located in both of the layers of cells, and send impulses to the nerve cells of the network, which are located at the base of the ectoderm layer. The impulses are transmitted either to muscle cells, which contract, or to gland cells, which secrete. The nerve net also coordinates the various activities of the tentacles in grasping prey and in swallowing it.

Earthworm

In contrast to the hydra, which has no central nervous system, the earthworm has a large *ganglion* or *brain* above the pharynx. It is connected by nerves to another large ganglion below the pharynx, from which a double ventral *nerve cord* extends back to the rear of the animal. In each segment, this nerve cord enlarges to form a double ganglion, which is connected with all parts of the segment by nerves. There are no eyes, ears, or nose. However, there are *sensory* cells in the skin that receive impulses of light and taste, and transmit them to the ganglia, which in turn stimulate muscles or glands. The rate of transmission of an impulse is slow, only about 5 feet per second, compared to over 300 feet per second in humans.

Insects

In insects, there is a *brain,* located between the eyes, above the esophagus. A pair of nerves connects it with a ventral double *nerve cord* that extends through the body. Some of the *ganglia* are fused, so that there are two ganglia in the head, three in the thorax, and five in the abdomen. The sense organs include a pair of *compound eyes,* three *simple eyes,* ears, and a pair of *antennae.* In addition, the mouth parts have sensory hairs for taste. There are also taste receptors on the feet. The compound eyes have numerous lenses and are sensitive to color. The simple eyes are sensitive to light but do not form images. The antennae are organs of touch, smell, and taste. The ears are simple sound-receiving organs, whose location varies in different insects; in the grasshopper, they are on the sides of the abdomen next to the third legs; in the katydid, they are on the upper part of the first legs. The brain serves chiefly to receive impulses from the sense organs in the head. It does not seem to control or coordinate the other organs of the body. Highly complex patterns of behavior are shown by the social insects, such as bees and ants. However, these patterns are inborn and instinctive, rather than based on intelligent self-direction or thought.

Fish

Like all vertebrates, fish have a *brain* located in the skull, and a dorsal spinal cord enclosed in the backbone. Beginning with the front end, the parts of the brain are as follows: a pair of *olfactory lobes*; a small double *cerebrum*; two *optic lobes,* which are the largest part of the brain; *a cerebellum*; *a medulla.* The *eyes* have no eyelids, being kept moist by the water; a large pupil allows a large amount of light to enter the eyes. The eyes are connected with the optic lobes. The nostrils are not used for breathing; they open into small pouches, which have nerve endings leading to the olfactory lobes, and provide the sense of smell. Fish cannot hear, although they have an inner ear for balance. The *lateral line,* which runs along the outside of the body from the head to the tail, is believed to be sensitive to stimuli of water-current pressure, and transmits impulses to the central nervous system through nerves.

Frog

The brain of a frog shows a larger *cerebrum* than that of a fish. It also has *olfactory lobes, optic lobes,* a *cerebellum,* and a *medulla.* The spinal cord extends through the backbone. Cranial and spinal nerves extend to the various parts of the body. The *autonomic nervous system* controls the internal organs. The *eyes* are large and bulge out of the head; each eye has a transparent eyelid in addition to a regular eyelid. The *ears* consist of a pair of eardrums in back of the eyes.

Other Vertebrate Brains

The brain of a bird has a relatively large pair of *optic lobes* for its well-developed sense of sight, and a large *cerebellum,* which helps coordinate the bird's movement and balance in flight. The *cerebrum* becomes progressively larger in fish, amphibia, reptiles, birds, and mammals. In mammals, it shows convolutions as well. Finally, in the primates it is larger than in any other order. Human beings have the largest cerebrum of all, accounting for their superior intelligence.

Comparison of the Nervous and Endocrine Systems

The nervous and endocrine systems together control and coordinate the various life activities of the organism (regulation). The two systems are similar in that they both (a) secrete chemical messengers, and (b) play a major role in maintaining homeostasis. The systems differ in that (a) the responses of nerves are much more rapid than those caused by endocrines, (b) the responses of nerves do not last as long as those caused by endocrines, and (c) nerve impulses are transmitted along neurons; hormones are carried in the plasma of the transport system.

SUMMARY OF THE NERVOUS SYSTEM

Structure		Function
Brain	1. Cerebrum	Thought; reasoning; memory; voluntary activities. Sensory areas for sensation. Motor areas for voluntary movement.
	2. Cerebellum	Coordination of muscles; balance
	3. Medulla	Controls breathing, heartbeat, and certain reflexes (blinking, sneezing, swallowing, salivary secretion)
Spinal cord		Pathway for impulses to and from brain; center of reflex actions
Autonomic nervous system		Regulates internal activities—heartbeat, peristalsis, kidney action, glandular secretion, size of bronchioles and blood vessels

Section Review

Select the correct choice to complete each of the following statements:

1. A nerve cell is known as a(n)   (A) axon   (B) axon sheath   (C) neuron   (D) trichocyst   (E) nerve

2. A part of a nerve cell that is not found in other tissue cells is the   (A) nuclear membrane   (B) matrix   (C) chromatin   (D) dendrite   (E) xylem

3. A sense organ is connected with the spinal cord by a(n)   (A) sensory neuron   (B) motor neuron   (C) efferent neuron   (D) interneuron   (E) connective neuron

4. A chemical produced in an axon is   (A) acetylcholine   (B) aster   (C) auxin   (D) EEG   (E) ECG

5. An area between neurons is called a(n)   (A) dendrite   (B) synapse   (C) terminal branch   (D) end brush   (E) axon sheath

6. A mass of cytons is called a   (A) cranial nerve   (B) synapse   (C) ganglion   (D) motor area   (E) nerve fiber

7. In the pathway of a reflex arc, the correct order of the structures involved (1-spinal cord; 2-sense organ; 3-motor neuron; 4-sensory neuron; 5-muscle) is   (A) 4—2—1—3—5   (B) 3—2—1—4—5   (C) 3—1—2—4—5   (D) 2—4—1—5—3   (E) 2—4—1—3—5

8. The spinal cord serves as the center of   (A) subconscious thought   (B) reflex actions   (C) habits   (D) tropisms   (E) synapses

9. All of the following functions are controlled by the cerebrum *except*   (A) intelligence   (B) memory   (C) sensations   (D) digestion   (E) voluntary movement

10. Muscular coordination is controlled by the   (A) spinal cord   (B) convolutions   (C) cortex   (D) cerebellum   (E) cerebrum

11. The breathing rate is controlled by the   (A) lungs   (B) medulla   (C) cerebrum   (D) cerebellum   (E) sensory area

12. The autonomic nervous system controls all of the following activities *except*   (A) secretion of digestive juice   (B) peristalsis   (C) sweating   (D) thought   (E) diameter of blood vessels

13. All of the following are examples of receptors *except*   (A) eyes   (B) ears   (C) cyton   (D) nose   (E) tongue

14. Visual purple is found in the   (A) retina   (B) cerebellum   (C) cornea   (D) sclerotic coat   (E) lens

15. The sense of balance is maintained with the help of the   (A) cochlea   (B) hammer   (C) anvil   (D) stirrup   (E) semicircular canals

Answer Key

1-C	4-A	7-E	10-D	13-C
2-D	5-B	8-B	11-B	14-A
3-A	6-C	9-D	12-D	15-E

Answers Explained

1. (C) Another name for a nerve cell is neuron.

2. (D) The dendrite is the branched, threadlike set of structures extending out from the cyton of a neuron.

3. (A) The sensory, or afferent, neuron transmits impulses from a sense organ to the spinal cord or brain.

4. (A) At the end of an axon, the end brushes (terminal branches) secrete chemicals known as neurotransmitters. One type of neurotransmitter is acetylcholine. It travels across a synapse to stimulate the dendrites of an adjacent neuron. In this way, a nerve impulse passes from one neuron to another.

5. (B) A synapse is an area where an impulse from one neuron passes from its axon across a tiny space to the dendrites of another neuron.

6. (C) A ganglion consists of a mass of cytons from many neurons, located on either side of the backbone. A ganglion is connected by nerves with the spinal cord on the one side and internal organs on the other side.

7. (E) In a reflex arc, sensations are received in a sense organ. The impulse travels along sensory neurons to the spinal cord. Here, the impulse crosses a synapse into a motor neuton, which stimulates a muscle.

8. (B) The spinal cord serves as the center of a number of reflex actions that take place without the action of the brain.

9. (D) Internal, involuntary functions, such as digestion, are controlled by the autonomic nervous system.

10. (D) The cerebellum is the part of the brain that coordinates the movement of the voluntary muscles so that they work together.

11. (B) The medulla is a part of the brain located directly above the spinal cord. It controls the breathing rate. When stimulated by increased amounts of carbon dioxide in blood, it sends impulses to the diaphragm and chest muscles that speed up the rate of breathing.

12. (D) The process of thought is centered in the cerebrum, the largest part of the brain.

13. (C) A cyton is the cell body of a neuron containing the nucleus and most of the cytoplasm. Spreading out from it in all directions are the branched dendrites.

14. (A) Visual purple is a light-sensitive pigment found in the retina of the eye. It requires vitamin A for its production.

15. (E) The semicircular canals are found in the inner ear. They are filled with liquid and have nerve endings connected with the cerebellum. They help maintain the sense of balance.

7.4   BEHAVIOR OF HIGHER ANIMALS

Lower organisms such as protozoa, the hydra, and the earthworm respond to stimuli *automatically;* that is, they do not think about them, nor are they taught about them. This type of behavior is said to be *inborn,* and *unlearned,* and is called a *reflex.* It is like a tropism in these respects. Higher animals also show reflex behavior and in some cases *instinctive* behavior, which, like reflexes, is inborn, automatic, and unlearned.

Animals with a well-developed nervous system show additional types of behavior that are acquired after birth and therefore are said to be *learned.* Examples are *conditioned responses* and *trial-and-error learning.* Humans are also capable of additional higher forms of acquired behavior, such as *habit formation* and *intelligent behavior.*

Unlearned Behavior

Reflex

If you accidentally touch a flame, your finger is automatically pulled back. The mechanism for this act was explained in the preceding section as follows: A sensory neuron carries the impulse from the receptor, at the end of your finger, to the spinal cord; here an interneuron receives it; and then sends it over a motor neuron to a muscle in your arm, which contracts and pulls your finger away. No thought is involved in the reaction. It is centered in the spinal cord. A moment later, you are aware of it, because impulses are sent up the spinal cord to the brain. Fortunately you do not have to think about such stimuli. The response is inborn and is a factor in your survival. Such a reflex protects you.

We show other reflexes, some of which may be centered in the brain: The pupil of the eye expands in dim light, and contracts in bright light; we start at a sudden noise; we blink when a cinder flies into the eye—and our eye tears; we sneeze when irritating dust enters the nose; the knee jerks when the knee cap receives a blow; we cough if some water accidentally "goes down the wrong pipe." A demonstration to show the automatic nature of the blinking reflex, for example, may be performed as follows:

1. Have someone hold a pane of glass in front of her face. Instruct her to look at the wad of paper you are about to throw at her.

2. Throw the paper at her eyes. Repeat several times. The person will blink each time, even though the glass protects her face.

A newborn baby shows many reflexes: he sucks a nipple, grasps a finger placed in his palm, cries, coughs, and starts at the sound of a sudden noise. As the baby grows older, he will change some of these inborn reflexes and will replace them with various types of learned behavior that he will acquire.

Instinct

A spider spins a web perfectly the first time. This is another example of an inborn, automatic unlearned act. This type of behavior is more complicated than a simple reflex. It consists of a series of reflexes, each one acting as a trigger for the next, and is known as an instinct. In this complicated type of unlearned behavior, the first reflex is probably the secretion of a silky thread from the spider's special glands. This leads to the attachment of the thread between two twigs. Another thread is then spun and attached, and so on, until the web is finished, each step leading to the next reflex in the series.

Other living things may also act by instinctive behavior, including those with a well-developed nervous system. Here are some examples: The caterpillar of a moth spins a cocoon during a certain stage in its life history. Honeybees store honey in the cells of a hive, which are constructed with great precision. The mud-dauber wasp collects mud and builds a nest containing chambers for its developing young.

It still remains a mystery how animals navigate when they migrate. Birds migrate many miles each fall to spend the winter in a warm climate, and then return to the same location in the spring. Salmon swim from the ocean back to the river where they were hatched. Eels, however, live in fresh water and migrate to the ocean to breed. Each fall, millions of monarch butterflies migrate from the United States to the mountains of western Mexico for the winter.

The migration of loggerhead turtles after they hatch from eggs off the coast of southern Florida has been studied by scientists who recently concluded that subtle differences in the earth's magnetic field are sensed by the turtles as they swim across the Atlantic Ocean toward Portugal and then back to Florida.

Each species of bird builds a particular type of nest, selecting from a variety of materials, such as twigs, thread, feathers, moss, mud, and hair. Beavers cut down trees by gnawing through their trunks, and then use the branches to build a complicated dam.

In all these cases, there are inborn pathways over which impulses travel in a series of reflex acts. Sometimes the instincts appear so complicated and well-directed that we are tempted to consider them as examples of intelligent, purposive behavior. But this is not so. The organization of a beehive with its workers,

soldiers, drones and queen, all specialized to perform specific jobs, is merely another example of the complex, inborn activity known as instinctive behavior.

Learned Behavior

Conditioned Reflex

Inborn behavior, such as a reflex, can be changed or conditioned so that a new or acquired type of behavior results. The Russian physiologist Ivan Pavlov (1849–1936) did this experimentally with dogs, for the first time. He noticed that saliva flowed when the dogs were being fed. This was a simple reflex: the stimulus was food; the response was the flow of saliva. For a number of days, Pavlov rang a bell at each feeding time. After a while, he observed that merely ringing the bell without food being present was sufficient to cause the flow of saliva. In other words, he had substituted a new stimulus to produce the original response. This was now a conditioned reflex; it was acquired or learned. Pavlov found that his dogs would also produce saliva in response to other stimuli, such as the flashing of a light. In fact, the dogs even associated with food the white-coated attendants who fed them, and would produce saliva when they saw the attendants. The conditioned reflex was found to be temporary, however, if it was not accompanied by the feeling of satisfaction that came with eating. When the bell was rung a number of times in the absence of food, less and less saliva was produced, until the secretion stopped altogether.

Conditioned Response

This results when a more complex form of behavior is changed. You teach a dog its name by saying "Spotty" each time you feed it. The dog will then come in response to the sound of "Spotty" even if you do not feed it, because you supply the element of satisfaction in another way; you play with the animal, or take it outside, or pet it.

Human beings also show conditioned responses by associating a new stimulus with an original one. In school, if your lunch period follows the math class, you will find yourself getting hungry during the math lesson. If on a certain day the math period occurs earlier, perhaps because of an assembly, you may be surprised to find yourself salivating long before lunchtime. In other words, you have associated food with math. Other types of conditioned responses are common. A child will probably develop a fear of mice if his mother screams at the sight of one, and will grow up associating mice with fear; his friend may play with mice without fear if raised in a different setting. A girl may tremble at the sight of a harmless snake for the same reason. A demonstration to show conditioning in a goldfish may be performed as follows:

1. Feed the fish sparingly, and tap the side of the aquarium with your finger at the same time. Continue to do this over a period of 2 weeks.

2. At the end of that time tap the aquarium without adding any food. The fish will be seen to rise to the surface of the water. It has associated the tapping with food, and now responds to it.

Trial-and-Error Learning

A form of learning that depends on trying out several ways of doing things, and then selecting the correct way, is learning by "trial and error." If you have a dozen similar keys and wish to open a closet door, you may try several wrong keys before choosing the correct one. The next day you may have to try over again until you find the right key. If you do this daily, however, after a while you will be able to select the right key at once. You have learned by trial and error.

Here are other examples of trial-and-error learning: It takes some time for a hamster or a white rat to learn the correct route through a maze, and to avoid the blind alleys; after a number of days of making mistakes it will go through in the least possible time. A baby learns to eat with a spoon by holding it in various ways, until she learns how to grasp the spoon properly with her fingers.

A demonstration to illustrate progress in learning by trial and error may be performed in this way:

1. Cut a letter "L" that is 1 inch wide out of a card, with the long arm 4 inches long and the short arm 3 inches long. Cut the letter diagonally into four uneven pieces as shown:

2. Time someone and see how long it takes him to put the four pieces together to form the letter.

3. Shuffle the pieces and time the person again. Do this a dozen times. Record the results in a simple table, and plot them on a graph.

It will be seen that this type of learning takes place unevenly. The second trial may take less time than the first; the third, however, may take a little more time than the second; the fourth less time than the third; and so on, until learning is complete and no errors are made.

Habit

A learned pattern of behavior that is so automatic that it requires no thought, and is performed quickly and efficiently, is known as a habit. It is as automatic as a reflex, but it is not inborn. A secretary can keyboard so well that she need not look at the keys. She can work rapidly and correctly because she has established the right nerve pathways for the impulses to travel automatically. How has she done this?

Steps in forming a habit There are three steps in habit formation:

1. *Desire.* First, the secretary had a desire to learn the habit. If a person is not interested in learning a skill, it will be difficult to acquire it; witness the slow progress of a reluctant piano pupil.

2. *Repetition.* Practice makes perfect. As the secretary was learning to keyboard, she practiced and repeated the steps many times. During this repetition, the correct nerve pathways were being established for the impulses involved.

3. *Satisfaction.* As in other forms of acquired behavior, there must be a feeling of satisfaction resulting from the habit. The secretary enjoyed the feeling of being able to keyboard successfully, and of doing her job properly.

Examples of other habits include using a word processor, tying your shoelaces, riding a bicycle, writing, driving an automobile, playing an instrument, brushing your teeth, getting dressed, and playing tennis. In these habits, there is no thought about the steps involved. They have become automatic.

Advantages of habit formation A habit is performed quickly and accurately. It is done the same way each time. It also frees your mind to do other things, so that you are more efficient. These advantages may be demonstrated in the habit of writing as follows:

1. Divide a paper into two columns. In the left-hand one, using the hand with which you ordinarily write, write your name as many times as you can in 30 seconds.

2. Now do the same in the other column, but use your other hand.

3. Compare the two columns. It is apparent that when you were writing the second column, you found yourself thinking about how each letter is formed, while in the first column you were thinking about the writing itself. Also note in the second column the awkward appearance of the letters, the unevenness of the writing, the slow pace, and so on.

Breaking a habit We may wish to break certain habits that we have formed, such as nail-biting or smoking. Since the nerve pathways are well established for these habits, it becomes necessary to break up the pattern that led to them.

1. First of all, you must now have the desire to break the habit.

2. You should concentrate on the action so that you do not find yourself unconsciously doing it.

3. As you overcome the habit, you should experience a feeling of satisfaction from the progress you are making.

4. You can lose the feeling of satisfaction in nail-biting by painting iodine over your nails so that you experience a bitter taste each time you attempt to bite them.

5. Substituting another stimulus is also effective: For smoking, instead of reaching for a cigarette, suck on a piece of candy; for nail-biting in girls, buy a simple manicure set and polish the nails.

6. In all these steps, it is important to allow no exceptions.

Intelligent Behavior

Much of our behavior is based on voluntary activity, and is directed by our thinking.

The Role of Memory

We have the ability to store memories and to recall them. How this is done is not understood at present. Apparently, neuron pathways in the cerebral cortex are established, which can be recalled after days, months, and even years. By putting together our experiences, memories, and reasoning ability, we are able to do abstract thinking and to create new ideas. If, after watching the trial-and-error learning demonstration of the letter "L," discussed earlier in this chapter, you now try to put the pieces together yourself, you make an interesting observation. You may not do the puzzle correctly and quickly the first time, but you will master it in much less time than the other person. Apparently, while he was putting the pieces together, you were noticing and remembering the correct arrangement. You were indirectly benefiting from his direct experience. When it was your turn, you were able to avoid some of his errors. Compare this with a hamster that watched a fellow hamster learn a maze. No matter how long the animal watched, it required just as long to go through the maze correctly.

Studies in *psychology*, the scientific investigation of behavior, have shown that some of the other higher animals, including mammals and especially primates, are endowed with a certain amount of memory and even intelligent behavior. A dog will remember its young master upon his return from a summer at camp, and will greet him with joy. The behavior of the chimpanzee has been studied extensively by Dr. Robert Yerkes and Dr. Wolfgang Köhler. In one experiment, a chimpanzee in a cage was allowed to become hungry. Some food was placed outside, just beyond his reach. When a stick was placed nearby, he used the stick to reach the food. To test his reasoning ability still further, the chimpanzee was confronted with another problem—the food was now moved beyond the length of the stick. A second short stick, which could be fitted into the first, was made available. After a number of vain attempts to reach the food with each stick, the chimpanzee happened to fit them together to make a long pole. He promptly proceeded to obtain the food without much difficulty. He thus showed both memory and reasoning ability.

Human Intelligence

Psychologists have devised tests that attempt to measure human intelligence. By means of a variety of questions they test such aspects of mental ability as verbal meaning, space perception, reasoning ability, use of numbers, memory, and speed of recognition. The *mental age*, MA, is obtained from such tests. This is compared with the *chronological age*, CA, to arrive at the *intelligence quotient*, or IQ. The following formula is used to find the IQ:

$$IQ = \frac{MA}{CA} \times 100$$

Thus, a 17-year-old boy (CA = 17) who solves the problems of 17-year-olds (MA = 17), has an IQ of 100. If his MA is that of 19-year-olds, his IQ is 112. The range of IQ in a large number of unselected individuals will include some above 130, some below 70, and most between 90 and 110. Results of intelligence tests must be interpreted carefully, however, because the score may be influenced by such factors as cultural background, amount of education, language facility, physical health, and emotional state.

Because a human has the most highly developed cerebrum of all the animals, he is superior to all of them. He also has the power of speech for communication; the opposable thumb enables him to write and use tools. He has developed devices by which he can exceed animals, even in physical abilities in which they excel: He can see better than an eagle, using a telescope; he can fly faster and higher than a bird, in an airplane; he can hear a longer distance than a dog, using the telephone; he can exceed the strength of an elephant, with his machines; he can travel faster than a horse, in an automobile; he can travel under water more quickly than a fish, in a submarine; he can overcome a lion, using a gun. In addition, humans have unlocked many of the secrets of nature, including the structure and the forces of the atom. The question now confronting humankind is whether humans will use this knowledge intelligently for the benefit of all living things, or will behave like lower animals that have been given power they cannot control, and let it lead to their downfall.

Memory and RNA

Memory is the ability to retain a particular bit of information and to recall it. When a white rat has learned to go through a maze, it has stored certain information in its brain; a week later, it can still do the maze correctly. Scientists have conducted experiments with RNA that seem to indicate that it may be involved in memory. They removed the brains from rats that had learned a particular task and separated out the RNA fraction. They then injected this RNA material into untrained rats and found that they took much less time to learn the task.

In other experiments with certain flatworms, planaria, it was found that they could be trained to avoid a particular area. After such training, these planaria were ground up and fed to untrained planaria. The latter were then observed to avoid the particular area. They had not learned this, but memory had apparently been transferred to them.

It is believed that memory information is stored in RNA. But RNA is not the end product; it is used for the production of certain proteins and polypeptides. Another experiment used rats that had learned to avoid a dark, electrified area in a lighted cage. Their brains were removed and injected into untrained rats, which then also avoided the dark area. One particular fraction of the brain, a polypeptide, which was given the name scotophobin (meaning fear of darkness), was injected into rats, and they, too, stayed away from the dark area.

It is not known how RNA and polypeptides or proteins are involved in memory. Not all scientists agree that the results show a connection between RNA and memory. More research remains to be done in this area.

The Biological Clock

Some plants and animals, including humans, are controlled by a type of inner biological clock, which causes them to adjust regularly over the period of a day. For example, humans show a rhythm in their body temperature. It is highest in the afternoon, and lowest in the early hours of the morning; 37°C is only a daily average.

Recent research by Dr. Charles Czeisler, a specialist in the study of the human biological clock, has indicated that light is the controlling factor in our daily rhythm. It acts on a part of the brain (the *suprachiasmatic nucleus*) which is connected to the eyes, and which influences the internal cycles of the body, causing such effects as degree of alertness; change in levels of adrenaline; and secretion of the hormone melatonin, from the pineal gland, which influences sleep. Using very bright lights, Dr. Czeisler was able to artificially change the sleep-wake cycles of astronauts preparing for the 1990 flight of the space shuttle Columbia.

Certain beach crabs on the ocean shore are active at low tide and stay in their burrows at high tide. Even when these crabs are moved inland to the Midwest, they still show the same rhythm.

Rodents such as mice and rats, as well as certain insects such as cockroaches, are active at night and sleep during the day. Birds show daytime activity.

Among plants, legumes show a type of sleep-movement rhythm. They fold their leaves and lower them at night; during the day, they open and raise them. When such plants are kept indoors under continuous darkness and constant temperature, they continue with the same daily rhythm of sleep-movement.

SUMMARY OF TYPES OF BEHAVIOR

Type	Characteristics	Examples
Tropism	Turning toward or away from a stimulus; inborn; automatic	Positive phototropism—leaves, stem, moth Positive geotropism—roots Negative geotropism—stem, leaves Positive hydrotropism—roots Positive thigmotropism—stems
Reflex	Inborn, unlearned, automatic; no thought involved	Knee jerk; eye pupil changes in dim and bright light; blinking; sneezing; pulling finger away from flame
Instinct	A series of reflexes; inborn, unlearned, automatic	Spider spinning a web; honeybee specialization; bird migration; bird nest-building; beaver dam-building

---

SUMMARY OF TYPES OF BEHAVIOR *(Continued)*

Type	Characteristics	Examples
Conditioned response	Learned; substitution of one stimulus for another	Dog salivating at a bell; dog learning its name; child learning fear of mice; fish responding to tapping on aquarium
Trial-and-error learning	Learned response acquired after many trials in which errors were gradually eliminated	Animal learning a maze; human learning a puzzle
Habit	Learned, automatic responses, developed through (1) desire, (2) repetition, (3) satisfaction	Typewriting, getting dressed, writing, tying shoelaces, roller-skating, driving a car
Intelligent behavior	Involves memory, reasoning, and deliberation	Benefiting from someone else's experience; learning a lesson; building a birdhouse; making a painting

---

Section Review

Select the correct choice to complete each of the following statements:

1. All of the following are examples of a reflex *except* (A) sneezing   (B) blinking   (C) thinking   (D) coughing   (E) contraction of the pupil of the eye

2. All of the following are characteristics of a reflex *except* that it is not   (A) inborn   (B) unlearned   (C) automatic   (D) acquired   (E) inherited

3. The spinning of a web by a spider is an example of (A) an instinct   (B) a conditioned reflex   (C) an automatic, acquired act   (D) a habit   (E) trial-and-error learning

4. The scientist who pioneered in research on conditioned reflexes was   (A) Schick   (B) Banting   (C) Pavlov   (D) Pasteur   (E) Reed

5. A conditioned reflex   (A) is inborn   (B) is unlearned   (C) is acquired   (D) occurs whenever saliva flows   (E) occurs whenever a bell rings

6. A hamster goes through a maze the first time by (A) instinct   (B) trial and error   (C) habit   (D) reasoning   (E) conditioned reflex

7. Playing the piano is an example of a(n)   (A) habit   (B) inborn activity   (C) instinct   (D) conditioned reflex   (E) reflex

8. All of the following steps are necessary for habit formation *except*   (A) frequent exceptions   (B) desire   (C) repetition   (D) satisfaction   (E) practice

9. Working out a mathematics problem is an example of   (A) an involuntary act   (B) a voluntary act   (C) an instinct   (D) a tropism   (E) a reflex

10. The IQ is a measure of a person's   (A) mental ability   (B) mental age   (C) chronological age   (D) verbal quotient   (E) space perception

Answer Key

1-C	3-A	5-C	7-A	9-B
2-D	4-C	6-B	8-A	10-A

Answers Explained

1. (C) Sneezing, blinking, coughing, and pupil contraction are all reflexes. There is no thought involved in a reflex; it is inborn and automatic. Thinking is not a reflex.

2. (D) Acquired behavior is learned. It may be a conditioned response, a result of trial-and-error learning, a habit, or intelligent behavior. Reflexes are not learned or acquired. They are inborn and automatic.

3. (A) An instinct is a series of reflexes, each one acting as a trigger for the next. It is a complex, inborn activity.

4. (C) The Russian physiologist Ivan Pavlov (1849–1936) changed the salivary reflex of dogs so that the animals would salivate at the ringing of a bell. He changed or conditioned the original reflex in which saliva flows at the sight of food.

5. (C) A conditioned reflex is acquired. It is not inborn, and results from a reflex that has been changed.

6. (B) The hamster tries several ways of getting through the maze and then selects the correct way.

7. (A) A habit is a learned pattern of behavior that has become so automatic that it requires no thought and is performed quickly and efficiently. A person who has learned to play the piano knows which fingers to place on the keys according to the notes on the sheet music.

8. (A) In order to form a habit and establish the right nerve pathways for the impulses to travel automatically, there should be frequent repetitions. Frequent exceptions would interfere with this process.

9. (B) Working out a math problem is not inborn, unlearned, or automatic. It requires voluntary participation.

10. (A) The mental age is obtained in an intelligence test. It is divided by the chronological age and multiplied by 100 to determine the intelligence quotient, or IQ.

7.5 THE HABITS OF TOBACCO, ALCOHOL, AND OTHER DRUGS

The habits of smoking, drinking alcohol, and using other drugs are so different from the usual run of habits that they require special attention because of their pronounced ill effects on the health and welfare of the body.

Tobacco

Boys and girls in their teens are often tempted to smoke. In their desire to appear grown up, they believe that smoking will make them seem more mature and will enhance their popularity.

Advertisements often help heighten this impression. Opposition to smoking by parents and other adults has sometimes been based on emotional appeals. There are many facts about the effects of smoking, however, that should be known to anyone interested in the habit.

Effects of Using Tobacco

Tobacco smoke contains nicotine, tars, carbon monoxide, and other substances irritating to the lungs. Nicotine is a poison so powerful that 60 milligrams would cause death if injected directly into the blood. In smoking an average cigarette, an individual absorbs about 1 or 2 milligrams of nicotine into the bloodstream, enough to produce certain effects on the body. It causes the blood vessels to constrict, resulting in a rise in blood pressure. The narrowing of the blood vessels leads to a lowering of the temperature in the fingers and toes by as much as 2.8°C. There is an irregular increase in the heartbeat. The breathing rate is also increased. The appetite is dulled, and the process of digestion is delayed. The smoker becomes breathless more easily.

Some of these effects are recognized by athletic coaches, who prohibit smoking by the members of their teams. Most doctors advise their patients to stop smoking. There is a high correlation between heavy cigarette smoking and coronary heart disease. In medical studies on this subject, death rates from coronary heart disease in middle-aged men were found to be 50–150 percent higher among heavy smokers than among those who did not smoke.

The tars in smoke are strongly suspected of being a cause of lung cancer. Scientists connected with the National Cancer Institute state that excessive cigarette smoking is one of the causes of lung cancer. A study by

Dr. Wynder and Dr. Hoffman of Sloan-Kettering Institute for Cancer Research concluded that the chances of a man over age 50 getting lung cancer are as follows:

1. Does not smoke—one chance in 29,411.
2. Smokes ½ –1 pack a day—1 chance in 1,686.
3. Smokes 1–2 packs a day—1 chance in 695.
4. Smokes more than 2 packs a day—1 chance in 460.

The dangerous effects of smoking on the cells of the body were proved in a study of a gene known as p53. This gene is essential to the body's health because it acts to suppress the unchecked growth of cells that lead to cancer. This *tumor suppressor gene*, as p53 is called, was found to be damaged by the chemical BPDE, a powerful carcinogen that is formed in the body from benzo(a)pyrene, one of the chemicals found in tobacco smoke. As a result, the p53 gene becomes incapable of checking excessive growth, and cancer results. Problems with the p53 gene are found in up to 70 percent of lung cancers.

In 1997, because of an increasing number of lawsuits brought against the tobacco industry based on diseases caused by smoking, the Leggett Tobacco Company finally agreed to acknowledge that:

1. smoking does indeed cause cancer;
2. smoking is addictive, and that it will include such a statement in future warning labels on cigarettes, and
3. it would refrain from marketing cigarette advertisements to teenagers.

The Federal Center for Disease Control recently announced that the number of adult smokers had declined to a new low of 26.5 percent. This is a dramatic decrease in the nation's smoking rate since 1964, the year the U.S. Surgeon General first warned about the connection between smoking and cancer, heart disease, and other health problems. At that time, 40 percent of the adult population smoked.

However, the number of women who smoke has increased to such an extent that lung cancer has now surpassed breast cancer as a leading cause of death among women. Also, more teenage girls smoke than do boys: the number of boys who smoke declined from 28 to 16 percent, while the number of girl smokers rose from 20.5 to 28.8 percent.

Effect of Smoking on Nonsmokers

In 1986, then-U.S. Surgeon General C. Everett Koop announced that nonsmokers who are exposed to cigarette smoke suffer an increased rate of lung disease. Children of smoking parents show an increase in bronchitis and pneumonia early in life. He stated that 80–90 percent of chronic lung disease in the United States is directly caused by cigarette smoking. Second-hand smoke has been found to almost double the risk of suffering a heart attack. To reduce this health threat, laws have been passed prohibiting smoking on commercial airline flights and in workplaces, elevators, building lobbies, and designated areas of restaurants.

With all this knowledge about the undesirable effects of smoking, people who have the habit often try to give it up. They usually have great difficulty doing so because they have become addicted to nicotine. However, those who do succeed invariably express satisfaction that they are through with it. They mention their improved appetite, keener sense of taste, better wind, and relief at being free from, rather than a slave to a bad habit. The publication ASH (Action on Smoking and Health) has been promoting a smoking-free world for the past 35 years.

Alcohol

When a person drinks liquor of any type, alcohol is absorbed rapidly in the stomach and small intestine, and is circulated through the body.

Effects of Alcohol

Alcohol has the following effects:

1. It is absorbed in relatively large amounts by the brain and nerves; it is not a stimulant, as is often claimed, but acts as a *depressant* on the nervous system, interfering with reflexes, judgment, and vision.
2. It is readily oxidized by the cells to release energy; however, unlike the nutrients, it does not furnish needed vitamins, proteins, and minerals, and may lead to deficiency symptoms resulting from a poor diet.
3. It relaxes the smooth muscles in the blood vessels, enlarging their diameter; this results in a greater flow of blood through them, causing a loss of body heat from the increased flow of blood through the skin, and giving an individual a flushed appearance.
4. In large quantities, alcohol causes loss of muscular coordination, mumbling, loss of memory, and unconsciousness.

Drinking is a major cause of automobile accidents. A driver who has had even one or two drinks has enough alcohol in the blood to affect the centers in the brain that control judgment, alertness, and coordination. Those who have had even more alcohol cannot think clearly and act quickly enough, especially at high speeds, to drive safely. Studies have shown that more than one-third of automobile accidents occurred among people who had alcohol in their bodies. The National Safety Council warns drivers: "It is unsafe to mix alcohol with gasoline."

Regular heavy drinkers of liquor have less resistance to disease. They are especially vulnerable to pneumonia. Because they neglect proper eating habits, they tend to become undernourished, and show a deficiency of the vitamin B complex. Another serious effect is cirrhosis of the liver. Life insurance company statistics indicate that heavy drinkers have a shorter life span.

Alcoholism

Alcoholism is the disease resulting from continued use of alcohol. It is considered to be a difficult disease to cure, since the individual shows an uncontrollable craving for liquor. Most alcoholics are emotionally maladjusted. It is thought that some type of diet deficiency may be responsible for their irrepressible urge to drink. The nationwide organization known as Alcoholics Anonymous (AA) consists of people who have successfully freed themselves of the habit, and who attempt to help others do likewise. The undesirable social effects of alcohol have led a number of states to supervise the sale of liquor carefully, especially to minors. Young people who drink liquor as a stimulus to a good time face the possibility of serious accidents while driving, or the consequences of forming a degrading lifelong habit.

Alcoholism also has damaging effects on people who come in contact with an alcoholic. It plays a role in more than half of all physical abuse in the home. Child abuse, in the form of neglect and malnourishment, is directly linked to alcoholism. In addition, children of alcoholics run a greater risk of becoming alcoholics themselves. Children of alcoholic mothers may be born with birth defects (*fetal alcohol syndrome*).

Other Drugs

Through the ages, doctors have used narcotics to deaden pain, lull the senses, and bring about sleep. Used under their care, these drugs have been very useful. In 1680, the physician Thomas Suydenham wrote: "Among the remedies which it has pleased Almighty God to give to man to relieve his sufferings, none is so universal and so efficacious as opium." However, narcotics are sometimes used improperly, inducing habit formation, and causing the user to become a drug addict. The chief drugs are opium, heroin, morphine, codeine; cocaine; marijuana; Ecstacy; the barbiturates; amphetamines; and the hallucinogenic drugs.

Opium

Opium is derived from the dried juice of the seed capsule of the opium poppy plant, which is raised largely in parts of Asia. In Oriental countries, opium smoking is the chief way of taking the drug. In this country, it is purified to form heroin, morphine, and codeine.

Heroin

This drug has the appearance of a white powder. It is sold illegally by dope peddlers ("pushers") who generally dilute it with harmless powder and sell it at enormous profits. Drug addicts usually start by sniffing it, and then find it loses its "kick" unless they "mainline" it, or inject it directly into a vein. In an effort to help "junkies" (heroin addicts) overcome their habit, the narcotic methadone is sometimes substituted under close medical supervision. As the treatment progresses, if the patient on methadone should take heroin, he notices no effect from it.

Morphine

A standard medical drug, morphine is used by doctors to relieve extreme pain, usually caused by accidents or terminal cancer. Doctors are aware of its habit-forming properties and use it sparingly.

Codeine

This milder narcotic is used in some medicines to relieve severe coughing, as well as to alleviate pain.

Cocaine

Cocaine is extracted from the leaf of the coca bush, which grows in the Andes Mountains of South America. It has been used for centuries by the Indians working in the rarefied atmosphere of the high altitude there, who chew the leaves to overcome fatigue. When the pure chemical ("coke," "snow") is taken by mouth, sniffed, or injected, it gives a temporary feeling of physical strength and intellectual stimulation. The user becomes restless and talkative. There seems to be a mistaken belief that cocaine is a relatively safe drug, but the fact is that chronic use results in nausea, loss of weight, insomnia, and occasional convulsions. It can cause respiratory failure and death within an hour. In 1990, more than half of all drug-overdose deaths involved cocaine. A cocaine addict who tries to give up the drug may feel intense craving for it, leading to a deep state of fatigue, mental depression, and possibly suicide. Prolonged sniffing can cause the septum of the nose to become perforated. Pregnant women who use cocaine run the risk of giving birth to babies who are addicted to the drug and who suffer physical defects.

"Crack" is a more potent, addictive, and dangerous form of cocaine. It comes in concentrated pellets and is smoked, rather than sniffed. When inhaled into the lungs, it passes into the bloodstream rapidly and produces a "high" quickly; however, a "low" can follow in 15–30 minutes, making the user crave more. Crack acts by interfering with the normal functioning of neurotransmitters at the synapses between neurons. Crack can cause inflammation of the lining of the heart and spasms of the arteries, as well as interfere with normal heart rhythm.

Marijuana

Marijuana is obtained from the hemp plant, *Cannabis,* and is commonly called "pot" or "grass." Cigarettes ("joints") are made from its leaves and may be the first step in drug addiction, as the users sometimes find they want other drugs, such as heroin, to enhance the effect. When taken regularly, marijuana induces hallucinations, distortion of images, depression, delusions, and difficulty in concentrating.

Since marijuana smokers tend to inhale deeply, the smoke damages the lung cells and lowers lung resistance to infection. One study showed that smoking five marijuana cigarettes a week was more damaging to the lungs than smoking six packs of cigarettes during the same period. Marijuana contains the mind-altering chemical THC (tetrahydrocannabinol), which is fat-soluble and accumulates in the brain; there, it has been found to lead to brain damage among chronic users. Marijuana is also suspected of causing chromosome damage. Although it does not cause physical addiction, as heroin does, it is nevertheless considered to be psychologically addictive.

Marijuana has been found to have some medical value. THC may be useful in overcoming the overwhelming nausea and vomiting sometimes experienced by cancer patients during chemotherapy. However, marijuana does not cure cancer. Another possible medical use is in the treatment of glaucoma, a disease of the eye that is caused by pressure within the eye, often leading to blindness. Because many glaucoma patients are elderly, however, the resulting increase in heart rate may diminish whatever usefulness the drug may otherwise have. The use of marijuana does not prevent glaucoma.

Barbiturates

These synthetic chemicals are not as strong as heroin but are just as vicious in their drug-addiction powers. The individual who starts out by taking them as sleeping pills may find it necessary to take stronger and stronger doses. Eventually, this person may become an addict and suffer from mental depression, emotional instability, and types of delirium. Addiction to the barbiturates is even more serious than addiction to morphine.

Amphetamines

Known as "pep" pills, amphetamines induce a state of alertness at first, so that the person does not feel the effects of fatigue. Truck drivers who drive for long periods, especially at night, have been known to take amphetamines to keep awake. However, they require stronger and stronger doses, since the body develops a tolerance to amphetamines. Various side effects begin to appear, such as headaches, dizziness, irritability, and decreased ability to concentrate. The person may eventually have hallucinations.

Hallucinogenic Drugs

These drugs, which include LSD (lysergic acid diethylamide), mescaline and psilocybin, heighten sensory perception to the point where images and thoughts are weirdly distorted. It is impossible to predict their effects. They are supposed to be mind-expanding, but they can result in temporary or permanent insanity. LSD has been shown to cause changes in the chromosomes, thus affecting heredity.

Ecstacy

A relatively new drug, Ecstacy, has become widespread in recent years. It has caused sudden death and permanent brain damage among young people. Ecstacy produces feelings of euphoria by triggering the release of the brain chemical serotonin, which regulates emotions, pain perception, sleep, appetite, and sexual function. In studies at Johns Hopkins University School of Medicine, scientists found a 20 to 60 percent reduction in serotonin levels among drug users. Another study reported the results of neuropsychological tests of users indicating that they had a slower reaction time in memory and learning and in focusing their attention.

Drug Addiction

A *drug addict* is a person who has developed the habit of taking drugs. He may start by smoking a marijuana cigarette for a thrill. After a few such experiences, as his body gets used to the drug, he feels the need for something stronger, and starts sniffing heroin. This, too, is soon too weak, and he goes on to inject the drug directly into a vein. While under the influence of the drug, he feels a sense of well-being. His problems appear to diminish in importance, and the world seems very pleasant. When the drug has worn off, however, he faces reality again and finds the problems are still there. Worse than that, he soon begins to crave the drug. He must have it to feel at ease again. If he runs out of money, he will steal and rob, if necessary, to buy the drug. He is at the mercy of the dope peddlers. If he tries to drop the habit, he finds he cannot. He experiences "withdrawal" symp-toms; he has sharp abdominal and leg pains, he cannot eat or sleep, he feels chilly, he is very restless, he has nausea and diarrhea, he loses weight. The major purpose of his existence seems to be to obtain narcotics for his daily needs. Ill health, a low standard of living, crime, and degeneracy result. The addict needs medical help.

Section Review

Select the correct choice to complete each of the following statements:

1. Tobacco smoke contains   (A) caffeine   (B) nicotine   (C) niacine   (D) quinine   (E) morphine

2. Smoking causes all of the following effects *except* (A) constriction of the blood vessels   (B) increase in blood pressure   (C) increase in breathing rate (D) regular decrease in heart beat   (E) lowering of temperature in fingertips

3. There is a high correlation between cigarette smoking and   (A) increase in appetite   (B) increase in process of digestion   (C) coronary heart disease   (D) rickets   (E) resistance to scurvy

4. Many scientists believe that cigarette smoking leads to   (A) an increase in skin pigmentation (B) resistance to colds   (C) a longer life span (D) baldness   (E) lung cancer

5. Alcohol serves as a stimulant to   (A) the brain (B) the nerves   (C) vision   (D) judgment (E) none of these

6. Alcohol is absorbed in relatively large amounts by the   (A) brain   (B) skin   (C) lungs   (D) legs (E) eyes

7. Alcoholism is a   (A) custom   (B) disease (C) preventive for pneumonia   (D) food substitute (E) socially desirable "crutch" for shy teen-agers

8. All of the following are drugs *except*   (A) heroin (B) morphine   (C) Ecstacy   (D) chloroform (E) marijuana

9. Opium is obtained from the   (A) cinchona tree (B) flax plant   (C) poppy   (D) bloodroot (E) Jack-in-the-pulpit

10. The drug addict who tries to drop the habit experiences   (A) withdrawal symptoms   (B) a craving for candy and other sweets   (C) sleepiness   (D) an increase in weight   (E) a sense of well-being

Answer Key

1-B	3-C	5-E	7-B	9-C
2-D	4-E	6-A	8-D	10-A

Answers Explained

1. (B) Nicotine, tars, carbon monoxide, and other substances are contained in tobacco smoke.

2. (D) Smoking causes an irregular increase in the heartbeat.

3. (C) Death rates from coronary heart disease in middle-aged men were found to be 50–150% higher among heavy smokers than among non-smokers.

4. (E) Many studies have linked lung cancer with smoking.

5. (E) Alcohol acts as a depressant on the nervous system, interfering with reflexes, judgment, and vision.

6. (A) After liquor has been absorbed by the stomach, it circulates through the body and is absorbed by the brain and nerves in relatively large amounts.

7. (B) Alcoholism is the habit resulting from continued use of alcohol. It is considered a difficult disease to cure, since the individual shows an uncontrollable craving for liquor.

8. (D) Chloroform is an anesthetic once widely used in surgery to put a person to sleep during an operation.

9. (C) Opium is obtained from the dried juice of the seed capsule of the poppy plant.

10. (A) A person who experiences withdrawal symptoms has abdominal and leg pains. The person also cannot eat or sleep, feels very chilly, is very restless, has nausea and diarrhea, and loses weight.

HOW WE FIGHT DISEASE

CHAPTER
8

8.1 STUDY OF BACTERIA

Bacteria (sing., bacterium) were first discovered in 1683 by the Dutch microscopist Anton van Leeuwenhoek. By means of his simple, homemade lenses, he was able to study various forms of bacteria and describe their activities.

Characteristics of Bacteria

A bacterium is a single-celled organism having a cell wall. Chromatin material is scattered throughout the cytoplasm, rather than being contained in a nucleus.

The average bacterial cell may be from 1 to 1.3 micrometers in length. (A micrometer, μm, is 1/1000 millimeter, or 1/25,400 inch.) A particle the size of a pinhead may contain as many as 8,000,000 bacteria. Giant bacteria, a million times larger than the average bacterium, were recently discovered in the intestines of surgeonfish.

Classifying Bacteria

Bacteria are classified according to their shape:

1. *Bacillus*(*i*)—rod-shaped. Example: *Bacillus anthracis*, the cause of anthrax.
2. *Coccus*(*i*)—spherical. Example: *Streptococcus scarlatinae*, the cause of scarlet fever.
3. *Spirillum*(*a*)—spirally curved. Example: *Vibrio comma*, the cause of cholera.

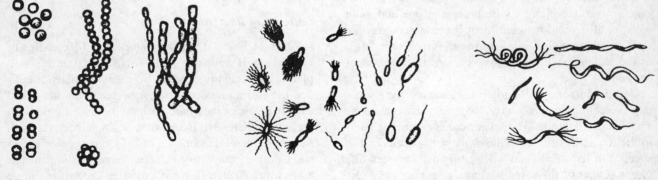

Cocci   Bacilli   Spirilla

Types of Bacteria

Cocci may be arranged singly or in groups. A single coccus is a micrococcus; cocci in pairs are diplococci; in groups of four, they are tetrads; in cubes of eight, sarcinae; in chains, streptococci; and in grapelike clusters, staphylococci.

Spore Formation

Certain bacteria have the ability to form a tough wall around their protoplasm. In this stage, they are known as *spores,* and can resist heat, low temperatures, dryness, and chemicals. When a spore germinates, the protoplasm emerges, grows, and multiplies. Spores may serve to carry bacteria through unfavorable periods; they may also constitute a stage in the life history of the bacteria. Few of the pathogenic (disease-producing) bacteria form spores. (Example: anthrax)

Reproduction

Bacteria multiply by the process of simple fission, during which they split in half. If conditions of food and temperature are good, they may multiply every 30 minutes. At this rate, one bacterium can produce millions within a 24-hour period, as can be seen from the following table covering just 12 hours:

Time		Number of Bacteria
30	minutes	2
1	hour	4
1½	hours	8
2	hours	16
4	hours	256
8	hours	65,536
10	hours	1,048,576
12	hours	16,777,216

Some Other Characteristics

Some bacteria move by means of hairlike projections, *flagella,* which may be present at one end, both ends, or all around the cell. Most bacteria need oxygen to live and are called *aerobic. Anaerobic* bacteria, such as the tetanus bacilli, can grow only in the absence of oxygen.

Certain bacteria are sometimes attacked by a virus parasite called *bacteriophage,* originally discovered by Twort and d'Herelle. The bacteriophage attaches itself to the bacterial cells and causes them to dissolve. It is specific in its action, and will distinguish between different strains of the same bacteria, attacking only certain types. For this reason, bacteriophage is used to type bacteria into various strains.

Growing Bacteria

Media

In the laboratory, bacteria may be grown and fed on various types of artificial culture media. The most common liquid culture medium is *nutrient broth,* containing beef extract, peptone, and water. This liquid is usually placed in test tubes covered with cotton plugs. When sterilized in an *autoclave,* or pressure cooker, for 20 minutes at 15 pounds of pressure, the liquid is clear. After bacteria are introduced, the liquid becomes cloudy. Other liquid nutrient media may contain milk, bile, stains, or glucose, depending on their specific uses.

Another, more common culture medium is solid *nutrient agar.* This is a mixture of beef extract, peptone, water, and agar-agar, a gelatinous seaweed extract. After it is sterilized, the melted nutrient agar is poured into either: (a) sterile test tubes, which are kept at an angle so that the agar may cool and harden at a slant, after the tubes are covered with a cotton plug, or (b) sterile petri dishes, in which agar form a solid layer about ⅛ inch thick. Other solid nutrient media may be made with gelatin, potato, blood, and milk.

Studying Bacteria Around Us

Sterile agar plates, that is, petri dishes containing nutrient agar, may be exposed to various conditions to show that bacteria are present everywhere. Thus, if there are seven plates to start with, six of them may be exposed as follows: A coin may be placed in one; a hair may be placed in another; a third may be exposed to the air for 5 minutes; freshly washed fingers may be touched to the surface; lips may be pressed against the surface; water from an aquarium may be added. The seventh agar plate should remain unopened, as the *control*; it is intended to show that all the dishes were sterile before they were exposed.

Colonies of Bacteria

After 24 hours in an incubator at 37°C, colonies of bacteria will appear on the surface of the exposed plates. Each colony contains so many millions of cells that the entire mass can readily be seen with the naked eye. It is sometimes hard to imagine that a colony ¼ inch in diameter originated from a single bacterial cell. A colony therefore consists of only one type of bacteria. Colonies may vary in size, shape, texture, luster, and color. Although most colonies are white, some appear red, as in the case of *Serratia marcescens*, or gray, yellow, green, or purple.

A Pure Culture

A pure culture consists entirely of one type of bacteria. It may be obtained by transferring bacteria from one of the colonies to the surface of an agar slant in a test tube. The transfer is accomplished with the aid of a platinum or nichrome wire needle, supported in a glass rod, in the following way:

1. Sterilize the needle by heating to red hot, both before and after using.
2. When cool, touch the end of the needle to the desired colony.
3. Remove the cotton plug from the test tube, and pass the mouth of the test tube through a flame to sterilize it.
4. Insert the needle into the bottom part of the test tube, without touching the sides, place it in contact with the surface of the agar slant, and draw it upwards to the top.
5. Replace the cotton plug, and sterilize the needle again.

These steps should be performed carefully, in order to avoid contamination by bacteria from the air or from other sources. After incubation for a day, a streak will be observed, with the naked eye, on the slant. This streak consists of the one type of bacteria that was deposited by the needle.

Microscopic Study of Bacteria

Hanging-Drop Method

It may be desired to observe the growth of bacteria directly. This may be done by using a hanging-drop slide, which can be prepared as follows:

1. Place a thin coat of Vaseline around the depression of a concave slide.
2. Heat a platinum or nichrome wire loop until red hot.
3. When cool, insert the loop into a liquid culture of bacteria, withdrawing a small quantity of it in the loop.
4. Touch the loop to the center of a clean cover slip, thus depositing a small drop of the culture.
5. Invert the cover slip over the center of the slide so that the drop of fluid hangs in the center of the depressed part of the slide.
6. Press the edges of the cover slip into the Vaseline to make an airtight seal that will prevent evaporation.
7. With the low-power objective of the microscope, focus on the edge of the drop. Then switch to high power, using the edge of the drop as a guide.

Living bacteria can be observed within the drop. A hanging-drop slide is also useful for observing motility, size, and shape of bacteria.

Staining Method

Bacteria can be observed more readily when they are stained with a weak solution of various aniline dyes, such as methylene blue or crystal violet. The following method may be used to stain bacteria:

1. Place a drop of water in the center of a slide.
2. Heat a straight wire needle until red hot.
3. When cool, touch the needle lightly to the desired colony.
4. Spread the bacteria evenly through the drop of water on the slide to form a thin film. Heat the needle again.
5. Allow the slide to dry in the air, and then pass it quickly through a flame three times. This fixes the bacteria to the slide.
6. Cover the bacterial film with the stain.
7. After 1 minute, wash the excess stain off with water.
8. Blot the slide dry with filter paper.

Under the microscope, the bacteria will be seen bearing the color of the dye. Special stains are employed to study specific features of bacteria, such as flagella, spores, and granules.

Gram's Stain

Probably the most important stain of all is Gram's stain, a differential stain which helps identify and classify bacteria. Bacteria that take a blue or violet color are called Gram-positive. Most cocci are Gram-positive. On the other hand, Gram-negative bacteria are stained pink. Most bacilli are Gram-negative, except for the diphtheria, tuberculosis, and spore-forming bacteria.

Helpful and Harmful Bacteria

Helpful Bacteria

Most bacteria are harmless. Some are quite useful to humans in a number of different ways.

1. *Soil bacteria.* The soil is teeming with numerous types of bacteria. Bacteria of decay attack and digest dead animal and plant remains in the soil, releasing chemical elements and compounds that enrich the soil. Soil composed largely of decayed plant material is known as *humus.*

2. ***Bacteria in the nitrogen cycle.*** Among the compounds formed during the bacterial decay of plant and animal proteins is ammonia. *Nitrifying bacteria* (*Nitrobacter* group) in the soil change ammonia to nitrates, which are valuable for protein synthesis by green plants. *Nitrogen-fixing bacteria (Rhizobum leguminosarum)* live within the swellings, or *nodules*, on the roots of plants of the legume family, such as beans, peas, alfalfa, and clover. These bacteria fix the free nitrogen of the air present in the soil into nitrates. In the free state, nitrogen cannot be used by plants, but as nitrates it can be taken up and used. The farmer recognizes the value of these bacteria by practicing crop rotation as a method of increasing the supply of nitrogen compounds in the soil.

3. ***Bacteria in the dairy industry.*** The flavor of butter depends to a great extent on bacterial activity. The cream used in making butter is inoculated with a culture of desirable lactic acid bacteria, which produce the desired flavor.

Swiss cheese, cheddar cheese, and several other types of cheese receive their flavor, texture, and color from the action of certain bacteria that are added to milk under carefully controlled conditions. The holes in Swiss cheese are produced as the result of gas formation while the cheese is still in the liquid condition.

Buttermilk, yogurt, and acidophilus milk are examples of fermented milk that has been treated with certain strains of bacteria, such as *Lactobacillus acidophilus.*

4. ***Making of vinegar.*** The word "vinegar" comes from the French and means "sour wine." The alcohol of wine or hard cider is changed to the acetic acid of vinegar by fermentation because of the action of acetic acid bacteria (*Acetobacter* group). The bacteria grow on the top of the liquid in a thick, jellylike layer called "mother of vinegar."

5. ***Silage fermentation.*** Chopped hay, cornstalks, beet leaves, and the like are packed by the farmer into the tall, cylindrical silo connected with the barn. The bacteria contained in the plant material carry on fermentation. This action results in an improvement in the flavor of the fodder, and at the same time preserves it against putrefaction.

6. ***Preparation of flax and hemp.*** Flax is the source of linen. The useful fibrovascular bundles or fibers in the flax stem are held together by certain "pectins." During the retting process, when the flax is placed in pools of water, these binding materials are removed by bacteria. After about 10 days, the fibers are removed, dried, and cleaned and are then ready for spinning. Hemp is prepared in the same way.

7. ***Other uses of bacteria.*** *Tobacco* leaves must be "cured" before they can be used; bacterial action produces the desired changes in the leaves. Meat that is cured has a more tender quality and flavor because of the action of bacteria. The production of *sauerkraut* from shredded cabbage, and *pickles* from cucumbers, depends on the fermentation action of bacteria. Natural *sponges* are prepared for commercial use by the action of bacteria in decaying and removing the protoplasmic material from the fibrous structure of the sponge.

Harmful Bacteria

Certain bacteria are harmful to human beings and their welfare.

1. ***Food spoilage.*** Many foods, such as meat, fish, poultry, milk, and cooked vegetables and fruits, spoil readily if allowed to remain at room temperature. Bacteria, yeasts, and molds find in these foods excellent conditions for growth. To protect our food, as well as our health, against these microorganisms, we use various methods of food preservation, such as the following:

- *Refrigeration.* Low temperatures inhibit the growth of bacteria. However, they do not kill many of them. When food is returned to normal temperatures, the bacteria emerge from the dormant state and begin to grow actively again.
- *Freezing.* The water in frozen foods is crystallized as tiny particles of ice and is therefore not available for the bacteria. Meat, fish, vegetables, and other foods may be kept preserved for a long time in the frozen condition.
- *Pasteurization.* This process of heating foods to temperatures that are high, but below the boiling point, in order to destroy harmful microorganisms was named after its originator, Louis Pasteur. He discovered that the spoiling of wine was due to the growth of undesirable bacteria. By heating the wine, he killed the bacteria without altering the taste. Milk is now pasteurized by heating to 145°F for ½ hour and then cooling rapidly; or, by the short-time method, it is heated to 162°F for 15 seconds and cooled rapidly. Either process destroys disease-producing bacteria of tuberculosis and typhoid fever, but not the harmless bacteria in milk. Pasteurization

is also used to preserve grape juice and apple juice. Pasteurization is not the same as sterilization, because many bacteria may still be present. During *sterilization*, all are killed.

- *Canning*. When food is canned, it is sealed and heated so that all the bacteria are killed. Unopened cans may therefore remain in good condition for many years. A sound can has concave ends. If the ends bulge, it is usually a sign that the contents have spoiled as the result of action by gas-forming bacteria. It is unsafe to open such cans because of the danger of botulism, caused by the deadly anaerobic *Clostridium botulinus*.
- *Dehydration*. Water is necessary for bacterial growth. When water is removed, food usually becomes powdery or flaky. In this condition, it occupies very little space, and may be shipped and stored readily with little danger of spoilage.
- *Salting*. Adding large quantities of salt to meat or fish helps preserve it by inhibiting the growth of bacteria. *Plasmolysis* of the protoplasm of the bacterial cells takes place when they are in contact with a concentrated solution of salt. The water of the protoplasm passes out of the bacterial cells, leaving the contents shriveled.
- *Other methods of food preservation*. *Smoking* meat helps prevent the growth of bacteria because of the chemicals present in the smoke; it also alters the flavor. Some *spices*, such as cinnamon and cloves, destroy bacteria, thus giving foods a keeping quality. Certain *chemicals*, such as sodium benzoate, are added directly to foods to preserve them. However, their use is strictly supervised under the terms of the Pure Food and Drug Act.

2. ***Dentrifying bacteria (Bacillus denitrificans)***. These soil bacteria play an undesirable part in the nitrogen cycle. They produce enzymes that convert ammonia, resulting from the breakdown of proteins by other soil bacteria, into free nitrogen. In this free state, nitrogen cannot be used by plants, and can therefore be considered as lost to them. This waste of nitrogen may be corrected by proper cultivation and aeration of the soil.

3. ***Bacteria and disease***. The greatest harm from bacteria comes from their effects on our health. The classic work of Pasteur and Koch helped establish the fact that certain diseases are caused by specific microorganisms. This is known as the *germ theory of disease*. This aspect of bacteria is discussed in detail in Section 8.2.

Some Famous Bacteriologists

Louis Pasteur (1822–1895)

Pasteur made many basic contributions to the field of bacteriology. He investigated the "disease" that caused wine to sour, and proved that souring was due to bacteria. When the silkworm industry in France was threatened by a disease of silkworms called *pebrine*, he showed that a microorganism was causing this disease. His solution of the silkworm problem stimulated him to turn his attention to other diseases, such as anthrax, chicken cholera, and rabies, for which he established successful principles of immunity.

Robert Koch (1843–1910)

Koch was a physician who became interested in the importance of bacteria in disease. He devised many of our present techniques for culturing, staining, sterilizing, and handling bacteria. He was the first to use gelatin for a solid culture medium, thus making it possible to isolate bacteria in pure culture. In 1876 he discovered the bacterium that causes anthrax in cattle. In 1884, he announced his discovery of the bacterium that causes tuberculosis in humans.

Koch used certain steps in identifying the tubercle bacillus. This procedure, which is the basis for proving that a certain disease is caused by a specific germ, is known as *Koch's postulates*:

1. The same microorganism is found in all cases of the disease.
2. It must be isolated and grown in pure culture.
3. The original disease is produced when the microorganism is injected into healthy animals.
4. The same microorganism is obtained from the diseased animals.

Joseph Lister (1827–1912)

Lister, an English surgeon, used Pasteur's discoveries about microorganisms as the basis for eliminating the terrible infections that accompanied surgery. He applied carbolic acid (phenol) directly to wounds, bandages, instruments, and the hands of the surgeon. This *antiseptic surgery* killed germs, but it was also harmful to the patient's tissues. Lister then introduced *aseptic surgery*, which is now used in all hospitals; the instruments, dressings, gloves, and so on are sterilized, and the operating rooms are kept free of germs—in other words, they are aseptic.

Section Review

Select the correct choice to complete each of the following statements:

1. Rod-shaped bacteria are called (A) bacilli (B) spirilla (C) cocci (D) plasmodia (E) streptococci

2. Bacteria may survive unfavorable conditions by (A) splitting in half (B) forming flagella (C) forming spores (D) forming a bacteriophage (E) forming tetrads

3. When bacteria are to be grown on an agar plate, they are (A) sterilized (B) pasteurized (C) stained (D) incubated (E) flamed

4. A type of glassware most commonly used for growing bacteria is the (A) flask (B) beaker (C) petri dish (D) delivery tube (E) thistle tube

5. Gram-positive bacteria are usually (A) disease resistant (B) antibiotic resistant (C) bacilli (D) stained pink (E) cocci

6. Nitrogen-fixing bacteria live on the roots of (A) peas (B) clover (C) alfalfa (D) legumes (E) all of these

7. Eight colonies of bacteria appeared on an agar plate after 24 hours; the number of individual bacteria present at the beginning was (A) 2 (B) 4 (C) 6 (D) 8 (E) 8 million

8. The flavor of all of the following is due to bacterial activity *except* that of (A) ice cream (B) buttermilk (C) Swiss cheese (D) butter (E) yogurt

9. Bacteria are useful in each of the following cases *except* the (A) making of linen (B) making of rayon (C) making of vinegar (D) production of sauerkraut (E) preparation of commercial sponges

10. All bacteria are killed by (A) refrigeration (B) food preservation (C) freezing (D) pasteurization (E) canning

11. The scientist who established the principles of immunity in anthrax and rabies was (A) Lister (B) Pasteur (C) Koch (D) van Leeuwenhoek (E) Harvey

12. The correct sequence of steps in Koch's postulates (1-injecting the germ to produce the original disease; 2-isolating the same germ in many cases; 3-obtaining the original germ from infected animals; 4-growing the germ in pure culture) is (A) 1—4—2—3 (B) 1—2—3—4 (C) 2—4—1—3 (D) 2—1—4—3 (E) 2—4—3—1

13. The basis of eliminating infections during operations is (A) pasteurization (B) dehydration (C) aseptic surgery (D) salting (E) cauterization

14. In practicing crop rotation, farmers rely on the desirable activity of (A) denitrifying bacteria (B) nitrogen-fixing bacteria (C) nitrifying bacteria (D) *Lactobacillus acidophilus* (E) acetic acid bacteria

15. All bacteria are classified as (A) algae (B) monera (C) parasites (D) saprophytes (E) symbionts

Answer Key

1-A	4-C	7-D	10-E	13-C
2-C	5-E	8-A	11-B	14-B
3-D	6-E	9-B	12-C	15-B

Answers Explained

1. (A) Bacilli are rod-shaped bacteria. Spirilla are spirally curved bacteria. Cocci are spherical bacteria. Plasmodia are protozoa that cause malaria.

2. (C) Certain bacteria have the ability to form a tough wall around their protoplasm. In this stage they are known as spores and can resist heat, low temperatures, dryness, and chemicals.

3. (D) Bacteria are grown in an incubator, where the temperature can be controlled, very often at 37°C.

4. (C) When a petri dish is used for growing bacteria, it is first sterilized. Then sterile, melted agar is added to form a layer about 1/8 inch thick.

5. (E) Most cocci are Gram-positive and take a blue or violet color with Gram's stain. Most bacilli are Gram-negative and take a pink stain.

6. (E) Nitrogen-fixing bacteria live in nodules on the roots of plants of the legume family, such as peas, clover, and alfalfa. They fix the free nitrogen of the air into nitrates, which can be used by the plants.

7. (D) A colony of bacteria on an agar plate grows from a single bacterium, which divides every 30 minutes to produce millions of bacteria in 24 hours.

8. (A) The flavor of some dairy products, including buttermilk, Swiss cheese, butter, and yogurt, is obtained by inoculating them with a culture of desirable bacteria. Ice cream is not prepared in this way and obtains its flavor from the ingredients used in making it.

9. (B) Bacteria are useful in many ways. However, rayon is not made through bacterial action. It is obtained from the artificial processing of cellulose.

10. (E) When food is canned, it is sealed and heated so that all the bacteria are killed. As long as the can remains unopened, no bacteria will appear.

11. (B) Louis Pasteur (1822–1895) weakened the bacteria causing anthrax and rabies and showed that they could protect an animal that was injected with them. In this way, the animals became immune to the diseases.

12. (C) By following these steps, Robert Koch (1843–1910) was able to identify the bacillus that causes tuberculosis.

13. (C) By means of aseptic surgery, Joseph Lister (1827–1912) was able to obtain a germ-free environment for surgical operations. The instruments, dressings, gloves, and so on are sterilized, and the operating room is kept clean with disinfectants.

14. (B) By planting legumes one year, between plantings of another crop such as corn, the farmer enriches the soil with nitrates produced by nitrogen-fixing bacteria that grow on the nodules of the legume plants.

15. (B) Bacteria are single-celled monera having a cell wall. They do not contain a nucleus, and their chromatin material is scattered throughout the cytoplasm.

8.2 GERMS VERSUS HUMAN BEINGS

Germs are microscopic organisms, such as bacteria, viruses, protozoa, and fungi, that cause disease in animals and plants.

How Germs Affect the Body

Human beings may be affected in a variety of ways by different germs.

1. *Production of toxins.* Toxins are poisons produced by germs. As the germs live on the tissues of the body, they give off these poisons. Toxins are powerful in minute quantities. It has been estimated that only 0.00023 gram (about ¼ milligram) of the toxin produced by the tetanus bacillus would be fatal to a full-grown man. The toxin of the botulism bacillus is about 100 times more powerful than this.

2. *Destruction of tissue.* Some germs settle in certain parts of the body and affect specific cells and tissues, as shown in the table below.

Disease	Part of Body Directly Affected
Amebic dysentery	Intestines
Malaria	Red blood corpuscles
Poliomyelitis	Nerves of spinal cord and brain
Rabies	Brain and spinal cord
Ringworm	Skin
Trichinosis	Skeletal muscles
Tuberculosis	Lungs, bone, or skin
Typhoid fever	Intestines, spleen

How Germs Enter the Body

Different germs may enter the body in various ways.

1. *Air.* Many germs are airborne and enter the body through the nostrils or the mouth. Examples: common cold, influenza, measles, mumps, pneumonia, tuberculosis.

2. *Water.* Water that appears clear may actually be contaminated with germs of typhoid fever, dysentery, or cholera.

3. *Food.* Germs may live on the nutrients in food and may be carried into the body when these foods are eaten. Examples: typhoid (milk, shellfish); tuberculosis (milk); trichinosis (undercooked pork); botulism (improperly canned food).

4. *Break in the skin.* Germs may be carried into the body when the skin is cut or broken. Examples: tetanus (punctured wound); skin infections; boils; rabies (bite of infected dog).

5. *Insect bite.* Some germs are injected into the body by the bite of an insect. Examples: malaria (*Anopheles* mosquito); yellow fever (*Aedes* mosquito); black plague (flea); African sleeping sickness (tsetse fly); typhus (body louse).

6. *Sexual transmission.* Some diseases are spread during sexual intercourse. Examples: syphilis, gonorrhea, genital herpes, AIDS. See Section 8.3.

How the Body Protects Itself Against Germs

The body is protected against germs in a number of ways:

1. *Lymphocytes* are types of white blood cells that help protect the body. They include *B cells* and *T cells*, which are formed in bone marrow. These cells develop an *immune response* to *antigen*. An antigen is a substance that invades the body, such as bacteria, viruses, and other types of foreign materials.

a. *B cells.* When a B cell in the bloodstream comes in contact with an antigen, it forms specific proteins called *antibodies* on its membrane that bind to the antigen. When this B cell reaches a lymph node, it divides into many *plasma cells*, each of which produces larger numbers of specific antibodies. These enter the bloodstream where they combine with the antigens, causing them to clump together. Large white blood cells called *macrophages* engulf and destroy the clumped antigens.

Other B cells are called memory cells because they are stimulated by a previous encounter with a particular antigen to respond to a new invasion of that same antigen months or years later. These memory cells produce a large number of specialized plasma cells with a stronger supply of antibodies that will resist the invading antigen. In this way they provide immunity to many infectious diseases, such as measles, polio, and smallpox.

b. *T cells.* T cells mature in the thymus gland, from which they get the name T. They do not produce free antibodies, as B cells do. Instead, they kill invading antigens directly. When specialized T cells, called *helper T cells*, meet an antigen, they become activated, and the antibodies on their surface attach directly to it. The activated helper T cells divide rapidly and stimulate the production of B cells and other T cells that bind to the antigen and destroy it by secreting toxins.

In addition, T cells can destroy cells that have been damaged by the invasion of a virus. T cells can also destroy cells that have been transformed by cancer. However, in the AIDS virus, the T cells are directly attacked, rendering them helpless to protect the body.

T cells may sometimes attack a transplanted organ, such as a kidney or the liver, that was surgically used to replace a diseased one. This could result in the organ being considered a foreign antigen and thus rejected shortly after the operation. In this case, the drug cyclosporin may be used to suppress the immune reaction. Often, close relatives are sought as donors of such organs because they may have more similar immune systems.

2. *Phagocytes* are the white blood corpuscles that engulf bacteria which have entered the tissues and the bloodstream. They are attracted to a break in the skin that has become infected. Pus is a thick liquid, consisting of lymph, white corpuscles, and bacteria, which may collect at such a place.

3. *Antibodies* are chemicals produced by the body, which act against germs or other antigens (foreign materials). Each is specific for a certain type of disease germ and is not effective against others.

Monoclonal antibodies are being used experimentally in studies of cancer, heart disease, and virus diseases. Monoclonal antibodies are produced by a new technique in tissue cultures wherein antibody-producing cells are fused with a certain type of cancer cell. The cancer part of the new hybrid cells keeps them growing indefinitely, while extraordinarily pure antibodies are being produced.

4. *Interferon* is a protein, produced by cells that are attacked by viruses, which protects neighboring cells against the germs. It was discovered in England, in 1956, by Dr. Alick Isaacs and Dr. Jean Lindemann. It offers great promise in fighting viruses because it is effective against not only one specific virus, but virtually all of them.

5. *Reticuloendothelial system (RES)* is an internal system of the body that clears foreign particles, such as germs, from the bloodstream. It consists of phagocytes and other cells that have the same function but are fixed in the lining of the blood vessels in such organs as the liver, thymus, lungs, spleen, bone marrow, and lymph nodes. It may also produce antibodies.

Providing Immunity Against Germs

Immunity is the ability of a body to resist disease. On the other hand, a person who becomes ill with a disease is *susceptible* to it. There are various types of immunity.

1. *Natural immunity* occurs when a person is born with the ability to resist a disease.

2. *Acquired immunity* results after a person either has recovered from a disease, or has been protected against it. Acquired immunity may be *active* or *passive*.

a. *Active acquired immunity.* Most often, a child who has had a case of the measles will not catch this disease again. While she was ill, she produced antibodies that helped her recover, and are now present in her body to protect her against further attacks of the germ. Since she acquired this type of germ immunity by actively producing the antibodies, this is known as active acquired immunity.

There are other ways to help the body acquire active immunity and to build a supply of antibodies. When you were vaccinated against smallpox, the mild virus of *cowpox* was scratched into the skin of your arm. This caused your body to produce a large enough supply of antibodies to protect you against smallpox for many years. In the Sabin vaccine against polio, the live virus, which has been weakened with chemicals, is swallowed to produce immunity.

Active immunity may also be acquired by inoculating a person with dead germs, as is done to protect against typhoid fever. This antigen stimulates the production of antibodies which kill any typhoid germs that may enter the body. The same principle is employed in the use of the Salk vaccine against poliomyelitis. Dr. Jonas Salk made the virus inactive by treatment with formaldehyde. Inoculation of the dead virus causes the formation of antibodies against the polio virus, thereby protecting the person against the disease.

Deadly toxins that are weakened with chemicals are called *toxoids*. If toxoid is injected, it stimulates the body to produce antitoxins against the specific disease. Active immunity may thus be acquired against diphtheria or tetanus. *Toxoid* has now largely replaced *toxin-antitoxin*, which was formerly used to stimulate the production of antitoxins by the tissues of the body. In toxin-antitoxin, the toxin stimulates the reaction, while, at the same time, the antitoxin protects the body against the toxin. Toxoids are now preferred because fewer injections are needed, and the danger of a sensitive serum reaction is avoided.

In all cases of active immunity, the body is protected for a long time, because its tissues have actively produced an abundant supply of antibodies.

b. *Passive acquired immunity.* To produce passive immunity, antibodies are injected into the person from some outside source, such as another person or a horse. During a case of diphtheria, for example, a doctor will inoculate a child with antitoxins, in order to neutralize the deadly toxins that are causing the symptoms. The child recovers without having produced his own supply of antibodies. In the case of a deep puncture wound, tetanus may be prevented by injecting antitoxins into the injured person. The antitoxins neutralize any toxins that may be produced by tetanus bacilli.

Passive immunity is temporary, lasting only a few weeks. The tissues of the body are not stimulated to produce antibodies.

Using Chemicals to Protect the Body

External Use

Germs on the outside of the body may be killed by various chemicals.

1. *Disinfectants* are powerful chemicals that can destroy not only germs but also spores. Examples: chlorine compounds, Lysol.

2. *Antiseptics* are chemicals that inhibit or prevent the growth of bacteria. They are less injurious to tissues than disinfectants. For this reason, we use them for a cut or break in the skin. Example: tincture of iodine.

Internal Use

Some chemicals or drugs are used internally to fight germs that have entered the body. The chief concern in using such chemicals within the body, of course, is that, while they are expected to kill the germs, they should not harm the living cells of the organism.

1. *Quinine* was originally used by the Indians of South America to treat malaria. They chewed the bark of the cinchona tree which contains the drug. Nowadays, *atabrine*, *chloroquine*, and *mefloquine* are largely used to treat the disease.

2. *Arsenic compounds* were used by Paul Ehrlich (1854–1915) to treat syphilis. His 606th experiment to discover a chemical treatment for this disease finally yielded a compound containing arsenic and bismuth that would not harm the body tissues. It was called "606," or *salvarsan*.

3. *Sulfa drugs* were first developed in the 1930s; the first of these so-called wonder drugs was *sulfanilimide*. Since then, hundreds of varieties have been made, including sulfathiozole and sulfadiazine. They are effective against bacteria that cause streptococcus infections, meningitis, pneumonia, boils, and blood poisoning. They interfere with the metabolism of these bacteria, causing them to die. Since sulfa drugs may have harmful effects on the body, they should be taken only under a doctor's care.

4. *Antibiotics* are chemical substances produced by living things, especially fungi, that stop the growth of germs.

• *Penicillin* was the first of the antibiotics to be discovered. In 1929, Alexander Fleming, a British doctor, observed that a green mold (*Penicillium notatum*) was accidentally growing in a petri dish containing staphylococcus colonies. Instead of discarding the contaminated dish, he investigated the clear bacteria-free zone immediately around the mold. He theorized that the mold was giving off a substance that destroyed the bacteria and, after additional investigation, named it penicillin. This drug was first used on a large scale to save the lives of wounded soldiers during World War II. It is valuable in treating pneumonia, venereal diseases, and infections caused by cocci. The Nobel Prize was awarded to Fleming, Howard Florey, and Ernst Chain for their work in developing penicillin.

- *Streptomycin* was obtained from soil fungi by Selman Waksman of Rutgers University, who also was awarded a Nobel Prize. This antibiotic is useful in combating tuberculosis, urinary infections, and other diseases.
- Other antibiotics are being developed all the time. *Chloromycetin* is useful against typhus and Rocky Mountain spotted fever, both caused by rickettsia germs. *Aureomycin, terramycin*, and *tetracycline* are just a few of the other useful antibiotics.
- Certain antibiotics (for example, *terramycin*) have been found to stimulate the growth of farm animals, and are now added to the food of chickens, pigs, and cows.
- Resistance to antibiotics and the sulfa drugs is sometimes developed by certain strains of germs. These immune germs survive and then increase in number. Doctors may then use *combinations* of the drugs to overcome the germs.

Allergic reactions are sometimes shown by people who have become sensitive to certain antibiotics. Because of the body's reaction to the antibiotics, these persons may develop hives or become quite ill after they have become protected against a specific disease germ. For this reason doctors now advise using antibiotics only when absolutely necessary.

5. *Isoniazid* and similar chemicals are very effective against the tuberculosis germs, and are often used in combination with streptomycin in the successful treatment of tuberculosis. Recently, another drug, *rifampin*, has been combined so successfully with isoniazid that the 18-month treatment period for tuberculosis has been cut in half.

Section Review

Select the correct choice to complete each of the following statements:

1. Poisons produced by bacteria are known as (A) toxoids (B) toxins (C) antitoxins (D) antibodies (E) toxin-antitoxins

2. A disease in which the red blood cells are attacked is (A) rabies (B) amebic dysentery (C) polio (D) trichinosis (E) malaria

3. Germs that enter the body in milk may cause (A) pneumonia (B) tetanus (C) mumps (D) tuberculosis (E) influenza

4. A disease caused by the bite of an insect is (A) typhus (B) typhoid (C) rabies (D) measles (E) botulism

5. All of the following protect the body against the entrance of germs *except* (A) tears (B) mucous membranes (C) ciliated cells (D) white blood cells (E) red blood cells

6. All of the following are examples of antibodies *except* (A) agglutinins (B) opsonins (C) lysins (D) antibiotics (E) antitoxins

7. An example of passive acquired immunity is (A) vaccination against smallpox (B) inoculating dead germs against typhoid (C) use of Salk vaccine (D) inoculation of toxoid against diphtheria (E) inoculation of antitoxin in case of a puncture wound

8. A drug obtained from the bark of a tree is (A) salvarsan (B) tincture of iodine (C) quinine (D) sulfanilimide (E) tetracycline

9. All of the following are antibiotics *except* (A) penicillin (B) streptomycin (C) aureomycin (D) terramycin (E) riboflavin

10. Penicillin was discovered by (A) Ehrlich (B) Fleming (C) Waksman (D) Koch (E) Metchnikoff

Answer Key

| 1-B | 3-D | 5-E | 7-E | 9-E |
| 2-E | 4-A | 6-D | 8-C | 10-B |

Answers Explained

1. (B) Toxins are poisons produced by germs as they live on the tissues of the body. Toxins are powerful in minute quantities.

2. (E) The red blood cells are attacked by the plasmodium protozoan that causes malaria.

3. (D) The tuberculosis bacillus may enter the body if a person drinks milk that has not been pasteurized properly.

4. (A) The germ that causes typhus may enter the body as the result of a bite by a body louse.

5. (E) Red blood cells contain hemoglobin, which combines with oxygen in the lungs and carries it to the cells of the body. Red blood cells are not involved in protecting the body from the entrance of foreign substances.

6. (D) Antibiotics are chemical substances produced by living things, especially fungi, that stop the growth of germs. Example: penicillin.

7. (E) In passive acquired immunity, antibodies are injected into a person from some outside source, such as another person or a horse. An injured person may be injected with tetanus antitoxins that neutralize any toxins that may be produced by tetanus bacilli in a puncture wound. In passive acquired immunity, the body does not produce its own antibodies.

8. (C) Quinine is obtained from the bark of the cinchona tree and is useful in treating malaria.

9. (E) Riboflavin, or vitamin B_2, is a coenzyme that plays a role in cellular respiration. It is found in milk, liver, kidneys, and other meat.

10. (B) In 1929, Alexander Fleming observed a clear zone around a green mold *(Penicillium notatum)* growing accidentally in a petri dish containing streptococcus colonies. He theorized that the mold was giving off a substance that destroyed the bacteria, and named it penicillin.

8.3 CONQUEST OF DISEASE

A disease is a condition in which the health of the body is impaired. Germs are responsible for many types of disease. These microorganisms may be bacteria, some fungi, viruses, protozoa, or worms. The successful conquest of disease germs has helped raise the average life expectancy of a child born in the United States to 75.1 years.

DISEASES CAUSED BY BACTERIA

Tuberculosis

This disease was formerly known as the white plague and was a chief cause of death in this country at the beginning of the century. Robert Koch (1882) was the first scientist to prove that it was caused by a specific germ, the tubercle bacillus. The bacilli may be spread through droplets in the air from coughing or sneezing; by using contaminated articles; in impure milk; and by direct contact with a person having the disease. The bacilli may infect the lungs where they form tubercles, little masses of germs surrounded by protective layers of calcium and fibers. Since 1985, there has been a dramatic increase in the number of TB cases, especially among people with AIDS.

Detection   The disease may be detected by:

1. Chest X ray, which helps reveal the presence of tubercles in the lungs.

2. Tuberculin test (patch test), in which tuberculin material obtained from the germs is placed in the skin. A red area a few days later indicates that the person is sensitized to the tubercle bacillus, and may now have, or once had, the active germs in the body. A positive result should be followed up with a chest X ray.

3. Sputum test, in which a sample of coughed-up sputum is examined for the presence of tubercle bacilli.

Treatment   Treatment for the disease includes:

1. Providing rest and good nutrition to the person having the disease. This is the classic treatment first developed by Dr. Edward Trudeau at the Lake Saranac sanatorium.

2. Drug treatment using streptomycin and the new isoniazid-rifampin combination. This has been so successful that many tuberculosis sanatoriums have now been closed, including the famous Trudeau sanatorium.

3. Pneumothorax, in which a badly infected lung is collapsed to permit the damaged tissue to repair itself.

Protection   Protective measures include:

1. BCG (Bacillus Calmette-Guérin) vaccination with weakened germs. This is a method of providing active immunity. Its value is controversial among doctors.

2. Proper health measures to prevent the spread of the germ. Examples: avoiding the use of common drinking cups or towels; covering a cough; no spitting; no kissing or handshaking with an infected person.

3. Pasteurization of milk. This is necessary because cows are susceptible to tuberculosis, and raw milk may contain the tubercle bacilli.

4. Proper rest and good nutrition. Overcrowding and slum conditions are favorite breeding grounds of the tuberculosis germ.

Typhoid Fever

Upon induction, all new members of the Armed Forces receive immunization against this disease. Impure water, milk, and food may bring the germ into the alimentary canal, where it multiplies and damages the intestines. It is carried out of the body with excretions, and may infect other people.

Detection   The *Widal test* is a useful method of diagnosis. A culture of typhoid germs is mixed with serum from a person's blood. If the person has typhoid, the agglutinins in his blood will cause the typhoid germs to clump together.

Protection and treatment   Protective measures include:

1. Vaccination with dead bacilli, which produces active immunity against the disease.
2. Inspection of food handlers.
3. Providing a good sewage system.
4. Proper purification of water.
5. Pasteurization of milk.
6. Not eating shellfish found in polluted water.
7. Inoculation of the sick person with the antibiotic chloromycetin.

Diphtheria

Until diphtheria was controlled, around the beginning of this century, it was a dread killer of children. The germs enter through the mouth and nose, and settle in the throat. They produce a thick membrane, which may close over the opening of the trachea and produce death by choking. Pierre Roux discovered in 1888 that the bacilli also produce a toxin that spreads throughout the body, causing high temperature, damaging tissues, and resulting in death. Emil von Behring showed in 1892 that antitoxins manufactured in the body of an animal such as a horse would protect children ill with diphtheria. This antitoxin in the serum neutralized the toxins produced by the germs.

Detection   The *Schick test* indicates whether a person is immune or susceptible to diphtheria. A small amount of the toxin is injected under the skin. If the person is immune, the area remains unaffected because the toxin is neutralized by the antitoxins in the body. A red spot indicates susceptibility.

Protection   Protective measures include:

1. A child already ill with diphtheria is inoculated with antitoxin. This confers passive immunity.
2. All infants are now inoculated with toxoid, which consists of toxin that has been weakened with chemicals. This stimulates the body to produce its own antitoxins, resulting in active immunity. As a result of this immunization, diphtheria has become a rare disease among treated children.

Tetanus

A person who sustains a puncture wound runs the risk of tetanus, or "lockjaw." The tetanus bacilli are anaerobic, and multiply deep within the injured area, away from oxygen of the air. Toxins are given off that cause the muscle to contract, and eventually result in death.

Immediate protection   Tetanus antitoxin is injected to neutralize toxins that might be produced by the tetanus bacilli.

Long-range protection   Toxoid is inoculated into healthy persons to confer active immunity.

Anthrax

Anthrax was a little-known disease affecting mostly cattle until it became a threat in 2001 as an agent of bioterrorism. Letters with a powdery substance containing spores of anthrax germs were delivered by mail to several prominent people, from an unknown source, in what became a nationwide scare. Five people died. Antibiotics were prescribed for those affected, but doctors cautioned about their indiscriminate use because of the risk that antibiotic-resistant strains of bacteria could develop.

DISEASES CAUSED BY VIRUSES

Viruses are ultramicroscopic: they were not visible until the perfection of the electron microscope. They do not grow on agar, as bacteria do, and must be cultured within living cells, such as those of fertile chicken eggs. They cause a number of diseases, for example, AIDS, smallpox, polio, rabies, yellow fever, the common cold, mumps, and measles. In 1977, the first successful use of a drug against a virus was announced. The drug, adenine arabinoside (ara-A), was used to treat cases of herpes virus encephalitis, a disease that destroys the brain. Antibiotics, such as penicillin, that conquer bacteria have not been effective against viruses. Viruses penetrate a cell and there replicate themselves. Ara-A is the first drug to enter a cell without harming it, as the drug attacked the virus.

Smallpox

Until the early part of the nineteenth century, smallpox was the cause of many deaths. Those who survived bore pockmarks on their faces. Dr. Jenner (1749–1823), an English physician, observed that milkmaids were immune to smallpox after an attack by cowpox, a similar but mild disease. He rubbed some of the material from the sores of a cow ill with cowpox into a slight scratch on a boy's arm. After developing a small sore, the boy was immune from smallpox. This was the first case of vaccination and the beginning of the conquest of the dread disease. In this process, the body develops active immunity against the mild virus of cowpox and the related powerful smallpox virus. Progress against smallpox has been so effective that in 1980 the World Health Organization declared it to be totally eradicated from the earth as an epidemic disease. Thus one of the world's worst scourges has finally been conquered.

Poliomyelitis

Infantile paralysis, or polio, mainly affects children, although adults may also contract this disease, as did Franklin D. Roosevelt before he became President. The three types of polio virus, Types 1, 2, and 3, are probably spread through the air, from person to person, although there is some evidence that they may also be spread through polluted water. Once in the body, the virus may damage the nerve cells of the brain and spinal cord that control the muscles. If the diaphragm and chest muscles are paralyzed, the patient must be kept breathing by means of an "iron lung" or respirator. Muscles of the arms and lungs stop functioning if the neurons that control them are damaged. A serious attack results in permanent crippling and possibly death.

It is believed that most adults have had a mild case of polio at one time or another, which produced symptoms like those of a cold. This probably resulted in the building up of antibodies, and may explain why most grownups are immune to the disease. Children do not have time to build up resistance to the polio virus, and as a consequence were the chief victims of the disease.

After many years of research, however, the tide against polio was finally turned, with the development of the Salk vaccine in 1953. The next year a mass inoculation experiment involving almost two million school children proved the effectiveness of the vaccine. It is now used as the chief means of protection against the disease. The vaccine perfected by Dr. Jonas E. Salk is composed of dead polio virus, which stimulates the production of antibodies when injected. A number of scientists, however, are convinced that the Sabin vaccine, consisting of live but weakened polio virus and taken by mouth, may be superior to the Salk vaccine in conferring active immunity.

Rabies

The bite of a mad dog (or other infected animal, such as a bat, wolf, or raccoon) may result in *hydrophobia*, commonly known as rabies. The germs present in the saliva of the animal enter the body through the broken skin, and find their way along the nerves to the brain and spinal cord. Death results unless the person is given treatment. The Pasteur treatment consists of a series of inoculations of weakened virus, each becoming progressively stronger, and stimulating the body to produce more and more antibodies against the virus. In recent years, a duck-embryo vaccine has been used instead of the Pasteur treatment, because it is safer and less painful.

An animal suspected of having rabies is usually isolated and observed for about 10 days. If it should die, its brain is examined for Negri bodies, the presence of which confirms the activity of the rabies virus. The incubation period of the virus varies from 2 weeks to 6 or more weeks, depending on how far the bite is from the brain.

Louis Pasteur developed the rabies vaccine by drying the spinal cord of infected animals. This treatment weakened the deadly virus so that it was no longer capable of causing the disease. However, it was able to stimulate the body to produce the valuable antibodies which gave protection against the disease. The rabies treatment was successfully used on a young boy for the first time in 1885.

A newly developed method of preventing the spread of rabies to humans is to make animals themselves immune to the disease. This is done by spreading food treated with a vaccine based on recombinant DNA. When animals eat this food, their bodies develop immunity to the rabies virus. This approach has been used successfully to control the spread of rabies among the coyotes in Texas, raccoons in New York, and red foxes in Belgium.

Hepatitis

Hepatitis is a virus infection of the liver. There are several forms of the disease. *Hepatitis A* is present in human feces and is spread through contamination of food and water. It causes a condition known as traveler's diarrhea, which threatens people who visit regions where there is inadequate sanitation. In 1995, a protective vaccine against hepatitis A was successfully developed.

Hepatitis B has been a serious threat to patients receiving a blood transfusion. The virus can be spread by sharing a toothbrush or razor. It has also been found in saliva, semen, and breast milk. This disease can now be controlled with a vaccine.

Recently, the drug alpha-interferon has been found to control the virus causing *hepatitis C,* the most common form of hepatitis. Interferon is a protein that the body makes in small quantities to fight virus infection. The present form of interferon was mass-produced by genetically altered bacteria (see Section 10.4).

AIDS

AIDS is a baffling, extremely serious disease. The acronym stands for "acquired immune deficiency syndrome"; the name refers to a severe breakdown in the body's immune system. A person with AIDS is vulnerable to a variety of infections and tumors that would normally be attacked by the body's white blood cells. As a result, the person becomes weak and dies.

Cause   The cause of AIDS is a virus. It was identified in 1984 by Dr. Luc Montagnier of the Pasteur Institute in Paris, who called it LAV-1, and by Dr. Robert C. Gallo of the National Cancer Institute, who named it HTLV-III. It is now known as HIV, or human immunodeficiency virus. It is believed that the AIDS virus kills a specialized type of white blood cell, called the helper T cell, which normally protects the body from infection by destroying foreign substances that enter it.

Transmission   The disease is believed to be passed through blood and semen, but not by casual contact, such as sneezing or using the same utensils. In the United States, certain groups of people have the greatest risk of getting the disease: male homosexuals, intravenous drug users who share contaminated needles, and, less commonly, recipients of contaminated blood transfusions. In Central Africa, where AIDS is prevalent, the disease is thought to be spread also by heterosexual contact. Heterosexual transmission is also becoming more common in the United States. Pregnant women who have AIDS transmit the disease to their newborn infants.

Protection   Considerable research is being conducted to develop a defense against the disease. In 1986, the drug *azidothymidine (AZT)* was used to interfere with the replication of the AIDS virus inside the body cells. A more recent advance has been the discovery that the virus produces an enzyme called *protease* that it needs to replicate. A new class of drugs called *protease inhibitors* has been developed to disable protease. The resulting virus is rendered defective and is not infectious. The latest treatment for AIDS now includes a *combination* of drugs consisting of protease inhibitors such as ritonavir or indinavir and AZT. This approach and many other similar approaches have been successful in reducing the amount of the AIDS virus present in the blood. It also reduced the death rate from the disease in New York City by 48 percent in 1997.

DISEASES SPREAD BY INSECTS AND TICKS

Serious diseases are transmitted by insects and ticks that inject germs into human beings as they suck blood from them. Thus they are quite different from the housefly, which may carry the germs of typhoid fever and tuberculosis on its feet and deposit them as it crawls over food.

Malaria

The female *Anopheles* mosquito receives the plasmodium, the protozoan cause of malaria, when it bites a malaria victim and sucks up some of his blood. In the mosquito's stomach, these plasmodia multiply by sexual reproduction, during which male- and female-type cells are produced. After these unite, the germs travel to the salivary glands of the mosquito.

The mosquito injects these germs the next time she bites a person, thus spreading the disease. Another phase of the plasmodium's life cycle now takes place. Each protozoan invades a red blood corpuscle and feeds on it. It multiplies rapidly until there are so many germs that the red corpuscle bursts. The protozoa then invade new corpuscles, and the process is repeated. Each time the red blood cells burst, toxins are released into the bloodstream, causing the person to have the characteristic chills and high fever of malaria. The destruction of so many blood cells also causes anemia.

Infected persons find relief by taking either quinine or some of the newer drugs, such as *atabrine, chloroquine,* or most recently, *mefloquine.* Researchers are trying to develop a vaccine to stimulate the body's immune system to produce antibodies that will kill the malaria germ.

The most effective methods of preventing malaria have been to control the *Anopheles* mosquito. Some ways of doing this are as follows:

1. Spray insecticides to kill the mosquito.
2. Drain swamps where mosquitoes breed.
3. Spread a thin layer of oil to suffocate the larvae "wrigglers," and pupae stages of the mosquito.
4. Stock ponds with small fish that feed on the mosquito larvae and pupae.
5. Remove rain barrels and other containers of standing water wherever mosquitoes may breed.
6. Screen windows and doors.

Yellow Fever

Another type of mosquito, the *Aedes*, was identified by Dr. Walter Reed at the end of the Spanish-American War as the carrier of the yellow-fever germ. Using human volunteers who allowed themselves to be bitten by the female mosquitoes, he conducted controlled experiments that definitely established this mosquito as the carrier of the germ. Dr. William Gorgas then waged a campaign of mosquito extermination that eliminated "yellow jack" from Panama, thereby permitting the successful construction of the Panama Canal. Previously, the French had been halted in their efforts to build a canal because of the enormous casualties inflicted by the *Aedes* mosquito.

Bubonic Plague

Fleas may carry the deadly *Bacillus pestis,* cause of the "black plague," among rats. When a person is bitten by one of these fleas, the germs are injected into his blood, thereby causing the disease. Ships docking in this country from the Orient are required to install large, circular shields on their hawsers to prevent any rats from coming ashore with germ-carrying fleas.

During the Middle Ages, approximately one-third of the population of Europe was wiped out by bubonic plague.

Typhus Fever

Typhus may be common in overcrowded, unsanitary living conditions where people do not change their clothing or wash very often. During World War II, it became a threat in bombed-out areas of Poland and Italy. The disease is spread by the infected body louse, which lives within the clothing and sucks the blood of its host. The germ is a *rickettsia*, an organism that is intermediate in size between a virus and a bacterium.

DDT powder dusted into the clothing halted a typhus epidemic when the American Army entered Naples during World War II. A vaccine composed of dead rickettsia has been used to confer active immunity. The antibiotic *chloromycetin* is useful against the rickettsias of both typhus and Rocky Mountain spotted fever. The latter disease is spread by a tick, a member of the spider group.

African Sleeping Sickness

Large regions of Africa have remained under-developed because of the prevalence of the tsetse fly, the carrier of African sleeping sickness. The germ is a protozoan, the *trypanosome,* which is transmitted by the fly when it bites a person.

The wild game of Africa, such as the various types of antelopes, are also attacked by the tsetse fly, and serve as a reservoir of the trypanosome.

Lyme Disease

Lyme disease received its name from the small town in Connecticut where an unusual number of cases among children were first observed in 1975. Since then, the disease has been reported in most of the states, especially in the Northeast and upper Midwest. It is also believed to exist in Europe, Asia, and Australia.

The disease is caused by a type of corkscrew-shaped bacterium called a spirochete (*Borrelis burgdorferi*), which is spread by the bite of a tiny deer tick about the size of a pinhead. The tick feeds on a number of animals, including deer, white-footed mice, and possibly birds.

How Spread	Disease Transmitted	Type of Germ
Anopheles mosquito	Malaria	Protozoan
Aedes mosquito	Yellow fever	Virus
Flea	Bubonic plague	Bacterium
Body louse	Typhus fever	Rickettsia
Tsetse fly	African sleeping sickness	Protozoan
Deer tick	Lyme disease	Bacterium

In most cases, the person who has picked up the tick in the woods, or even on the family lawn near a wooded area, develops a skin rash measuring over an inch in size, which has a bull's-eye pattern with a clear center. Later on, if the disease is untreated, there may be fever, headache, pain in the joints, arthritis, or cardiac disorders.

Lyme disease can be successfully treated with antibiotics, especially if detected early. As protection, people in suspected areas are advised to wear light-colored clothing with long sleeves, and pants that are tucked into high socks. Adults should examine themselves, children, and even pets immediately upon coming indoors.

If a tick is found attached to the skin, it should be carefully removed with fine-nosed tweezers, by grasping the head as close to the skin as possible and pulling straight out; the body should not be squeezed, since this could cause the tick to inject its fluid into the skin. An antiseptic should then be applied to the skin. A new vaccine was announced in 1997.

SEXUALLY TRANSMITTED DISEASES

Venereal diseases (VD) are largely spread through sexual intercourse between people who are infected with the germs that cause the diseases. The infection of two such diseases, syphilis and gonorrhea, generally takes place through the mucous membranes of the reproductive organs or the mouth. A pregnant woman who has one of these diseases may give birth to a deformed, blind, or dead baby. As a precaution, it is common practice in hospitals to treat the eyes of newborn babies with silver nitrate.

Syphilis

The cause of syphilis is a spirochete, a type of bacterium. Sores on the skin (chancres) are the first sign of disease. Eventually, brain damage and insanity result. Some states require a blood test (Wassermann test) before a marriage certificate is issued. Paul Ehrlich's successful search for a chemical compound—salvarsan—to fight syphilis was a first victory against the disease. Penicillin is the latest weapon against it.

Gonorrhea

This social disease is caused by the gonococcus bacteria. It is characterized by running sores on the membranes of the infected organs. Eventually arthritis and heart disease result. The sulfa drugs and penicillin are used to treat the disease.

Genital Herpes

The cause of this previously little known venereal disease is the herpes simplex virus, which is similar to the type that causes cold sores or fever blisters. The disease became more common as a result of the increase in sexual activity during the 1980s. Tiny painful blisters appear on the genital (sex) organs. When they burst, they pour out millions of infectious virus particles. In about 10 days, the sores heal but the virus retreats to nerves in the lower spinal cord to remain for life. Under conditions of stress, the blisters may return without warning. There is no cure at present.

Chlamydia

Another previously little known venereal disease is caused by a tiny bacterium that can reproduce only inside living cells. Chlamydia has become so widespread that it is now recognized as the commonest venereal disease. It infects the urinary and genital tracts and can lead to infertility. The lymph glands in the groin become swollen. The disease may lead to pneumonia in people with poor immune systems. Another complication is conjunctivitis, an eye inflammation. The disease is usually treated with the antibiotic tetracycline.

Prevention of Sexually Transmitted Diseases

According to the American Social Health Association, some methods of protection against sexually transmitted diseases are as follows: abstain from having sex; (for the male) use a condom; limit the amount of sexual activity; practice a high degree of selectivity in sex relations; maintain good personal hygiene. In view of the increasing rate of sexually transmitted disease among young people, it has been suggested that there be an extended sex education program for them.

DISEASES CAUSED BY FUNGI

Athlete's Foot

A parasitic fungus is responsible for this common foot condition. The spores may be picked up in swimming and gymnasium areas used by people with infected feet. The spores then germinate in the moist areas between the toes, causing irritation of the skin. The best protection is to dry the toes thoroughly with the aid of a foot powder. Chlorinated foot baths leading to swimming pools help to retard the spread of the spores.

Ringworm

Circular infected patches on the skin, or lesions of the scalp, may be caused by one or more species of fungus. This type of infection may be transmitted by wearing apparel or by contact with scales or hairs from lesions. General sanitary measures can prevent the infection from spreading.

DISEASES CAUSED BY WORMS

Trichina worm

Poorly cooked pork may contain tiny living roundworms known as trichina worms. When such food is eaten, the worms pass through the intestinal wall into the bloodstream. They travel to the voluntary muscles of the legs, arms, diaphragm, and face, where they form hard-walled cysts around themselves. This infection produces pain, fever, nausea, and often more serious effects. Prevention of *trichinosis* depends on proper cooking of pork products to at least 150°F in order to kill any worms that may be present. Some foreign countries, and some of our states, require that all garbage fed to pigs be precooked, in order to prevent them from getting the infection. Besides humans and pigs, other hosts of the trichina worm are rats and bears.

Hookworm

The tiny, round worms that cause hookworm disease may occur in the moist soil of rural regions of the South, wherever unsanitary conditions prevail. They enter the body by boring through the feet of people walking barefoot. Entering the bloodstream, they travel to the lungs, where they damage the tissue, and then pass up the bronchial tubes and the trachea to the throat. Here they are swallowed and finally reach the small intestine, where they hook onto the wall. They live on the blood of their victim and cause bleeding, producing anemia and listlessness. They reproduce rapidly, and their eggs pass out of the intestines with the wastes to infect the soil.

Treatment consists of feeding such drugs as tetrachlorethylene to remove the worms from the intestinal tract. Preventive measures include the wearing of shoes and the installation of proper sanitary disposal systems for human wastes.

Tapeworm

Undercooked beef or fish may contain tapeworm larvae. In the intestine, they attach themselves to the wall by means of hooks and suckers. As they feed on the food, they grow rapidly, sometimes reaching a length of 20 feet. They produce millions of eggs, which pass out of the body with the wastes, and may be swallowed by grazing cattle.

Treatment with drugs makes the tapeworm loosen its hold on the intestinal wall so that it can be excreted from the body. Careful meat inspection and thorough cooking of meat are important preventive measures.

HEART DISEASES AND CANCER—DISEASES NOT CAUSED BY GERMS

In 1900, the chief causes of death in this country were the infectious diseases. Since that time, these diseases have been so successfully studied and treated that they are no longer dreaded threats. Today, the chief causes of death are diseases not caused by germs—heart diseases and cancer.

Heart Diseases

As the leading cause of death, the heart diseases pose a challenge both to the research scientist trying to discover the predisposing causes, and to the average member of our fast-moving civilization to learn to live sensibly in view of currently known facts.

Prevention The following advice is being offered by more and more doctors as a precaution against heart troubles:

1. Limit the ingestion of fats to a maximum of 30 percent of the calorie intake. Recent evidence indicates that the deposit of *cholesterol*, a fatty substance, on the linings of the arteries (*atherosclerosis*) may lead to high blood pressure, and a greater chance of heart attack from the blocking of the coronary arteries that feed the heart. Saturated fats in butter, cream, and animal fats that remain solid at room temperature tend to raise the level of cholesterol in the blood. However, unsaturated fats such as liquid vegetable cooking oils tend to lower this level. Recent studies have shown that many teenagers have too high a cholesterol level in their blood, making them prime candidates for a heart attack later in life.

2. Quit smoking, cut down on the number of cigarettes smoked, or—best of all—do not start smoking in the first place. There is extensive evidence that smoking contributes to or accelerates the development of coronary heart disease. The American Heart Association has stated that death rates from coronary heart diseases in middle-aged men are 50–150 percent higher among heavy cigarette smokers than among nonsmokers.

3. Keep your weight down to normal. Extra weight produces a strain on the heart since blood must be pumped to and from the excess tissue. For every excess pound, the body builds nearly a mile of capillaries. This unnecessary load makes the heart work harder and over a period of time weakens it. A review of mortality records of one life insurance company showed that, for every 100 men of normal weight who died at the age of 45 years, 139 men who were definitely overweight died at the same age.

4. Exercise mildly every day. This helps to maintain muscle tone, lymphatic circulation, and good heart condition. Walking is considered good exercise. By contrast, strenuous "week-end" athletics may be harmful if a person is not accustomed to regular exercise. The heart specialist who treated former President Eisenhower after his heart attack, while still in office, stated that daily bicycle riding, swimming, or other mild form of exercise is essential for a healthy heart condition.

5. Relax and get plenty of rest. Do not develop the habit of worrying or becoming anxiety-ridden.

Types of heart diseases There are various conditions that produce heart disease.

1. *Coronary heart disease.* As a person grows older, the walls of the arteries lose their elasticity and become thicker. This "hardening of the arteries" (*arteriosclerosis*) causes an increase in the blood pressure. The coronary arteries that supply blood to the tissues of the heart itself are also affected. When this occurs, the heart muscle may not receive adequate blood, and severe pains of the chest may result during a heart attack.

2. *Coronary thrombosis.* A small blood clot (*thrombus*) may block one of the coronary arteries, thereby shutting off the supply of blood to a part of the heart and causing a heart attack, or *myocardial infarction*. This part of the heart dies, leaving scar tissue.

3. *Angina pectoris.* When the coronary arteries have become narrowed, the heart may not receive an adequate blood supply, leading to an oxygen shortage in the cardiac muscle. This may result in sharp pains radiating from the chest to the left shoulder and arm.

4. *Hypertension* (*high blood pressure*). Prolonged strenuous living without sufficient periods of relaxation may be one of the causes of hypertension, or high blood pressure. Other causes may be overweight and high salt use. The continuous strain on the heart may cause it to become enlarged, and finally to stop beating.

5. *Infections of the heart.* An attack of scarlet fever or a streptococcus infection in young children may result in *rheumatic fever*, causing a weakening of the valves of the heart and pains in the joints. Such children may need special care for many years before they can outgrow the condition. Infections of the heart lining may also occur among people who are suffering from venereal disease.

Cancer

Cancer is now the second leading cause of death. Its cause is unknown and has been one of the baffling mysteries confronting medical research. It is a wild growth of cells that starts in any part of the body. A *benign tumor* is an abnormal growth that stops growing and is harmless. A *malignant tumor* is a harmful growth that continues to grow and invades surrounding tissues and organs. *Metastases* are cancer cells that spread to other parts of the body by means of the lymph and circulatory systems, and continue to multiply there. At first, a cancerous growth may be undetected or may appear as a painless lump. It is for this reason that an educational campaign has been sponsored by the American Cancer Society to alert everyone to certain symptoms that should be reported to a doctor at once.

Seven danger signals of cancer None of the following should be ignored:

1. A persistent sore that does not heal.
2. A lump or thickening, especially in the lip or breast.
3. Irregular bleeding from any of the body openings.
4. A change in the size or color of a wart or mole.
5. Persistent indigestion.
6. Continuous hoarseness or difficulty in swallowing.
7. Marked changes in normal bowel habits.

Carcinogens Cancer-causing agents are known as *carcinogens*. They may be chemical substances, radiations such as X rays or sunlight, and possibly viruses. About two dozen or more chemical substances are known to cause cancer.

The first record of carcinogen dates back to Sir Percival Potts in England some 200 years ago. He studied the relatively high rate of scrotum cancer among chimney sweeps, and concluded that their condition resulted from exposure to *soot* in chimneys.

Other substances now suspected of being carcinogenic are *asbestos*, especially among workers in the asbestos industry, as a cause of cancer of the lungs, stomach, colon, and rectum; *trichloroetheylene*, a degreasing agent used in dry-cleaning and metalworking industries; *dioxins* in herbicides such as 2,4,5-T (phenoxacid) and Agent Orange, used to defoliate trees in the Vietnam War (1965–1971); and *aflatoxin*, produced by a fungus infesting moist peanuts, grains, and vegetables, as a cause of liver cancer.

How carcinogens cause normal cells to become cancerous is being studied in relation to DNA, the genetic material of living things. It is believed that cancer-causing chemicals bind to the DNA molecule and distort it, leading to errors in its replication. As a result, the genetic instructions for the new generations of cells may be changed, resulting in the uncontrolled growth of cancer cells. It is now believed that most carcinogens produce *oxygen radicals*; that is, these molecules contain oxygen, have an extra electron, and react actively with DNA. If oxygen radicals bind to the DNA molecule, the function of DNA may be changed, leading to cancer cell formation. Fortunately, there are antioxidant compounds that serve as strong anticarcinogens because they deactivate the oxygen radicals. Some examples are carotenoids found in many green and yellow vegetables, vitamin E, ascorbic acid, and selenium.

Tests for cancer Most cancers in their earliest stages do not cause pain and so go undetected. Fortunately, a number of specific procedures have been developed that help to reveal whether cancer is present in a particular organ. If treated early, the disease can be cured more readily.

1. A *mammogram* is an X ray of a woman's breast that can often reveal, at an early stage, a lump that may be cancerous.

2. *Breast self-examination* by a woman herself can sometimes detect the presence of a lump that may be cancerous.

3. A *Pap (Papanicolau) test* is performed by a doctor, who painlessly removes a few surface cells from a woman's cervix with a swab. The cells are then examined with a microscope to determine whether they are normal, precancerous, or cancerous.

4. A *sigmoidoscopy* or *colonoscopy* helps a doctor look for cancer in the colon and rectum, using a thin, lighted instrument.

5. A *guaiac stool (hemoccult) test* can be used to find unseen blood in stool samples, using a simple kit supplied by a doctor to be used at home. A positive result may indicate cancer of the colon.

6. Prostate cancer often develops in men after the age of 50. A doctor can conduct a *digital rectal exam (DRE)* to feel for the presence of abnormal lumps in the prostate. A *prostate specific antigen (PSA) test* is also available. If there are unusual findings in either or both of these tests, a *transrectal ultrasound test* can scan the prostate for possible tumors.

7. Bladder cancer may be detected by *examining cells* that are shed normally in the urine for the presence of a mutated form of a particular gene called p53.

Treatment of cancer If detected early enough, cancer can usually be cured. Much progress has been made in early detection and treatment. Every year, about 40,000 Americans are now saved who would have died of cancer 15 years ago. Four out of every ten cancer patients are now being cured. The effective methods of treatment are (1) surgery; (2) radiation from X rays or radioactive elements; and (3) chemotherapy.

Surgery involves the removal of the cancerous tissues. If a surgeon is undecided about the malignancy of the tissue, a *biopsy* is performed. While the patient is on the operating table, a small piece of tissue is removed and quickly examined microscopically by a pathologist. If typical cancer cells are present, the surgeon is notified within a matter of minutes and the operation proceeds.

Cancer cells are more sensitive than normal cells to *radiations* of X ray, radium, or radioactive isotopes, and can be destroyed by them without harm to the rest of the body. The amount of the dosage and the time of exposure are carefully controlled. In some cases, radioactive cobalt is injected directly into cancerous tissue. Cancer of the thyroid gland is treated by having the patient drink radioactive iodine. The concentration of this iodine in the thyroid gland serves as a source of destructive rays to kill the cancerous cells.

Chemotherapy involves the use of drugs that destroy the cancerous cells. These chemicals are so powerful that they often produce uncomfortable side effects (nausea, vomiting, diarrhea, hair loss) that are largely temporary. During treatment, patients are usually able to continue with their normal routines.

Laetrile is another name for the chemical amygdalin, which is found naturally in bitter almonds and the kernels of apricots, peaches, and plums. Some people believe it to be effective in curing cancer. However, little or no evidence has thus far been found to support such claims.

Importance of diet According to the American Cancer Society, diet can play a role in reducing the risk of cancer. It links high-fat diets to cancers of the colon, breast, and prostate, and recommends cutting fat consumption by 30 percent by reducing the intake of fatty meats, whole milk products, butter, cooking oils, and fats. It also urges a reduced intake of salt-cured, salt-pickled, and smoked foods. It recommends fruits, foods high in fiber, members of the cabbage family, other vegetables, and whole grains as offering protection against cancer.

Recent advances in the fight against cancer
Certain cancers seem to run in families. An example is breast cancer. Two genes, BRCA1 and BRCA2, have been identified in breast cancers. When mutations in these genes do occur, there appears to be a greater chance that cancer will develop. Colon cancer is another example, especially among Ashkenazi Jews. A gene for this condition has recently been identified.

Ultraviolet-B radiation in sunlight has been identified as a carcinogen for squamous-cell carcinoma, a common skin cancer. The deeply penetrating wavelengths of ultraviolet-B appear to damage a gene called p53, which normally prevents excessive cell growth. It is for this reason that p53 is known as a *tumor suppressor gene*. Without the protection of the p53 gene, squamous skin cells divide without check, causing a cancerous growth.

Scientists have found that a human cell growing in a test tube divides only about 50 to 100 times before dying. Each time it divides a small section at the end of the chromosomes, called a *telomere*, is lost. This continued process leads to the ultimate breakdown of chromosomes and the death of the cell. It is theorized that cancer cells produce a chemical, *telomerase*, which repairs the breakdown of the broken chromosomes, permitting these cells to keep on growing endlessly. A gene for telomerase production was discovered in 1997 in four kinds of tumors. One research approach is to develop a drug to shut down the telomerase repair system in cancer cells. However, some scientists are not convinced that the information about telomeres is correct.

Another advance has been the identification of cancer genes, or *oncogenes*. These are genes in the body that may be affected by radiation or other carcinogens and become activated to cause cancer. When an oncogene is thus affected, there may be a change in the sequence of nucleotides in its DNA, causing a change in its genetic message.

It has been found that melanoma, the most dangerous form of skin cancer, can be detected by a blood test. A protein called tumor antigen 90, or TA90, which shows up mostly in cancer tissue, can be traced by a simple blood test that checks for antibodies to TA90. If TA90 is present after the cancer has been surgically removed, it means the cancer still exists elsewhere in the body. In these cases further surgery, chemotherapy, or radiation may be necessary.

OTHER DISEASES NOT CAUSED BY GERMS
Nutritional Diseases

Some diseases are caused by a lack of proper amounts of nutrients in the body.

1. *Anemia.* Failure of the body to build hemoglobin, found in red corpuscles, because of a lack of iron.
2. *Beriberi*. A nervous disorder accompanied by loss of energy, due to lack of thiamin (vitamin B_1).
3. *Bleeding*. Failure of the blood to clot; may be caused by lack of vitamin K.
4. *Goiter* (endemic). A swelling of the thyroid gland brought on by lack of iodine in the diet.
5. *Night blindness*. Inability to see well in dim light, caused by lack of vitamin A, which is needed in the production of visual purple in the retina of the eye.
6. *Pellagra.* Inflammation of the skin and tongue, accompanied by mental breakdown, caused by lack of niacin.
7. *Rickets.* A deformity of the bones of children in which the bones remain soft, because of lack of either calcium or vitamin D, or both.
8. *Scurvy.* A condition of hemorrhages, bleeding gums, and swollen joints brought on by lack of ascorbic acid (vitamin C).
9. *Sterility.* Failure to produce young, caused, in rats, by lack of vitamin E.
10. *Tooth decay*. Recent evidence points to the lack of fluorine in the diet as a major cause; calcium, phosphorus, and vitamin D are also necessary for tooth formation.
11. *Xerophthalmia*. Drying condition of the eye and epithelial tissue, caused by lack of vitamin A.
12. *Genetic diseases*. These are discussed in Section 10.3.

Endocrine Diseases

Some diseases are caused by improper functioning of the endocrine glands.

1. *Cretinism.* A condition of mental and physical retardation in children, caused by lack of thyroxin.
2. *Myxedema.* Extreme mental and physical sluggishness and overweight in adults, brought on by a lack of thyroxin.
3. *Exophthalmic goiter.* A condition of nervous activity, protruding eyeballs, and underweight, brought on by excessive production of thyroxin.
4. *Rheumatoid arthritis*. A disease characterized by swollen, inflamed joints; one cause seems to be lack of cortisone and/or ACTH.
5. *Dwarfism.* Stunted growth, caused by an undersecretion of the hormone of the pituitary gland that controls the growth of the long bones.
6. *Gigantism.* Excessive growth, caused by an oversecretion of the hormone of the pituitary gland that controls the growth of the long bones.
7. *Acromegaly*. Abnormal enlargement of the hands, feet, and face in adults, caused by excessive secretions of the growth hormones of the pituitary gland.

8. *Addison's disease.* A condition in which the skin becomes bronze-colored, accompanied by loss of weight and weakness, caused by lack of cortin.

9. *Diabetes.* Inability of the body to utilize and store glucose, caused by lack of insulin.

MISCELLANEOUS DISEASES

Another group of diseases are caused in various ways.

1. *Allergies.* A variety of conditions, including asthma, hay fever, skin rashes, sneezing, and coughing, caused by sensitivity to various substances called *anti-gens.* An antigen is a foreign substance that enters the body and causes it to produce antibodies against the antigen. When the person is subsequently exposed to the antigen, an antigen-antibody reaction occurs, in which the white blood cells release histamines. These chemicals bring on the uncomfortable symptoms of the allergy. Various antihistamine drugs are now available to bring relief to the allergy sufferer.

2. *Hemophilia.* Bleeder's disease, which is hereditary and due to a recessive, sex-linked gene.

3. *Silicosis.* Occupational disease of miners caused by the collection of fine sand particles in the lungs.

4. *Radiation sickness.* Burns and weakness caused by overexposure to nuclear radiation. Death may result.

SUMMARY OF DISEASES

Disease	Cause	How Spread	Prevention	Treatment
AIDS	Virus	Homosexual and heterosexual transmission; contaminated needles; contaminated blood transfusion	Avoidance of causes; use of condoms; educational program	No cure at present; AZT and protease inhibitor
Anthrax	Bacillus	Contact with spores or infected animals	Avoidance of causes	Antibiotics; vaccine
Bubonic plague (black plague)	Bacillus	Bite of flea from infected rodent	Control of rats; vaccine	Antiplague serum
Diphtheria	Bacillus	Airborne; contact	Active immunization with toxoid; Schick test	Inoculation of antitoxin
Hepatitis	Virus	Contaminated food, water; contaminated blood transfusion; personal contact	Sanitary practices	Vaccine; alpha-interferon
Hookworm	Roundworm	Walking barefoot on infected soil	Wearing shoes; proper sanitary facilities	Drugs to eliminate the worm
Lyme disease	Spirochete	Bite of deer tick	Avoidance of tick areas; personal body scrutiny	Antibiotics; vaccine
Malaria	Plasmodium (protozoan)	Bite of infected female *Anopheles* mosquito	Insecticides and other methods of mosquito control	Quinine, chloroquine, atabrine, mefloquine
Measles	Virus	Airborne; contact	Vaccine of weakened virus; isolation	Bedrest
Pneumonia	Coccus or virus	Droplet infection	Maintenance of high resistance; vaccine	Penicillin, sulfa, immune serum

SUMMARY OF DISEASES (Continued)

Disease	Cause	How Spread	Prevention	Treatment
Poliomyelitis (infantile paralysis)	Virus	Suspected: contact, airborne, polluted water	Either Sabin vaccine—weakened live virus, or Salk vaccine—dead virus; isolation	Rest; muscle therapy
Rabies	Virus	Bite of rabid animal	Pasteur treatment or duck-embryo vaccine before symptoms appear; immunization of dogs with vaccine	No specific treatment; horse serum for immediate protection
Smallpox	Virus	Contact; airborne	Vaccination with weak virus; isolation	None
Tetanus	Bacillus	Puncture wound	Toxoid immunization	Antitoxin injection
Trichinosis	*Trichinella* worm (roundworm)	Eating poorly cooked pork	Thorough cooking of pork; feeding steamed garbage scraps to pigs	No specific treatment
Tuberculosis	Bacillus	Droplet infection; contact; contaminated food and articles; feet of housefly	Chest X ray; tuberculin test; sputum analysis; BCG vaccine; tuberculin testing of cows; milk pasteurization; good bodily health	Rest; good food; combination of streptomycin and isoniazid-rifampin; pneumothorax
Typhoid fever	Bacillus	Polluted water and contaminated food; feet of housefly; carrier (human)	Immunization with dead bacteria; water and food purification; inspection of food handlers; fly control	Chloromycetin
Typhus fever	Rickettsia	Bite of infected body louse	Body cleanliness; vaccine of dead virus	Chloromycetin
Whooping cough	Bacillus	Droplet infection; contact	Isolation of patients; active immunization with vaccine of dead bacilli	Convalescent serum; terramycin or aureomycin in early stages
Yellow fever	Virus	Bite of infected female *Aedes* mosquito	Insecticides and other methods of mosquito control; vaccine of weak virus	No specific treatment

Section Review

Select the correct choice to complete each of the following statements:

1. All of the following are types of germs *except* (A) bacteria (B) viruses (C) protozoa (D) toxins (E) worms

2. All of the following are useful methods of providing protection against tuberculosis *except* (A) pneumothorax (B) use of Schick text (C) BCG (D) rest (E) pasteurization of milk

3. A chest X ray is useful in detecting a case of (A) diphtheria (B) typhoid (C) tetanus (D) tuberculosis (E) typhus

4. Oysters grown in polluted water may contain the germs of (A) typhoid (B) typhus (C) smallpox (D) yellow fever (E) scurvy

5. A test for syphilis is the (A) Wassermann test (B) Widal test (C) Dick test (D) Schick test (E) Heaf test

6. Antitoxins against diphtheria are obtained from (A) the weakened virus (B) horses (C) a toxoid (D) *Penicillium notatum* (E) people who are naturally immune

7. A disease caused by bacteria is (A) smallpox (B) rabies (C) polio (D) the common cold (E) tetanus

8. A person being vaccinated against smallpox receives (A) live cowpox virus (B) dead cowpox virus (C) live smallpox virus (D) dead smallpox virus (E) weakened smallpox virus

9. A disease in which the nerves are damaged is (A) silicosis (B) tetanus (C) rheumatic fever (D) polio (E) yellow fever

10. Viruses can be seen with the aid of the (A) oil immersion microscope (B) synchotron (C) electron microscope (D) spectroscope (E) ultracentrifuge

11. A disease in which the brain is affected is (A) bubonic plague (B) pellagra (C) gonorrhea (D) rabies (E) xerophthalmia

12. The *Anopheles* mosquito spreads (A) typhoid (B) malaria (C) typhus (D) African sleeping sickness (E) amnesia

13. All of the following are effective ways to control mosquitoes *except* (A) spray insecticides (B) spray antitoxin (C) drain swamps (D) remove rain barrels (E) screen windows and doors

14. AIDS may be spread by all of the following *except* (A) a shared needle used in taking drugs (B) a contaminated blood transfusion (C) semen (D) a promiscuous male homosexual (E) casual contact

15. The black plague is transmitted by a (A) cockroach (B) mosquito (C) fly (D) flea (E) mad dog

16. Hepatitis C can be treated effectively with (A) liver injections (B) alpha-interferon (C) chloromycetin (D) quinine (E) AZT

17. Two diseases caused by protozoa are (A) malaria and yellow fever (B) malaria and African sleeping sickness (C) yellow fever and African sleeping sickness (D) malaria and pneumonia (E) yellow fever and pneumonia

18. All of the following diseases are caused by worms *except* (A) trichinosis (B) hookworm (C) tapeworm (D) ringworm (E) liver fluke

19. Eating undercooked pork may lead to (A) hookworm (B) trichinosis (C) pellagra (D) scurvy (E) athlete's foot

20. The leading cause of death at the present time is (A) tuberculosis (B) cancer (C) heart disease (D) myxedema (E) pneumonia

21. All of the following are advisable as precautions against heart disease *except* (A) exercise mildly every day (B) reduce smoking (C) increase ingestion of fats (D) relax (E) keep weight down to normal

22. Rheumatic fever may result   (A) from an attack of chicken pox   (B) from an attack of measles   (C) from a dislocated joint   (D) in weakening of the valves of the heart   (E) in weakening of the long bones

23. Effective methods of cancer treatment include the use of all of the following *except*   (A) cosmic rays   (B) X rays   (C) surgery   (D) radium   (E) radioactive isotopes

24. A contagious disease is   (A) anemia   (B) cretinism   (C) acromegaly   (D) cancer   (E) tuberculosis

25. Lyme disease is caused by the bite of a   (A) male deer   (B) female deer   (C) deer tick   (D) deer mouse   (E) deer fly

Answer Key

1-D	6-B	11-D	16-B	21-C
2-D	7-E	12-B	17-B	22-D
3-D	8-A	13-B	18-D	23-A
4-A	9-D	14-E	19-B	24-E
5-A	10-C	15-D	20-C	25-C

Answers Explained

1. (D) Toxins are harmful chemical substances given off by bacteria, such as the tetanus and the botulism bacilli.

2. (B) The Schick test indicates whether or not a person is immune to diphtheria.

3. (D) A chest X ray can reveal the presence of tubercles in the lungs.

4. (A) Shellfish, such as oysters, growing in polluted water may contain bacteria that cause typhoid.

5. (A) The Wassermann test is a blood test that helps detect the presence of syphilis, a serious venereal disease that can affect the health of not only the infected person but also that of any children born to that person. Some states require the Wassermann test as a requirement for the issuance of a marriage license.

6. (B) Von Behring in 1892 first showed that antitoxins manufactured in the body of a horse could protect children ill with diphtheria.

7. (E) Tetanus is caused by bacilli. The other diseases mentioned in the question are caused by viruses.

8. (A) A person vaccinated against smallpox receives live cowpox virus and develops active immunity against the relatively mild virus of cowpox, as well as the related powerful smallpox virus.

9. (D) The polio virus may damage the nerve cells of the brain and spinal cord that control the muscles, leading to paralysis of part of the body.

10. (C) Viruses are ultramicroscopic and were not visible until the perfection of the electron microscope.

11. (D) The bite of a mad dog, or other infected animal, may cause rabies. The germs present in the saliva of the animal enter the body through the broken skin and find their way along the nerves to the brain and spinal cord.

12. (B) The female *Anopheles* mosquito spreads malaria. The mosquito picks up the protozoan (plasmodium) cause of malaria when it bites a malaria victim and sucks up some of his blood. When that mosquito bites the next person, it injects these germs, spreading the disease.

13. (B) The most effective way of preventing malaria from being spread is to control the breeding of mosquitoes. Spraying antitoxin would not be effective against mosquitoes.

14. (E) AIDS is not spread by such examples of casual contact as sneezing or sharing utensils.

15. (D) Fleas transmit the black plague when they bite a person and inject the germs into her blood.

16. (B) Alpha-interferon is a protein that is produced by the body to fight a virus infection. It is now being mass-produced by bacteria specially altered through genetic engineering.

17. (B) Malaria is caused by a plasmodium, a type of protozoan spread by the bite of an infected *Anopheles* mosquito. African sleeping sick-

ness is caused by a type of protozoan called a trypanosome and is spread by the bite of an infected tsetse fly.

18. (D) Ringworm is an infection of the skin caused by a parasitic fungus.

19. (B) Poorly cooked pork may contain tiny trichina worms. When such food is eaten, the worms pass through the intestinal wall into the bloodstream. They travel to the voluntary muscles of the legs, arms, diaphragm, and face, where they form cysts around themselves. This infection produces pain, fever, and nausea.

20. (C) The leading cause of death in the United States today is heart disease, a disease not caused by germs. By comparison, in 1900 the chief causes of death were infectious diseases caused by germs.

21. (C) As a precaution against heart disease the intake of fats should be reduced to a maximum of 30 percent of the calorie intake to reduce the risk of high cholesterol levels. Deposits of cholesterol, a fatty substance, on the lining of the arteries may lead to a fatal blockage of the coronary arteries that feed the heart.

22. (D) A streptococcus infection in young children may cause rheumatic fever, leading to a weakening of the valves of the heart and pains in the joints.

23. (A) Cosmic rays come to the earth from outer space. They are not used to treat cancer.

24. (E) Tuberculosis is a contagious disease caused by the tubercle bacillus, a rod-shaped type of bacteria.

25. (C) The deer tick carries the germ, a spirochete, which it injects into the blood when it bites a person. Since the deer tick is very tiny, only about the size of a pinhead, it can easily be overlooked by the affected person, who then develops the symptoms of Lyme disease.

HOW LIVING THINGS REPRODUCE

CHAPTER

9

9.1 REPRODUCTION AMONG LOWER ORGANISMS

One of the properties of protoplasm is its ability to reproduce more of its kind. In one-celled organisms, the entire cell reproduces. In higher organisms there are specific structures that specialize in reproducing the individual.

Spontaneous Generation

At one time it was believed that living things could arise from nonliving matter. Here are some examples of this idea: Frogs and mice originate from mud; threads of cotton turn into worms; flies develop from decaying meat. This belief in spontaneous generation was disproved for the first time by Francesco Redi during the seventeenth century. He conducted the following experiment to determine whether flies could develop from decaying meat:

1. He allowed meat to decay in each of three jars. He left the first jar opened, covered the second jar with gauze, and sealed the third with parchment.
2. For several weeks, he observed each jar. In the first jar, flies flew in and out, and maggots appeared on the meat. In the second jar, flies flew around the gauze and formed their maggots on top of it, but there were no flies inside. No flies were present around the third jar.

3. He decided that, in the first jar, flies were attracted by the odor of decaying meat and reproduced on it. In the second jar, they could not reach the meat and reproduce on it. The sealed third jar did not attract them at all. Therefore, he concluded, flies originate from other flies, and not from decaying meat.

Later, the discovery of bacteria by Antonie van Leeuwenhoek led to the belief that these microorganisms were simple enough to originate spontaneously from favorable food conditions. Lazarro Spallanzani in the eighteenth century, and Louis Pasteur and John Tyndall in the nineteenth century, conducted conclusive experiments to disprove this idea. Spallanzani and Pasteur showed that there were no bacteria in sealed flasks of broth that had been heated, but that, when the flasks were opened, bacteria of the air entered and multiplied. Pasteur drew out the necks of flasks in a pronounced curve, and left the ends open; bacteria had no way of entering, and the broth remained sterile. It was definitely decided that living things can come only from previously existing living things.

Although there is no spontaneous generation of present forms of living things, scientists postulate that the simplest form of life on earth probably arose billions of years ago when conditions were different, from combinations of complex molecules, and gradually evolved into life as we now know it.

Asexual Reproduction

Reproduction in which there is only one parent is known as asexual reproduction. Examples are (1) fission, (2) budding, and (3) spore formation.

Fission

After a one-celled organism, such as an ameba, grows to its maximum size, it reproduces as follows: The nucleus divides in two; equal amounts of cytoplasm collect around each nucleus; the cell splits into two daughter cells that are equal in size. Each cell is independent; it grows to its maximum size, then splits again by binary fission. Other organisms that reproduce in this way are the paramecium, pleurococcus, spirogyra, and bacterium. In the paramecium, both the macronucleus and the micronucleus divide in half and are distributed to each of the two cells.

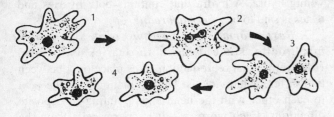

Fission in the Ameba

Budding

Yeast cells are microscopic fungi that are useful in the baking and brewing industries, because they carry on *fermentation* of sugar. During this process, they produce carbon dioxide and alcohol. When a yeast cell reproduces, it forms a *bud*; the nucleus splits; a small amount of cytoplasm gathers around one of the nuclei; and a cell wall separates it to form a little cell attached to the large one. The two cells are unequal in size. The bud remains attached to the larger cell. In time, it will grow and produce other buds.

The small animal, hydra, also reproduces by budding. When it is growing vigorously, a projection forms on its outer layer of cells, elongates, develops tentacles and a mouth, and remains attached to the parent animal. After a while, cells grow across its base and separate it from the parent.

Spore Formation

If bread is kept in a moist, dark place, mold usually develops on it, unless the bread was made with calcium propionate, a chemical that retards the growth of molds. The bread mold (*Rhizopus*) consists of a mass of threads called the *mycelium*. Under the low power of a microscope, it can be seen to be made of stemlike structures called *hyphae*. At the end of a hypha, a ball-like swelling, the *sporangium*, or spore case, forms. As it

develops, the nucleus divides many times. Each one of the nuclei becomes surrounded by a small amount of cytoplasm and a tough cell wall. Hundreds of these tiny cells, or *spores*, are produced, and when the wall of the sporangium bursts, they are scattered about. Under favorable conditions of moisture, food, and subdued light, each spore can germinate to form a new mold. Until then, the spores are resistant to extremes of temperature and to drying. They have been recovered from the stratosphere, more than 7 miles up, and have germinated successfully when brought down and incubated.

Other fungi, including yeast, mushrooms, and wheat rust, also reproduce by *sporulation*, or spore formation. In the blue-green mold from which penicillin is made (*Penicillium*), the spores are produced in little chains at the ends of fingerlike projections from the hypha. Mosses and ferns also produce spores in the asexual phase of their life history. In mosses, the spores are found in capsules at the end of separate stalks; in ferns, they are usually located in the small round spots on the underside of the leaves.

Regeneration

If a hydra loses some of its tentacles, it grows new ones. If it is cut into several pieces, most of the pieces will grow into complete hydras. This ability to grow back a missing part, or to develop an entire organism from a part, is known as *regeneration*. A crab or a lobster can grow back a new claw. If a starfish is cut into several pieces, each piece can regenerate into a new organism, especially if it contains parts of the central disc. Many experiments in regeneration have been performed with the small flatworm *Planaria*, which is about ¼ inch long. If cut in half, it will regenerate into two new worms. Interesting results may be obtained by cutting it in different ways. In vertebrates, the power of regeneration is reduced. A salamander or a lizard can grow back a new tail. A human being can regenerate new skin in a cut, but cannot replace a limb, or part of one.

Sexual Reproduction

Two parents, or two parent cells, are involved in sexual reproduction. Two cells or gametes unite to form a single cell, the zygote, which develops into a new individual. The two gametes are alike in the type of sexual reproduction called *conjugation*; they are unlike in *fertilization*.

Conjugation

Conjugation may be studied in the spirogyra, bread mold, and paramecium.

1. *Spirogyra.* The cells of a spirogyra are arranged in a long thread or filament. Each cell reproduces asexually by binary fission, resulting in lengthening of the filaments. In the fall of the year, as the filaments lie parallel to each other in the pond, they undergo conjugation. Projections grow out toward each other from the cells of adjoining filaments. When they touch, their ends are dissolved by enzymes, forming a connecting tube between the cells. Now all the protoplasm of the cells in one filament moves through the tubes into the adjoining cells, and unites with them. The cells that move across are called *active gametes*; those that remain stationary, *passive gametes*; they are alike in size and shape. A single cell, the *zygote*, results from the union of the two gametes. A tough, resistant wall forms around it, and it is now a *zygospore*. Under the microscope, the two adjoining filaments can be seen with connecting tubes between the adjoining cells; one filament has only empty cells, while the other contains dense zygospores. During the winter, the protoplasm remains protected within the heavy covering of the zygospores. In the spring, the wall of each breaks open, and a new cell emerges; it grows and reproduces asexually by binary fission to become a long filament. Scientists have succeeded in making conjugation proceed artificially by varying the concentration of chemicals in the water.

2. *Bread mold.* Besides forming spores asexually, bread mold may also reproduce by conjugation if two strains, called *plus* and *minus*, are growing near each other. A thread from one strain will grow toward a thread from the other. When they touch, a cell is formed at each end; these cells are the *gametes*; they are alike in size and shape. The separating wall dissolves, and the protoplasm of both gamete unites to form a *zygote*. A zygospore results when a heavy wall grows around it. After a period of time, the wall of the zygospore breaks and a new hypha grows from it; it forms a sporangium and produces spores asexually. Conjugation occurs only when there are two different strains, a plus and a minus; it will not take place between two like strains.

3. *Paramecium.* During conjugation, two paramecia lie next to each other, and unite at their oral groove regions. Their macronuclei disintegrate, and their micronuclei undergo a series of complicated changes. A portion of each micronucleus goes into the adjoining paramecium and fuses with the opposite micronucleus. The paramecia then separate and divide actively by binary fission. The exchanged portion of the nucleus serves as a gamete. The resulting organisms contain nuclear material from the two original paramecia.

Fertilization

In fertilization, the uniting gametes are different from each other in size, shape, and activity. Fertilization is the chief method of sexual reproduction among animals and plants.

1. *Hydra.* In addition to reproducing asexually by budding, the hydra reproduces sexually, usually in the fall. A large bulge forms on the body wall; it is called the ovary and will contain one large cell, the *egg*, which is well supplied with food. Another bulge on the body wall of either the same animal or another one, called the testis, forms and contains many small *sperm* cells. The egg is the female gamete or sex cell; the sperm is the male gamete or sex cell. The testis opens and releases the sperm cells into the water. One of them enters the egg and unites with it to form the *zygote*, or *fertilized egg*. This divides a number of times to form an embryo, or developing hydra. It drops away from the parent hydra and forms a hard membrane. In the spring, the embryo completes its development into an independent young hydra. A hydra that contains both ovaries and testes is said to be a *hermaphrodite*.

2. *Earthworm.* The earthworm is also a hermaphrodite, since it has both ovaries and testes. However, sexual reproduction takes place only between two different individuals. During mating, they lie next to each other with the head ends pointing in opposite directions. Each worm receives sperm cells from the other in special sacs called sperm receptacles. After this exchange, the worms separate. A swelling near the middle of the worm secretes a ring of mucus, which glides toward the front of the worm. As it moves over the opening containing the eggs, a number of eggs pass into it; farther along, sperm cells come out. The eggs are fertilized by the sperm in this mucous ring, or cocoon, which finally comes off the worm at the head end and is left in the soil. The fertilized eggs or zygotes develop into young earthworms, which move about and then leave the cocoon.

3. *Insects.* Insects are either male and have testes, or female with ovaries. During mating, the male introduces sperm into a sac, called the sperm receptacle, in the female's body, where they are stored until egg-laying time. As the eggs pass by this sac on the way out of the female insect's body, they are fertilized by the sperms. Each egg has a large amount of stored food, on which the developing embryo feeds. Some insects, such as the butterfly, go through four stages in their life history; egg → larva (caterpillar) → pupa (chrysalis) → adult. This is known as complete metamorphosis. Others, such as the grasshopper, have only three stages: egg → nymph → adult. This is known as incomplete metamorphosis.

4. *Mosses.* Lower plants also reproduce by fertilization. Mosses develop from asexual spores that are produced in spore cases. Each spore germinates on the

moist ground and produces a small green filament (*protonema*). This becomes the moss plant. Egg cases, each containing an egg, are formed on the upper part of the plant. Sperms are produced nearby, on the same plant or on other plants, and move through the moisture to the egg cases, where they fertilize the eggs. From the fertilized egg, a new "plant" develops, with a stalk and a spore case, deriving its nourishment from the original moss plant beneath it. There are thus two stages in the life cycle of the moss: the spore-making stage, or *sporophyte*, which is asexual and consists of the stalk and spore case; and the gamete-producing stage, or *gametophyte*, which is the moss plant and in which fertilization occurs. This alternation of a sporophyte phase and a gametophyte phase is known as *alternation of generations*.

5. *Ferns*. In ferns, an alternation of generations also occurs. Spores are produced asexually on the undersurfaces of the leaves, in rows of little spots. Each of these spots contains a number of *sporangia*, or spore cases. When these are ripe, they burst, scattering the spores. A spore will germinate in a moist place and, instead of developing into a new fern, forms a small, heart-shaped plant, the *prothallus*. This is the gametophyte; it is less than $1/2$ inch in size and can carry on photosynthesis. It forms egg cases with egg cells in them and sperm cells. The sperms swim in the moisture to the egg cases, and one sperm will unite with an egg to form a fertilized egg. The new fern plant, or sporophyte, develops from this fertilized egg. It forms leaves and then produces more spores asexually. Thus, the spore-forming stage, or fern plant, is asexual; it alternates with the gamete-forming stage, or prothallus, which is the sexual phase of the fern's life history.

Section Review

Select the correct choice to complete each of the following statements:

1. All of the following scientists helped disprove the idea of spontaneous generation *except* (A) Redi (B) Spallanzani (C) Trudeau (D) Pasteur (E) Tyndall

2. Most protozoa reproduce by (A) budding (B) fission (C) sporulation (D) parthenogenesis (E) regeneration

3. Two organisms that reproduce by budding are the (A) hydra and yeast (B) spirogyra and yeast (C) hydra and spirogyra (D) pleurococcus and yeast (E) hydra and pleurococcus

4. All of the following organisms produce spores *except* (A) bread mold (B) *Penicillium* (C) paramecium (D) moss (E) fern

5. A lobster can grow back a new claw by the process of (A) sporulation (B) metamorphosis (C) parthenogenesis (D) budding (E) regeneration

6. The uniting cells in sexual reproduction are called (A) zygotes (B) zygospores (C) spores (D) gametes (E) buds

7. Two organisms that reproduce by the process of conjugation are the (A) spirogyra and ameba (B) paramecium and ameba (C) spirogyra and bread mold (D) ameba and bread mold (E) ameba and yeast

8. A zygote is a cell that (A) results from the union of two gametes (B) unites with another zygote to form a gamete (C) unites with another zygote to form a zygospore (D) results from the union of two zygospores (E) results from the union of a gamete and a zygospore

9. The simplest method of reproduction is (A) conjugation (B) fertilization (C) fission (D) sporulation (E) parthenogenesis

10. During the life history of a fern, the fern plant is (A) the sporophyte stage (B) the gametophyte stage (C) the prothallus stage (D) the structure that produces egg cells and sperms (E) developed from a spore

Answer Key

1-C	3-A	5-E	7-C	9-C
2-B	4-C	6-D	8-A	10-A

Answers Explained

1. (C) Edward Trudeau developed a treatment of tuberculosis that includes rest and good nutrition for a patient at a sanitarium. He was not involved in research on the theory of spontaneous generation.

2. (B) Most protozoa divide in half to produce two equal cells, a method of reproduction called fission.

3. (A) Yeast reproduces by forming a little cell, called a bud, attached to the original cell. Hydra reproduces by budding when it forms a small hydra attached to the original animal.

4. (C) Paramecium reproduces by fission, in which it splits in half, or by conjugation, in which two paramecia lie next to each other and exchange portions of their micronuclei. Paramecia do not form spores.

5. (E) Regeneration is the ability to grow back a missing part or to develop an entire organism from a body part. Lobsters, starfish, and some other animals exhibit regeneration.

6. (D) Cells that unite to form a single cell, the zygote, are called gametes. In conjugation, the two uniting gametes are similar. In fertilization, they are different from each other in size, shape, and activity (e.g., in humans, the male gamete—a small, comma-shaped motile spermatozoan—unites with the female gamete—a comparatively large, ovoid, nonmotile ovum, or egg).

7. (C) In bread, mold, and spirogyra, similar gametes unite to form a zygote in the process of conjugation. A zygospore results when a heavy wall grows around the zygote.

8. (A) A zygote is a single cell that results from the union of two gametes. The zygote is also known as the fertilized egg.

9. (C) During fission, an organism, such as an ameba, splits into two cells that are equal in size. Fission is the simplest method of reproduction.

10. (A) A fern develops from a fertilized egg on the prothallus. The new fern plant produces many spores asexually, and thus is the sporophyte stage of the life cycle.

9.2 REPRODUCTION IN HIGHER PLANTS

Flowers are the specialized organs of reproduction among the higher plants, or angiosperms. These plants are classified into various families largely on the basis of the structure and arrangement of the flowers.

Reproduction by Means of Flowers
Structure of a Flower

Although different flowers vary in their appearance and structure, they tend to have the same basic parts:

1. *Petals.* These are the brightly colored parts that make up the most attractive feature of flowers. The group of petals in their circular arrangement or whorl is called the *corolla*.

2. *Sepals.* These are the small, green, leaflike structures that form an outer layer at the base of the petals. At one time, they enclosed the flower bud. Together, they make up the *calyx*.

3. *Stamens.* Within the petals are many stamens, each of which consists of a slender stalk, the filament; and the enlarged part on top, the *anther*, or pollen case, which contains the powdery pollen.

4. *Pistil.* Located in the center of the flower, within the stamens, the pistil contains an enlarged portion at the bottom, the *ovary*. Extending above is the slender *style*, at the top of which is the sticky *stigma*. The ovary contains one or more small structures, the *ovules*. When the flower matures, these ovules will become the seeds; the surrounding ovary will become the fruit.

The stamens and pistils are the male and female organs and are referred to as the essential organs of the flower, since they are directly concerned with reproduction. In some plants they are located in separate flowers. Thus, in corn, the stamens are borne in *staminate* flowers at the top of the plant, and make up the *tassel*. The pistils are found farther down on the plant, in *pistillate* flowers, making up the ear; the mass of their stigmas and styles makes up the *silk*. In the poplar family, which includes poplar and willow trees, the staminate and pistillate flowers are formed on separate trees, in structures called *catkins*. However, most plants have flowers that contain both stamens and pistils and are said to be *perfect* flowers.

Pollination

The transfer of pollen from the stamen to the stigma is known as pollination. If this takes place within the same flower, as in the pea or bean, it is called *self-pollination*;

if pollen is transferred from the anther of one flower to the stigma of another flower, it is called *cross-pollination.* Cross-pollination may be accomplished by various agents, that is, insects (bees, especially important in cross-pollinating apple trees); wind (corn and other grasses; gymnosperms); and human beings (to improve varieties of flowers—this is also known as *artificial pollination*). Insects, as well as the hummingbird, are attracted to flowers to obtain the sugary liquid called *nectar* formed at the base of the flower. In this process of food getting, the pollen that becomes deposited on the insects' bodies is distributed to other flowers.

Fertilization

Pollination is the first step in the fertilization of a flower. When the pollen grains are deposited on the stigma, they each begin to germinate, and form a *pollen tube.* A pollen grain is a single cell; after it has matured in the anther, it usually contains two nuclei. Now, as the pollen tube grows down through the style toward the ovule, one of these nuclei divides to form two nuclei. These two are the *sperm nuclei,* or male gametes; the other is the *tube nucleus.*

Within each ovule, one of the cells enlarges and is called the *embryo sac.* Its nucleus divides several times to form eight nuclei; one is the *egg nucleus;* two others unite to become the *endosperm nucleus;* the other nuclei are not significant.

As the pollen tube reaches an ovule, it enters through a tiny opening called the *micropyle,* and the sperm nuclei pass into the embryo sac. Double fertilization now takes place. One of the sperm nuclei unites with the egg nucleus to form a *fertilized* egg, or *zygote.* The other sperm nucleus unites with the *endosperm nucleus.* The tube nucleus disintegrates.

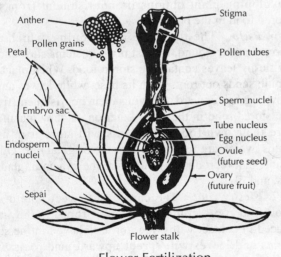

Flower Fertilization

Formation of the Seed

Following double fertilization, the ovule becomes a seed. The fertilized egg undergoes numerous divisions to form an embryo, or baby plant. At the same time, the fertilized endosperm nucleus divides to form many large cells in which food is stored for use by the embryo later on, when the seed germinates. The outer cells of the ovule surrounding the embryo sac become the covering layer of the seed, or *testa.*

Parts of a Seed

These parts can be seen by studying a large seed, such as the lima bean. It should first be soaked in water for several hours. On the outside of the testa, along the narrow end of the seed, a scar, or *hilum,* is easily seen; it shows the point of attachment of the seed to the ovary wall, or the pod. Near it, the tiny opening is the micropyle, through which the pollen tube entered the ovule. If the bean is split open, it can be seen to have two large *cotyledons,* each of which contains stored food. The remaining part of the embryo contains the *epicotyl* and its tiny plumules, which make up the upper part of the stem and leaves of the plant, respectively, and the *hypocotyl,* which will develop into the lower part of the stem and the roots.

Formation of the Fruit

After fertilization, while the ovule is maturing into a seed, several other changes are taking place in the flower. The petals, sepals, stamens, style, and stigma shrivel and usually drop off. The ovary enlarges and becomes the fruit. In the apple, additional fleshy tissue is added to the ovary, providing the edible part. The dried-up remains of the sepals, styles, and stamens can be seen at the bottom of an apple. A pea pod is also a fruit, since it is a matured ovary and contains seeds; the remains of the sepals can usually be seen at one end of the pod, and the dried-up style and stigma at the other. The tomato, cucumber, and squash are also fruits, having developed from the ovary of a flower. A kernel of corn is considered to be a fruit containing one seed, since it developed from an ovary, in which there was one ovule; an ear of corn is thus a collection of many fruits.

In order for an ovule to become a seed and for an ovary to mature into a fruit, pollination and fertilization must take place. Unless the male and female gametes unite, a fruit does not "set," and the seeds do not develop. Recent research with chemical growth-promoting substances, however, has shown that ovaries can be made to develop into fruits artificially. *Gibberellin* and *auxins* are two such substances. When they are sprayed on a young tomato flower, the ovary grows and develops into a seedless fruit.

Seed Dispersal

Plants have various adaptations by which seeds can be scattered or dispersed.

1. Fleshy fruits such as apples, cherries, and watermelons are eaten by humans, birds, and other animals, and the seeds then discarded far from the original plant.

2. The wind scatters the seeds of the dandelion and milkweed, each being equipped with a silken parachute; the seeds of maple, ash, and elm trees have broad blades by which they are blown about in the wind.

3. Some seeds, such as those of the cocklebur, have small hooks by which they become attached to passing animals.

4. The seeds of the witch-hazel tree and the touch-me-not plant are popped away from the plant by the explosion of the fruit as it dries up.

5. The coconut can be carried to distant shores by floating in the water.

If there were no seed dispersal, all the seeds would germinate in one restricted area around the mother plant. They would then be so crowded that there would not be enough light, space, water, and minerals for them all to survive. When the seeds are scattered about, there is a better opportunity for many of them to develop into new plants.

Seed Germination

After a seed has been formed, its embryo usually goes through a resting, or dormant, period, during which it can endure conditions of drying up and freezing weather. In the spring, with warmth and moisture available, the seed begins to sprout, or *germinate*. The embryo grows, obtaining its nourishment from the stored food in the seed. Roots develop first, to anchor the seedling and to obtain moisture. The stem and leaves grow up above the ground, and the young plant begins to make its own food. By differentiation, the various tissues of the plant develop.

Vegetative Propagation

In some cases, flowering plants can reproduce without the use of flowers, from stems, leaves, or roots. Since these are the vegetative parts of the plant normally used for nutrition, this method of reproduction is known as *vegetative propagation*. It is a form of asexual reproduction, since it involves only one parent.

Methods of Vegetative Propagation

Vegetative propagation can occur in eight ways:

1. *Cutting, or slip.* A stem may be cut and removed from a geranium or coleus plant, and placed in moist sand or in water. After a period of time, this cutting, or slip, will develop roots at the cut end of the stem. *Plant growth hormones* are now used to speed up the rate of root formation or cuttings. The cut end of the stem is dipped into a solution, or the powder itself, of the growth hormone. In this way, woody cuttings that normally do not produce roots have been induced to form them. If the leaves of some plants, such as the bryophyllum and begonia, are placed on moist soil, new plants begin to develop from various points on the leaves; these are *leaf cuttings*.

2. *Runner.* A horizontal stem of the strawberry plant, called a runner, will grow along the ground. A short distance from the main plant, a new plant will develop, with roots that anchor it to the ground, and with a stem with leaves. In time, this new plant will also form runners.

3. *Layering.* In blackberry and forsythia plants, some of the branches bend over and come into contact with the ground. They take root at these places, and new stems and leaves develop into independent plants.

4. *Rhizome.* In some plants such as ferns, the snake plant, and the Canada thistle, a thick underground stem, called a rhizome, at intervals sends up stems that produce leaves. Each of these becomes an independent plant.

5. *Tuber.* The white potato is really a fleshy, modified underground stem, known as a tuber. There are numerous buds, or *eyes*, on it, that can develop into new plants with roots, stems, and leaves. A farmer plants a potato crop by cutting up potatoes in such a way that each piece has at least one eye. The young developing plant obtains its nourishment from the stored food in the potato.

6. *Bulb.* The tulip and onion bulbs, which are formed underground, consist of short stems surrounded by fleshy leaves containing stored food. When planted, a bulb sends out roots from its base, while leaves and a flower grow upward, deriving their nourishment from the stored food in the bulb. Later on, as they manufacture their own food, the plants will store it in new bulbs, which are formed next to the original one. Other plants that reproduce by bulbs are the hyacinth, gladiolus, and lily.

7. *Fleshy root.* The carrot and sweet potato are modified roots containing stored food. If they are placed in contact with water, or are planted in the soil, stems and leaves will sprout upward, and roots will grow down.

8. *Grafting.* Humans propagate certain desirable types of woody plants by grafting. The first step is to select and cut a stem or bud of the desired type of plant. This is the *scion*; it is then attached to a closely related tree, the *stock*, on which an equal-sized branch has also been cut and notched. The place of attachment is securely bound and covered with melted wax, to prevent drying out or infection. The graft is performed in such a way that the *cambium*, or growing layer, of the scion is in contact with that of the stock. In some cases, a bud is used as a scion, and is inserted under a slit in the bark of the stock. Soon the graft takes, and the scion produces the desired fruit or flower, while obtaining its water and minerals from the stock. It is possible to buy an apple tree that will bear several different varieties of apple, such as Delicious, Macintosh, Winesap, Northern Spy, and Baldwin. Scions of each of these varieties were grafted onto the stock. The seedless orange is propagated by grafting, as are many desirable varieties of grapes, oranges, lemons, peaches, and plums. In all these cases, the scion must be closely related to the stock on which it is grafted.

Values of Vegetative Propagation

Vegetative propagation offers three advantages:

1. The plants are of the same type as the parents. They do not vary, as may usually be the case in sexual reproduction, where the characteristics of two parents are inherited.

2. Plants are reproduced much more quickly and in larger numbers than if they were grown from seeds.

3. Seedless fruits such as oranges and grapes can be maintained and propagated.

Section Review

Select the correct choice to complete each of the following statements:

1. Pollen is produced in the part of the flower called the (A) corolla (B) calyx (C) stigma (D) anther (E) style

2. A fruit develops from the part of a plant called the (A) embryo sac (B) ovary (C) pollen tube (D) egg nucleus (E) endosperm

3. The ripened ovule of a pea plant is called the (A) seed (B) pistil (C) stamen (D) micropyle (E) essential organ

4. The structure that includes all the others is the (A) ovary (B) ovule (C) style (D) pistil (E) stigma

5. The transfer of pollen from the stamen to the stigma of the same flower is known as (A) cross-pollination (B) self-pollination (C) fertilization (D) germination (E) seed dispersal

6. The female gamete of a flower is formed in the (A) pollen grain (B) pollen tube (C) stigma (D) style (E) embryo sac

7. When a bean seed germinates, its embryo obtains nourishment from the (A) stored food in the pistil (B) stored food in the ovary (C) stored food in the seed (D) stored yolk in the seed (E) minerals in the sepals

8. A plant that reproduces by runners is the (A) geranium (B) onion (C) lily (D) bryophyllum (E) strawberry

9. A tulip reproduces vegetatively by forming a (A) tuber (B) bulb (C) cutting (D) slip (E) rhizome

10. A tuber is an underground (A) stem (B) root (C) leaf (D) nodule (E) flower

11. Vegetative propagation is a form of (A) sexual reproduction (B) asexual reproduction (C) conjugation (D) fertilization (E) vegetative pollination

12. If a branch of a seedless orange is grafted to a tree that produces oranges with seeds, the (A) stock will bear both seedless and seed oranges (B) stock will bear only seedless oranges

(C) scion will bear only seedless oranges
(D) scion will bear only seed oranges   (E) scion will bear both seedless and seed oranges

13. A bean seed contains all of the following *except*
(A) a seed coat   (B) an epicotyl   (C) a hypocotyl
(D) a hypha   (E) cotyledons

14. All of the following are considered to be fruits *except* the   (A) tomato   (B) cucumber   (C) squash
(D) carrot   (E) corn

15. All of the following are advantages of asexual reproduction *except*   (A) the offspring are the same type as the parents   (B) cross-pollination is simpler with large flowers   (C) there can be many offspring   (D) reproduction is faster   (E) seedless fruits can be propagated

Answer Key

1-D	4-D	7-C	10-A	13-D
2-B	5-B	8-E	11-B	14-D
3-A	6-E	9-B	12-C	15-B

Answers Explained

1. (D) The anther is the enlarged part of the stamen that produces the pollen.

2. (B) The pistil contains an enlarged portion at the bottom, the ovary. When the flower matures, the ovary becomes the fruit.

3. (A) The ovule is contained within the ovary. Following double fertilization, the ovule develops into the seed

4. (D) The pistil is located in the center of the flower, within the stamens. The pistil contains the ovule-containing ovary and the slender, tube-like style, topped by the sticky stigma. Along with the stamens, the pistil is considered an essential organ of the flower, since it is directly concerned with reproduction.

5. (B) When pollen is transferred from the stamen to the stigma of the same flower, the process is called self-pollination. If it is transferred from the stamen to the stigma of another flower, the process is called cross-pollination.

6. (E) Within the ovule, one of the cells enlarges to form the embryo sac. Its nucleus divides several times to form eight nuclei; one is the egg nucleus, or female gamete.

7. (C) The two large cotyledons in a bean seed contain stored food that the embryo uses for nourishment when the seed germinates.

8. (E) A horizontal stem of the strawberry plant, called a runner, grows along the ground. A short distance from the main plant, a new plant develops.

9. (B) A bulb, which is formed underground, consists of a short stem surrounded by fleshy leaves containing stored food. When planted, a bulb sends out roots from its base, while leaves and a flower grow upward, deriving their nourishment from the stored food in the bulb.

10. (A) A fleshy, modified underground stem is known as a tuber; a white potato is really a tuber.

11. (B) Vegetative propagation, a form of asexual reproduction, involves only one parent that can reproduce by a vegetative part of the plant—the stem, leaf or root.

12. (C) In grafting, the desired branch—in this case, that of the seedless orange—called the scion, is attached to a closely related tree—in this case, that which produces oranges with seeds—called the stock. The scion obtains water and minerals from the stock but retains its hereditary material—for seedlessness.

13. (D) The hypha is a stemlike structure in molds, such as the bread mold.

14. (D) The carrot is an enlarged root that stores food.

15. (B) Since cross-pollination involves the fusion of male and female gametes to produce a fertilized egg in the ovule, it cannot be considered a form of asexual reproduction.

9.3 REPRODUCTION IN HIGHER ANIMALS

The higher animals, or vertebrates, are alike in having special reproductive organs (gonads) that are devoted to perpetuating the species. These gonads produce two types of substances: (1) sex hormones and (2) gametes.

Secondary Sexual Differences

The male sex hormone, *testosterone*, and the female sex hormone, *estrogen*, are produced when male and female animals mature. They are responsible for the appearance of secondary sexual differences between them. Thus, male birds have bright plumage and a singing voice, while the females have dull feathers and rarely sing. Among deer, the male has large antlers and is aggressive; the female lacks antlers and is gentle. The lion has a mane; the lioness does not. In humans, the male develops hair on his face and becomes deep-voiced; the female does not grow hair on her face, continues to have a high-pitched voice, and develops breasts.

The Reproductive Process

Production of Gametes

The male gonads are called *testes*, or *spermaries*. They have special cells that form the male gametes, or sperm. In addition, they produce male sex hormones. In humans, a pair of testes is located in an outpocketing of the body wall known as the *scrotum*. The temperature here is 1–2°C lower than the body temperature, providing the best conditions for sperm manufacture and storage. Several glands, including the *prostate*, secrete a liquid that serves as a transport medium for the sperm. This liquid and the sperm it carries are known as the *semen*.

Each sperm is a microscopic structure in which there are a tail, or *flagellum*, made of cytoplasm, and a head containing the nucleus. Sperms move by lashing the flagella. Sperm cells are produced in extremely large numbers. They leave the body of the male through the urethra and the penis.

The *ovaries* form the female sex hormones as well as produce the female gametes, which are also known as eggs, or *ova*. Each ovum is a single cell containing a nucleus, cytoplasm, and stored food. In practically all of the vertebrates except the mammals, there is a considerable amount of stored food in the eggs. As a result, the eggs can readily be seen with the naked eye. They reach their largest size among the birds. After the eggs are formed in the ovaries, they enter the oviducts.

Fertilization and Cleavage

When mating takes place, many sperms swim to the egg cells and surround them. One sperm cell penetrates the membrane surrounding an egg, and unites with its nucleus. A fertilization membrane forms around the egg and keeps the other sperms out. The union of sperm and ovum to form a *fertilized egg*, or *zygote*, is known as *fertilization*. Soon afterwards, the fertilized egg undergoes a series of cell divisions, developing into an embryo that has two, then four, then eight, and then more cells, as each cell divides.

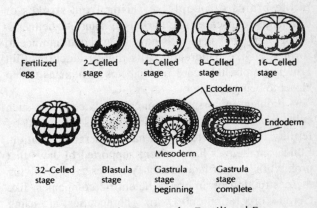

Early Development of a Fertilized Egg

This series of cell divisions is known as *cleavage*. In many animals, the embryo next takes the appearance of a hollow ball of cells one cell thick, known as the *blastula*.

Gastrulation

One part of the blastula then grows inward (gastrulation), forming a second layer. This is the cuplike *gastrula*. The outer layer is called the *ectoderm*; the inner, the *endoderm*. A third or middle layer then develops between these two; it is known as the *mesoderm*.

Differentiation

Up to this point, by cleavage, the gastrula, consisting of many cells quite similar to each other, has been produced. It is approximately equal in size to the original zygote. From this point on, the embryo elongates as the three layers of cells begin to *differentiate*, or form the different tissues and organs, as follows:

1. The *ectoderm*, or outer layer—epidermis of skin; nervous system.
2. The *endoderm*, or inner layer—lining of alimentary canal, trachea, and lungs; pancreas; liver; bladder.

3. The *mesoderm*, or middle layer—muscles; connective tissue, including bone and cartilage; blood and circulatory system; kidneys; reproductive organs.

The organs of the body are developed from tissues derived from one or more of these layers. No organ originates exclusively in any one of the layers.

Stem cells recently have been identified in the early embryonic stages as having the ability to differentiate into many different types of tissue cells including neurons, and those of heart, kidney, and liver. They are called stem cells because all other cells stem from them. When grown artificially in the laboratory, such cells have shown some promise of replacing damaged cells in the brains of people suffering from strokes or degenerative diseases like Parkinson's and Alzheimer's. Also, embryonic stem cells have been shown in mice to form islet-like pancreatic cells that produce insulin, bringing a possible cure for diabetes in humans a step closer to being realized.

A problem in pursuing stem cell research is that such cells come from either human embryos discarded by fertility clinics or fetuses that have undergone abortion. Ethical challenges have arisen, supported by laws that prohibit the use of federal money for research involving the destruction of embryos. Scientific companies, like Geron Corporation and Osiris Therapeutics, are using private funds to investigate methods of mass production of stem cells.

Much research on *adult stem cells* has yielded promising results. Although they were thought to give rise only to specific types of tissue, adult muscle stem cells have been shown to generate blood cells, and adult bone marrow cells have formed types of brain and bone tissue. Adult stem cells are also claimed to be isolated from fat tissue.

Another type of stem cell has been isolated from the placenta, or afterbirth. It is believed that this type of stem cell is the equivalent of human embryonic stem cells which can differentiate into cartilage, nerves, the lining of blood vessels, and other tissues of the adult body.

Reproduction in Fish, Frogs, and Birds

Reproduction in Fish

In most fish, fertilization takes place in the water, outside of the body. The female sheds many eggs, called *roe*, at spawning time. In some species such as the perch, 100,000 eggs are laid by one fish. The male deposits *milt*, a liquid containing countless sperms, over the eggs. The sperms move about and fertilize the eggs. Eggs and sperm that do not unite fail to develop,

and disintegrate. After cleavage, an embryo called a *fry* develops. It is nourished by a large *yolk* sac attached to the lower part of its body and containing the stored food of the egg. As the fry grows, the yolk sac disappears, and the young fish begins to hunt for its own food. Despite the large number of eggs produced by one female, comparatively few of the offspring reach adulthood; there is a constant struggle for existence against various predatory animals, as well as adverse environmental conditions.

In some fish, including certain tropical fish, fertilization takes place in the oviduct, within the female's body. The developing fertilized eggs may be retained for varying lengths of time before being laid. In "live-bearers" such as the guppy, the young fish develop entirely within the female's body, obtaining their nourishment from the stored food in the eggs. Most fish pay no attention to their offspring. A few, such as the stickleback and the sea horse, however, take special care of the eggs.

Reproduction in Frogs

In the spring, the female frog lays a mass of eggs, which have been formed in her ovaries and which pass to the outside through the oviduct. As the eggs leave, the male frog deposits sperm over them. In a very short time, a clear, jellylike material around each egg swells and holds the mass of eggs together. Fertilization occurs when a sperm unites with each egg to form a zygote. Cleavage and differentiation take place, but instead of a young frog being formed, a *tadpole* is produced. It has external gills for breathing and a tail for movement; there are no legs. At first, it is nourished by the stored food of the egg. Soon internal gills replace the external gills. The tadpole feeds on algae and green plants. It has a two-chambered heart.

After several weeks or months, the tadpole goes through a series of changes called *metamorphosis*, and becomes a frog. In the bullfrog, this takes place over a 2-year period. First, hind legs and then forelegs appear. The tail gradually shrinks. Lungs form and the gills disappear. The heart becomes three-chambered. The digestive system becomes modified for an animal diet, and the frog leaves the water and lives on land. Like most fish, frogs and other amphibia give no care to their offspring.

Parthenogenesis

Although in most animals an egg will not develop unless it has been fertilized by sperm, some lower animals reproduce by *parthenogenesis*, in which eggs develop without sperm. This is true of aphids, or plant lice, water fleas, and others. In bees the drone, or male, develops from an unfertilized egg.

Artificial parthenogenesis has been accomplished by scientists experimenting with the eggs of sea urchins and frogs. Jacques Loeb stimulated frog eggs to go through cleavage and eventually to form frogs without fertilization by sperms; he used various stimuli, such as pricking the membrane with a needle, and treatment with salt solutions or acids. Dr. Gregory Pincus was successful in producing "fatherless rabbits" by removing the ova from female rabbits, treating the ova with salt solutions, and implanting them in other female rabbits. The baby rabbits that developed were females and were subsequently mated to produce normal offspring.

Reproduction in Birds

A hen's egg consists of the following parts: (a) an outer shell; (b) a shell membrane within the shell; (c) albumen, or white of the egg; (d) the central yolk, which is the ovum, or "true egg," and contains, in addition to stored food, the tiny living material, including a nucleus and cytoplasm. A bird's egg is therefore a tremendous single cell. The ovary contains many ova in various stages of development. One ovum at a time, with its yolk mass, leaves the ovary and enters the single oviduct. It takes a little more than half a day for the finished egg to be laid; during this time, various glands in the wall of the oviduct secrete, in turn, the albumen, shell membrane, and shell.

Fertilization in birds and other land mammals is internal. During mating, sperms are introduced into the oviduct, which provides a moist environment for the sperms to move. They move to the upper end, where one of them fertilizes the ovum as it enters the oviduct from the ovary. The embryo begins to develop at once, and is in the late gastrula stage when the egg is laid. If fertilization does not occur, the egg that is laid will not develop into a bird. The mother bird incubates the eggs for a period of time (21 days in chickens) by sitting on them and providing the warmth needed for the embryo to develop.

During this time, the embryo is being nourished by the yolk. It is surrounded by a set of membranes that provide a favorable environment for development. These membranes include (a) the *chorion*, which lines the shell, surrounds the other membranes, and serves as a moist membrane for the exchange of gases; (b) the *allantois,* which collects metabolic wastes from the embryo and, with the chorion, serves to exchange oxygen and carbon dioxide; (c) the *amnion*, which contains the amniotic fluid surrounding the embryo and protecting it from shock; (d) the *yolk sac,* around the yolk, containing blood vessels that bring food to the embryo. Reptile eggs have a similar set of membranes.

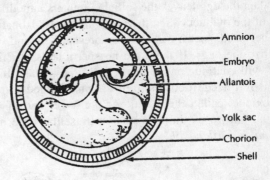

Embryonic Membranes in a Bird's Egg

When mature, the baby bird pecks its way out of the shell. In some birds, such as the chicken, the young chick is able to run about and obtain its own food a few hours after hatching. In the songbirds, however, the young are helpless, with little body covering to protect them. The mother bird makes countless trips to feed them with insects and worms; she continues to sit on the nest to keep them warm. After about 2 weeks, the young birds are sufficiently developed to learn how to fly and obtain their own food.

Reproduction in Mammals

Mammals give birth to their young alive. Their embryos develop and are nourished internally; they are said to be *viviparous,* as contrasted with *oviparous* animals, which lay eggs that develop outside the female's body (birds, amphibia, fish). Some fish, such as guppies, and some snakes retain the eggs in their oviducts until they hatch; the embryos are nourished by the food contained in the eggs. These animals are called *ovoviviparous.*

Development of the Embryo

The mammal embryo develops internally, nourished by nutrients obtained from the mother's bloodstream. The ovum, or egg, has practically no stored food. Consequently, it is microscopic. The ova of a mouse and an elephant are about the same size. As in other animals, the eggs are produced in the ovaries. In some mammals, such as rabbits, pigs, mice, and dogs, a number of eggs are formed at one time. In other mammals, including humans, elephants, and cows, usually only one egg is produced at a time. In humans, there are two ovaries, located in the lower portion of the abdomen. They produce eggs in cavities called follicles. The release of an egg from a follicle is called *ovulation.* A female baby is born with all the eggs she will ovulate as a woman.

When the egg leaves the follicle, it enters a modified part of the oviduct called the *Fallopian tubes*. If mating has occurred, sperm cells will be present here and will fertilize the eggs. If there are no sperm cells, the tiny eggs will continue out of the body of the female. If an egg is fertilized, it comes to rest in an enlarged part of the oviduct called the uterus. At the lower end of the uterus, known as the *cervix*, is a muscular tube, the *vagina*, or *birth canal*.

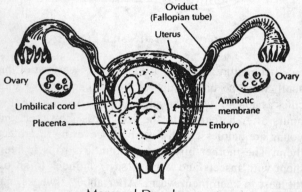

Mammal Development

In the uterus, the embryo (known as a *fetus* after it begins to take form) receives its nourishment from the mother through the *placenta*. This consists of a set of thick membranes composed of tissue of both the parent and the embryo, lying in intimate contact with each other. These membranes are well supplied with many capillaries connected to the separate blood systems of both individuals. As blood flows through the two different sets of capillaries, food and oxygen diffuse from the mother's bloodstream into the embryo's bloodstream. The blood circulates out of the placenta through the *umbilical cord* to the embryo, whose heart keeps the blood circulating to its cells. Wastes such as carbon dioxide and urea are sent from the embryo through the umbilical cord to the placenta, where they diffuse into the mother's bloodstream. There is no direct connection between the bloodstream of the mother and that of the embryo. Transport is accomplished by diffusion and active transport. During its development, the fetus lies suspended in the amniotic fluid, surrounded by the *amnion*. The fluid provides a watery environment and protection against shock.

Assisting Human Fertilization

One reason why some women cannot conceive is that their husbands are unable to produce active gametes, that is, they are sterile. One way of overcoming the problem has been through *artificial insemination*. Frozen sperm from a donor can be stored in a sperm bank at a temperature of −196°C for many years.

A doctor implants such sperm into the Fallopian tubes of the woman. Here fertilization occurs, and a normal child will be born. This has been an established procedure for more than 80 years.

A newer approach to human reproduction was the delivery of a "test tube" baby in England by Dr. Patrick C. Steptoe and Dr. Robert G. Edwards in 1978. The mother had been unable to have a baby for 12 years. Through a delicate surgical procedure, an egg was removed from a follicle in her ovary. It was placed in an incubator, and then a few drops of sperm were added to the petri dish containing the egg. The egg was thus fertilized and began to divide. When the eight-cell stage was reached, it was implanted into the uterus. Here the fetus developed and a normal baby was born. This type of *in vitro* (in a test tube, or similar container, outside the body) fertilization has been repeated many times since then, but the surgical techniques involved are very delicate.

The Human Menstrual Cycle

The human female has a reproductive cycle in which hormones from the hypothalamus, the pituitary gland, and the ovaries interact. Under the influence of the hypothalamus, the pituitary gland releases hormones that stimulate the development of a follicle in the ovary and stimulate ovulation. The ovaries produce hormones that start the thickening of the uterus wall and also control the production of hormones by the pituitary gland and the hypothalamus. The uterus secretes hormones that affect the production of ovarian hormones. This interplay of hormones is an example of a negative feedback mechanism.

These hormones influence the progression of the various stages in the menstrual cycle, which is the series of changes that take place in the preparation of the uterus wall for the receipt of the embryo. If the egg has not been fertilized, menstruation occurs. The stages in the menstrual cycle usually take about 28 days. They may be described as follows:

1. *Follicle stage* (10–14 days). The follicle in the ovary is ripening, and the wall of the uterus is beginning to develop a rich supply of blood vessels.

2. *Ovulation.* In the middle of the cycle, the egg bursts out of the follicle.

3. *Corpus luteum stage* (4–5 days). The corpus luteum is the yellowish material that forms in the follicle after ovulation. During this stage, the lining of the uterus becomes spongy and thick with glands and blood vessels. If the egg has been fertilized, implantation of the embryo in the uterus wall takes place, about 6–10 days after fertilization, and the next stage does not occur.

4. *Menstruation* (2–6 days). If the egg was not fertilized, the thickened lining of the uterus breaks down and is shed, along with the microscopic egg, through the vagina.

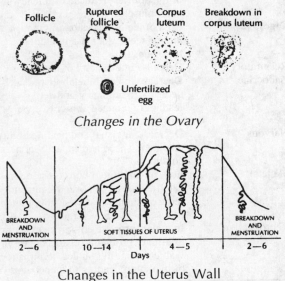

Changes in the Ovary

**Changes in the Uterus Wall
The Menstrual Cycle**

The menstrual cycle starts at puberty, or sexual maturity, anywhere between the ages of 9 and 18. It is temporarily suspended during pregnancy. It ends after about 35 years, at menopause, around the age of 50. The duration of the monthly cycle may vary considerably from one female to another and may be interrupted by illness.

Birth

The period of the embryo's development, or *gestation*, varies in different mammals: mouse, 20 days; cat, 9 weeks; human, 9 months; elephant, 20 months. At the end of that time, the young individual is born. The *navel* marks the place where the umbilical cord was attached. The young mammal is nourished on milk produced by the *mammary glands*. Some mammal young (e.g., the horse) are able to move about freely shortly after birth. Others (e.g., human, cat) are helpless for a period of time before they can walk and feed themselves. Mammals provide extended care for their young.

Nonplacental mammals The duckbill platypus and the spiny anteater are unusual Australian mammals that lay eggs. When the young are born, they are fed on milk. Another group of Australian mammals, the marsupials, have a pouch in which the young are placed after they are born in an undeveloped condition; the kangaroo and the koala bear are examples. The only marsupial found outside Australia is the opossum. Although pouched animals have internal development, they do not develop a placenta and the embryo does not receive direct nourishment from the parent in this stage.

Multiple births In human beings, two children, or twins, may sometimes be born at the same time. They may be either fraternal or identical twins.

Fraternal twins develop from two eggs that were produced and fertilized at the same time. They are nourished from two separate placentas. They may be of the same sex or of different sexes. They are as alike as other members of the same family.

Identical twins develop from a single fertilized egg that started to undergo cleavage, and then split into two masses of cells, each of which developed into a separate organism. They are nourished from one placenta. They are exactly alike and are of the same sex.

It has been shown that twins appear in one out of 85 births; triplets have a frequency of $1:(85)^2$; and quadruplets appear in the ratio of $1:(85)^3$. In triplets, three sep-

SUMMARY OF VERTEBRATE REPRODUCTION

Class	Number of Eggs	Where Fertilization Occurs	Where Development Occurs	Care of Young
Fish	Thousands or millions	Usually external	Usually external	Practically none
Amphibia	Hundreds or less	External	Water	None
Birds	One to a few	Internal	Internally for about ½ day; then ex-external incubation	Several weeks
Mammals	One to a few	Internal	Internal entirely, except for egg-laying mammals and marsupials	From a few weeks in mice up to 20 years in humans

arate eggs may be fertilized at the same time; or two eggs may be fertilized, with one forming identical twins and the other producing a fraternal triplet. It is also possible for identical triplets to develop from one fertilized egg; the developing egg mass splits in two, and then one of these splits again. Identical quadruplets probably originate from a fertilized egg that splits into two cell masses, which then split again, to form four similar offspring. The five Dionne sisters born in Canada in 1934 are thought to be identical quintuplets, originating from one fertilized egg that split to form two cell masses; each of these then split again to form four cell masses, and one of these split again to form five all together.

Life Span and Aging

Varying Life Spans

Living things live for varying lengths of time. The sequoia tree and the bristlecone pine tree may be over 4,000 years old. The average life spans of some other organisms are shown in the following table:

Organism	Average Life Span (years)
Human (in U.S.)	76.1
Margined tortoise	152
Indian elephant	70
Eagle owl	78
Crown pigeon	16
Cat	14
Dog	10
Mayfly	1 day

It is not known what determines the life span of a species.

Physical and Mental Changes with Aging

The human life cycle includes a number of stages: birth → infancy → childhood → youth → maturity → old age → death. The signs of advancing age usually appear gradually. When the hair of a young man begins to thin out because of a loss of hair follicles, it signals the onset of baldness. Another gradual sign of advancing age is the appearance of wrinkles on the face as the yellow elastic fibers of the connective tissue in the skin become less elastic. The ability of the aging body to produce important enzymes and antibodies is reduced, often times leading to various disorders.

Changes in other parts of the body due to aging are summarized in the following table:

System	Changes Due to Aging
Circulatory	Increase in blood pressure; hardening of the arteries; reduced efficiency of heart muscle
Digestive	Loss of teeth; recession of gums around teeth; decrease in number of taste buds
Respiratory	Decrease in elasticity of lungs
Endocrine	Decrease in basal metabolism rate; decrease in adrenal gland activity leading to reduced ability to cope with stress
Nervous	Decrease in visual and hearing ability; decline in speed and capacity of nerve condition
Skeletal	Loss of calcium in bones (osteoporosis); calcification of joints and ligaments; reduced stature resulting from flattening of discs between vertebrae
Reproductive	In males, prostate enlargement leading to urinary problems; also decrease in viable sperm. In females, termination of ovulation, leading to menopause

Alzheimer's Disease

As some people age, they may tend to experience normal forgetfulness. However, in the condition known as Alzheimer's disease, severe mental deterioration develops which leads to memory loss, mental confusion, and personality changes. The incidence of the disease increases greatly with advancing age.

According to one research theory, beta-amyloid peptides secreted by the brain cells are soluble and are normally cleared away. If, however, they became insoluble, they lead to the formation of large *plaques* that damage the brain cells. Recently, scientists have identified the enzymes beta-secretase and gamma-secretase, which play key roles in the buildup of these plaques. They are now searching for agents that can block these enzymes.

Progress has been reported on development of a vaccine that reverses some of the damaging effects of Alzheimer's. The vaccine consists of a laboratory-made protein called AN-172, which attaches to amyloid in brain plaques. This prepares the amyloid for removal by naturally occurring brain cells that scavenge for unwanted materials. In one case, experimental mice treated with the vaccine showed improvement in navigating a maze.

Other research has identified that the neurotransmitter *acetylcholine*, which is important in memory function, is deficient in the disease. In addition, the hormone estrogen may have a positive influence on memory. Four genes, including one known as ApoE, have also been associated with the disease.

Although there is presently no known way to prevent Alzheimer's disease, there are medications to slow its progress, such as tacrine, donepezil, and metrifonate. Scientists believe that a combination of these treatments may be effective in the future.

The Aging of America

With the rising life expectancy of people in the United States, the number of older people has increased steadily. At the present time, more than 12 percent of the population is over 65 years of age. In 1900, this was true of only about 4 percent. It is estimated that in the year 2000 there was 16.9 percent, or about 45 million people, in this age group. It is also estimated that the group aged 80 and above will grow the fastest in the future. These people now number about 7 million, but by the year 2050 their number will probably reach 25 million.

The relatively new science of *gerontology* deals with aging and the special problems of older people. Similarly, in medicine, *geriatrics* deals with the health, diseases, and care of the aged. Although the United States has had a "youth-oriented" culture from its earliest days, more and more attention is now being paid to utilizing the abilities and meeting the needs of senior citizens. Advancing age is a fact of life, but feeling old can be a state of mind. When Bernard Baruch reached his 85th birthday, he remarked: "To me, old age is always 15 years older than I am."

Section Review

Select the correct choice to complete each of the following statements:

1. The head of a sperm consists largely of (A) a flagellum (B) cytoplasm (C) yolk (D) a nucleus (E) stored food

2. One effect of the female sex hormone is the (A) bright plumage of birds (B) dull plumage of birds (C) singing voice in birds (D) antlers of the deer (E) lion's mane

3. The largest egg is present in a (A) mouse (B) horse (C) whale (D) sparrow (E) cow

4. The proper sequence of steps (1-gastrula; 2-zygote; 3-egg; 4-blastula; 5-two-celled stage) in embryo development is (A) 5—2—3—4—1 (B) 3—5—2—4—1 (C) 3—2—4—1—5 (D) 3—2—5—4—1 (E) 3—2—5—1—4

5. The ectoderm develops into the (A) circulatory system (B) nervous system (C) liver (D) muscles (E) bones

6. A young fry develops during the life cycle of a (A) fish (B) frog (C) bird (D) hydra (E) lobster

7. The egg-producing structure of an animal is called the (A) oviduct (B) ovum (C) ovary (D) ovule (E) ovipositor

8. A young tadpole is different from a frog in having (A) four legs (B) a three-chambered heart (C) a tail (D) lungs (E) a large, sticky tongue

9. The development of an egg without being fertilized by a sperm is known as (A) metamorphosis (B) regeneration (C) differentiation (D) parthenogenesis (E) incubation

10. The only part of a chicken's egg not formed in the oviduct is the (A) albumen (B) yolk (C) shell membrane (D) shell (E) shell pigment

11. The embryo of a mammal develops in the structure called the (A) ureter (B) uterus (C) urethra (D) pistil (E) ovary

12. The embryo of a cat receives its food (A) from the egg (B) from the yolk supply (C) through the micropyle (D) through the placenta (E) from the mother's small intestine

13. Fraternal twins (A) may be of the opposite sex (B) are always of the same sex (C) develop from one egg fertilized by two sperms (D) develop from two eggs fertilized by one sperm (E) are nourished from one placenta

14. A mammal embryo obtains oxygen from   (A) its mother's bloodstream   (B) its lungs   (C) the amniotic fluid   (D) oxidation of its food   (E) its diaphragm

15. Testes are structures that produce   (A) spores   (B) sperm   (C) seeds   (D) ova   (E) testa

16. The release of an egg from a follicle is called   (A) fertilization   (B) cleavage   (C) metamorphosis   (D) osmosis   (E) ovulation

17. All of the following are membranes in a bird egg *except* the   (A) chorion   (B) allantois   (C) amnion   (D) yolk sac   (E) albumen

18. The human fetus is surrounded by   (A) liquid   (B) air   (C) solid cells   (D) a fertilization membrane   (E) nitrogenous wastes

19. All of the following organs produce hormones involved in the reproductive cycle *except* the   (A) testes   (B) pituitary   (C) pancreas   (D) ovary   (E) uterus

20. The correct sequence of stages in the menstrual cycle is   (A) ovulation–follicle stage–corpus luteum stage–menstruation   (B) corpus luteum stage–ovulation–follicle stage–menstruation   (C) follicle stage–ovulation–corpus luteum stage–menstruation   (D) menstruation–ovulation–corpus luteum stage–follicle stage   (E) fertilization–cleavage–differentiation–ovulation–menstruation

Answer Key

1-D	5-B	9-D	13-A	17-E
2-B	6-A	10-B	14-A	18-A
3-D	7-C	11-B	15-B	19-C
4-D	8-C	12-D	16-E	20-C

Answers Explained

1. (D) The sperm is a microscopic structure containing a flagellum made of cytoplasm and a head containing the nucleus.

2. (B) In birds the female sex hormone estrogen induces the production of dull-colored feathers and inability to sing.

3. (D) The egg of a sparrow has much stored food, making it larger than the egg of a mammal, which has practically no stored food.

4. (D) After the egg is fertilized and becomes a zygote, it undergoes a series of divisions known as cleavage. First the zygote becomes a two-celled stage of the embryo; then it appears like a hollow ball of cells called the blastula, and finally, the gastrula.

5. (B) The ectoderm is the outer layer of the gastrula: during differentiation, it develops into the nervous system.

6. (A) The embryo of a fish, called the fry, is nourished by a large yolk sac attached to the lower part of its body, and containing the stored food of the egg. As the fry grows, the yolk sac is absorbed.

7. (C) The ovary produces eggs, which contain a considerable amount of stored food in practically all vertebrates except mammals.

8. (C) The young tadpole has a tail for movement, but no legs. It breathes by gills and feeds on algae and plants.

9. (D) In a few cases, an egg produces new individuals, without being fertilized by a sperm, in a process known as parthenogenesis. Examples: aphids, water fleas.

10. (B) The ovum leaves the ovary containing a large mass of yolk and tiny living material, including a nucleus and cytoplasm. All the other parts of a chicken's egg are added as the ovum passes through the oviduct.

11. (B) The embryo of a mammal develops in an enlarged part of the oviduct called the uterus.

12. (D) While developing in the uterus, the embryo of a cat receives its nourishment through the placenta. As food circulates through the bloodstream of the mother, it diffuses into the embryo's bloodstream.

13. (A) Fraternal twins develop from two eggs that were fertilized independently of each other. These twins can be as alike or as different as any other members of the same family.

14. (A) The placenta consists of tissues of both the mother and the embryo, lying in intimate contact with each other. Oxygen diffuses from the mother's bloodstream into the embryo's bloodstream.

15. (B) The testes are the male gonads; they produce the male gametes called sperm.

16. (E) The ovary produces eggs in cavities called follicles. The process by which an egg leaves a follicle is called ovulation.

17. (E) Albumen is the white of an egg; it is secreted by glands in the oviduct around the yolk as the egg passes through the tube.

18. (A) During its development, the fetus is surrounded by the amniotic fluid, which is enclosed by the amnion.

19. (C) The pancreas produces the hormone insulin, which stimulates the liver to store excess glucose as glycogen, and the hormone glucagon, which increases the blood level of glucose by causing the liver to release it.

20. (C) The menstrual cycle is the series of changes that takes place in the uterus wall to prepare for the reception of an embryo. The egg develops in a follicle in the ovary and is released in ovulation. The follicle then forms a yellowish body, the corpus luteum. However, if the egg has not been fertilized, the thickened lining of the uterus breaks down and is shed, along with the microscopic egg, through the vagina during menstruation.

HOW LIVING THINGS INHERIT TRAITS

CHAPTER

10

10.1 THE BASIS OF INHERITANCE

Mitosis

A newly born individual contains features of both of its parents. The fertilized egg of a rabbit develops into a rabbit, not a cat; the fertilized egg of an apple flower develops into another apple tree, not an oak. The reason is that the hereditary material present in the fertilized egg is distributed equally to all the cells of the new individual. The nuclear division by which this takes place is mitosis.

Phases of Mitosis

Mitosis can be observed when a fertilized egg divides during cleavage, and during the growth of any part of an organism, such as the tip of a root. The steps of mitosis can be observed in the following stages:

1. *Interphase.* Before undergoing mitosis, the nucleus contains a mass of *chromatin* material and a nucleolus. In animal cells, there is a small body, the *centrosome*, in the cytoplasm outside the nucleus. The centrosome contains a rodlike particle, called the *centriole*, which occurs in pairs.

2. *Prophase.* The chromatin material becomes visible as short, stubby rods called *chromosomes*. These have already split lengthwise during the interphase, but remain attached together. This self-duplication is

known as *replication*. Each member of a double chromosome is called a *chromatid*. The chromatids are held together by a small structure called a *centromere*. The nuclear membrane disappears, and the chromosomes are distributed through the cytoplasm. The nucleolus also disappears. In animal cells, the centrosome divides in two; the two new centrosomes separate to opposite sides of the cell. Fine threads of cytoplasm called the *spindle* become attached to the centromere of the chromosomes from the centrioles. The radiating mass of spindle threads, together with the centrosome from which they spread out, is called the *aster*. In plant cells, these spindle fibers radiate from concentrated parts of the cytoplasm equivalent to the centrosomes.

3. *Metaphase.* The chromosomes become lined up in the center of the cell. The split between the chromatids appears more pronounced. The centromeres replicate.

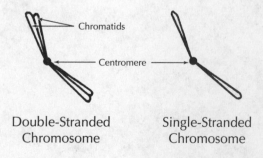

Double-Stranded Chromosome   Single-Stranded Chromosome

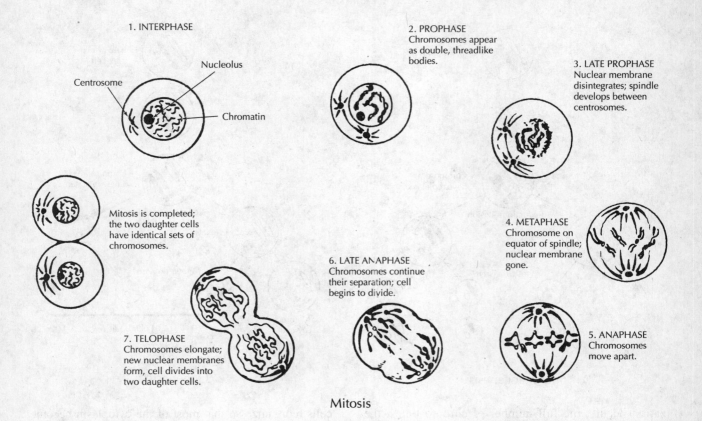

1. INTERPHASE

Nucleolus

Centrosome

Chromatin

2. PROPHASE
Chromosomes appear as double, threadlike bodies.

3. LATE PROPHASE
Nuclear membrane disintegrates; spindle develops between centrosomes.

Mitosis is completed; the two daughter cells have identical sets of chromosomes.

4. METAPHASE
Chromosome on equator of spindle; nuclear membrane gone.

6. LATE ANAPHASE
Chromosomes continue their separation; cell begins to divide.

7. TELOPHASE
Chromosomes elongate; new nuclear membranes form, cell divides into two daughter cells.

5. ANAPHASE
Chromosomes move apart.

Mitosis

4. *Anaphase.* The split chromosomes begin to separate and to move toward the opposite sides of the cell. The number of chromosomes at each end remains the same as the number contained in the original cell.

5. *Telophase.* The chromosomes at each end of the cell collect to form a nucleus containing chromatin. A nuclear membrane forms. The spindle fibers disappear. The cell membrane *indents* in the middle of the cell to divide the cell in two. In plant cells, however, a *cell plate* forms in the middle of the cell, on which particles of cellulose are deposited, to form a cell wall.

After a period of time, each cell may divide again. Regardless of how many diversions take place, however, the resulting cells have the same number of chromosomes as the original. This is possible because *each chromosome always splits lengthwise into two identical chromosomes during mitosis.*

Chromosomes

Each species of plant or animal has a characteristic number of chromosomes in every one of its cells. Corn has 20; pea, 14; cotton, 52; frog, 26; fruit fly, 8; gypsy moth, 62; human, 46. In all cases, the chromosomes occur in pairs. The chromosomes in a pair are said to be *homologous*; one of the pair originated in the male gamete; the other, in the female gamete. Each pair may differ somewhat from the others in size and shape. When the human fertilized egg containing 46 chromo-

somes splits, each of the 2 resulting cells receives 46 (or 23 pairs) of chromosomes. Adult humans continue to have 46 chromosomes in every one of their cells, except the gametes.

Genes

Chromosomes carry the genes, which determine the hereditary characteristics of an individual. The genes are arranged lengthwise along a chromosome. They split each time the chromosome splits, so that the daughter cells always receive the same number as the parent cell. Dr. Thomas Hunt Morgan first suggested the existence of genes when he stated the gene theory in 1910. Since then, it has been found that the principal substance of which genes are made is a complex chemical called DNA, or deoxyribonucleic acid. The DNA molecule, which is large enough to be studied with the electron microscope, appears to be coiled like a spring, or helix, and consists of two spiral chains of atoms. It is discussed in detail in Section 10.4.

Meiosis

The gametes, or eggs and sperm, are the only cells that do not have the same number of chromosomes as all the other cells of the organism. When they are formed, they receive half the number of chromosomes, the *monoploid*, or *haploid*, number. Then, when fertil-

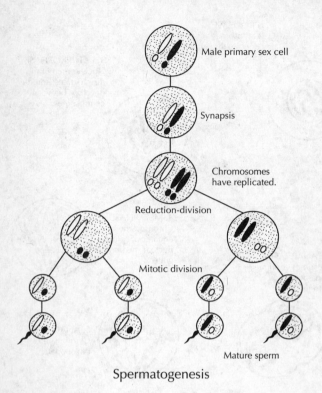

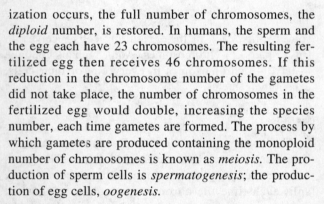

Spermatogenesis

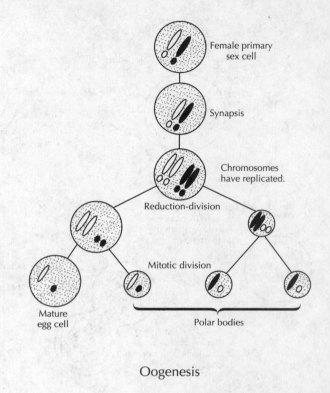

Oogenesis

ization occurs, the full number of chromosomes, the *diploid* number, is restored. In humans, the sperm and the egg each have 23 chromosomes. The resulting fertilized egg then receives 46 chromosomes. If this reduction in the chromosome number of the gametes did not take place, the number of chromosomes in the fertilized egg would double, increasing the species number, each time gametes are formed. The process by which gametes are produced containing the monoploid number of chromosomes is known as *meiosis*. The production of sperm cells is *spermatogenesis*; the production of egg cells, *oogenesis*.

Spermatogenesis

The testes contain *primary sex cells;* these are the cells that will eventually become the sperm cells. They have the diploid number of chromosomes. When these cells are mature, the homologous chromosomes come together in pairs. This process is called *synapsis*; sometimes, during the pairing, the chromosomes twist about each other and exchange parts. This is called *crossing-over*. Each of the chromosomes also replicates. The chromosomes line up in the center of the cell, the homologous members separate, and the cell divides. Each daughter cell now has the monoploid number; this division is known as *reduction-division*. Following this, the two cells split again, but this time by mitosis, in which the replicated chromosomes separate. There are now four cells, each with the monoploid number. These

cells reorganize so that most of the cytoplasm becomes fashioned into a tail, and they become sperm cells.

Oogenesis

The ovary contains *primary sex cells* with the diploid number. When each cell matures, the homologous chromosomes come together in pairs, as in spermatogenesis. During this pairing, or synapsis, the chromosomes may twist about each other and crossing-over occurs. Each of the chromosomes also replicates. The chromosomes line up in the center of the cell, the homologous members separate, and the cell divides by *reduction-division*. The two daughter cells are unequal in size, one being practically the size of the original cell and the other being a tiny *polar body*. Each has the monoploid number. They divide again, this time by mitosis, in which the replicated chromosomes separate. The larger cell again divides unequally to form a large cell and a tiny polar body, while the other polar body forms two more. The large cell is the egg; the three polar bodies disintegrate.

Fertilization

As the result of meiosis, sperm cells and ova containing the monoploid (n) number of chromosomes are produced. When a sperm fertilizes an egg, their nuclei unite, to form a zygote having the diploid ($2n$) number. The chromosomes then are in homologous pairs again, one of each pair originating in the sperm, the other in the egg. The species number of chromosomes is thus restored. By cleavage and differentiation, a new individual having the diploid number in each of its cells is produced. Each offspring has one set of chromosomes from the male parent and one set of chromosomes from the female parent. The hereditary characteristics of the individual are thus produced.

In flowers, the stamens have diploid cells that produce pollen containing monoploid nuclei. A mono-ploid egg nucleus is contained in the ovule. Following pollination, the pollen tube grows into the ovule. The sperm nucleus fertilizes the egg nucleus, resulting in a zygote with the diploid number of chromosomes. As the ovule develops into the seed, the zygote divides mitotically to produce an embryo with the diploid number. By differentiation and growth, a new plant is produced in which each cell has the diploid number of chromosomes.

Section Review

Select the correct choice to complete each of the following statements:

1. Nuclear division is known as (A) binary fission (B) mitosis (C) differentiation (D) cleavage (E) fertilization

2. The rodlike structures that appear during nuclear division are known as (A) spindle fibers (B) asters (C) nucleoli (D) chromosomes (E) centrosomes

3. Chromatin is to chromosomes as DNA is to (A) daughter cells (B) mitosis (C) genes (D) maturation (E) the electron microscope

4. The chromosome number of human gametes is (A) 12 (B) 23 (C) 46 (D) higher for eggs than sperms (E) higher for sperms than eggs

5. If the sperm cell of a fruit fly has 4 chromosomes, then the number of chromosomes in each body cell is (A) 2 (B) 4 (C) 6 (D) 8 (E) 16

6. Of the following cells, the only one to have the monoploid number of chromosomes is (A) skin (B) muscle (C) nerve (D) connective (E) ovum

7. The diploid number is restored as the result of (A) differentiation (B) fertilization (C) cleavage (D) reduction-division (E) maturation

8. Polar bodies are formed during (A) phototropism (B) oogenesis (C) spermatogenesis (D) the prophase stage (E) fertilization

9. During meiosis, the chromosome number (A) is doubled (B) remains the same (C) is reduced (D) becomes diploid (E) becomes tetraploid

10. Reduction-division occurs during the process of (A) cleavage (B) differentiation (C) fertilization (D) meiosis (E) parthenogenesis

11. When a chromosome duplicates, it forms a pair of (A) centrosomes (B) spindles (C) chromatids (D) centromeres (E) nucleoli

12. The pairing of homologous chromosomes is called (A) meiosis (B) synapsis (C) reduction-division (D) cell plate (E) centriole

13. Spindle threads are attached to the chromosomes at the (A) asters (B) mitochondria (C) prophase (D) interphase (E) centromeres

14. In replication, a chromosome (A) splits (B) separates (C) twists about another (D) lines up in the center of the cell (E) reduces

15. In flowers, the monoploid number is found in the (A) seed (B) fruit (C) stigma (D) pollen (E) epicotyl

Answer Key

1-B	4-B	7-B	10-D	13-E
2-D	5-D	8-B	11-C	14-A
3-C	6-E	9-C	12-B	15-D

Answers Explained

1. **(B)** During nuclear division, or mitosis, a nucleus divides in two, with each of the resulting cells receiving the same number of chromosomes as the original.

2. **(D)** During nuclear division, the chromatin material becomes arranged as short, stubby rods called chromosomes.

3. **(C)** The chromosomes are made of chromatin material; the genes, of DNA (deoxyribonucleic acid).

4. **(B)** When human gametes, containing 23 chromosomes each, unite in fertilization, they form a zygote, or fertilized egg, containing 46 chromosomes.

5. **(D)** When a fruit-fly sperm, containing 4 chromosomes, unites with an egg containing 4 chromosomes, the resulting fertilized egg has 8 chromosomes. As the egg divides repeatedly by mitosis, each of the body cells will have 8 chromosomes.

6. **(E)** The ovum receives the monoploid (haploid) number of chromosomes as the result of the process of meiosis.

7. **(B)** When the gametes, each of which contains the monoploid number, unite in fertilization, the fertilized egg will contain the diploid number.

8. **(B)** During oogenesis, the female primary sex cell divides into two unequal cells; the smaller one is the polar body. The two cells divide unequally once more to form a large cell and a small polar body; the other polar body forms two more. The large cell is the ovum; there are three polar bodies, which disintegrate.

9. **(C)** During maturation, or meiosis, the diploid chromosome number of the body cells becomes reduced to the monoploid number of the gametes.

10. **(D)** During meiosis, the primary sex cell divides into two cells, each of which has the monoploid number of chromosomes. This division is known as reduction-division.

11. **(C)** When a chromosome duplicates, it becomes two chromatids that are held together by the centromere. During the metaphase stage of mitosis, the split between the two chromatids appears more pronounced.

12. **(B)** During meiosis, the homologous chromosomes come together in pairs. This process is called synapsis. The chromosomes may twist about each other. Each of the chromosomes also replicates.

13. **(E)** During the late prophase stage, spindle threads become attached to the centromeres of the chromosomes.

14. **(A)** During replication, a chromosome splits lengthwise into two equal chromosomes.

15. **(D)** In flowers, the stamens have diploid cells that produce pollen containing monoploid nuclei.

10.2 INHERITANCE IN PLANTS AND ANIMALS

The science of heredity is called *genetics*. It is one of the newer fields of science, having made its start in 1900. However, 35 years earlier, an Austrian monk named Gregor Mendel (1822–1884) described a series of experiments that laid the foundations of our knowledge of heredity. Unfortunately, his paper, "Experiments in Plant Hybridization," did not receive recognition until after his death.

Mendel's Experiments

Mendel studied inheritance in garden pea plants. He selected seven contrasting characteristics and traced them from one generation to the next. He was careful to study only one of these traits at a time, thus avoiding the complex situation that frustrated previous scientists who had tried to trace the inheritance of many traits at a time, and had failed.

Mendel started out by establishing *pure* lines. He obtained *pure tall* plants by allowing the flowers of tall plants to self-pollinate for several generations. These plants were about 6 feet tall. In like manner, he obtained *pure short* plants; they were only about 1 foot tall. He then cross-pollinated the two types of plants. Pollen from tall plants was placed on the stigma of flowers on short plants; of course, Mendel first removed the stamens of the latter flowers, in order to prevent self-pollination. The flowers were then covered with paper bags to keep stray pollen from being introduced. The reverse type of cross-pollination was also

TABLE I

Pure Parents (P₁)			First Generation (F₁)
1. Tall	×	short	All tall
2. Smooth seeds	×	wrinkled seeds	All smooth seeds
3. Yellow cotyledons	×	green cotyledons	All yellow cotyledons
4. Colored seed coat (gray or brown)	×	white seed coat	All colored seed coat
5. Inflated pods	×	constricted pods	All inflated pods
6. Green pods	×	yellow pods	All green pods
7. Axial flowers	×	terminal flowers	All axial flowers

performed, in which pollen from short plants was placed on the stigmas of flowers on tall plants. Mendel collected the seeds and planted them. Would the plants be intermediate in height, short, or tall?

Law of Dominance

Much to Mendel's intense interest, all the seeds developed into plants that were as tall as the original plants. None of these first-generation (F₁) offspring appeared short, despite the fact that one of the parents had been short. These plants were called *hybrid* tall, since they had come from parents that were different in height, as compared with pure tall plants, whose parents had both been tall. When Mendel conducted similar experiments for the other characteristics, he obtained similar results; only one of the characteristics appeared, as is shown in Table I.

He called the characteristic that appeared *dominant*; the one that did not appear was called *recessive*. Mendel stated his results in the *Law of Dominance*: When organisms having pure contrasting traits are crossed, their offspring will show only one of the traits; the trait that appears is called the dominant trait, and the trait that does not appear is called the recessive trait.

Law of Segregation

Mendel next turned his attention to the hybrid tall plants, which were indistinguishable in appearance from the pure tall plants. He crossed hybrid tall plants, collected their seeds, and in the next, or second, generation (F₂) he observed both tall and short plants. Although the recessive trait seemed to have disappeared in the first generation, it now reappeared in the second generation. He obtained similar results when he crossed plants that were hybrid for the other characteristics. When he counted all the plants, he obtained the results shown in Table II.

Mendel observed that there were three times as many dominant offspring as recessive. Then, when he crossed the short plants in the F₂ with each other, he obtained only short plants. When he crossed the tall plants, he noticed that some gave rise only to tall plants; others gave rise to a 3:1 ratio of tall and short plants. The same results were obtained for the other characteristics. On the basis of all these data, Mendel summarized his findings in the *Law of Segregation*: When large numbers of hybrids are crossed, the factors segregate and then recombine to produce dominant and recessive offspring in a ratio of 3:1.

Genes and Inheritance

Modern-day knowledge of genes permits us to explain Mendel's results and to predict the possible offspring of various crosses. Homologous pairs of chromosomes have pairs of genes called *alleles*. Pure tall plants have two genes for tallness (*TT*) in every cell; one gene was contributed by the sperm, the other by the egg, and the fertilized egg therefore received two genes for tallness. Pure short plants likewise have two genes for shortness (*tt*). The capital letter, *T,* is used for the

TABLE II

F₁ Hybrid Parents	F₂ Results				Ratio in F₂
1. Tall	787	Tall	– 277	short	2.84 : 1
2. Smooth seeds	5,474	Smooth	– 1,850	wrinkled	2.96 : 1
3. Yellow cotyledons	6,022	Yellow	– 2,001	green	3.01 : 1
4. Colored seed coat	705	Colored	– 224	white	3.15 : 1
5. Inflated pod	882	Inflated	– 299	constricted	2.95 : 1
6. Green pod	428	Green	– 152	yellow	2.82 : 1
7. Axial flower	651	Axial	– 207	terminal	3.14 : 1
Totals	14,949	Dominant	– 5,010	recessive	2.98 : 1

dominant allele; the small letter, *t*, for the recessive allele. The method of transmission of the genes in the first generation may be depicted as follows:

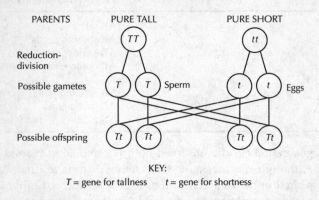

PARENTS   PURE TALL   PURE SHORT

Reduction-division

Possible gametes   Sperm   Eggs

Possible offspring

KEY:
T = gene for tallness   *t* = gene for shortness

Results: Appearance (phenotype)—all tall hybrid
Gene makeup (genotype)—100% *Tt*

Genes in the First Generation

The hybrid tall plants have a gene for tallness and a gene for shortness in every cell (*Tt*). Although the recessive gene for shortness is present, its effect is masked by the dominant gene for tallness. Such an individual possessing two unlike genes of an allelic pair is said to be *heterozygous* for the trait. The parents of the heterozygous tall plants are said to be *homozygous* for tallness and shortness since they possess like allelic genes, *TT* and *tt*. The (F₂) results may be depicted as follows:

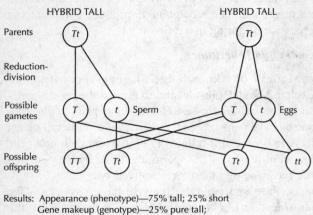

HYBRID TALL   HYBRID TALL

Parents

Reduction-division

Possible gametes   Sperm   Eggs

Possible offspring

Results: Appearance (phenotype)—75% tall; 25% short
Gene makeup (genotype)—25% pure tall;
50% hybrid tall; 25% pure short (1 : 2 : 1)

Genes in the Second Generation

In appearance, or *phenotype,* there are three times as many tall as short plants, or 75 percent tall and 25 percent short. The genetic makeup, or *genotype*, reveals a ratio of 1:2:1, or 25 percent pure tall, 50 percent hybrid tall, 25 percent pure short. In this example, a sperm with a gene for tallness may combine either with an egg containing a gene for tallness or with an egg containing a gene for shortness. Similarly, a sperm with a gene for shortness may combine with either type of egg. When many hybrid plants are crossed, both types of sperms will fertilize both types of eggs. The larger

the number used, the closer will the results be to the ideal ratio of 1:2:1. If only a few individuals are used, however, the actual results will rarely give this ratio. *Chance* determines which sperms and which eggs will unite. A demonstration to illustrate how chance operates may be performed as follows:

1. Flip two pennies at the same time, and observe whether the results are two heads, a head and a tail, or two tails.

2. Tally the results under three columns labeled TT, TH, HH.

3. Flip both coins together 100 times, and tally the results.

4. Summarize the results. It will be seen that the ratio is approximately 1:2:1. Compare with the results of flipping the coins only 4 times. There may or may not be a ratio of this type, because of the small number of cases involved.

Using the Punnett Square

A convenient method of working out different genetic crosses and the possible offspring involves the use of the Punnett square.

EXAMPLE 1:

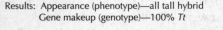

HYBRID TALL   ×   HYBRID TALL

Possible gametes

KEY:
T = tall   *t* = short

The possible male gametes are placed at the top of the Punnett square, one over each box; the possible female gametes are placed at the left side of the square, one alongside each of the boxes. Each box now represents the type of zygote resulting from the union of the male gamete, indicated at the top, and the female gamete, indicated at the side:

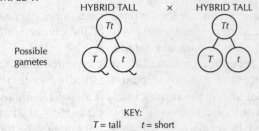

	T	*t*
T	*TT*	*Tt*
t	*Tt*	*tt*

Possible results:
50% *Tt*, hybrid tall
25% *TT*, pure tall
25% *tt*, pure short

EXAMPLE 2:

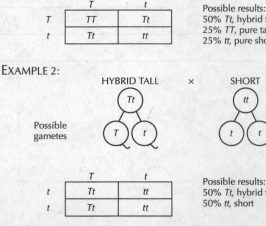

HYBRID TALL   ×   SHORT

Possible gametes

	T	*t*
t	*Tt*	*tt*
t	*Tt*	*tt*

Possible results:
50% *Tt*, hybrid tall
50% *tt*, short

EXAMPLE 3:

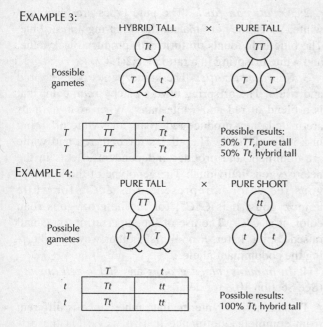

Possible gametes

	T	t
T	TT	Tt
T	TT	Tt

Possible results:
50% TT, pure tall
50% Tt, hybrid tall

EXAMPLE 4:

PURE TALL × PURE SHORT

Possible gametes

	T	t
t	Tt	tt
t	Tt	tt

Possible results:
100% Tt, hybrid tall

Law of Independent Assortment (Unit Characters)

Mendel was interested in determining whether characteristics are inherited independently from each other, or together. He started with plants that were pure for two contrasting traits: Tall plants with Yellow seeds x short plants with green seeds. The first-generation plants were hybrid for both characteristics, or dihybrids; they appeared Tall and had Yellow seeds. When these dihybrids were crossed, Mendel found that the characteristics were assorted independently, giving four different types of individuals: Tall Yellow, Tall green, short Yellow, short green, in a ratio of 9:3:3:1. Mendel stated this idea in his *Law of Independent Assortment, or Unit Characters*: Characteristics are inherited independently from each other, and are not affected by each other. We now know that the genes for each of the characteristics studied by Mendel were, by coincidence, located on separate chromosomes, thus making this law possible.

We may study dihybrid inheritance on the Punnett square as follows:

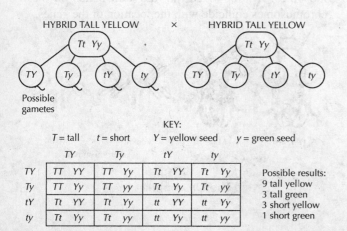

HYBRID TALL YELLOW × HYBRID TALL YELLOW

Possible gametes

KEY:
T = tall   t = short   Y = yellow seed   y = green seed

	TY	Ty	tY	ty
TY	TT YY	TT Yy	Tt YY	Tt Yy
Ty	TT Yy	TT yy	Tt Yy	Tt yy
tY	Tt YY	Tt Yy	tt YY	tt Yy
ty	Tt Yy	Tt yy	tt Yy	tt yy

Possible results:
9 tall yellow
3 tall green
3 short yellow
1 short green

Inheritance in Other Organisms

Since Mendel's day, the heredity of many other plants and animals has been carefully studied. In general, Mendel's findings were also found to apply to them, as is shown in Table III.

TABLE III

Organism	Dominant Trait	Recessive Trait
Cattle	Hornlessness	horned
Guinea pig	Black fur	white fur
Mice	Pigmented coat	white coat (albino)
Chickens	White feathers	pigmented feathers
Wheat	Late ripening	early ripening
Corn	Yellow grain	white grain
Pea	Colored flower	white flower
Barley	Beardless	bearded

Test Cross ("Back Cross")

An organism that is heterozygous has the same appearance, or phentotype, as the pure dominant. To determine whether a dominant-appearing individual is homozygous or heterozygous, a test cross, in which the unknown is crossed with a recessive, is performed. There are two possible results: (1) if any of the offspring are recessive. it indicates that the unknown was heterozygous; (2) if none of the offspring is recessive, it indicates that the unknown probably was homozygous. This can be illustrated with a black guinea pig that may be either heterozygous black or homozygous black. It is mated with a white one.

(1) If a white offspring is born, it shows that the black guinea pig was heterozygous. Thus:

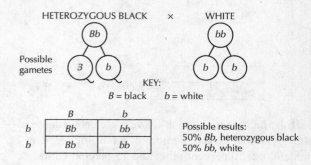

HETEROZYGOUS BLACK × WHITE

Possible gametes

KEY:
B = black   b = white

	B	b
b	Bb	bb
b	Bb	bb

Possible results:
50% Bb, heterozygous black
50% bb, white

(2) If all the offspring are black, there is a good possibility that the black guinea pig was homozygous. Thus:

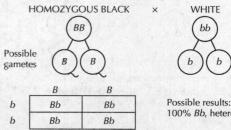

HOMOZYGOUS BLACK × WHITE

Possible gametes

	B	B
b	Bb	Bb
b	Bb	Bb

Possible results:
100% Bb, heterozygous black

Intermediate Inheritance

In some living things, there are characteristics that are neither dominant nor recessive. Heterozygous offspring are intermediate in appearance between their homozygous parents. Such intermediate inheritance, known as incomplete dominance, or codominance, is illustrated in the following examples:

1. ***Flowers of the Japanese four-o'clock and the snapdragon.*** A red flower crossed with a white flower will result in pink flowers. (Since neither red nor white is dominant, all the genes are designated by capital letters: RR = red, WW = white, RW = pink.)

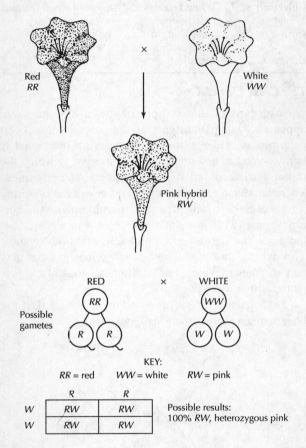

Red
RR

×

White
WW

Pink hybrid
RW

RED   ×   WHITE

RR   WW

Possible
gametes

R   R   W   W

KEY:

RR = red WW = white RW = pink

	R	R
W	RW	RW
W	RW	RW

Possible results:
100% RW, heterozygous pink

If the heterozygous pink flowers are crossed, the offspring possess all three colors in a ratio of 1:2:1, or 25 percent red, 50 percent pink, and 25 percent white.

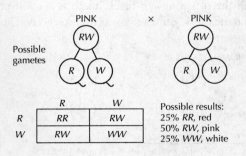

PINK   ×   PINK

RW   RW

Possible
gametes

R   W   R   W

	R	W
R	RR	RW
W	RW	WW

Possible results:
25% RR, red
50% RW, pink
25% WW, white

2. ***Andalusian fowl.*** The pure types are black and white. When they are mated, the offspring appear blue. The blue individuals, in turn, will produce black, blue, and white offspring in a ratio of 1:2:1.

3. ***Shorthorn cattle.*** The homozygous types are red and white. The offspring have a patchy *roan* color that is a blend of red and white hairs. When roan animals are mated, they produce red, roan, and white offspring in a ratio of 1:2:1. The alleles for both red and white are dominant and produce independent effects in the heterozygous individual. The genotype of the homozygous red color is represented as $C^R C^R$; for white homozygous, it is $C^W C^W$; for the heterozygous roan color, it is $C^R C^W$. The use of the superscript (the small raised capital letter, R or W) is another way of indicating the codominant allele.

4. ***In humans, blood groups*** and ***sickle-cell anemia*** (See Section 10.3 for details.)

To summarize: Intermediate inheritance is different from complete dominance in two ways: (1) the offspring have a different phenotype from homozygous parents that have opposite alleles; (2) the cross of heterozygous individuals gives a phenotype ratio of 1:2:1, rather than 3:1.

Morgan's Work

Drosophila, or fruit fly

Probably the most useful of all organisms in genetic research has been the diminutive *Drosophila melanogaster*, the fruit fly. Among its advantages are the following: it breeds, in large numbers, every 10 days; it is easily stored in half-pint bottles; it has only four pairs of clearly distinguishable chromosomes in each cell. Thomas Hunt Morgan received the Nobel Prize for his development of the gene theory, based on research with *Drosophila*. He and his co-workers drew up chromosome maps showing the locations of various genes on the chromosomes. Certain cells of the salivary glands have subsequently been shown to contain *giant chromosomes* about 100 times larger than normal chromosomes. It is believed that the dark bands on these chromosomes coincide with the location of the genes.

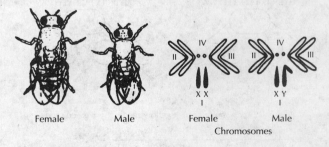

Female   Male   Female   Male
Chromosomes

The Drosophila Fly

Linkage

Organisms have many genes distributed along the length of the chromosomes, like beads on a string. When a gene for one characteristic is inherited, many other genes on the same chromosome are also inherited with it. This is known as linkage. In *Drosophila,* for example, the genes for black body and short wings are linked, or inherited, together. This type of inheritance, therefore, is an exception to Mendel's Law of Independent Assortment. This is possible because Mendel happened to study characteristics whose genes are located on different chromosomes.

Crossing-Over

Morgan also observed that during synapsis, in the first part of meiosis, the chromatids of homologous chromosomes sometimes twisted about each other, and then separated, exchanging adjacent parts. This resulted in sets of genes crossing over from one chromosome to another, and producing new combinations of characteristics that were inherited together. The frequency with which this occurred was also the basis for drawing up the chromosome maps of the genes.

Mutations

Gene Mutations

Genes sometimes change, altering the inheritance of the characteristics they control. Such a change is known as a mutation. The term was first used by Hugo DeVries (1848–1935) during his study of sudden changes in the evening primrose. Mutations have been found in many organisms, including white eyes, vestigal (very short) wings, and other traits in *Drosophila*; albinism (lack of all pigment) in white mice, frogs, and corn; hornlessness in cattle; seedlessness in oranges and grapes. Most mutations are probably harmful, since they upset the balance among the genes and change the ability of the organism to survive in the original environment. This causes the organism to die out (*lethal* effect). However, some mutations are useful and may improve the organism's chances of survival or its value to humans. The cause of these mutations is not known, although it is suspected that cosmic rays from outer space may be responsible for the occasional change in a gene.

Mutations were produced artificially in 1927 by Herman J. Muller, who exposed *Drosophila* flies to X rays. This treatment affected the genes in the reproductive organs, and resulted in offspring that showed changed characteristics. Muller received the Nobel Prize for this work. Other animals and plants have also shown mutations as the result of X ray exposure. Research has indicated that mutations may also be brought about by other *mutagenic agents,* such as ultraviolet rays, mustard gas, colchicine, and radioactive isotopes. Dr. Muller and other scientists have predicted that atomic warfare and possibly also atomic tests may cause undesirable mutations to appear among human beings. For this reason, as well as others, the birth rate and infant statistics in Hiroshima and Nagasaki were under observation by scientists.

Chromosomal Mutations

A change in the number of chromosomes, or in their structures, is known as a chromosomal mutation. One example is *nondisjunction*. Sometimes chromosomes fail to separate from one another during meiosis, resulting in gametes with slightly more or less than the normal monoploid number. After fertilization, the new individuals may have more or less than the normal 2n chromosome number, often leading to various defects.

Another example is *polyploidy*, in which an organism may have more than one entire set of chromosomes, leading to the triploid (3n), tetraploid (4n), etc., condition. A chemical, *colchicine*, which is used to treat gout, has been found to cause the chromosome number of plant cells to double. Its effect is due to its interference with mitotic division, preventing the chromosomes from spreading apart after they have split. This results in the cells having twice the normal number, or the tetraploid number, of chromosomes. Colchicine treatment has been used to produce giant marigold flowers.

Other types of chromosome changes have been studied, which also change the characteristics of organisms. In wheat and in jimson weed, for example, when tetraploid and diploid plants were crossed, the resulting offspring had the triploid number of chromosomes. In other cases, pieces of chromosomes become lost, or become attached to other chromosomes, during maturation. This changes the number of genes and alters the heredity of the offspring accordingly.

Genes and the Environment

How Genes Produce Their Effects

Although the mechanism of gene inheritance has been thoroughly studied, relatively little has been learned about how the genes produce their effects. The following facts are known: Frequently, several pairs of genes (*multiple alleles*) act together in affecting a particular characteristic; thus, in mice, gray coat color depends on the presence of two pairs of genes, while eye color of *Drosophila* is determined by 50 genes. Genes probably produce their effects by causing the production of enzymes that are needed to bring about certain chemical changes; thus, it has been shown that rabbits with black fur possess an enzyme that stimulates the production of black pigment, while white rabbits do not form this pigment; in another case, that of the mold *Neurospora,* a gene controls the production of an enzyme that permits the mold to synthesize a particular member of the vitamin B complex. A pair of genes may affect more than one trait; for example, *Drosophila* flies with vestigial wings have a shortened life span, indicating that other parts of the body are also affected by the particular gene called vestigial.

Interaction of Heredity and Environment

That the proper environmental condition is needed for genes to exert their effects can be demonstrated in the following manner: Germinate some corn seedlings in the dark, and some in the light. You will observe that, although all the seedlings have genes for the formation of chlorophyll, the ones kept in the dark will remain white. The control seedlings germinated in the light will show the normal condition of green color. If the first group of seedlings are now transferred from the dark to the light, they will become green.

A certain type of *Drosophila* inherits a condition of curly wings at a temperature of 25°C. When these flies are raised at a temperature of 16°C, however, their wings have the normal straight appearance. If the flies are mated, the offspring will continue to have straight wings at this temperature. However, if the temperature is raised to 25°C, the offspring will have curly wings.

The Himalayan rabbit inherits a color pattern in which the animal is white, with black ears, nose, paws, and tail. However, if some of the white fur is shaved off and an ice pack applied to the spot, the new fur that grows in will be black. If the fur on a paw is shaved off, and the paw kept warm in a wrapping, the fur grows back white. Apparently, the genes produce white fur on the warm parts of the body, and black fur on the extremities, which have a lower temperature.

Identical Twins

Identical twins have the same sets of genes, since they originated from one fertilized egg. If such twins are separated from birth and raised under different environments, this would amount to a controlled experiment as to the comparative effects of heredity and environment. Professor H. H. Newman of the University of Chicago conducted studies on a number of identical twins that had been separated in this way and found that (1) the two individuals resembled each other greatly, showing that their genetic makeup controlled their physical appearance; and (2) in some cases, the IQ was higher for an individual raised in a more favorable background, indicating that environmental factors help in the expression of native intelligence.

Section Review

Select the correct choice to complete each of the following statements:

1. The foundations for the study of heredity were first laid by   (A) Morgan   (B) Mendel   (C) Mendeleeff   (D) Muller   (E) Metchnikoff

2. In breeding experiments, paper bags are placed over flowers to   (A) keep them warm   (B) keep them from blowing away   (C) prevent self-pollination   (D) keep stray pollen away   (E) protect the flowers from excessive sunlight

3. The Law of Dominance is illustrated in the garden pea by   (A) homozygous tall x heterozygous tall   (B) heterozygous tall x heterozygous tall   (C) homozygous tall x homozygous tall   (D) pure short x pure short   (E) homozygous tall x pure short

4. In pea plants, all of the following are examples of dominant traits *except*   (A) wrinkled seeds   (B) tallness   (C) yellow cotyledons   (D) axial flowers   (E) inflated pods

5. When hybrids are crossed, the genotype ratio of the offspring is   (A) 1:1   (B) 3:1   (C) 1:2:1   (D) 4:1   (E) 3:2

6. If a pair of hybrid black guinea pigs are mated, and there are four offspring, their appearance may be   (A) all black   (B) 3 black, 1 white   (C) 2 black, 2 white   (D) 1 black, 3 white   (E) any of the above

7. A cross in which three-fourths of the offspring appear dominant is   (A) *Tt* x *TT*   (B) *TT* x *tt*   (C) *Tt* x *tt*   (D) *Tt* x *Tt*   (E) *TT* x *TT*

8. When Mendel crossed hybrids, he derived the (A) Law of Dominance (B) Law of Genes (C) Law of Incomplete Dominance (D) Law of Unit Characters (E) Law of Segregation

9. The possible gametes of *TTYy* will contain the genes (A) *TY, Ty* (B) *TT, Yy* (C) *TT, TY* (D) *TT, Ty* (E) *TT, YY*

10. A test cross is performed (A) only with hybrids (B) only with pure types (C) to determine whether an organism is heterozygous or homozygous dominant (D) only between recessives (E) only between homozygous dominants

11. In the four-o'clock flower, cross-pollinating pink flowers gives (A)100% pink (B) 100% red (C) 100% white (D) 50% pink, 25% red, 25% white (E) 50% red, 50% white

12. *Drosophila* is useful for all of the following reasons *except* (A) it breeds every 10 days (B) it produces many offspring (C) it has no linked characteristics (D) it has four pairs of chromosomes (E) it has giant chromosomes in its salivary glands

13. A change in genes is called a (A) genotype (B) phenotype (C) cross-over (D) back-cross (E) mutation

14. In peas, if 50% of the offspring are short, and 50% are tall, the parents were probably (A) *TT* x *tt* (B) *Tt* x *tt* (C) *Tt* x *TT* (D) *Tt* x *Tt* (E) *TT* x *TT*

15. Genes are not found in pairs in a(n) (A) sperm cell (B) fertilized egg (C) muscle (D) epithelial cell (E) zygote

16. In nondisjunction, chromosomes fail to (A) replicate (B) reduce (C) separate (D) fertilize (E) recombine

17. The exchange of genes between homologous chromosomes is called (A) polyploidy (B) chromatids (C) crossing-over (D) test cross (E) roans

18. All the following are mutagenic agents *except* (A) X rays (B) colchicine (C) carbon dioxide gas (D) mustard gas (E) ultraviolet rays

19. Union of a gamete having the monoploid number of chromosomes and a gamete having the diploid number results in a zygote that is (A) monoploid (B) diploid (C) triploid (D) tetraploid (E) none of these

20. In the Himalayan rabbit, low temperatures cause the growth of (A) white fur (B) black fur (C) white paws (D) white nose (E) black genes

Answer Key

1-B	5-C	9-A	13-E	17-C
2-D	6-E	10-C	14-B	18-C
3-E	7-D	11-D	15-A	19-C
4-A	8-E	12-C	16-C	20-B

Answers Explained

1. (B) Gregor Mendel (1822–1884) laid the foundation for the study of heredity when he conducted a number of experiments on garden pea plants.

2. (D) By covering the flowers with paper bags, stray pollen was prevented from introducing factors that were not part of the experiments.

3. (E) The Law of Dominance states that, when organisms containing pure contrasting traits are crossed, the offspring will show only one of the traits, the dominant one; the trait that does not appear is called recessive. In a cross of homozygous tall with pure short, only tall (heterozygous) garden peas result.

4. (A) Smooth seeds are dominant to wrinkled seeds.

5. (C) A cross of hybrids yields a genotype ratio of 1:2:1, in which the genes are distributed as 25% homozygous dominant, 50% heterozygous dominant, and 25% homozygous recessive.

6. (E) Large numbers of offspring are needed to give the ideal ratio. With a small number, the actual ratio will rarely be the expected one.

7. (D) When hybrids are crossed, the ratio shows 25% homozygous dominant, 50% heterozygous dominant, and 25% homozygous recessive. In short, 75%, or three-fourths, will appear dominant.

8. (E) The Law of Segregation states that, when large numbers of hybrids are crossed, the factors segregate and then recombine to produce dominant and recessive offspring in a ratio of 3:1.

9. (A) This illustrates the Law of Independent Assortment, which states that characteristics are inherited separately from each other. When reduction division of *TTYy* takes place, the genes separate into *TY* and *Ty* gametes.

10. (C) An organism that is hybrid has the same appearance as the pure dominant. In a test cross, a dominant is crossed with a recessive. If any of the offspring are recessive, it indicates that the organism was heterozygous, or hybrid. If none of the offspring is recessive, it indicates that the unknown probably was homozygous dominant, or pure dominant.

11. (D) Pink flowers are hybrid. Their offspring appear in the ratio of 25% red, 50% pink, and 25% white. This illustrates incomplete dominance.

12. (C) Linked characteristics are inherited together because their genes are on the same chromosome. Example: the genes for black body and short wings are linked.

13. (E) Genes sometimes change, altering the inheritance of the characteristics they control. Such a change is called a mutation.

14. (B) The heterozygous tall plants, *Tt*, produce two types of gametes, one bearing *T*, the other bearing *t*. The homozygous recessive short plants, *tt*, produce only gametes containing *t*. When these unite with gametes containing *T*, only *Tt* heterozygous tall plants result. However, if they unite with gametes containing *t*, only *tt* homozygous short plants result. By the laws of chance, 50% of each result.

15. (A) Pairs of genes are separated when the pairs of chromosomes that contain them go through reduction-division during meiosis. A sperm cell will therefore receive only one gene of a pair.

16. (C) In nondisjunction, chromosomes sometimes fail to separate from one another during meiosis. As a result, the gametes may have slightly more or less than the normal monoploid number. Defects may result in a new individual who has more or less than the normal $2n$ chromosome number.

17. (C) During synapsis in the first part of meiosis, the chromatids of homologous chromosomes may twist about each other and then separate, exchanging adjacent parts. This results in sets of genes crossing over from one chromosome to another and producing new combinations of characteristics.

18. (C) Mutagenic agents cause mutations and change genes, altering the inheritance of the characteristics they control. Carbon dioxide gas has not been shown to be such an agent.

19. (C) When a gamete with the monoploid number of chromosomes, n, unites with a gamete having the diploid number, $2n$, the result is a zygote that has a combination of both n and $2n$, or $3n$, the triploid number.

20. (B) The genes of the Himalayan rabbit produce white fur on the warm parts of the body and black fur on the extremities, which have a lower temperature.

10.3 HUMAN INHERITANCE

The laws of heredity apply to humans as well as to other living things. However, relatively little is known about human inheritance for the following reasons: (1) about 25 years are required for one generation; (2) small numbers of offspring are produced; (3) planned experiments are not possible; (4) description of ancestors on the basis of records or memory is not too reliable.

Inherited Traits

Eye Color

It is known that brown eyes are dominant to blue. Therefore, blue-eyed parents have only blue-eyed children. Brown-eyed parents may have blue-eyed children if both are heterozygous for brown eyes. However, if both are homozygous for brown eyes, or if one parent is heterozygous and the other is homozygous, there will be no blue-eyed children. This can be shown as follows:

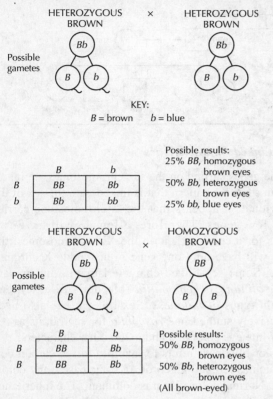

However, it is well to remember that we are considering only possibilities. If there are four children in a family, do the above ratios apply? The answer is—possibly, but not necessarily. When both parents are heterozygous for brown eyes, there may actually be two or more blue-eyed children—or none at all. Four is a small number, statistically speaking, and the law of chance decides which combination of gametes will take place. The different shades of eye color, from dark brown to light gray, lead to the conclusion that several genes are involved in the determination of eye color.

Other Human Traits

Some other traits among human beings are summarized in the following table:

Dominant	Recessive
Dark hair	light hair
Curly hair	straight hair
Nomal pigmentation	albinism
Free ear lobe	attached ear lobe
White forelock	normal hair color
Rh-positive blood factor	Rh-negative blood factor
Normal color vision	color blindness
Bitter taste reaction to PTC (phenylthiocarbamide)	no taste reaction to PTC

The blood types A, B, AB, and O are inherited in predictable ways, based on the action of three alleles. The allele for type A, I^A, and the allele for type B, I^B, are both dominant to the third allele, i. Although not dominant to each other, I^A and I^B may appear together in type AB. When both are absent, type O results. The possible genotypes for the various blood groups may be summarized as follows:

Blood Group	Genotype
A	$I^A I^A$; or $I^A i$
B	$I^B I^B$; or $I^B i$
AB	$I^A I^B$
O	ii

Rh-positive blood factor is dominant to Rh-negative. When a father is Rh positive and the mother is Rh negative, the developing child is usually Rh positive. As mentioned in Section 6.3, this leads to complications later because the mother builds antibodies against the Rh-positive blood of the developing child. This is serious if the mother becomes pregnant again, but a vaccine has been developed that, when injected into an Rh-negative mother, protects the fetus from developing Rh blood disease.

A demonstration to show inheritance of taste reaction to the chemical phenylthiocarbamide (PTC) may be performed as follows:

1. Distribute a piece of PTC-impregnated paper to several people, and have them chew it. (PTC taste papers may be obtained from the American Genetic Association, Washington, D.C.)

2. Note that some will experience an extremely bitter taste; others will report no taste sensation whatever.

The tasters may be able to test the reactions of members of their family. Since the ability to taste PTC is dominant, there will be more tasters than nontasters.

Tracing Inheritance

The inheritance of a human trait for several generations is traced on a pedigree chart as shown in Table IV.

Although the mother and father have brown eyes, two of their three children have blue eyes. Their blue-eyed daughter, Mary, married blue-eyed Charles and had blue-eyed children. The mother and father's brown-eyed son, John, married brown-eyed Anne and had only brown-eyed children. It is obvious that the mother and father are heterozygous for brown eyes. Since Mary and her husband are recessive for blue eyes, all their children have blue eyes. On the other hand, son John and his wife are brown-eyed and have five brown-eyed children; since five is a small number, statistically speaking, it is not possible to state whether they are homozygous or heterozygous for brown eyes.

TABLE IV

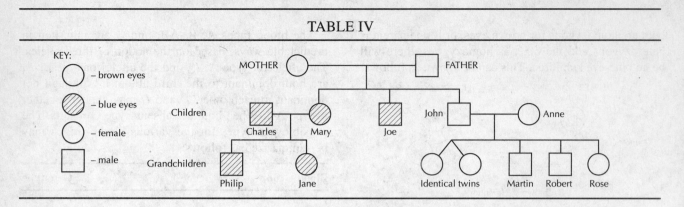

KEY:
- ◯ – brown eyes
- ◯ (hatched) – blue eyes
- ◯ – female
- ▢ – male

Sex Inheritance

As mentioned earlier, chromosomes occur in pairs that are alike in males and females.

The Sex Chromosomes

One pair of chromosomes, however, has been identified as the sex chromosomes. The other chromosome are called *autosomes*. Thus, of the 46, or 23 pairs, of chromosomes in the cell, 22 pairs are autosomes. One pair are the sex chromosomes. In females, they are alike and are known as X chromosomes, or as XX. In males, they are different, one being smaller than the other. The latter is known as the Y chromosome; the other, as X. Males thus have XY chromosomes. When gametes are formed, these chromosomes separate by reduction-division into separate cells. All eggs have one of the X chromosomes. A sperm cell, however, may have either an X or a Y chromosome. This makes two kinds of combinations possible:

1. If a sperm with an X chromosome fertilizes an egg, the offspring will have XX chromosomes and will be a female.
2. If a sperm with a Y chromosome fertilizes an egg, the offspring will have XY chromosomes and will be a male.

Recently, a gene on the Y chromosome was discovered that is responsible for the male characteristics. It becomes active in the seventh week of pregnancy, causing the fetus's immature sex organs to develop into testes. Without it, the sex organs develop into ovaries.

From this, it follows that the sex of a child is determined at the time of fertilization and also that the sperm determines whether a child will be male or female.

Sex chromosomes have also been identified in *Drosophila* and other organisms. In birds and moths, however, the male has two identical sex chromosomes (ZZ), while the female has two unlike chromosomes (ZW).

Sex-Linked Characteristics

The genes carried on the sex chromosomes are inherited in a certain pattern that is linked with the sex of the individual. It appears that the Y chromosome carries very few genes. Therefore, a female will have two genes for a characteristic in the XX chromosomes; the male will have only one gene in his single X chromosome. Some sex-linked characteristics in humans are *color blindness, hemophilia* (bleeder's disease), and certain types of *baldness*. Sex linkage has also been extensively studied in *Drosophila* for such traits as red and white eye color and about 200 other characters.

Color blindness refers to inability to see certain colors such as red and green or, in some cases, any colors at all. Normal color vision is dominant. The inheritance of color blindness may be traced as follows:

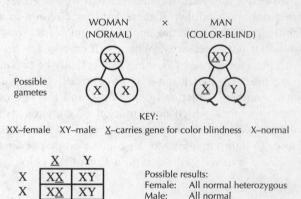

KEY:
XX–female   XY–male   X̲–carries gene for color blindness   X–normal

	X̲	Y
X	XX̲	XY
X	XX̲	XY

Possible results:
Female: All normal heterozygous
Male: All normal

The children of a color-blind man and a normal (homozygous) woman are thus normal. The daughters, however, are heterozygous, or *carriers*. The effect of one of these females bearing children to a normal male may be shown as follows:

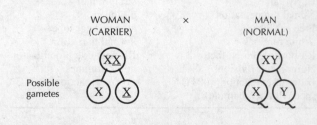

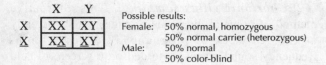

	X	Y
X	XX	XY
X	XX	XY

Possible results:
Female: 50% normal, homozygous
          50% normal carrier (heterozygous)
Male: 50% normal
          50% color-blind

In this case, there is a 50–50 chance that the sons will be color-blind. It can be seen that there is more likelihood that color-blind people will be males; there are relatively few color-blind females. The same pattern of inheritance applies to hemophilia.

Inheritance of Defects

Gene Defects

Some inherited diseases are known to be caused by gene defects:

1. *Hemophilia* is inability of the blood to clot. As a result, a hemophiliac may bleed to death. This, too, is a recessive trait. Its inheritance follows the same pattern as that for color blindness.

2. *Phenylketonuria* (PKU) is a condition of feeble-mindedness caused by the lack of a dominant gene that normally leads to conversion of the amino acid phenylaline into the amino acid tyrosine. The presence of the homozygous recessive genes leads to failure to produce a liver enzyme that causes this conversion. Phenylaline accumulates and causes brain damage. Fortunately, a simple PKU blood test has been developed for infants, which identifies the condition. The disease can be corrected by feeding the baby a special formula containing very little phenylaline.

3. *Sickle-cell anemia* is also inherited as a recessive trait. When the oxygen supply is low, the red blood cells take on a sickle shape and do not pass readily through the capillaries. This blocks the flow of blood, causing swelling and pain. The abnormal blood cells tend to break up. The homozygous condition usually leads to early death. It has been found that the hemoglobin molecules differ from normal hemoglobin by only a single amino acid out of a total of several hundred. There is a substitution of the amino acid *valine* for the normal *glutamic acid* in the hemoglobin; otherwise the two hemoglobin molecules are identical. Apparently, a mutation occurred in the allele for glutamic acid in the hemoglobin molecule.

Sickle-cell anemia is common in certain parts of Africa and is now carried by a number of Afro-Americans. In heterozygous individuals the disease may be present in mild form. It has the advantage of providing greater resistance to malaria than usual. In this respect, the defective gene has survival value and has continued to be inherited.

4. *Tay-Sachs disease* is a fatal condition that affects very young children. It is caused by the lack of a dominant gene that normally provides for the presence of an enzyme, hexosaminidase-A, which assists in the breakdown of fatty substances and other cell products. The homozygous recessive genes cause the enzyme to be missing. Fatty substances build up in the cells of the newborn child, particularly in the brain, leading to the loss of most normal body functions. The disease primarily affects Jewish individuals of central and eastern European extraction. There is no known cure. People who are considering marriage and whose families have a record of the disease are usually advised to consult their doctor. A carrier of the Tay-Sachs gene can be identified by a simple blood test.

5. *Thalassemia*, or *Cooley's anemia*, is due to inability of the body to produce red blood cells with a normal amount of hemoglobin. As a result, the red blood cells are misshapen and rapidly destroyed. The child becomes anemic at the age of 1 or 2 years. The liver and spleen become enlarged. The structure of the skull and the bones of the face are typically affected, and growth and development are poor. The only treatment at present is to give blood transfusions regularly. The disease is relatively common among people of Italian or Greek descent who trace their ancestry to the Mediterranean region. It is due to a recessive gene; in other words, the homozygous recessive condition will lead to the disease. Heterozygous individuals, or carriers, are completely normal. As in the case of Tay-Sachs disease, if two carriers marry, they run a risk of one in four that a child will have thalassemia.

6. The genes that cause such crippling, inherited diseases as *Lou Gehrig's disease (amyotrophic lateral sclerosis, ALS)* and *Huntington's disease*, a gene that causes *colon cancer*, and a gene responsible for 40 percent of *heart attacks* have recently been identified.

Chromosomal Aberrations

Through the recent perfection of techniques for growing human cells in tissue culture, it has been possible to examine the chromosomes with increasing detail. Certain diseases, or other conditions affecting humans, have been identified as being caused by abnormalities, or *aberrations,* in some of the chromosomes. *Down syndrome*, in which there is physical and mental retardation, has been traced to the presence of an extra chromosome in chromosome pair number 21, resulting from nondisjunction. Such an individual has a total of 47 chromosomes. Recently a gene on chromosome 21 has been identified as producing a protein in the brain that leads to the production of Down syndrome.

Another condition resulting from an extra chromosome is *Klinefelter's syndrome*, where a male has XXY chromosomes, and is underdeveloped and mentally retarded. Other variations in chromosomes are receiving increased attention as important factors in many physical and mental diseases and in behavior disorders.

Prenatal Testing

It is now possible to remove and examine the cells of a developing fetus to determine if it is carrying chromosomes or genes for an inherited abnormality. The technique is called prenatal testing, because it deals with an unborn fetus that is still developing in the uterus. While the procedures are relatively simple, there are some doubts about their complete safety. Two such methods are described below.

1. *Amniocentesis.* In this relatively simple and safe procedure, a doctor inserts a thin, hollow needle into the uterus of an expectant mother during the fourth month of pregnancy and withdraws a few cubic centimeters of the amniotic fluid. This is the liquid that surrounds the fetus and contains free cells that are normally shed by the developing fetus. The liquid is whirled in a centrifuge, and the cells are concentrated in a mass. They are carefully grown in nutrient culture and then studied for biochemical or chromosome defects.

The number and shape of the chromosomes are inspected. If they are defective, this fact helps determine whether they might lead to abnormalities in the baby. Thus, the presence of an extra chromosome in chromosome number 21 may lead to Down syndrome.

This process is especially being followed in families where there is a history of a genetic defect that could affect a child mentally or physically. About 100 different genetic disorders are being diagnosed through this technique. If it reveals a defect, as happens in about 5 percent of the cases, the pregnant woman may be given the benefit of special advice called *genetic counseling*. As a result, she may decide to terminate the pregnancy. The couple may try again later to have a healthy baby.

2. *Chorionic villus sampling.* This alternative method of prenatal diagnosis provides results as early as the 9th week of pregnancy. It utilizes high-frequency sound waves, recorded on a screen, to help guide the insertion of a thin, hollow tube through the cervix into the uterus of a pregnant woman. Cells from the developing fetus are then suctioned out from extensions of the chorion, a membrane that surrounds the fetus. These cells have the same genetic makeup as the fetus itself and can be analyzed for defects such as Down syndrome, PKU, thalassemia, and sickle-cell anemia. However, not all doctors consider this procedure to be safe.

Karyotype

A karyotype is a photograph or diagram of the chromosomes of a cell arranged in homologous pairs. It is prepared by staining the chromosomes in a cell and photographing them. The picture is then cut into sections, and the chromosomes are matched together in pairs according to length. Each pair is given a number. The pairs may differ considerably from one another in size of chromosomes and in position of centromeres.

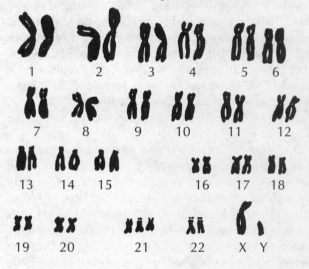

A Human Karyotype

In the accompanying illustration of a human karyotype, the 23 pairs of chromosomes are clearly visible. The fact that the last pair of sex chromosomes consists of a very short Y chromosome and a medium-length X chromosome indicates that the individual is a male. In a female the sex chromosome would be XX.

This karyotype also reveals an abnormality: chromosome pair 21 has an extra chromosome. This condition occurred when chromosomes failed to separate from one another during meiosis. The resulting gametes contained one more than the haploid number, so that,

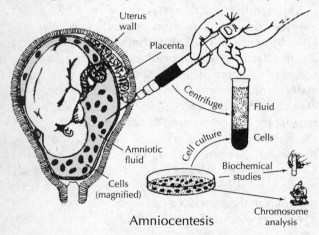

Amniocentesis

after fertilization, the new individual has one more than the normal *2n* number. The presence of three chromosomes in pair 21 results in *Down syndrome*, a serious abnormality marked by mental and physical retardation.

Eugenics

Eugenics is the science of improving the human race by applying our knowledge of genetics. It was founded by Sir Francis Galton (1822–1911). Eugenics does not have much appeal among modern biologists because human heredity is too complex for specific guidelines to be followed. The subject is included here for its historical interest only. Eugenicists have based their recommendations on a study of pedigrees over a number of generations in such families as the Edwards, Darwins, Jukes, and Kallikaks. However, their conclusions have not been definitive because of the environmental factors involved.

Section Review

Select the correct choice to complete each of the following statements:

1. If two parents who are hybrid for brown eyes have four children, the eye colors of the children may be (A) all brown   (B) 3 brown, 1 blue   (C) 2 brown, 2 blue   (D) 1 brown, 3 blue   (E) any of the above combinations

2. All of the following human traits are recessive *except* (A) curly hair   (B) albinism   (C) Rh-negative blood   (D) color blindness   (E) light hair color

3. In the accompanying pedigree of a family, in which brown eye color is designated as ◯, and blue eye color as ●,   (A) Alice and Joseph are pure for brown eyes   (B) Mary and Charles are hybrid for blue eyes   (C) Martin and Robert may be hybrid or pure for brown eyes   (D) John and Anne are pure for brown eyes   (E) Philip and Jane are hybrid for blue eyes

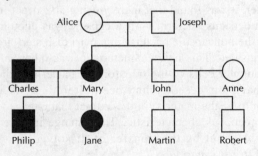

4. The human sperm cell contains   (A) 46 chromosomes   (B) 44 chromosomes and 2 X chromosomes   (C) 44 chromosomes, and 1 X and 1 Y chromosome   (D) 23 chromosomes   (E) 23 chromosomes, and 1 X and 1 Y chromosome

5. If a man who is color-blind marries a woman who is pure normal for color vision, the chances of their sons being color-blind are   (A) 0   (B) 100:1   (C) 50:50   (D) 75:25   (E) 25:50:25

6. A son with hemophilia will most likely result from parents represented as (Key: X–chromosome with gene for normal; X̲–chromosome with gene for hemophilia)   (A) XX × X̲Y   (B) XX × XY̲   (C) XX × X̲Y̲   (D) XX × XY   (E) XX̲ × XY

7. The blood-group alleles $I^A i$ are for blood group (A) O   (B) A   (C) B   (D) AB   (E) M

8. PKU is caused by the lack of a gene for   (A) a hormone   (B) brain cells   (C) an enzyme   (D) liver cells   (E) intelligence

9. The scientist who founded eugenics was   (A) Eijkman   (B) Gorgas   (C) Jukes   (D) Galton   (E) Dalton

10. The hemoglobin of people with sickle-cell anemia differs from normal hemoglobin by one   (A) lipid   (B) monosaccharide   (C) disaccharide   (D) fatty acid   (E) amino acid

Answer Key

1-E	3-C	5-A	7-B	9-D
2-A	4-D	6-E	8-C	10-E

Answers Explained

1. (E) A person who is hybrid for brown eyes contains the genes for both brown and blue eyes. Four is a small number, statistically speaking, and the law of chance decides which combination of gametes will take place. A ratio can be predicted only when large numbers are involved.

2. (A) In humans, curly hair is dominant to straight hair.

3. (C) Brown eyes are dominant to blue eyes. Both Martin and Robert are brown-eyed and could be either hybrid or pure for brown eyes. Their parents, John and Anne, are likewise brown-eyed. John could be hybrid for brown eyes because his parents, Alice and Joseph, are brown-eyed, but have a blue-eyed child, Mary, as well. This indicates that Alice and Joseph must have a recessive gene each, and are hybrid. John could also be hybrid, and this could be true of his two sons, Martin and Robert.

4. (D) The diploid number in humans is 46. By reduction-division, the sperm cells each receive half the number, or 23 chromosomes.

5. (A) The color-blind man has only one gene for color blindness on the X chromosome; the Y chromosome is practically empty of genes. When his Y chromosome unites with the X from the pure normal woman, the resulting zygote will have no genes for color blindness.

6. (E) The mother, as represented by X<u>X</u>, is normal, but a carrier for hemophilia. During reduction-division, she can form eggs with either X or <u>X</u> chromosomes. If an <u>X</u>-chromosome egg is fertilized by a sperm containing a Y chromosome, the zygote, that is, <u>X</u>Y, will develop into a son with hemophilia.

7. (B) The blood types are based on three alleles, I^A, I^B, and i. The allele for type A, I^A, is dominant to allele i.

8. (C) In PKU (phenylketonuria), a condition of feeble-mindedness results from the presence of homozygous genes that lead to failure in the production of an enzyme in the liver. This enzyme brings about the conversion of phenylaline, an amino acid, into another amino acid, tyrosine. Lack of the enzyme causes the accumulation of phenylaline, resulting in brain damage.

9. (D) Sir Francis Galton (1822–1911) founded eugenics, the science of trying to improve the human race by applying knowledge of genetics.

10. (E) In sickle-cell anemia, there is a substitution of the amino acid valine for the normal glutamic acid in the hemoglobin molecule. Otherwise the two hemoglobins are identical.

10.4 MODERN GENETICS

Evidence About DNA

In recent years, evidence has accumulated showing that the nucleic acid DNA (deoxyribonucleic acid) makes up the gene and determines heredity. Some of this evidence is based on transformation, bacteriophage activity, and transduction.

Transformation

The first indication of DNA action was found in the pneumococcus bacteria that cause pneumonia. These bacteria cells have a covering or capsule around them. Otherwise, they are similar to other pneumococci that do not have capsules and that do not cause the disease. In the investigations of Frederick Griffith in 1928, the harmless bacteria were injected into mice along with dead bacteria of the capsule type. The mice developed pneumonia! When the live bacteria were withdrawn from the sick mice and examined, they were found to have capsules. They had been transformed by something from the dead bacteria.

Later, it was found that merely using an extract from pneumococcus bacteria with a capsule was enough to make the noncapsule bacteria in a petri dish start growing capsules. This new characteristic was inherited by the daughter cells of the transformed bacteria, indicating that their heredity had been altered. DNA was shown to be the material in the extract that changed the pneumococcus characteristic. This change in heredity brought about by the transfer of dissolved DNA is called *transformation*.

Bacteriophage Activity

A bacteriophage is a virus that attacks bacteria. Like many other viruses, its structure consists of a central core containing DNA surrounded by a protein covering. To determine whether both of these parts or only one of them attacks a bacterial cell, a classic experiment was performed in which the DNA core was tagged with radioactive phosphorus and the protein coat with radio-sulfur. When the virus attacked a bacterial cell, the DNA entered through the tail end, which became attached to the cell; the empty shell of the virus was left behind, and was shown to contain the radiosulfur.

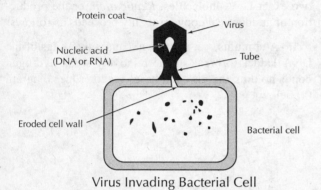

Virus Invading Bacterial Cell

Once inside, the DNA replicated to form more DNA molecules. The radioactive phosphorus was shown to be in the bacterial cells. The DNA molecules formed coats of protein around themselves to complete the structure of many additional new viruses. These were released when the bacterial cell burst. Thus, it was shown that the hereditary material of a virus enters a cell as DNA and redirects its metabolism to form additional DNA molecules with protein coverings.

Some viruses, including the tobacco mosaic virus, may contain RNA (ribonucleic acid) in their core, instead of DNA.

Transduction

It was found that in transduction, when a virus attacks a bacterial cell and reproduces itself at the expense of the cell, some of the DNA of the bacterial cell may be incorporated into the viruses that are formed. Then, when the bacterial cell bursts, these viruses attack other bacteria, and bring with them the DNA from the original bacteria. This is another indication that the genetic material carried from bacteria to bacteria is DNA.

The DNA Molecule

DNA Structure

The DNA molecule is composed of thousands of smaller units called *nucleotides*. A nucleotide consists of three parts: (1) a five-carbon sugar, deoxyribose; (2) phosphate; and (3) a nitrogen base, which may be either adenine, thymine, guanine, or cytosine. Thus, there are four types of DNA nucleotides, depending on which nitrogen base they contain.

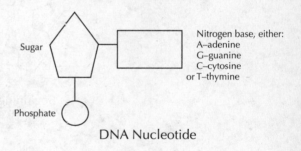

DNA Nucleotide

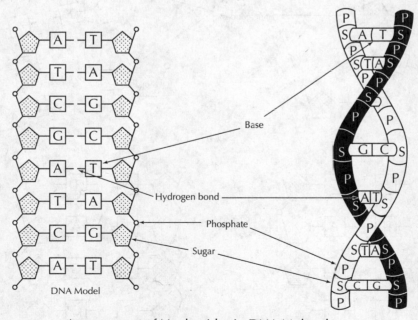

Arrangement of Nucleotides in DNA Molecule

Watson-Crick Model

The arrangement of the nucleotides was described in 1953 by James D. Watson and Francis H. C. Crick in their now-famous model of DNA. It was based on X ray diffraction data supplied by Rosalind Franklin and Maurice H. F. Wilkins. Watson, Crick, and Wilkins received the Nobel Prize in 1962. According to the model, the DNA molecule has the following characteristics: (1) there is a ladder-type organization; (2) it is twisted in a double spiral, or helix; (3) the upright parts are made of deoxyribose sugar, S, and phosphate, P; (4) the rungs are made of paired nitrogen bases; (5) these bases are always paired as follows: adenine with thymine, A-T; and cytosine with guanine, C-G; (6) there is a relatively weak hydrogen bond between each of these bases; (7) a single molecule of DNA is made up of thousands of these nucleotides arranged in spiral fashion.

DNA Replication

DNA molecules are present in the chromosomes. When the chromosomes replicate during mitosis and meiosis, each DNA molecule also replicates, to form two exact DNA molecules. *Replication*, or the production of an identical copy, probably proceeds as follows:

1. A stimulus, yet unknown, causes the rungs of the DNA ladder to break at the relatively weak hydrogen bonds holding the nitrogen bases together [see diagram (a) below].

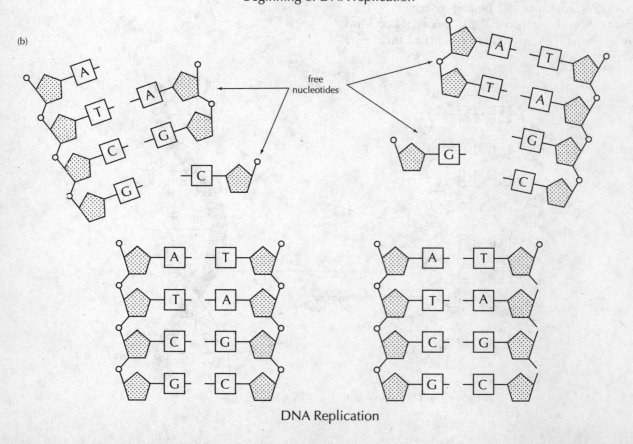

(a)

Beginning of DNA replication

(b)

free nucleotides

DNA Replication

2. The spiral ladder is now present as two half-ladders.

3. There are numerous free nucleotides present in the cell, which move into position to form a bond with the matching nucleotide of the half-ladder [see diagram (b)].

4. The bases are matched, so that adenine (A) bonds with thymine (T), and cytosine (C) with guanine (G).

5. Gradually a new upright portion is added to the half-ladder to complete two sections of a new spiral ladder [see diagram (c)].

6. Each half-ladder has now added an exact duplicate of the other, giving two identical DNA molecules.

RNA (Ribonucleic Acid)

DNA contains the hereditary code for the various characteristics of an organism. It also directs the formation of specific enzymes that are involved in the cell's activities. DNA is located in the nucleus of the cell, and sends its instructions for making enzymes into the cytoplasm through the nucleic acid RNA (ribonucleic acid).

RNA, like DNA, is composed of nucleotides. However, it shows these differences from DNA: (1) its five-carbon sugar is ribose, not deoxyribose; (2) RNA has the same nitrogen bases, adenine (A), cytosine (C), and guanine (G), but its fourth base is uracil (U) instead of thymine (T); (3) the RNA molecule is a single strand, while the DNA molecule has a double strand or, in some cases, a single strand.

There are three different types of RNA in the cell:

1. *Messenger RNA* (mRNA) carries the code from DNA in the nucleus to the ribosomes in the cytoplasm.

2. *Transfer RNA* (tRNA) picks up amino acid molecules in the cytoplasm and transfers them to the ribosomes. There are more than 20 kinds of tRNA, one for each kind of amino acid.

3. *Ribosomal RNA* (rRNA), located in the ribosomes, lines up the amino acids in the sequence dictated by the mRNA molecule for the formation of proteins.

Protein Synthesis

Protein synthesis consists of the building up of complex proteins from combinations of amino acids. This takes place in the ribosomes, which contain most of the RNA of the cell. The formation of enzymes, which are proteins, probably takes place in the following manner:

1. A portion of a DNA molecule, which is a gene, serves as a template, or pattern, for the synthesis of messenger RNA from free RNA nucleotides in the nucleus.

2. Some of the weak hydrogen bonds between the nucleotides break, and a portion of the DNA strands separate.

3. The free RNA nucleotides line up next to the appropriate DNA nucleotides—cytosine (C) with guanine (G), and uracil (U) with adenine (A) [see diagram (a) on the following page]. RNA contains a uracil instead of a thymine base. The process of transferring the coded information from DNA to the new strands of messenger RNA is called *transcription*.

4. The newly formed RNA molecule, which is a "reverse copy" of the DNA that produced it, separates from the DNA strand, moves through a pore in the nuclear membrane, and enters the cytoplasm. The messenger RNA molecule now carries the genetic code, which consists of triplets of nucleotides called *codons*.

5. The new messenger RNA becomes located in a ribosome; the order of arrangement of its nucleotides was determined by the arrangement of nucleotides on the DNA molecule.

6. Small transfer RNA molecules pick up specific amino acid molecules in the cytoplasm, and line up with messenger RNA.

7. Each transfer RNA molecule has a group of three nucleotides called an *anticodon*, and its bases fit the appropriate codon with its three bases of the messenger RNA; for example, if transfer RNA has an anticodon containing the sequence UGC, it will fit in with the messenger RNA codon ACG; U (uracil) always bonds with A (adenine), and C (cytosine) always bonds with G (guanine) [see diagram (b)].

8. The arrangement of the nucleotides on messenger RNA dictates the order in which the amino acids are lined up and bonded together into polypeptide chains.

9. A new protein is formed by dehydration synthesis in this way, ready to be used by the cell [see diagram (c)].

10. The transfer RNA separates, and messenger RNA is available to dictate the synthesis of more protein molecules. This process of turning the instructions from mRNA into protein in the ribosome is called *translation*.

The Genetic Code

The sequence of the nucleotides in a DNA molecule is important in determining the formation of proteins, through its directions to messenger RNA. Thus, if the order of nucleotide bases is adenine-guanine-thymine-cytosine, the complementary arrangement in messenger RNA is uracil-cytosine-adenine-guanine. This dictates the type of transfer RNA, and the particular amino acid that it carries, that will line up on a part of messenger RNA. All living things contain DNA, and the same

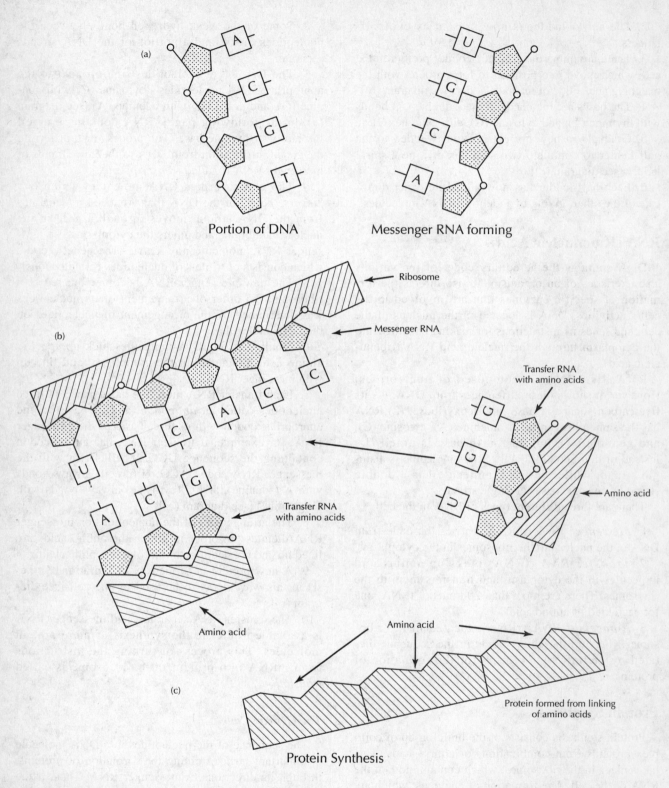

(a)

Portion of DNA

Messenger RNA forming

(b)

Ribosome

Messenger RNA

Transfer RNA
with amino acids

Amino acid

Transfer RNA
with amino acids

Amino acid

(c)

Amino acid

Protein formed from linking
of amino acids

Protein Synthesis

four types of nucleotides. However, the DNA of a human is different from the DNA of a maple tree. The arrangement of the nucleotides on the DNA molecule is thought to be responsible for the difference.

The genetic code, based on a four-letter alphabet, is as follows:

DNA—A, T, C, G

RNA—A, U, C, G

These letters are arranged in multiples of three in the formation of the various amino acids. Each such combination of three bases in a messenger RNA molecule is known as a *codon*. Here are some examples of amino acids and the codons that lead to their formation:

Amino Acid	RNA Codon
Phenylaline	UUU
Alanine	CCG, UCG
Histidine	ACC
Tryptophan	UGG
Lysine	AAA, AAG, UAA

Gene Action

The genes, which are made up of DNA, have two principal types of action: (1) they pass on copies of themselves by replication, to all the cells of an individual, following the formation of the zygote; (2) they control the activities of the cell by producing specific enzymes.

How Genes Act

The current *one gene-one enzyme* hypothesis is based on the idea that a single gene governs the synthesis of a specific enzyme in the cell. An example discussed in Section 10.3 deals with the disease PKU, in which a simple recessive gene affecting the production of a certain enzyme in the liver results in a form of feeblemindedness. Lately, the expression *one gene-one polypeptide* hypothesis has come into use because it has been determined that a single enzyme may be composed of several polypeptides; the synthesis of each polypeptide is governed by a different gene.

Experimental evidence of single action was provided by Dr. George W. Beadle and Dr. Edward L. Tatum. They exposed the mold *Neurospora* to X rays. The spores of the mold were then grown in test tubes containing a simple culture medium. Some of the spores did not germinate, indicating that mutations had been induced by the X rays. Apparently these new spores were now unable to produce a vital amino acid from the culture medium.

The scientists then added one of the different amino acids to each of 20 test tubes of the medium. The spores were found to grow in one of these, the test tube containing the amino acid arginine. This led to the conclusion that the X ray treatment had affected a single gene that was normally responsible for the production of an enzyme involved in the synthesis of arginine. The mold was no longer able to form arginine, and could not grow unless it was supplied artificially. Additional mutants were also discovered that required other amino acids, such as ornithine and citrulline. The Nobel Prize was awarded to Beadle and Tatum in 1958.

Gene Mutations

A mutation may occur as the result of a slight change in the DNA molecule. This change may affect just one nucleotide. Yet, as a result, there may be a change in the structure of a single enzyme. This, in turn, may affect an important reaction dealing with the life of the organism.

As stated previously, the ability of *Neurospora* to grow depends on its ability to synthesize certain amino acids. The loss of ability to make arginine, in the experiments of Beadle and Tatum, for example, prevented the mold from growing. Such a mutation would be harmful, and the organism would not be able to survive. However, other mutations could enhance an organism's ability to compete in the struggle for existence.

Mutations can also result from a change in the sequence of the nitrogen bases in the DNA molecule, as well as from additions or subtractions in these bases.

In sickle-cell anemia, the hemoglobin molecule differs from normal hemoglobin by only a single amino acid out of a total of over 300. The sickle-cell type of hemoglobin has the amino acid valine instead of the normal amino acid known as glutamic acid. It is thought that this change occurred as a gene mutation, changing one base in the genetic code for glutamic acid (UAG) into the code (UUG) for valine.

DNA Fingerprinting

Since 1985, DNA "fingerprinting" has been used to solve crimes, determine paternity, and reunite missing relatives. Only minute quantities of blood, semen, hair cells, or other tissues are needed to provide enough DNA to be analyzed in a series of steps, including a technique called gel electrophoresis. Basically, although all DNA has the same chemical structure, the *sequence* of the *base pairs* of *nucleotides* is unique to each individual. Scientists can differentiate people by studying the segments of small DNA sequences.

Cytoplasmic Inheritance

Although inheritance is directed by DNA in the nucleus, evidence indicates that there may be some forms of inheritance in the cytoplasm. Chloroplasts and mitochondria appear to divide during cell division. The chloroplasts have been found to contain small amounts of DNA. This leads to the possibility that they may contain separate genes that are regulated by the genes in the nucleus.

The chloroplasts in a plant appear to be inherited from the cytoplasm that accompanies the egg nucleus. If the female parent has leaves that are pale green or are variegated with blotchy areas of color, the descendents will have leaves of the same appearance.

Another example of cytoplasmic inheritance occurs in a certain paramecium, *Paramecium aurelia*. It produces a "killer" strain, which secretes into the water a poison that kills strains of paramecia sensitive to it. When the two strains, killer and sensitive, were crossed in conjugation, some unexpected results were obtained. The descendents of killers remained killers, and the descendents of sensitives remained sensitives. Under normal conditions, the descendents would have been heterozygotes, and either killers or sensitives. It was concluded that the killer trait is determined by the parent cytoplasm. Later, it was found that the secretion of the killer substance is due to particles in the cytoplasm that contain DNA and are self-duplicating.

Manipulating Genes

Cloning

A single, mature cell of an organism contains a complete set of genes in its chromosomes. Can such a cell yield a new individual? Dr. F. C. Steward, in the 1950s, planted tiny bits of carrot slices in flasks containing a sterile nutrient solution. Normal carrot plants, which were genetic copies of the original carrot, developed in the flasks. Such identical copies of an organism that grow from single cells without fertilization are called *clones*.

Another example of cloning was demonstrated in frogs by J. B. Gurdon. After destroying the nucleus of an unfertilized egg, he inserted the nucleus from a tadpole's intestinal cell into the egg. A new tadpole developed that was genetically identical to the tadpole that had donated its intestinal cell nucleus. In 1984, bits of genes taken from dried muscle tissue of an extinct zebralike animal in a museum, the quagga, were cloned. This was the first time gene fragments of a vanished animal species had been extracted and reproduced.

The cloning of a mammal, a sheep called Dolly, was announced in 1997 by a British scientist, Dr. Ian Wilmut. He removed a cell from the mammary gland of a sheep and introduced its DNA into an egg from another sheep, after removing that egg's own DNA. He then fused this egg cell with the adult mammary cell. The fused cells carrying the adult DNA began to grow like a perfectly normal fertilized egg to form an embryo. He transplanted the embryo into the uterus of another sheep where it developed normally. When the newborn lamb was examined, it was shown to be a clone of the adult sheep that first supplied the DNA.

Cloning has been performed in cows, pigs, goats, mice, and, in 2002, a cat. But fewer than 3 percent of all cloning efforts have succeeded. Apparently the rapid reprogramming in cloning can introduce random errors into the clone's DNA, interfering with embryo development. There is caution among scientists, religious leaders, and others regarding possible attempts to extend cloning to humans. Responding to recommendations from the National Bioethics Commission, President Clinton acted, one week after news about the cloning of Dolly, to institute a ban on the use of any federal money to support research on the cloning of humans.

Genetic Engineering

As scientists have learned more and more about modern genetics, they have begun to manipulate DNA and genes. This has led to the new field of *genetic engineering*. Its purpose is to direct or control the action of genes. It may be accomplished by inserting new genetic material or replacing certain genes in a cell, thus making fundamental changes in an organism's characteristics.

Recombinant DNA   Recombinant DNA is DNA combined from two different organisms to produce characteristics not found in nature. The technique for doing this involves the use of special enzymes called "restriction enzymes," which have the ability to split a DNA strand. The fragments of DNA have sticky ends, and when they touch a strip of DNA from another organism, they adhere to it.

The common intestinal bacterium of humans and other animals, *Escherichia coli* (*E. coli*, for short), is generally used in such research. Its structure is relatively simple. It has one chromosome containing a large, regular DNA molecule and two smaller rings of DNA called *plasmids*. It is into these plasmids that sections of DNA from other organisms have been spliced. Dr. Paul Berg, the principal pioneer in developing recombinant DNA, was awarded the Nobel Prize in 1980.

Early in this research, sections of DNA from the eggs of the South African clawed toad, *Xenopus*, were combined with the DNA of *E. coli*. These bacteria, multiplying with animal genes in them, can be regarded as man-made organisms. After being replicated through 100 generations of bacteria, the toad genes were still intact in them.

Suggestions had been made that genes that make insulin could possibly be transplanted into *E. coli* bacteria by the technique of recombinant DNA, or gene splicing. Since these bacteria usually reproduce at least every 60 minutes or so, each one could multiply to over 8 million cells in only 24 hours. Large quantities of insulin could be produced in a relatively short time by such bacterial "minifactories."

In fact, by 1980, not only human insulin, but also human interferon and human growth hormone, were being produced in the laboratory. In 1984, a vaccine for hepatitis B was developed for the first time through the use of recombinant DNA. It is believed safer than the vaccine made from donated human blood. Other possibilities that recombinant DNA technology may allow include the production of bacteria that will manufacture disease-fighting antibodies, vitamins, and antibiotics.

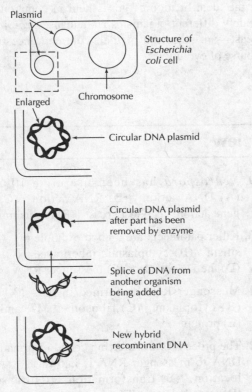

Plasmid

Structure of *Escherichia coli* cell

Chromosome

Enlarged

Circular DNA plasmid

Circular DNA plasmid after part has been removed by enzyme

Splice of DNA from another organism being added

New hybrid recombinant DNA

Formation of Recombinant DNA

Other examples of genetic engineering In 1986, a synthetic vaccine was developed against the hepatitis B virus (see Section 8.3). This vaccine does not require the use of live germs or human blood in its production. It is made by inserting a gene from the hepatitis B virus into yeast cells. The yeast cells then produce a protein from the surface of the virus, which triggers the body to produce antibodies, thereby resulting in immunity. It is hoped that using the coats of viruses will similarly provide synthetic vaccines against other virus diseases, such as AIDS and herpes. Experimental research is also producing a promising synthetic vaccine against the malaria germ, which is a protozoan.

Here are some recently announced results: A gene has been inserted into plants to reduce the saturated fat in vegetable oil. Steps are being taken to produce coffee beans that have a lower caffeine content. Gene-altered cattle have been experimentally produced, using genes from other species, including human genes; the scientists expect to speed the growth of the cattle and make them leaner. A gene that controls growth in one species of fish, rainbow trout, has been inserted into another fish species, the common carp, making the carp grow significantly faster than usual. Human genes have been introduced into plants with promising results: potatoes manufacture serum albumin, a human blood protein effectively used in surgery; tobacco plants produce antibodies and also a sun screen containing melanin, a natural skin pigment.

The Human Genome Project

The genome is the sum total of all the genes contained in a haploid set of chromosomes in each cell. In humans, it had been estimated that there are between 50,000 and 100,000 such genes. The Human Genome Project was initiated in 1988 to map the positions of all these genes in their proper sequence on the chromosomes of a cell. At the time, this ambitious project, involving thousands of scientists, was expected to take at least 15 years, at an eventual cost of $3 billion. By June 2000, however, the project was declared practically completed by two groups of scientists involved in its study, namely the government-sponsored Human Genome Project, under the direction of Dr. Francis S. Collins, and the private company Celerica Genomics, headed by J. Craig Venter.

Knowing the exact location of the 3 billion pairs of base nucleotides on the DNA molecules that make up the chromosomes is considered by scientists to be only the first step in understanding human genetics. What remains in the future is to determine just how the genes function and how they interact with each other. One of the benefits will undoubtedly be a more precise understanding of the defective genes that cause inherited diseases. This should lead to improved methods of treatment and prevention of such diseases.

A surprising outcome is the discovery that humans contain a total of only about 30,000 to 40,000 genes. Furthermore, humans are 99.99 percent identical in their genes. The .01 percent variation accounts for the differences between individuals with such traits as eye color, facial appearance, musical aptitude, intelligence, and disease susceptibility. More than a million spots on the DNA sequence called SNPS (single nucleotide polymorphisms) change a particular nucleotide and make each person unique.

Other Genome Progress

Of great economic value is the recent decoding of the genome of rice, a plant that provides food for half the world's population. This breakthrough opens possibilities for development of varieties that are more nutritious and more resistant to drought and insects.

Also, another plant genome project has succeeded in mapping the genes of the simple weed *Arabidopsis thaliana*. Because this weed has a relatively small genome of 25,498 genes and reproduces every six weeks, it is considered to be an excellent model for increasing knowledge of basic life processes in plants, and ultimately for contributing to the improvement of crops.

Within the last few years, the complete genomes have been successfully deciphered for other living things, including the mouse, *Drosophila*, roundworm, yeast, and a number of bacteria. Among the latter, the entire genome of the bacteria that cause human ulcers, *Heliobacter pylori*, has been identified. Also, the genome of a deadly strain of bacteria known as *E. coli* 015:47, which was recently decoded, has revealed that this type of otherwise harmless bacteria picks up genes from other bacteria and viruses.

One of the interesting discoveries in genome research is the identification of genes held in common by such widely different organisms as humans, *Drosophila*, and yeast—an indication of their common ancestry in the course of evolution.

Section Review

Select the correct choice to complete each of the following statements:

1. The first indication of DNA action was found in (A) protozoa (B) bacteria (C) algae (D) *Drosophila* (E) corn

2. The change in heredity brought about by the transfer of dissolved DNA is called (A) transduction (B) transformation (C) transpiration (D) trichinosis (E) trypanosome

3. Bacteria may be attacked by viruses called (A) cocci (B) bacilli (C) spirilla (D) bacteriophages (E) staphylococci

4. The central core of a virus is to its covering as (A) protein is to DNA (B) DNA is to protein (C) lipid is to DNA (D) DNA is to lipid (E) lipid is to protein

5. Of the following, the combination of bases that cannot be present in DNA is (A) AUCG (B) ATCG (C) ATTA (D) CGTA (E) GCAT

6. During replication of DNA, a nucleotide base that would bond with cytosine is (A) adenine (B) thymine (C) uracil (D) cytosine (E) guanine

7. RNA is different from DNA in that it has (A) a ribose sugar (B) an extra adenine base (C) an extra thymine base (D) a double strand (E) no similar nitrogen bases

8. All of the following are involved in the steps of protein synthesis *except* (A) messenger RNA (B) transfer RNA (C) ribosome (D) centriole (E) DNA code

9. *Neurospora* has been useful in illustrating (A) DNA action (B) RNA action (C) single-gene action (D) test cross (E) replication

10. Killer paramecia are a good example of (A) parasitism (B) cytoplasmic inheritance (C) meiosis (D) the DNA code (E) messenger RNA

11. Messenger RNA is formed in the (A) nucleus (B) cytoplasm (C) ribosome (D) amino acid (E) protein

12. The process of transferring coded information from DNA to messenger RNA is known as (A) translocation (B) transformation (C) transcription (D) transduction (E) transpiration

13. A sequence of three bases on messenger RNA is called (A) tRNA (B) rRNA (C) a codon (D) an anticodon (E) a double helix

14. DNA is artificially combined from two different organisms in the process of (A) cytoplasmic inheritance (B) gene mutation (C) clone formation (D) DNA replication (E) recombinant DNA

15. All of the following are examples of genetic engineering except the (A) production of a synthetic vaccine against hepatitis B by yeast cells (B) development of a strain of bacteria to break down oil spills (C) production of human antibodies in tobacco plants (D) production of the Shasta daisy (E) artificial creation of a gene

Answer Key

1-B	4-B	7-A	10-B	13-C
2-B	5-A	8-D	11-A	14-E
3-D	6-E	9-C	12-C	15-D

Answers Explained

1. (B) The first indication of DNA action was found in pneumococcus bacteria that have a capsule around them and cause pneumonia. DNA was found to transform harmless pneumococci that do not have a capsule into harmful bacteria containing a capsule.

2. (B) Transformation was brought about when the heredity of noncapsule pneumocci was changed to make them grow their own capsules. This was induced by subjecting them to an extract of capsulate pneumococci. DNA was shown to be the material in the extract that changed the pneumococcus characteristics.

3. (D) When a bacteriophage virus attacks bacteria, its DNA replicates to form many viruses, which are released when the bacteria burst.

4. (B) The central core of a virus contains DNA. The outside covering of the virus consists of protein. The arrangement of the nucleotides in DNA determines the arrangement of the nucleotides in messenger RNA. This, in turn, dictates the order in which the amino acids will be assembled into proteins.

5. (A) RNA contains uracil (U), which DNA does not contain [DNA contains thymine (T) instead]. Otherwise, RNA and DNA contain the same three other bases: adenine (A), cytosine (C), and guanine (G).

6. (E) During replication of DNA, numerous free nucleotides present in the cell move into position to form a bond with the matching nucleotides of DNA. In the process, guanine always bonds with cytosine.

7. (A) RNA has the five-carbon sugar, ribose, not deoxyribose, which is present in DNA.

8. (D) The centriole is present as a pair of structures within the centrosome in animal cells and is active during nuclear division. It is not active in protein synthesis.

9. (C) *Neurospora* is a mold that can synthesize certain compounds when grown on a culture medium. If a mutation occurs and a gene is changed, an enzyme is not produced and the amino acid arginine is not made. The use of *Neurospora* for research led to the formation of the one gene-one polypeptide theory by Beadle and Tatum.

10. (B) The secretion of a poison into the water by killer paramecia kills certain strains of sensitive paramecia. The secretion of the killer substance was found to be due to particles in the cytoplasm that contain DNA and are self-replicating.

11. (A) After messenger RNA is formed in the nucleus, it passes into the cytoplasm through pores in the nuclear membrane.

12. (C) During the process of transcription, the genetic code of DNA is transferred to molecules of messenger RNA. A segment of DNA separates into two strands. One of these strands serves as a template, or pattern, for RNA nucleotides to line up with the complementary DNA nucleotides.

13. (C) A codon is a combination of three nucleotides in messenger RNA that directs the formation of the various amino acids in building proteins.

14. (E) Recombinant DNA is a form of genetic engineering in which DNA segments from one organism are transferred artificially into the DNA of another species. This process produces characteristics not found in nature. A common organism used in this research is a bacterium, *Escherichia coli*.

15. (D) The Shasta daisy was produced by Luther Burbank, who hybridized three types of daisies. Genetic engineering is a relatively new field of modern genetics in which new DNA is inserted into a cell, or certain genes are replaced, to produce fundamental changes in an organism.

10.5 **IMPROVING THE SPECIES**

As early humans began to domesticate animals and raise plants, they gradually improved them to serve their purposes. Modern humans now apply their knowledge of genetics toward improving the various species for the purpose of breeding better plants and animals. Probably the largest research center for this purpose is maintained by the U.S. Department of Agriculture at Beltsville, Md.

Breeding for Better Plants and Animals

Aims

Plants and animals are bred for the following purposes:

1. *Improved yield.* For example, dairy cattle today produce twice as much milk as at the beginning of the century. Prize hens lay almost an egg a day, which is about three times the yield of 50 years ago. The productivity of other domesticated animals and plants has likewise been increased over the years.

2. *Improved quality.* For example, the butterfat content of milk has been improved. The strength of the fiber in cotton has been increased. Although the weight of hogs has been raised, there has been an increase in the leanness of pork.

3. *Increased resistance to disease.* Kanred wheat has been developed, which is resistant to the fungus parasite known as the wheat rust. An improved variety of cattle immune to Texas fever can endure the intense summer heat and insect pests of the Gulf Coast area.

4. *New varieties for special purposes.* Special breeds of dogs, larger and more attractive daisies, faster racehorses, new fruits such as the plumcot and pink grapefruit, and showy varieties of tropical fish have been bred for purposes of attractiveness, uniqueness, speed, endurance, special utility, and so on.

Methods of the Breeder

Several methods have been used in breeding these better animals and plants. In some cases, new types have actually been made to order.

1. *Selection.* The oldest method was simply a matter of selecting the best type for breeding in each generation. The various pure breeds of dogs have been produced in this way, each being selected for a particular purpose. The best of the desired types were then *inbred,* or mated to one another, in order to maintain the line. We now know that this method essentially resulted in selecting the best combinations of genes, and keeping them by inbreeding. Selection may sometimes give extreme results, as in the case of the Boston bulldog, which has been selected for its flat face to such an extent that it has difficulty in breathing through its nose. As an example among plants, cotton plants resistant to the fungus wilt disease have been improved by selection. In practicing selection among plants, self-pollination is carried on so that no other traits will be introduced. Eventually, the type is fixed by selection, and pure lines are established that will breed true.

2. *Hybridization.* By hybridization, or crossbreeding, the breeder combines desirable qualities of two different organisms into one. An example of this is the improvement of the shorthorn cattle in the Southwest. These animals were useful for their good beef qualities, but they became sickly as a result of Texas fever. They were crossbred with Brahman cattle imported from India; these cattle did not have the same good beef qualities, but were immune to Texas fever. The offspring showed various qualities: good beef but susceptible; poor beef but immune; poor beef and susceptible; good beef and immune. The last type was selected and bred for several generations until the Santa Gertrudis breed, containing both desired qualities, was developed.

Luther Burbank (1849–1926) is probably the best-known plant breeder. As part of his work, he created new plants by combining qualities of different existing plants: the plumcot was the result of crossing the plum and the apricot; the white blackberry resulted from a cross of the Lawton blackberry with a wild pale type; the Shasta daisy came from a cross of three daisies: the American, which is sturdy, the English, which has large petals, and the Japanese, which is brilliantly white.

One result of hybridization is *heterosis,* or hybrid vigor. In this case, the hybrid surpasses the parents. The outstanding example is hybrid corn, produced from two separate inbred pure strains that are cross-pollinated. Hybrid corn combines the desirable genes from both parents. It is more vigorous and produces more high-quality ears of corn. The yield from hybrid corn has been increased about 35 percent. Because of its high yield, the introduction of hybrid corn into the devastated countries of Europe at the close of World War II helped avert a famine.

The mule is an example of a hybrid of superior vigor and strength resulting from a cross of two species, the horse and the donkey. The cattalo, the hybrid of the bison and domestic cattle, also has unusual vigor. In both of the above cases, however, the hybrids are usually sterile. Additional curiosities of little economic value are the zebroid, a cross between the zebra and the horse, and the tiglon, resulting from a cross between a tiger and a lion.

3. *Saving the best genes.* A breeder who develops a useful variety values the outstanding individuals in it because they contain the best genes. A prize-winning racehorse will continue to be valuable even after it has grown too old to race; it is used for breeding other racehorses. One of the most famous, and most valuable for this purpose, was Man o' War; his offspring continued to win many prizes. A purebred bull that carries genes for good milk production is worth thousands of dollars. Records are kept of purebred dogs and are passed on to new owners so that they may know the pedigrees of their pets. The American Hereford Cattle Breeders

Association keeps a registry of purebred Hereford cattle. The American Guernsey Cattle Club, founded in 1877, maintains a record of purebred Guernsey dairy cows.

A practical way to make use of useful genes is in the technique of *artificial insemination.* The sperm of prize bulls, for example, is stored and frozen. It is used to inseminate or fertilize prize cows, and is available long after the death of the bull.

4. *Useful mutations.* Mutations, or gene changes, that occur spontaneously in domesticated plants or animals may introduce new favorable features, for which breeders are always on the lookout. For example, the seedless or navel orange originated in Brazil. It was introduced into California in 1873 and has been propagated by grafting ever since. The pink grapefruit is another mutant that has been maintained by grafting. The nectarine is a peach with a smooth skin that arose as a mutation. A very useful mutation among cattle is the polled or hornless condition, which is a dominant characteristic. The present polled Hereford strain of cattle was bred by mating horned Hereford cows to mutant polled bulls. The Ancon breed of sheep originated from a mutant that had very short legs.

5. *Genetic engineering.* By inserting genes into plants, as mentioned in the discussion of genetic engineering at the end of the preceding section, scientists are producing new, improved varieties. For example, corn, tomato, and tobacco plants have received a bacterial gene that forms a natural protein lethal to rootworms, budworms, and bollworms; it causes the digestive systems of these parasites to disintegrate, thus eliminating the need to use pesticide chemicals to kill them. Cucumber plants are being treated with a virus gene that produces a protein which protects the plants against cucumber mosaic disease. Rice, which is the most important food crop in many countries, especially in Asia, has come under intense research to study the genome of its 12 chromosomes; DNA is being experi-

mentally inserted into rice cells whose walls have been removed, with the expectation that rice plants with resistance to a serious tungro-virus blight will develop.

Advantages of Plant Breeding

The breeding of plants offers certain advantages over animal breeding: (1) plants reproduce in large numbers; (2) there is thus a greater chance of favorable mutations appearing; (3) by vegetative propagation (grafting, cutting, bulbs, etc.), the same genetic makeup can be obtained in the new individuals, thus making them true to type; (4) by vegetative propagation, more plants can be obtained in a shorter time; (3) inbreeding is easily practiced by simply self-pollinating the flowers.

Native American Plants and Animals

The explorers and early settlers in this country came across plants and animals they had never seen before. The turkey is a native American bird; Benjamin Franklin suggested using it as our national symbol instead of the eagle. Columbus found corn being grown by the Indians in Cuba and on the mainland of South America. Potatoes were brought back to Europe by the early Spanish explorers. Tomatoes, beans, squash, and pumpkins also originated in the New World.

Importance of the Environment

The best results of improved breeding are of little value unless the proper environment is provided. A prize cow will not give a record-breaking supply of milk if she is poorly fed or becomes sick. The yield from choice hybrid corn seeds will be low unless proper fertilizer and soil conditions are provided by the farmer. In short, the best results are obtained by a combination of the right genes and a good environment.

Section Review

Select the correct choice to complete each of the following statements:

1. All of the following are aims of the breeder *except* (A) improved yield (B) improved quality (C) increased susceptibility to disease (D) increased resistance to disease (E) new varieties

2. A new variety of fruit produced by hybridization is the (A) seedless grape (B) seedless orange (C) seedless grapefruit (D) plumcot (E) Delicious apple

3. In practicing selection among plants, the breeder carries on (A) self-pollination (B) cross-pollination (C) crop rotation (D) strip cropping (E) soil fertilization

4. Inbreeding is carried on by (A) crossing unrelated animals (B) crossing related animals (C) cross-pollinating flowers (D) hybridizing plants (E) using hydroponics

5. Combining the desirable qualities of two different organisms into one is referred to as (A) inbreeding (B) self-breeding (C) pure-line breeding (D) hybridization (E) conjugation

6. All of the following organisms resulted from cross-breeding *except*   (A) Ancon sheep   (B) the cattalo   (C) Santa Gertrudis cattle   (D) the mule   (E) the Shasta daisy

7. An advantage of the plant breeder over the animal breeder is that the former can use   (A) vegetative propagation   (B) mutations   (C) hybridization   (D) alternation of generations   (E) inbreeding

8. Two plants that originated in the New World are   (A) corn and wheat   (B) wheat and tomato   (C) tomato and rice   (D) rice and corn   (E) corn and tomato

9. A useful animal mutation is the   (A) long-horned Texas steer   (B) Boston bulldog   (C) Guernsey cow   (D) Hereford cattle   (E) polled cattle

10. One result of hybridization is   (A) a pure line   (B) heterosis   (C) gene change   (D) inbreeding   (E) an improved environment

Answer Key

1-C	3-A	5-D	7-A	9-E
2-D	4-B	6-A	8-E	10-B

Answers Explained

1. (C) One of the aims of the breeder is to increase resistance to disease. Example: Kanred wheat, which is resistant to wheat rust.

2. (D) Luther Burbank (1849–1926) created the plumcot by crossing the plum and the apricot.

3. (A) In selection, the most desired types are inbred to maintain the same genes. In self-pollination, no other traits are introduced.

4. (B) Inbreeding involves crossing related animals to select the best combination of genes for the desired traits. Hybridizing, or cross-breeding, introduces other traits.

5. (D) In hybridization, the desirable traits of two or more different animals or plants are combined into one variety. The Shasta daisy came from a cross of three daisies—the American, which

is sturdy; the English, which has large petals; and the Japanese, which is brilliantly white.

6. (A) Ancon sheep originated from a mutant that had very short legs.

7. (A) By vegetative propagation, the same genetic makeup can be obtained in the offspring, thus making them true to type.

8. (E) Corn and tomatoes were brought back to Europe by the early explorers of the New World.

9. (E) Polled cattle are hornless. This condition arose as a mutation and is useful because the cattle are less dangerous.

10. (B) Heterosis is hybrid vigor. The mule is an example of a hybrid of superior vigor and strength, resulting from a cross of two species, a horse and a donkey.

HOW LIVING THINGS HAVE CHANGED

CHAPTER

11

11.1   THE RECORD OF PREHISTORIC LIFE

Paleontology is the study of prehistoric life as revealed by fossils. The record shows that these earlier organisms were different from those of today; in particular, they were simpler.

Rocks and Fossils

Rock Formation

Geology is the study of the earth and its rocks. It is estimated that the earth is about 4 billion years old. Originally, it was extremely hot, with a surface of molten material. After it cooled, three main types of rocks were formed:

1. *Igneous rock* was the earliest type of rock, being formed from the molten material. An example is granite. Igneous rock is still being formed today from the molten lava of volcanoes.

2. *Sedimentary rock* is formed from sediment that is carried into the sea by rivers. This sediment results from the effects of erosion and weathering in breaking up rocks into smaller particles, and eventually forming soil. Under the pressure of water over the ages, the material in the sediment becomes cemented together, turning it into sedimentary rock. In this way, sandstone was formed from sand, shale from clay, and limestone from mud or the remains of shells or coral. Sedimentary rocks usually appear in layers or strata.

3. *Metamorphic rock* is formed from previously existing igneous or sedimentary rocks whose structure has been transformed by great pressure, heat, and other agents: an example is marble.

Fossil Formation

A *fossil* is the remains of a living thing that existed long ago. Fossils are practically always found only in sedimentary rocks, since they would be destroyed by the high temperature of igneous rocks and the crushing pressures of metamorphic rocks. Fossils may be formed in a number of ways:

1. *Petrifaction.* An animal or woody plant that is covered with undisturbed sediment under water for thousands of years gradually becomes petrified, or turned into stone. This happens as minerals replace the hard parts of the bones or wood, while the sediment itself is changing into sedimentary rock. The petrified bones of dinosaurs and the petrified wood of trees were formed in this way. Subsequent changes in the earth's surface and erosion have brought them within reach of paleontologists.

2. *Imprints.* Footprints or leafprints that were left in soft mud or sand have been preserved as imprints. The soft soil dried up and later changed into sedimentary rock. Dinosaur tracks are a good example. Leaf and stem imprints are sometimes found in coal. Coal itself was formed from the remains of trees under the surface of water, following thousands of years of pressure.

3. *Casts or molds.* Animals such as snails were buried in mud under water for ages. As the organism decayed, its place was taken by minerals, leaving a cast of the original animal in the surrounding sedimentary rock.

4. *Freezing.* The preserved bodies of prehistoric mammoths have been found in frozen Siberia, where the intense cold prevented bacterial decay from taking place.

5. *Tar pits.* The La Brea tar pits in Los Angeles have yielded the skeletons of prehistoric animals, including the saber-tooth tiger, which were trapped in the sticky material.

6. *Amber.* Prehistoric ants, wasps, and other insects, and even protozoa, bacteria, algae, and pollen, were trapped in the sticky resin of evergreen trees, which gradually hardened into amber. Their bodies were protected from bacteria and the air and did not decompose.

Ages of Rock and of Fossils

There are several ways of determining the ages of rock and of fossils:

1. *Radioactive elements.* Uranium is a radioactive element that gradually changes into lead. By analyzing the proportions of uranium and lead in a rock, it is possible to calculate its age. On this basis, the age of the earth has been estimated to be in the vicinity of 4 billion years. Another radioactive element, C-14, an isotope of carbon, has a half-life of 5,760 years. That is the period of time in which it loses half of its radioactivity. Using C-14, scientists have been able to compute the ages of wood, charcoal, and bone remains of early human beings and their culture.

2. *Position of strata.* Since sedimentary rock is laid down in layers, or strata, it is evident that the lowest strata are the oldest. Fossils found in them are older than fossils located in upper strata.

3. *Rate of sedimentation.* It has been estimated that some types of sediment turn into sedimentary rock at the rate of 1 foot in 900 years. Thus, by measuring the thickness of a stratum, geologists can determine its age.

4. *Saltiness of the ocean.* Another method of estimating the age of the earth is to analyze the salinity of the ocean. It continues to increase as rivers flow into it carrying dissolved salt.

Geologic Eras

By studying the various rock strata and the fossils in them, geologists have drawn up a history of the earth. It is divided into six *eras,* each of which is subdivided into *periods.*

1. *Azoic era.* The earth was cooling and the mountains and the oceans were being formed. There was no life at this time. This era lasted for about 1–2 billion years.

2. *Archaeozoic era.* During this era, the simplest forms of life probably appeared. Some rocks of this era contain carbon, an indication that living things probably existed. This era lasted for over 600 million years.

3. *Proterozoic era.* Traces of algae, bacteria, shelled protozoa, and the burrows of worms from this era have been found. Fossils are rare because the organisms did not have hard parts. Life existed only in the water. This era lasted about 600 million years.

4. *Paleozoic era.* There are abundant fossils of animals and plants from this era. In the earliest (Cambrian) period, there were sponges, jellyfish, and corals. Unique lobster like forms, *trilobites,* appeared, and then became extinct. In later periods, mollusks arose, followed by vertebrates. Fish became established, and amphibians, air-breathing vertebrates, developed.

Primitive reptiles began to appear. Insects, including giant dragonflies with wingspreads of over 2 feet, flourished. A fossil discovered in Argentina may have been a common ancestor of the great dinosaurs. Named *Eoraptor,* it lived about 225 million years ago, and was as large as a medium-size dog. Mosses and ferns grew on land. The Paleozoic era, which included the age of the coal-forming plants, the Carboniferous period, lasted about 400 million years.

5. *Mesozoic era.* This is the "Age of Reptiles," during which the huge dinosaurs appeared. Among them were the herbivorous *Brontosaurus,* the carnivorous *Tyrannosaurus rex,* and the armored *Stegosaurus.* They eventually became extinct, possibly because the climate changed, or because their eggs were destroyed by the newly risen mammals that began to appear toward the end of this era. Another theory regarding the extinction of dinosaurs is based on the unusual abundance of the element *iridium* found in the narrow band of rock separating the top layer of rock at the end of the Mesozoic era from the next rock layer at the beginning of the Cenozoic era. Since iridium is 1,000 times more abundant in meteorites from space than in earth rocks, it is possible that an asteroid may have smashed into the earth. The resulting collision would have thrown up a dense cloud of dust that might have blocked most of the sunlight reaching the earth for a few years. During this time, photosynthesis could have been prevented long enough to disrupt the food chain, with disastrous consequences to dinosaurs.

The fossil bird *Archaeopteryx* appeared toward the close of the era. It was about the size of a crow and had birdlike features, such as feathers and wings; it also had reptilian features: numerous teeth in the beak, an

HOW LIVING THINGS CHANGED

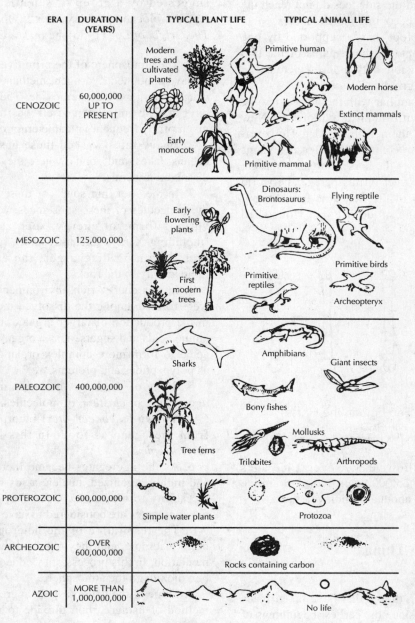

ERA	DURATION (YEARS)	TYPICAL PLANT LIFE	TYPICAL ANIMAL LIFE
CENOZOIC	60,000,000 UP TO PRESENT	Modern trees and cultivated plants; Early monocots	Primitive human; Modern horse; Extinct mammals; Primitive mammal
MESOZOIC	125,000,000	Early flowering plants; First modern trees	Dinosaurs: Brontosaurus; Flying reptile; Primitive birds; Primitive reptiles; Archeopteryx
PALEOZOIC	400,000,000	Tree ferns	Sharks; Amphibians; Giant insects; Bony fishes; Mollusks; Trilobites; Arthropods
PROTEROZOIC	600,000,000	Simple water plants	Protozoa
ARCHEOZOIC	OVER 600,000,000	Rocks containing carbon	
AZOIC	MORE THAN 1,000,000,000	No life	

Geologic History of the Earth

elongated tail containing about 20 vertebrae, and claws at the end of the wings. Evergreen trees also made their appearance at this time. The Mesozoic era lasted about 125 million years.

6. *Cenozoic era.* About 60 million years ago, the present "Age of Mammals" began. Most of the modern mammals and flowering plants began to appear. Rather complete fossil records that have been found show the various changes that have taken place in such mammals as the horse, camel, elephant, and pig since the beginning of the Cenozoic era. The fossil history of the horse, for example, reveals the following record:

- *Eohippus,* the dawn horse, lived during the earliest period of the Cenozoic era. It was about the size of a cat, had simple teeth, and had four toes on its front legs and three toes on its hind legs.
- In higher rock layers, fossils of a somewhat larger animal are found, with a foot consisting of a large middle toe and two smaller toes on each side of it that just reach the ground.

- In still higher strata, the fossils show an animal about the size of a pony, in which the middle toe of each foot is larger, and the side toes do not reach the ground. The teeth are becoming more complex.
- Modern horse has a foot that is supported by one large toe and that contains the remains of the two side toes as splints, visible only in the skeleton. The teeth have a complex grinding surface.
- The Indians were unfamiliar with the horses that the Spaniards brought with them to explore the New World, and regarded them with wonder. However, the ground under them contained the fossil remains of the many ancestors of the horse. For some reason, the animal became extinct in this part of the world during the last stages of the Cenozoic Era. After it was reintroduced, the modern horse flourished throughout the continent.

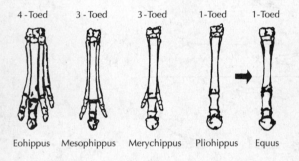

| 4 - Toed | 3 - Toed | 3 - Toed | 1 - Toed | 1 - Toed |

Eohippus   Mesophippus   Merychippus   Pliohippus   Equus

Evolution of the Forelimb in the Horse

The single toe of the modern horse (*Equus*) is the sole remainder of the four toes present in *Eohippus*.

Within the last million years, several ice ages occurred, the last about 25,000 years ago. A primitive type of human appeared about a million years ago.

Evolution of Living Things

Process of Evolution

The changes in living things that have taken place during the various eras since the earliest beginnings of life are known as *organic evolution*. As described above, modern living things have developed from simpler organisms of the past. This process of change has taken place over countless millions of years. It is still going on.

Origin of Life

In the early history of the earth, conditions were very different from those of today. The earth itself may have been a molten mass of rock. As it cooled, conditions become established for the emergence of a primitive form of life.

Heterotroph hypothesis   An explanation of how life originated is offered in the heterotroph hypothesis, suggested by a group of scientists, including the Russian biochemist Alexander I. Oparin. In his book, *The Origin of Life* (1936), he makes these points:

1. The atmosphere of the primitive earth is supposed to have contained hydrogen, methane (CH_4), ammonia, water vapor, and carbon dioxide. There was little or no oxygen. The atmosphere itself was formed from gases given off by frequent volcanic eruptions.

2. Heavy rains washed these gases into the early oceans, lakes, and ponds, where they were mixed with dissolved minerals.

3. In the "hot thin soup" of these bodies of waters, the molecules of these substances were acted upon by various forms of energy, such as solar radiation (including X rays and ultraviolet radiation), cosmic rays, lightning discharges, the earth's heat, and radioactivity in the rocks.

4. In this energy-rich environment, chemical bonds were formed among the dissolved molecules, resulting in the production of larger organic molecules, such as amino acids and sugars. These organic molecules interacted to form more complex organic molecules, such as polypeptides and proteins.

5. Some of the large, complex molecules formed *aggregates*, or *clusters*, of molecules. These aggregates (also known as *coacervates*) incorporated molecules from the ocean as food. In this form, they were *heterotrophs*.

6. As the aggregates became increasingly complex and highly organized, nucleic acids were formed, with the ability to reproduce. Once they could reproduce, the aggregates are considered to have been alive.

7. The respiration of the heterotrophs must have been anaerobic, and they obtained their energy by fermentation. In this process, they added quantities of carbon dioxide to the atmosphere.

8. Some of the heterotrophic aggregates developed a method of using carbon dioxide to manufacture their own food, and so became pioneer *autotrophs*.

9. Through the food-making activities of the autotrophs, oxygen was added to the atmosphere. Some of the autotrophs and heterotrophs developed methods of using oxygen to obtain energy from food, in the beginning of aerobic respiration.

10. The evolution of the first primitive forms of life to the complex organisms of today took place gradually over a period of 2 or more billion years.

Laboratory experiments of Dr. Stanley Miller, working under the guidance of Nobel Prize winner Harold Urey, led to the artificial production of amino acids. He boiled water continuously in a flask containing methane, ammonia, hydrogen, and water vapor. These materials were subjected to electric discharges. After a week of duplicating the earth's primitive conditions in this way, Miller found that a number of organic molecules, including amino acids, had been formed in the flask.

Other Explanations   Some scientists believe that deep sea vents of volcanoes under the ocean, which were discovered only 20 years ago, created the conditions for life to emerge. This idea is based on the fact that ancient forms of primitive microbes, classified as *Archaea,* have been found thriving in the superheated waters around these undersea volcanoes. An analysis found that the heated water was rich in chemicals containing carbon

monoxide, hydrogen sulfide, and metal sulfides. In experiments with such chemicals, acetic acid was produced, demonstrating a natural way of combining carbon atoms, which is the basis of organic compounds. A study of ancient remains in Australian rocks that are 3.2 billion years old holds that the primitive ancestor of *Archaea* consisted of a loosely knit association of cells that exchanged their genetic information.

Also under consideration   Rocks from Mars that were found on earth as meteorites in 1997 are thought to contain evidence of primitive life. Also comets, which have been shown to carry complex compounds, may have dropped them on earth in its early history, to start producing primitive life.

Section Review

Select the correct choice to complete each of the following statements:

1. The study of prehistoric life is called   (A) eugenics (B) morphology (C) paleontology (D) abiosis (E) physiotherapy

2. It is currently estimated that the age of the earth is about   (A) 1 million years   (B) 4 million years (C) 1 billion years   (D) 4 billion years   (E) 1 light-year

3. Fossils may be found in (A) limestone (B) sandstone   (C) shale   (D) none of these   (E) all of these

4. The replacement of bone or wood by minerals is known as   (A) amber formation   (B) transmutation   (C) petrifaction   (D) vulcanization (E) metamorphosis

5. A prehistoric animal preserved in tar pits was the (A) brontosaurus   (B) tyrannosaurus   (C) trilobite (D) coelacanth   (E) saber-tooth tiger

6. An animal whose body was preserved for thousands of years in ice was   (A) the triceratops (B) the mammoth   (C) the prehistoric ant (D) the giant grizzly   (E) *Eohippus*

7. The most reliable method of estimating the age of the earth is through the study of   (A) volcanic ash (B) igneous intrusions   (C) living volcanoes (D) uranium-lead deposits   (E) craters on the moon

8. A form of animal life that probably arose after fishes was   (A) sponges   (B) molluscs   (C) trilobites   (D) amphibians   (E) jellyfish

9. The Mesozoic era is referred to as the age of reptiles because during it   (A) dinosaurs first appeared (B) dinosaurs became extinct   (C) reptiles reached their greatest development   (D) reptiles destroyed all other types of vertebrates   (E) reptiles were the only living things on earth

10. *Eohippus* is considered to be an ancestor of the (A) camel   (B) horse   (C) elephant   (D) pig (E) whale

11. The heterotroph hypothesis was proposed by (A) Darwin   (B) Oparin   (C) DeVries   (D) Lamarck   (E) Metchnikoff

12. According to the heterotroph hypothesis, the correct sequence in the origin of life was   (A) heterotrophs — aggregates — autotrophs — organic molecules (B) organic molecules — aggregates — heterotrophs — autotrophs   (C) organic molecules — autotrophs — heterotrophs — aggregates   (D) aggregates — autotrophs — organic molecules — heterotrophs (E) heterotrophs — autotrophs — organic molecules — aggregates

13. The remains of *Archaeopteryx* indicate that birds are most closely related to   (A) flying insects (B) flying mammals   (C) flying fish   (D) reptiles (E) amphibia

14. The number of toes on each leg modern horse walks on is   (A) 1   (B) 2   (C) 3   (D) 4   (E) 5

15. All of the following statements about evolution are true *except*   (A) living things have changed   (B) modern living things developed from simpler organisms   (C) evolution has taken place over many millions of years   (D) evolution is still going on   (E) evolution has ceased

Answer Key

1-C	4-C	7-D	10-B	13-D
2-D	5-E	8-D	11-B	14-A
3-E	6-B	9-C	12-B	15-E

Answers Explained

1. (C) Paleontology is the study of fossils, which reveal a record of prehistoric organisms.

2. (D) Geologists estimate that the earth is at least 4 billion years old.

3. (E) Limestone, sandstone, and shale are examples of sedimentary rock, the type of rock in which most fossils are found. It is formed from sediment, which, under the pressure of water over the ages, became cemented into rock.

4. (C) During petrifaction, minerals replace parts of bones or wood as they lie undisturbed under water for thousands of years.

5. (E) The La Brea tar pits in Los Angeles have yielded the remains of the saber-toothed tiger and other prehistoric animals that were trapped in the sticky material.

6. (B) The intense cold prevented bacterial decay and preserved the bodies of prehistoric mammoths in Siberia.

7. (D) Uranium is a radioactive element that gradually changes into lead. By analyzing the proportions of uranium and lead in a rock, it is possible to calculate the age of the rock. On this basis, the age of the earth has been estimated to be about 4 billion years.

8. (D) Amphibian fossils have been found in rock that was formed after fish had appeared.

9. (C) During the Mesozoic era reptiles reached their greatest development and were the dominant animal life.

10. (B) *Eohippus,* the dawn horse, was about the size of a cat and had four toes on its front legs and three toes on its hind legs. It lived during the earliest period of the Cenozoic era.

11. (B) A Russian biochemist, Alexander I. Oparin, in 1936, proposed the heterotroph hypothesis to explain how life originated on the earth.

12. (B) According to the heterotroph hypothesis, molecules in the "hot, thin soup" of the bodies of water on early earth were acted on by various forms of energy to form bonds, resulting in the production of larger organic molecules that aggregated, eventually forming heterotrophs and later autotrophs.

13. (D) *Archaeopteryx* had birdlike features, such as feathers and wings. It also had reptilian features, such as numerous teeth in the beak, an elongated tail containing vertebrae, and claws at the ends of the wings.

14. (A) The modern horse has a foot that is supported by one large toe and that contains the remains of two side toes as splints, visible only in the skeleton.

15. (E) Evolution is the change that takes place in living things. It has occurred in the past, is going on at present, and will continue in the future.

11.2 **EVIDENCES OF EVOLUTION**

There are many lines of evidence that establish the fact that animals and plants have changed through the ages, progressing from simple to more complex forms.

Types of Evidence

Fossil Evidence of Evolution

The record of prehistoric life discussed in the preceding section has revealed the following points:

1. The oldest layers of rock contain only the simplest fossils.
2. The adjacent higher strata reveal similar fossils, but also new and slightly more complex organisms.
3. Successively more recent fossils show continuing complexity, from simple invertebrates to more advanced invertebrates, to vertebrates, to mammals.
4. Fossils of the most complex and contemporary plants and animals are present only in the most recent strata of rock.
5. There are rather complete fossil series, extending over millions of years, for animals such as the horse and the elephant, showing step-by-step progressive changes from a primitive type to the modern animal of today.
6. The relationship between two separate classes of animals, reptiles and birds, is shown by *Archaeopteryx,* which has characteristics of both classes.

Evidence from the Study of Heredity

Mutations in plants and animals are going on all the time. Humans have been able to make use of some of these gene changes in breeding better varieties. In fact, by means of radiations from isotopes and X rays, some animals and plants have been made to develop mutations artificially. The use of chemicals such as colchicine has also changed the chromosome number, resulting in new types. In addition to these sudden changes in the characteristics of animals and plants, other changes have resulted from the breeding of better and newer varieties. Improved fruits, vegetables, cattle, and so on that did not exist two generations ago are now well established.

Evidence from Comparative Anatomy

Comparative anatomy is the comparative study of the structures of different organisms. The arm of a human or an ape, the flipper of a whale, the leg of a cat, the wing of a bird, and the leg of a frog are clearly used for different purposes. Yet the skeletal structures of these forelimbs reveal a similar bone arrangement,

with the same names being applied to their parts. Such similar structures, which are not necessarily alike in function, are called *homologous* structures. Similarities are also evident when comparisons are made of the brains and nervous systems, the muscular systems, the digestive systems, and the circulatory systems of various vertebrates.

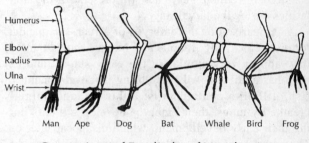

Humerus
Elbow
Radius
Ulna
Wrist

Man   Ape   Dog   Bat   Whale   Bird   Frog

Comparison of Forelimbs of Vertebrates

These similarities are explained by the presence of similar genes, which all of these animals have inherited from some distant, common vertebrate ancestor. Since the time of that common ancestor, changes have occurred, resulting in the evolution of new types of animals. However, these present-day animals still have enough genes in common to retain the similarities noted.

Evidence from Vestigial Structures

Vestigial structures are useless parts of animals that no longer have any function. In humans, some examples are the appendix, scalp and ear muscles, third eyelid (nictitating membrane) in the inside corner of the eye, and coccyx, or tailbone. In herbivorous animals, the appendix is used for digestion; many mammals use their scalp and ear muscles, and some rare people are able to wiggle their ears; amphibians, reptiles, and birds have a functional third eyelid, which covers the entire eye. The presence of vestigial organs indicates descent from an ancestor that once used them. Although they are no longer functional, genes are still present for them, and they appear in the animal's body.

Here are two other examples of vestigial organs: In the whale, small, vestigial hipbones indicate descent from vertebrate ancestors that had hipbones and hind legs. In 1994, the fossil remains of a whale that had legs and feet and walked on land were found in Pakistan. This whale lived about 50 million years ago and is named *Ambulocetus*. The modern horse has splints of two former toes on either side of the large toe.

Evidence from Embryology

Embryology is the study of the development of the embryo. The larvae of certain molluscs and annelids are remarkably similar in having a ciliated band around them. In both phyla the larvae, which are called *trochophores,* possess other similar structures. Beyond this stage, differences appear and development proceeds along different lines. Because of these early similarities, however, it is probable that the two phyla had a common ancestor. Although many changes have occurred since then, genes are still present for the similarities during the larval stage.

A comparison of the embryos of a fish, salamander, tortoise, chick, pig, rabbit, and human shows that (1) they appear very similar in the early stages, and (2) each has a tail and gill slits during its development. In fish, gill slits contain gills; although land animals breathe by lungs, their embryos develop gill slits during their early development, which disappear later on. The embryos of chick and human have tails, which disappear. The various embryos become less and less alike as they continue to develop.

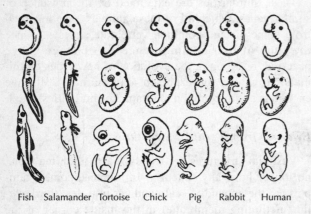

Fish Salamander Tortoise Chick Pig Rabbit Human

Embryological Development of Vertebrates

The similarity of the embryos in the early stage indicates that these vertebrates are descended from a common ancestor, and that they still have some similar genes. The more closely related animals are, the greater the similarity; the less closely related they are, the earlier in their development will differences appear. The "theory of recapitulation" attempted to summarize these observations of embryonic development by stating that, as the individual organism develops, it repeats the evolutionary development of the species—in other words, "Ontogeny recapitulates phylogeny." However, not all scientists agree with this theory.

Evidence from Classification

A "tree of life" arrangement of animals and plants, based on their similarities, gives us the system of classification described in Chapter 13 of this book. Like animals are grouped together in groups. Within a genus, for example, animals have many characteristics in common, indicating a common descent. Branching out from genus to family, to order, to class, and to phylum, the animals have fewer and fewer similarities, but enough to group them together. The further apart the phyla are, the greater the differences. Nevertheless, they all show a relationship and a progression from simple, one-celled organisms to more complex forms of life.

Evidence from Geographic Distribution

Geographic distribution refers to the fact that plants and animals tend to spread in all directions until they are stopped by barriers, such as climatic conditions, oceans, mountains, and deserts. If they become isolated for a long period of time as the result of such barriers, and cannot interbreed with other plants and animals, they will eventually become different from them. These differences come about as the result of mutations that occur from time to time, and the random recombination of genes. The longer the period of separation, the greater are the differences.

These effects of isolation are illustrated in the native mammals of Australia, which are all pouched, such as the kangaroo, koala, and wombat. At one time, Australia was connected with the mainland of Asia and the primitive pouched mammals of the time wandered freely between the two areas. After Australia was isolated by the ocean as the result of geologic changes, its pouched mammals continued to evolve to the present types. On the Asian mainland, however, the pouched mammals evolved into mammals that gave birth to their young. The two lines of development within and outside Australia occurred over a long period of time, demonstrating the ever-changing nature of life.

Similarly, Charles Darwin found distinct differences among the birds and other animals that inhabited the various Galapagos Islands, which are about 600 miles west of Ecuador. There were 13 different species of finch, a small bird. Some lived in trees and ate insects; others lived close to the ground and sought seeds. Their beaks were different in shape and size, varying from thin, pointed beaks to broad, heavy ones. These species resembled each other more than they resembled the related species on the mainland of South America. This was so because migration from the continent was rarer than migration from one island to another. The longer apart the animals had been, the more pronounced the differences between them.

Evidence from Comparative Biochemistry

The similarities among living things that are illustrated by the facts of comparative anatomy are strikingly substantiated by biochemical and physiological resemblances. The *precipitin test* establishes a delicate scale of blood relationships. The test is performed as follows:

1. A small amount of human serum is injected into a rabbit, stimulating it to produce antibodies against human serum.
2. The sensitized rabbit serum is mixed with human serum.
3. A white precipitate forms.
4. If this sensitized rabbit serum is mixed with serum from a chicken, there is no reaction and the liquid remains clear.

It has been found that a precipitate also forms in decreasing degrees when the serums of a chimpanzee, an orang-outan, and a gorilla are added to rabbit serum that has been sensitized to human serum. In like manner, the serums of a dog and a wolf show percipitations with serum sensitized to dog serum. The precipitin test also links birds with reptiles, whales and porpoises with cows and pigs, and horseshoe crabs with scorpions. The geological records indicate similar relationships in these various cases.

Cytochrome *c* is a protein that acts as an electron carrier in cellular respiration. It is made up of a chain of 104 amino acids. When the sequence of these amino acids was compared for different organisms, it was found that they varied from each other. Monkeys were found to differ from humans by only one molecule in this respect. By comparison, the molecular structure of cytochrome *c* in humans differed from these other organisms: rabbit-9, duck-11, horse-12, turtle-15, tuna fish-21, yeast-45, and bread mold-48. Although cytochrome *c* is active in the respiration of all these organisms, the closer their relationship, the greater the similarity in molecular structure. This suggests their evolutionary relationship.

Other biochemical similarities point to close relationships among organisms. Thus, antibodies for diphtheria and for tetanus, which are produced by a horse, may also be used in the human body. Various hormones produced by cattle, such as insulin and ACTH, are used by doctors to treat diseases in humans. Among plants, the potato plant is attacked by the potato beetle, which will not eat the leaves of any other cultivated plant. However, it will eat the leaves of wild plants that are related to the potato.

Evidence from Molecular Biology

The cells of all living things are basically similar. Whether bacterium, oak tree, starfish, or human, they all possess DNA, which contains the genetic code for the life activities of the organism. Through the processes of mitosis and meiosis, plants and animals pass DNA on to all the cells of the organism and to the next generation. The nucleotides containing guanine, cytosine, adenine, and thymine are universally present. The process of protein synthesis, utilizing messenger RNA and transfer RNA, occurs in practically all organisms. Such evidence points to a common ancestry of all living things.

Evidence from a "Living Fossil"

An unusual fish, the *coelacanth,* was caught in the depths of the Indian Ocean between the island of Madagascar and Africa. This fish was believed to have died out 70 million years ago. It is considered a link between fish and amphibians. Several living specimens of this "living fossil fish" have been found.

Conclusion

A study of all the evidence leads to the inescapable conclusion that evolution has taken place over the ages and is continuing at the present time. The various lines of evidences are listed below:

Fossils
Heredity
Comparative anatomy
Vestigial structures
Embryology
Classification
Geographic distribution
Comparative biochemistry
Molecular biology
"Living Fossils"

Section Review

Select the correct choice to complete each of the following statements.

1. All of the following provide evidence for evolution except (A) eugenics (B) heredity (C) comparative anatomy (D) fossils (E) vestigial structures

2. The flipper of a whale is most similar in structure to the (A) claw of a lobster (B) claw of a crab (C) tentacle of an octopus (D) trunk of an elephant (E) arm of a human

3. An example of a vestigal structure is the (A) horse's foot (B) human appendix (C) human brain (D) dog's ear muscles (E) cat's claws

4. The presence of gill slits in a rabbit embryo indicates that (A) rabbits breathe by gills in the gastrula stage (B) rabbits descended from amphibians (C) vertebrates have a common ancestor (D) acquired characteristics can be inherited (E) the theory of regeneration may be true

5. Animals with the greatest number of similarities are grouped together in a (A) phylum (B) family (C) genus (D) class (E) order

6. Early in its embryonic development, a chicken has (A) gills (B) fur (C) hair (D) a tail (E) fins

7. A native animal of Australia is the (A) koala (B) rabbit (C) mongoose (D) dormouse (E) horse

8. Darwin studied similarities among species of birds on the (A) island city of Venice (B) Thousand Islands (C) islands of the West Indies (D) Galapagos Islands (E) islands of Langerhans

9. The precipitin test helps establish evidence for evolution in the field of (A) anatomy (B) comparative biochemistry (C) embryology (D) plant and animal breeding (E) geographic distribution

10. Progression from simple to complex forms is summarized in the term (A) sedimentation (B) petrifaction (C) evolution (D) catastrophism (E) erosion

Answer Key

1-A	3-B	5-C	7-A	9-B
2-E	4-C	6-D	8-D	10-C

Answers Explained

1. (A) Eugenics is the study of improving the human race by applying our knowledge of genetics.

2. (E) The skeletal structures of the whale flipper and the human forelimb reveal similar bone arrangements, with the same name being applied to the parts. Such similarities are explained by the presence of similar genes inherited from some distant, common vertebrate ancestor.

3. (B) A vestigial structure is a useless part of an animal that has no function. It indicates descent from some ancestor that probably once used the structure.

4. (C) Although the rabbit breathes by lungs, its early embryo develops gill slits that later disappear. This indicates descent from some ancestor with gills.

5. (C) Within a genus, animals have many characteristics in common, indicating a common descent.

6. (D) The chicken embryo has a tail in its early development, which later disappears. This indicates descent from some distant ancestor and the presence in the chicken of remaining genes.

7. (A) The koala is a pouched mammal found exclusively in Australia.

8. (D) When Charles Darwin visited the Galapagos Islands, he found many species of finches, small birds, each with specific characteristics that were different from those of finches on the mainland of South America.

9. (B) The precipitin test establishes a delicate scale of blood relationships among animals. When human serum is injected into a rabbit, the rabbit produces antibodies against human serum. When this sensitized rabbit serum is mixed with human serum, a white precipitate forms. There is no such reaction when the rabbit serum is mixed with chicken serum. However, when it is mixed with the serum of a chimpanzee, a precipitate forms. This indicates a close evolutionary relationship and descent from a common ancestor for humans and chimpanzees.

10. (C) Many lines of evidence establish the fact that animals and plants have changed through the ages, progressing from simple to more complex forms.

11.3 THEORIES OF EVOLUTION

That evolution has taken place appears to be an incontrovertible fact. However, *how* it took place is open to speculation. There are several theories that attempt to explain how changes have occurred among plants and animals.

Lamarck's Theory of Use and Disuse

Jean Lamarck (1744–1829) was one of the earliest biologists to recognize that living things have changed. In 1809, he attempted to explain how evolution has taken place by his theory of use and disuse.

Main Points of the Theory

The theory includes the following points:

1. When an animal uses an organ to adapt itself to its environment, that organ becomes well developed and enlarged. New organs arise according to the needs of the organism.
2. If an animal does not use an organ as it adjusts to its environment, the organ is undeveloped and remains small.
3. These *acquired characteristics are inherited,* so that the offspring will have either a well-developed organ that has been used, or a smaller organ that has not been used. After many generations, the descendants will be changed as the result of inheriting well-developed organs, or as the result of the disappearance of unused organs.

Lamarck explained the development of the long neck of the giraffe as being the result of constant stretching to eat the leaves of trees. The cumulative inheritance of slightly greater neck length in each generation resulted in the unusually long neck of today. According to this theory, fish found in the waters of caves are blind because their ancestors did not use their eyes; after many generations of inheriting weak eyes, they eventually became blind.

Evidence against Lamarck's Theory

Lamarck's theory is not accepted because of the lack of experimental evidence to substantiate his claims for the inheritance of acquired characteristics. His basic error was that he gave purpose to evolutionary change. August Weismann (1834–1914) conducted an experiment to test this idea. He cut off the tails of 22 generations of mice but found that the size of the tail in subsequent generations was not affected. He stated that changes in the body cells *(somatoplasm)* are not passed on to the next generation. Only changes in *germplasm*, consisting of the reproductive organs and their gametes, are inherited in the next generation. Weismann formulated the theory of *"continuity of germplasm"* to summarize his ideas that germplasm is passed on from one generation to the next, but that the somatoplasm portion of an individual dies and is not passed on. According to him, germplasm is immortal.

The chief present-day support for Lamarck's theory came from the Russian agriculturist Trofim Lysenko, who claimed that environmental effects are inherited. Practically all other scientists remain unconvinced that acquired characteristics can be inherited.

Darwin's Theory of Natural Selection

Charles Darwin (1809–1882) spent 25 years gathering facts about evolution before he finally published his epoch-making *The Origin of Species* in 1859. While a young man in 1831, he had traveled to many parts of the world as the naturalist on the ship *Beagle*. The observations he made at that time led him to realize that evolution had taken place. When he first announced his theory, many people were not ready to accept the idea of evolution. Since then, however, it has had profound effects on most fields of biological, philosophical, and social thought. By coincidence, a similar theory was proposed to Darwin by Alfred Wallace, whereupon Darwin presented both theories to the world of science.

Main Points of the Theory

Darwin's theory includes the following points:

1. *Overproduction.* Plants and animals reproduce in such large numbers that the earth would soon be covered with them if they all survived. The roe of a single codfish may consist of 10 million eggs; the common oyster may shed as many as 80 million eggs in a season; protozoa could, in a few weeks, produce a mass of offspring many times the size of the earth. Darwin estimated that a pair of elephants, the slowest breeding animals known, would have 19 million descendants in 800 years if all survived.
2. *The struggle for existence.* The tendency toward overproduction is checked by the struggle for existence carried on by living things. They compete for food, space, water, and other requirements in the environment. This rivalry occurs not only among members of the same species, but also among different species. The organisms that cannot survive die out. Of

all the eggs produced by a codfish, only two need develop to maturity if the number in the ocean is to remain constant.

3. *Variation.* All living things differ in size, color, strength, speed of reactions, and countless other ways. Some individuals are therefore better equipped in the struggle for existence than others, because they possess more favorable variations.

4. *Survival of the fittest by natural selection.* The organisms that are most fit for a particular environment will survive; the others will die out. Sometimes, the animal that is a little faster, or a little stronger, or a little larger, will survive; the slower, weaker, or smaller animal will be killed off by nature. However, the strongest or largest do not always survive, as is seen in the case of the dinosaurs. Fitness is comparative, depending on the requirements of the environment. On some islands that have strong winds, there are very few winged insects; insects with wings are blown out to sea and perish, leaving the "weaker" ones without wings as those most fit to survive in that environment. When an animal is introduced into a new country, it may be extremely successful because of the lack of natural enemies. Thus, the rabbit overproduced in immense numbers after it was brought to Australia. The Japanese beetle has likewise become a menace to cultivated plants in the United States.

5. *The origin of new species.* By the gradual accumulation of favorable variations, generation after generation, the various species eventually changed. When the variations were unfavorable, the species died out.

Darwin's theory explains the development of the long neck of the giraffe in this way: The short-necked ancestor of the giraffe produced many offspring whose necks varied slightly in length. Those with slightly longer necks had the advantage of being able to feed on the leaves of trees. In the struggle for existence, they were the fittest and survived. In each succeeding generation, the longer necked animals continued to survive. Gradually, only giraffes with long necks came to be present on the earth.

Difficulties of the Theory

Darwin was the first to recognize certain limitations of his theory, and included a full chapter, Chapter VI, of his famous book on this subject. He could not distinguish between variations that are inherited and those that are not; the science of genetics was still 40 years away. Also, the mere act of selecting the fittest does not in itself create new variations; natural selection operates only after the variations have appeared. Another problem is that minor variations do not seem important enough in the first establishment of new organs; an organ would have to be of some size before it could be significant in the survival of the organism.

Other Contributions to Our Understanding of Evolution

De Vries' Theory of Mutations

One answer to the question of how variations arise was supplied by Hugo De Vries (1848–1935), a Dutch botanist. In 1901, he stated that new species arise as a result of sudden changes, called mutations, in their hereditary makeup. He based his conclusions on a study of variations in the evening primrose. His theory added to Darwin's theory by clearing up the nature of variations and their role in the change of species. The main points of his theory are as follows: (1) mutations occur that may be beneficial or harmful to a species; (2) by natural selection, the beneficial mutants survive, while the harmful ones die out; (3) new species arise by the accumulation of useful mutations.

The mutation theory would explain the development of the present-day giraffe with the long neck as follows: At one time, the ancestor of the giraffe had a short neck. As the result of a mutation, one of its offspring was born with a slightly longer neck. It had the advantage of being able to reach the leaves of trees, and thus survived. This mutation was passed on to succeeding generations. Again, a mutation occurred, producing an offspring with a neck that was still a little longer than the others. Over millions of years, many such mutations resulted, giving rise to giraffes with longer and longer necks, which were better fitted to survive.

Punctuated Equilibrium Theory

In more recent years, Niles Eldredge and Stephen Jay Gould have suggested updating the theory of Darwin. They believe that some species (e.g., trilobites) remained unchanged for long periods, during which time there was comparative genetic stability. Then there were short periods during which new species were rapidly formed. Such short periods were followed once again by long periods of relative equilibrium. This theory of punctuated equilibrium attempts to explain the relatively rapid appearance of a new species that can be seen on the geological time table. It thus is different from Darwin's theory, which is based on the concept of gradualism—that organisms change gradually during the ages.

Genetic Basis for Evolution

The development of genetics has given us greater insight into the process of evolution. Today, the basis for variations is explained by gene mutations, chromosome mutations, and the recombination of genes. New understanding of how species change has come through the study of population genetics.

Population Genetics

Our early knowledge of genetics was obtained by tracing the inheritance of genes in individuals. Now, studies are made of the distribution of genes (the *gene frequency*) in a population.

In biological language, a *population* includes all the members of a species inhabiting a certain location. Examples are the paramecia in a pond; a grove of pine trees; dandelions in a lawn; people in a town. All of the individuals in the population share the same *gene pool*. The gene pool is considered to be the sum total of all the genes in the population. These genes are exchanged as the individuals interbreed, to give many recombinations.

The Hardy-Weinberg principle The two scientists for whom this law is named studied populations and arrived at the conclusion that the gene pool of a population remains constant from generation to generation under the following conditions: (1) the population is large; (2) there are random matings; (3) there are no new factors such as mutations or migration. Under these theoretical conditions, a species would tend to remain practically the same. In other words, this principle applies to populations that are in equilibrium for their alleles.

The genotypes in a population in which there have been random matings distribute themselves as follows:

Genotype	Frequency
AA	p^2
Aa	$2pq$
aa	q^2

In this relationship, p = the frequency of the dominant allele *(A)*; q = the frequency of the recessive allele *(a)*; p^2 = the frequency of the homozygous dominant *(AA)*; q^2 = the frequency of the homozygous recessive *(aa)*; $2pq$ = the frequency of the heterozygous *(Aa)* individuals in the population.

The possibilities of the various matings can be shown in the usual way:

	A (p)	a (q)
A (p)	AA (p^2)	Aa (pq)
a (q)	Aa (pq)	aa (q^2)

The total genotypes in the population can be indicated by the following expression, in which 1 represents 100 percent of the population:

$$p^2 + 2pq + q^2 = 1$$

Suppose we consider the frequency of genes for brown *(B)* and blue *(b)* eye color. Assume that 64 percent of a given population has brown eyes and 36 percent has blue eyes. Can we determine how many of the brown-eyed people are homozygous or heterozygous?

We know that the frequency for gene *b* is expressed as *q*. If $q^2 = 0.36$, the square root of $q = 0.6$. Since *q* is the frequency of the recessive allele, we can say that 60 percent of the alleles are recessive, *b*. The other 40 percent are dominant, *B*; *p* therefore = 40 percent.

The possibilities can be expressed as follows:

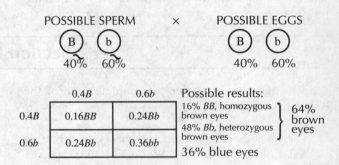

In other words, of the 64 percent who are brown-eyed, 16 percent are homozygous for brown eyes, and 48 percent are heterozygous for brown eyes.

Population changes Gene pools tend to be unstable, because the theoretical conditions needed under the Hardy-Weinberg principle are not always met. As a result, changes occur in the population. If environmental conditions change, some of the genes may provide an advantage, such as protective coloration. By natural selection, the individuals that have the favorable genes will survive and reproduce. Soon there will be a higher proportion of individuals with these genes, and a lower proportion with the unfavorable alleles.

Isolation When members of a population are isolated into smaller groups, they are prevented from breeding freely with each other. In such small populations, changes in the gene pool take place more readily. Random breeding establishes a new gene balance in each group that may be different from the gene balances in the others. Any mutations will probably be different in each of the groups, giving rise to distinct characteristics.

Development of New Species

As a result of the interaction of many factors, new species arise, which will not be able to interbreed with the original species. Some of the factors leading to the emergence of new species may be listed as follows: (1) variations—the individuals in a population share the gene pool, and reflect many variations; (2) changes in existing environment—the traits with high survival value will increase in frequency; (3) migration to new environments—under new conditions, the gene pool changes as there is a smaller group of individuals for random mating; (4) natural selection—the individuals with favorable characteristics for the particular environment will pass their assortment of genes along to their offspring; (5) isolation—when groups are isolated, they are not able to breed at random, and differences become accentuated on both sides of the barrier. The Kaibab squirrels on the north rim of the Grand Canyon are similar to the Abert squirrels on the south rim but have become different, probably because of the separation between them. Another type of isolation is biological, caused by differences in breeding time, as between two groups of toads; their populations are kept apart, leading to the development of two species.

As a group of organisms adapts to a new environment, they may change in different directions. The evolutionary pattern that takes place as the organisms fit into the different environmental niches is known as *adaptive radiation*. The finches on the Galapagos Islands illustrate this. Some evolved into insect eaters and live in trees. Others became seed eaters and live close to the ground. Within each type there were additional variations.

Recently, environmental conditions for house flies changed and they adapted accordingly. The introduction of the insecticide DDT was effective in killing practically all that were exposed to it. However, a few flies had a genetic makeup that made them immune, and they survived. DDT became less and less effective as more and more of these resistant flies survived and reproduced. Thus, a new strain of flies developed, with different genes for survival.

A similar situation resulted in the development of new strains of bacteria resistant to antibiotics. The occurrence of mutations led to the survival of strains of staphylococcus that were not killed by penicillin.

Section Review

Select the correct choice to complete each of the following statements:

1. The first scientist to present a theory of evolution was (A) De Vries (B) Weismann (C) Lamarck (D) Darwin (E) Lysenko

2. The theory of use and disuse was based on (A) gene changes (B) chromosome changes (C) changes brought about by the environment (D) mitotic changes (E) mutations

3. According to Lamarck, evolution occurred as the result of (A) natural selection (B) the theory of recapitulation (C) overproduction (D) blending inheritance (E) inheritance of acquired characteristics

4. According to Weismann, change in (A) somatoplasm are inherited (B) germplasm are inherited (C) somatoplasm and germplasm are inherited (D) body cells are inherited (E) body cells and germplasm are inherited

5. The name of Darwin's famous book is (A) *The Beagle* (B) *The Continuity of Germplasm* (C) *Brave New World* (D) *The Origin of the Earth* (E) *The Origin of Species*

6. All of the following are parts of Darwin's theory *except* (A) metamorphosis (B) survival of the fittest (C) natural selection (D) struggle for existence (E) variation

7. In order for the population of codfish to remain constant, the number of eggs produced by one codfish that must survive is (A) 1 (B) 2 (C) 4 (D) 1 million (E) 10 million

8. In the struggle for existence, the animals that survive are always the (A) largest (B) strongest (C) fastest (D) heaviest (E) fittest

9. Most scientists believe that species have changed as the result of   (A) use and disuse   (B) inheritance of environmental variations   (C) mutations   (D) changes in somatoplasm   (E) heterosis

10. The scientist who proposed a revision of Darwin's theory was   (A) Redi   (B) Banting   (C) Mendel   (D) De Vries   (E) Ehrlich

11. A population in an area includes   (A) all the living things   (B) all members of the same phylum   (C) all members of the same species   (D) all the living things in relation to their physical environment   (E) all the autotrophs and heterotrophs

12. According to the Hardy-Weinberg principle, the gene pool may remain stable if there are   (A) random matings   (B) many mutations   (C) frequent migrations   (D) selected matings   (E) random mutations

13. In the Hardy-Weinberg principle, $p^2 + 2pq + q^2 = 1$, q^2 represents the frequency of the   (A) homozygous dominant   (B) heterozygous dominant   (C) heterozygous recessive   (D) homozygous recessive   (E) blended genes

14. If, in a given population, 36 percent of the people have blue eyes, the percentage of the recessive allele is   (A) 24   (B) 36   (C) 40   (D) 60   (E) 64

15. All of the following are factors in the development of new species *except*   (A) variation   (B) asexual reproduction   (C) sexual reproduction   (D) isolation   (E) natural selection

Answer Key

1-C	4-B	7-B	10-D	13-D
2-C	5-E	8-E	11-C	14-D
3-E	6-A	9-C	12-A	15-B

Answers Explained

1. (C) In 1809, Jean Lamarck (1744–1829) attempted to explain, by his theory of use and disuse, how evolution takes place.

2. (C) According to the theory of use and disuse, an animal adapts itself to its environment and an organ that is used increases in size, one that is not used remains small, and such changes are inherited.

3. (E) According to Lamarck, evolution occurred as a result of the inheritance of acquired characteristics.

4. (B) August Weismann (1834–1914) formulated the theory of continuity of germplasm to summarize his ideas that germplasm is passed on from one generation to another, but that the somatoplasm of an individual dies and is not passed on.

5. (E) Darwin published *The Origin of Species* in 1859, after many years of study and observation, to explain how evolution takes place.

6. (A) Metamorphosis is a term that describes the changes that take place in the life history of an animal—for example, in a butterfly, the changes from the fertilized egg to the larva to the pupa and finally to the adult butterfly.

7. (B) Although a single codfish may produce millions of eggs, all do not survive. Only two need to survive to maturity if the number of cod in the ocean is to remain constant.

8. (E) Fitness is comparative, depending on the requirements of the environment. On a windy island, the "weaker" insects without wings are most fit to survive; those with wings would be blown out to sea.

9. (C) Mutations are sudden changes that take place in the hereditary makeup of an organism. If they are favorable, such mutations are inherited and passed on to the offspring, eventually giving rise to a new species.

10. (D) Hugo DeVries (1848–1935) proposed a theory to explain that variations that arise through mutations are inherited, if they are beneficial. By natural selection, the beneficial mutants survive, while the harmful ones die out. The accumulation of useful mutations leads to development of new species.

11. (C) All the individuals in the population share the same gene pool. The genes are exchanged as the individuals interbreed. Example: a grove of pine trees.

12. (A) The Hardy-Weinberg principle states that the gene pool of a population remains constant from generation to generation if there are random matings in a large population and if no new factors such as mutations or migration are introduced.

13. (D) The equation represents the total genotype in a population; p^2 represents the frequency of the homozygous dominant; q^2, the frequency of the homozygous recessive; $2pq$, the frequency of all the heterozygous individuals in the population.

14. (D) Blue eyes are homozygous recessive. If q^2 represents the frequency of the homozygous recessive genes, $q^2 = 36$ percent, or 0.36. The square root of this is $q = 0.6$, or 60 percent. From the equation we know that q represents the frequency of the recessive allele.

15. (B) In asexual reproduction, the same set of genes is passed on to the offspring. Since there are no inherited variations, the individuals do not differ. Therefore none is better fitted than any others to survive by natural selection. The species remains constant.

11.4 HUMAN DEVELOPMENT

Humans, like other living things, have changed through the ages. The first types of humans appeared about 1 million years ago. Humans have many similarities to the apes, indicating the existence of a common ancestor. Relatively few fossils of humans have been found; hence our knowledge of human evolution is not complete. In addition to the few fossil fragments, early humans are studied from their cultural remains, including tools and weapons (artifacts), refuse heaps, carvings, and paintings in caves. Present-day *Homo sapiens* probably did not come upon the scene until about 50,000 years ago.

Early Humans

Homo erectus

This name is now applied to two early types of humans originally called Java man and Peking man. In 1891, at Trinl, Java, Dr. Eugène Dubois found the top of a skull, a left thighbone, and some molar teeth that were recognized as belonging to a primitive type of human. The eyebrow ridges were prominent. The brain had a capacity of about 940 cubic centimeters, much smaller than that of a modern human (about 1,500 cubic centimeters) but larger than that of an ape (less than 580 cubic centimeters). Java man (originally called *Pithecanthropus erectus*, "the ape man that walked erect") had a heavy neck and a receding chin. He lived about 500,000 years ago. In 1926 and 1929, a number of fossil remains of a similar type were found near Peking, China; Peking man was originally named *Sinanthropus pekinensis*. It is believed that he had the power of speech, used fire, and made crude tools.

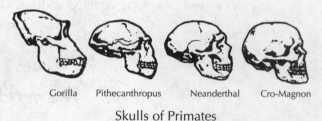

Gorilla   Pithecanthropus   Neanderthal   Cro-Magnon

Skulls of Primates

Neanderthal Man

Homo neanderthalensis probably appeared about 150,000 years ago, and became extinct about 25,000 years ago. He had projecting eyebrow ridges, a low retreating forehead, a massive jaw, and thick, heavy bones. His brain capacity, however, was about the same as a modern human's. His average height was about 5½ feet. Many fossil remains have been found in various parts of Europe. Neanderthal man's later tools were chipped and polished, indicating a rather advanced form of culture. In 1997, a study was made of DNA extracted from the bones of a Neanderthal who lived at least 30,000 years ago. The results showed that Neanderthals were a different species from modern humans and did not give rise to present-day *Homo sapiens*.

Cro-Magnon Man

Many skeletons have been found of Cro-Magnon man, who lived about 50,000 years ago. He was of slightly taller stature than modern humans, with a brain capacity of 1,550 cubic centimeters, a little larger than that of today. He had a high forehead and a well-developed chin. Cro-Magnon man's tools were made of polished stone, bone, and ivory. He sewed skins for clothing. He has left us paintings of deer, bisons, and mammoths on the walls of caves in the southern part of France. Believed to be the same species as modern *Homo sapiens,* he overlapped the more primitive Neanderthal man and may have led to the latter's extinction.

Other Finds

The excavations of the late Dr. Louis Leakey, his wife, Mary, and their sons, Richard and Jonathan, in the Olduvai Gorge of northern Tanzania, have yielded the fossil remains of other types of primitive humans. *Zinjanthropus* (1959) is supposed to have lived 1 million years ago and used stone tools. *Homo habilis* (1961) may have lived 1.75 million years ago. The fossils indicate that he was small-brained, walked erect, and used tools. In 1967, the Leakeys reported the discovery of a still older type of human, *Kenyapithecus,* in Kenya, whose age was established by radioactive potassium-argon dating as being 19–20 million years. Another very old set of fossil remains represented a primitive type called *Proconsul,* that may have lived 20–30 million years ago. In 1979, Drs. Donald C. Johanson and Tim White announced their discovery, in Ethiopia, of *Australopithecus afarensis*; they stated that this primate, which was nicknamed Lucy, lived about 3.5 million years ago and was the common ancestor of humans and a related type of primate that died out.

Some Controversial Questions

Humans and the Apes

Are humans descended from the apes? No. Scientists believe that both are descended from a common ancestor. When the skeletons of humans and apes are compared, there is close similarity of the bones. According to the results of the precipitin test (see Section 11.2), humans are biochemically similar to the orangutan and the chimpanzee. The sequence of nucleotides in the DNA molecules is very similar, and the proteins are almost identical.

One group of scientists holds that humans and chimpanzees have a common evolutionary ancestry. Another school of scientists believes that the orangutan, not the chimpanzee, and humans have a more modern ancestor. It is possible that a primate named *Ramapithecus* that lived about 10 million years ago may be one of the common ancestors from which both apes and humans have descended.

The Mother of All Humans

A controversial theory about human origin, developed from the study of molecular biology, states that present-day humans can trace the inheritance of the DNA in their mitochondria (mtDNA) back to a single woman ("Eve") who lived in Africa about 200,000 years ago.

Although most human DNA is located in the nucleus of the cell, the mitochondria also contain some DNA. Humans receive all their mitochondria from their mothers and none from their fathers. When the sperm unites with an egg during fertilization, only its head, which contains the nucleus, enters the egg. The rest of the sperm, which contains mitochondria, does not enter. As a result, the offspring receives only the mtDNA that was present in the egg from the mother. By contrast, the nuclear DNA is inherited equally from both parents. According to the theory based on these facts, the mtDNA can be traced back from mother to maternal grandmother and beyond that for generations to an ancient female ancestor who was the first member of the future *Homo sapiens* species.

Modern Humans

There are now over 5 billion human beings on earth, all belonging to the same species, *Homo sapiens.* They are similar in five respects: (1) the ability to walk erect, (2) having a well-developed cerebrum, (3) having a hand with the thumb opposite the other fingers, (4) having the power of speech for communication, and (5) other physical and physiological characteristics. Although they all show variation in physical makeup, they are more similar than different. On the basis of some of their differences, notably skin, eye, and hair pigmentation, hair texture, stature, skull shape, and facial features, modern humans are classified by some scientists into three main stocks: Caucasoid (white), Negroid (black), and Mongoloid (yellow). Each of these may be further classified into *races*.

Racial Characteristics

Anthropology is the study of human origin and classification. One of the characteristics studied is the *cephalic index,* representing the shape of the head. This is obtained by measuring the width of the head, and dividing the result by the length of the head. Heads with a cephalic index of 75 or less are considered long, or *dolichocephalic,* while those about 80 are broad, or *brachycephalic.*

Another characteristic that helps to distinguish races is blood group. The four blood groups, A, B, AB, and

O, are found in practically all races. A person with type A blood, for example, can be black, Chinese, English, or Navajo Indian. The various races differ, however, in the relative numbers of persons who possess each of the four blood groups. A blood transfusion may be made as conveniently between two type A persons of different races as between two members of the same race. In fact, it would be fatal for a person of one race to receive blood from a member of the same race if the blood types were different.

Confusions about Race

Confusion is often created by considering a race as being pure. Because of migrations and intermarriages, the genes for the various characteristics are carried in varying quantities by many individuals of different races. The Nordic race, for example, is found in the northern part of Europe, especially in the Scandinavian countries. Yet one finds many individuals there who do not possess all of the characteristics of the Nordic race. A study of the Swedish army revealed that only about 11 percent of the men were tall, blond, and blue-eyed. Many Nordics may be short, have dark hair, and have round heads. To avoid the confusion arising from the use of the word "race," some anthropologists suggest replacing it with the expression *racial type*. This would help convey the idea that races are not pure and that a person is classified on the basis of having an average of certain characteristics. There may be so much variation within racial types that there are Hindu Caucasoids who are darker than some Negroids, and there are Chinese who are taller than some Nordics.

Another source of confusion sometimes arises from the tendency to consider racial type and national boundary as being related. Thus, Italy is considered as consisting of Mediterranean racial types. However, there are parts of northern Italy in which many of the people have Nordic characteristics and others have Alpine traits. Also, there is no Italian race; the people who live in Italy have the characteristics of many racial types.

Similarly, other cultural traits, such as religion and language, have no connection with racial types. There is no Jewish race, since Judaism is a religious conviction. Also, there is no Latin race, based on languages derived from Latin, such as French, Spanish, and Italian. There is no Aryan race, either, since the term "Aryan" also refers to language.

Racial differences have led some people to believe that some groups are superior to others. However, there is no scientific evidence to support this idea; all groups seem to have equal average mental ability. Since the various characteristics are inherited independently of each other, there can be no linkage of intelligence with hair color, stature, eye color, or any other trait. When human beings from various parts of the world are compared, it is evident that they are similar in practically all of their characteristics (i.e., they all have a four-chambered heart, an appendix, a large cerebrum, two kidneys, 206 bones in the adult skeletal system, etc.), and differ in only a few ways.

The future of humankind may depend on the ability of *Homo sapiens,* man the wise, to cooperate with his fellow members of the human race.

Section Review

Select the correct choice to complete each of the following statements:

1. The reason for human similarities to the ape is that (A) humans descended from apes (B) apes descended from humans (C) a mutation of the apes became human (D) a mutation of humans became the apes (E) humans and apes had a common ancestor

2. Our knowledge about primitive humans is based on all of the following *except* (A) widespread fossil remains (B) refuse heaps (C) carvings (D) paintings in caves (E) artifacts

3. The first types of humans appeared about (A) 25,000 years ago (B) 100,000 years ago (C) 1 million years ago (D) 10 million years ago (E) 100 million years ago

4. A type of human that appeared after Neanderthal man was (A) Java man (B) Peking man (C) Heidelberg man (D) Cro-Magnon man (E) *Pithecanthropus erectus*

5. *Homo sapiens* is thought to have been derived from (A) Heidelberg man (B) *Sinanthropus pekinensis* (C) *Pithecanthropus erectus* (D) Cro-Magnon man (E) Peking man

6. All of the following are true of Cro-Magnon man *except* (A) his tools were made of polished stone (B) his brain capacity was smaller than that of modern humans (C) he sewed skins for clothing (D) he made paintings of animals on the walls of caves (E) he was slightly taller than modern humans

7. Humans are probably descended from (A) the chimpanzee (B) the orangutan (C) a hybrid form of the chimpanzee and the orangutan (D) the ape (E) an ancestor common to humans and apes

8. Human beings are similar in all of the following ways *except* (A) they walk erect (B) they have a well-developed cerebrum (C) they have an opposable thumb (D) they have the power of speech (E) they have the same amount of skin pigmentation

9. The study of human origin and classification is known as (A) eugenics (B) demography (C) anthropology (D) paleontology (E) physiology

10. A Navajo Indian with type A blood can receive a transfusion of type A blood (A) from another Navajo Indian (B) from a Mexican (C) from a black (D) from an American cowboy (E) from all of the above

Answer Key

1-E	3-C	5-D	7-E	9-C
2-A	4-D	6-B	8-E	10-E

Answers Explained

1. (E) It is believed that humans and apes have many similarities because both are derived from a common ancestor, whose descendants developed along different lines to become the species of today.

2. (A) Relatively few fossils of primitive humans have been found; hence our knowledge of human evolution is not complete.

3. (C) On the basis of fossil records, the closest estimate is that the first types of humans made their appearance about 1 million years ago.

4. (D) Cro-Magnon man overlapped Neanderthal man and may have led to the latter's extinction.

5. (D) Cro-Magnon man has so many characteristics of modern humans that he is thought to be of the same species.

6. (B) The brain capacity of Cro-Magnon man was about 1,550 cc, a little larger than that of today's humans.

7. (E) Although humans are biochemically similar to the chimpanzee and the orangutan, scientists believe that humans and apes are descended from a common ancestor.

8. (E) Human beings differ in the amount of skin pigmentation, from white to black.

9. (C) Anthropology is the study of human origin and classification. The cephalic index, representing the shape of the head, is one of the characteristics studied.

10. (E) Although the racial groups differ in the relative numbers of individuals having the different blood groups, a blood transfusion can be made between any two type A persons, regardless of background.

THE BALANCE OF LIFE

CHAPTER
12

12.1 THE WEB OF LIFE

Humans and the other living things on earth live in an entangling relationship with each other. They do not exist in isolated fashion. They are interdependent, and each forms a strand in the web of life. *Ecology* is the study of living things in relation to each other and to their environment.

Relationships Among Living Things

Structure of Life

The pattern of life may be said to be organized according to the following arrangement:

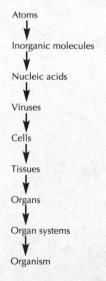

Atoms
↓
Inorganic molecules
↓
Nucleic acids
↓
Viruses
↓
Cells
↓
Tissues
↓
Organs
↓
Organ systems
↓
Organism

Organisms do not live alone; a dynamic equilibrium exists in the interactions of organisms with other organisms and with their environments. The structure of life for an individual may therefore be shown in its wider relationships as follows:

Organisms
↓
Populations
↓
Communities
↓
Ecosystems
↓
World biomes
↓
Biosphere

Population

All the members of a species inhabiting a given location comprise a population. The members of a species are more or less alike and are capable of breeding with one another. Examples are sunfish in a lake, trees in a pine grove, people in a city.

Community

All the plant, animal, and other organism populations interacting in a given environment comprise a community. There are aquatic communities, which inhabit a pond, a river, a lake, or an ocean. There are also terrestrial communities in a field, a desert, a cave, or a forest.

Ecosystem

An ecosystem is a self-sustaining, dynamic community of plants and animals in relation to their physical environment. In such a system, matter is transferred, and used over and over again in a series of cycles, for example, water, carbon, oxygen, hydrogen, and nitrogen. Light, temperature, and a flow of energy are also involved in the interrelationships of organisms with their physical surroundings and themselves.

The living things in an ecosystem affect each other and depend on each other in a variety of ways, in food chains and food webs.

World Biome

A biome is a major ecological grouping of organisms on a broad geographical basis. It is determined largely by the major climate zones of the world. The land biomes are classified according to the climax vegetation they contain. Aquatic biomes, however, are not classified on this basis because the distribution of plants is not fixed. Since more than 70 percent of the earth's surface is covered by water, aquatic biomes are a widespread influence on the content of life.

Biosphere

The biosphere is the thin layer of life at the surface of the earth, including the biologically inhabited soil, water, and air. It is also thought of as including a system of relationships between the living things and the materials and energy surrounding them.

Living Things and the Environment

Physical Factors in the Environment

The nonliving (*abiotic*) factors that affect living things include light, temperature, water supply, oxygen supply, minerals, pH, and soil or rock (substratum). Light is needed by green plants for food-making. The depth in the ocean to which plants can grow is limited to the area of penetration of sunlight below the surface.

Temperature affects the metabolism of living things, and consequently determines their geographical distribution. The various climatic zones have different varieties of living things that are best adapted to live at their temperatures.

Plants may be classified according to their ecological need of water as follows:

1. *Hydrophytes*—plants that live in water or in marshes. They usually have large leaves that float on the water, and have reduced root systems. Examples: water lilies, water weeds, cattails.

2. *Xerophytes*—plants that live in dry conditions. They have deep root systems, much water-storage tissue, a thick epidermis, and modified leaves or spines. Examples: cactus, sagebrush, yucca tree.

3. *Mesophytes*—plants that live on a medium amount of water. They have a well-developed system of roots, stems, and leaves. Examples: plants of field and forest, including grasses, maple trees, and clover.

Practically all living things need oxygen. The depth to which organisms can live in the soil is limited by the penetration of air. Likewise, the supply of oxygen in the ocean waters is a limiting factor that determines the depth at which marine organisms can survive. The mineral content of the soil and water has an important influence on the nature of the living things that can grow in a particular location. Likewise, the pH of soil affects the growth of plants; lime must be added by gardeners and farmers to counteract the effects of soil that is too acid.

The substratum is the soil or rock in which organisms live. Clay, sand, and rocky soils have different communities that can live in them. Soil containing much organic matter is teeming with organisms, as compared to sand.

Cycle of Materials

Water as well as such important elements as nitrogen, carbon, oxygen, hydrogen, and phosphorus are constantly being cycled between organisms and their environment. These elements pass from inorganic sources to organic forms and back again to inorganic forms. Much of this cycling is accomplished through the actions of decomposers, scavengers, and saprophytes.

The Nitrogen Cycle

The chief features of the nitrogen cycle, illustrated on the following page, are as follows:

1. During protein synthesis, green plants bind nitrogen into their protoplasm.

2. Animals eat plants and synthesize animal protein from plant protein.

3. Plants and animals produce nitrogenous wastes. When they die, their bodies are used as sources of energy by decomposers, bacteria of decay.

4. The nitrogen of the nitrogenous wastes and the proteins is released and converted into nitrates by decay bacteria and nitrifying bacteria; nitrogen-fixing bacteria also produce nitrates from gaseous nitrogen. The nitrates are made available to plants for protein synthesis.

5. Denitrifying bacteria return molecular nitrogen to the atmosphere.

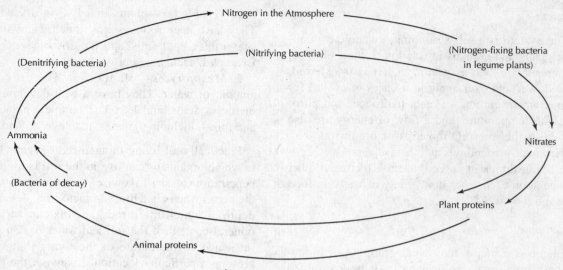

The Nitrogen Cycle

The Carbon-Hydrogen-Oxygen Cycle

The elements carbon, hydrogen, and oxygen are present in all living things. By the process of respiration, carbon dioxide and water are released. When plants and animals die, carbon dioxide and water are given off through the action of decomposers such as bacteria of decay and fungi. When green plants carry on photosynthesis, they take in carbon dioxide and water and give off oxygen. In the burning of coal, petroleum, and wood, oxygen is used up and carbon dioxide and water are given off. Through these and other activities, the elements carbon, hydrogen, and oxygen are kept in circulation between the biotic (living) and the abiotic (nonliving) environment.

The Phosphorus Cycle

Elements such as phosphorus and calcium are cycled from soil, rock, or water to organisms and back again. The cycle for phosphorus is typical. This element is present in compounds that become dissolved in water. Plants take in the phosphates, and animals eat plants. Animal wastes include some of these phosphates, which are returned to the soil or to the ocean. When the plants and animals die, decomposers break them down through decay, and release the phosphorus compounds, which are returned to the soil or ocean for further circulation.

Food Relationships

Practically all living things depend for their existence on green plants, which can make their own food by photosynthesis. Green plants are autotrophs and are considered to be *independent organisms*. Organisms that cannot synthesize their own food are heterotrophs, and are dependent.

Saprophytes are organisms that live on dead organic matter, for example, bacteria of decay, mushrooms. *Herbivores* are plant-eating animals, such as the cow and horse. *Carnivores,* or meat-eating animals, eat the flesh of other animals, which directly or indirectly obtained their food from plants. Carnivores are of two types: (1) *predators,* which kill other animals (lion, hawk), and (2) *scavengers,* which eat dead organisms that they did not kill (vulture, jackal, snail). The snapping turtle is an example of an animal that may be both a predator and a scavenger.

Omnivores are animals that eat both plants and animals (human, rat). When plants and animals die, their remains are used as food by smaller organisms, *decomposers,* such as bacteria of decay and fungi.

Food Chain

A food chain represents the different links along which food is passed from one organism to another. It starts with green plants, which can make their own food in the presence of sunlight. Sunlight serves as the source of energy for photosynthesis to take place. The green plant may then serve as a source of food for herbivores, which, in turn, are eaten by carnivores, and for decomposers, which bring about decay in animals, as well.

The links in a food chain may be illustrated by referring to the well-known jingle by the 18th-century satirical writer Jonathan Swift:

> Big fleas have little fleas
> Upon their back to bite 'em,
> And little fleas have lesser fleas
> And so, ad infinitum.

Aquatic food chain In the oceans, great numbers of algae synthesize carbohydrates, and then convert them into proteins. Protozoa and minute invertebrates feed upon the algae. Together, they make up the mass of microscopic life called *plankton*, which is found floating near the surface of oceans and inland waters. Small fish eat the plankton and, in turn, are eaten by larger fish. Tiny crustaceans less than an inch in size, called *krill*, also feed upon the algae. Krill is the principal food of the largest whales. Eventually, the food chain extends from microscopic algae to whales, sharks, and humans. Without the aquatic algae and green plants, there could be no aquatic animals.

Terrestrial food chain A terrestrial food chain may involve the following sequence: green plants, such as corn, make food → mice feed on the plants → snakes prey on the mice → hawks eat snakes as well as mice → bacteria of decay and fungi live on all of the organisms when they die.

Classification in a food chain The various links in a food chain may be classified as follows:

1. *Producers*—green plants are the producers, since they synthesize organic compounds that serve as food.
2. *Consumers*—organisms that feed directly on green plants are called *primary,* or *first-order,* consumers. Animals of this type are known as herbivores. *Secondary,* or *second-order,* consumers are carnivores with sharp tearing teeth, which prey on the primary consumers. There may also be *third-order consumers,* which are predators on the secondary consumers, and frequently on the primary consumers.
3. *Decomposers*—these organisms break down the wastes and dead bodies of producers and consumers into simpler compounds. These compounds are cycled, or returned to the environment, where they are used over again by other living things.

The sequence in the terrestrial food chain mentioned above can thus be rewritten in this way: green plants (producers) → mice (first-order consumers) → snakes (second-order consumers) → hawks (third-order consumers on snakes; second-order consumers on mice) → decomposers.

Food Web

Food chains overlap into food webs, because there are many kinds of producers, which may be eaten by different kinds of primary consumers (insects, rabbits, etc.); these herbivore consumers may shift from one plant to another. Carnivores, which live on herbivores, may likewise shift from one source of prey to another,

and may be of different types (frogs, foxes, owls). An ecological balance is achieved when the numbers of the various populations are kept at relatively stable levels in relation to each other.

Symbiosis

Many organisms live intimately together in close associations that may or may not be beneficial to them. Although the relationships are not always clearcut, the types of symbiosis may be described as follows:

1. *Commensalism*—one organism is benefited and the other is not affected. Examples: Barnacles live on the hide of a whale, obtaining a habitat as well as a means of transportation. The remora fish attaches itself to the bottom of a shark by means of a suction pad on its head, and feeds on food scraps left over by the shark.
2. *Mutualism*—both organisms mutually benefit from living together. Examples: A lichen is made up of both a fungus and an alga; the fungus provides moisture, while the alga makes food by photosynthesis. Nitrogen-fixing bacteria live in the nodules on the roots of legume plants such as clover; the bacteria convert nitrogen into nitrates, while the plants provide a habitat and nutrition for the bacteria.
3. *Parasitism*—one organism, the parasite, attaches itself to another, the host, and benefits at the latter's expense. Examples: The athlete's foot fungus grows on human skin. Tapeworms and other types of worms attach themselves in the intestines of animals and absorb digested food. The lamprey eel, a primitive fish without jaws, has a circular, sucking mouth lined with hooklike teeth, by means of which it attaches itself to trout and other fish and sucks their blood. Disease germs, such as viruses and bacteria, live within the body of an organism and may cause serious illnesses. Among green plants, the mistletoe carries on its own photosynthesis. However, it absorbs water and minerals through roots that it embeds in the host tree's branches. It is thus considered to be partially parasitic.

Niche

The specific environment of a particular species is known as its *niche*. If two different species occupy the same ecological niche, they will compete with each other for food and reproductive sites. The species that reproduces faster will eliminate its competitor. As a result, one species is established per niche in a balanced community. All the members of that species use the same kinds of food, and occupy similar reproductive sites.

The giant saguaro cactus of the Southwest harbors two kinds of owls, the elf owl and the screech owl, in holes made in its stem by woodpeckers. Although both species of owl may occupy adjoining holes, they are not in the same niche. The elf owl eats mainly small insects. The screech owl eats larger insects, as well as mice and scorpions; it also breeds earlier in the year. Thus, these two species occupy different niches.

Energy Flow

Green plants have the ability to store the radiant energy of the sun as chemical energy in the bonds of the organic compounds they synthesize. When green plants are eaten as food, this energy is taken into the consumers and used by them for their life activities. The various pathways by which the food energy is transmitted make up the energy system of the food web.

There is a decrease in the total amount of energy as it is passed along from the producer to the consumers. Each member of the food web uses up some of the energy as it carries on its various metabolic activities. In the sequence of energy transfers, the amount of usable energy runs down. In addition, since the consumers must seek and find their prey, they may not always be successful in obtaining food. In other words, the various members of the food web pass on less energy than they received.

Pyramid of Energy

Since there is a loss of energy at each feeding level, it can be seen that the producers contain the greatest amount of energy. This amount decreases at each level. Since consumers are generally larger than the animals they live on, many organisms at the lower feeding levels are required to furnish sufficient food energy for a single organism at a higher feeding level. This idea is referred to as the pyramid of numbers.

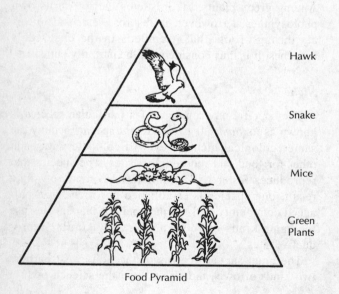

Hawk

Snake

Mice

Green Plants

Food Pyramid

Biomass Pyramid

Biomass refers to the total amount of organic matter. Since there is a decrease in the amount of energy at each successive feeding level, less biomass can be supported at each level. In other words, the total mass of consumers in a particular biomass is less than the total mass of the producers.

Because of the shortage and high cost of oil, many countries are attempting to obtain large fractions of their energy from biomass. In 1980, experts in bioenergy from 76 countries met in Atlanta for the first World Congress and Exposition on Bioenergy. Some of the progress reported is as follows:

Brazil is obtaining alcohol from the fermentation of sugar cane and is constructing 250,000 cars employing only alcohol as a motor fuel. Sweden, which has no coal, oil, or natural gas, is planning to shift to its large forest areas for wood as the prime source of energy. It plans to grow fast-rotation trees, such as willow and birch, that can be harvested every 3–5 years. The trees are mowed down and wood is collected in winter. The next spring, new shoots arise from the stumps, so the tree does not have to establish a new root system.

China is developing the production of biogas methane from biomass human and animal wastes. The United States is developing processes to obtain alcohol from grain, and to reduce trees to small chips in seconds for the same purpose. One problem with the harvesting of plants for bioenergy purposes is that the production of humus in the topsoil may be reduced by the removal of these plant products. This could lead to poor crop yield and erosion.

Ecological Succession

Communities change as the conditions affecting them change. This may be illustrated by the history of a pond. A small pond may be covered with water lilies, which make up the dominant species. Other types of plants may also grow in it, including *Elodea* and the bladderwort. Frogs, tadpoles, snakes, water beetles, and small fish may also inhabit the pond.

As the plants in the pond die, their remains build up until they form a semisolid base at the edge of the pond. Cattails and other marsh plants begin to grow there. Mosses and ferns soon invade the area. As the amount of soil increases, a new community develops, which will be replaced later on by shrubs and willow trees. Other low plants, shrubs, and trees, including goldenrod, wild carrot, and sumac, succeed this community. The amount of available light is changed as poplar and pine trees replace the earlier community. The soil becomes increasingly enriched and changed.

This orderly sequence of communities that replace each other in a given area is known as ecological succession. It may take a few years, or several hundred years, for the pond to disappear, depending on the circumstances of size, source of water, location, surrounding topography, and so on. The replacement of communities continues until a climax community is reached.

Climax Community

In the order of succession of communities in the pond area, a community of oak and hickory trees is next. Other species are also present, but these are dominant. As the amount of shade increases in the forest, however, oak and hickory seedlings have difficulty surviving, and maples and beeches appear, to become the successful types. Their seeds are able to sprout and to survive, thus resulting in a permanent or final community. This now is a *climax* community. There is an equilibrium between the various living things in it, and it may continue indefinitely. Each species reproduces at a rate that maintains itself at about the same number from year to year.

A climax community receives its name from the dominant types of plants that characterize the situation. Besides the maple-beech climax, which is the typical climax in the northeastern part of the United States, other climaxes are cypress-white cedar in wet areas of Louisiana; beech-magnolia in the South; tall grasses on the prairies; scrub oak on a mountaintop; pine forests in New Jersey.

If a climax community is destroyed by humans through unlimited lumbering practices, or by fire, or by a hurricane, a succession of communities starts all over again. The maple-beech forest does not come up at once. Low shrubs and trees will be the pioneer dominant forms for many years until they are eventually replaced by the succeeding types of tall trees, leading once more to the maple-beech climax, perhaps a century later.

Often, the climax community is not replaced. Climax communities may vary in the same area. In New York State, for example, hemlock, beech, and maple trees often comprise the climax community at high elevations. At lower elevations, the climax trees are oak and hickory.

Role of Pioneer Organisms

The succession that ends in a climax community can be traced back to the pioneer organisms that first populated a given area. Pioneer organisms, such as lichens, first appear on bare rock. They help build up soil by breaking down the surface of the rock and releasing its minerals. Spores of mosses that are blown around may land in such a very thin layer of soil. As mosses grow, their remains help build up the soil. After a while, other plants will appear, and the succession of communities is on its way.

Biogeography

World Biomes

Through the study of biogeography, which deals with the distribution of plants and animals in the various areas of the world, a greater understanding can be reached of the major biomes. A biome is a major ecological grouping of organisms.

Terrestrial Biomes

The major plant and animal associations on land are determined by the chief climate zones of the earth. These climate zones are distinguished from each other on the basis of factors such as temperature, amount of rainfall, and amount of solar radiation. There are also modifications caused by local land and water conditions. The land biomes take their characteristics and names from the climax vegetation in the area. Since green plants are the producers of food, the major plant associations on the earth determine the kind of animals that will inhabit a particular locality. The major land biomes can be classified as follows:

1. *Tundra* is the treeless region in the far North. It has an arctic climate of severe cold, with a long, dark winter, and a mild summer having continuous daylight. The underlying part of the soil is permanently frozen. The upper part thaws out temporarily during the summer, to form numerous bogs and ponds. Mosses, lichens, including reindeer moss, and small plants grow actively during the summer. Birds, the snowshoe hare, caribou, flies, and polar bears are representative of the animal life.

2. *Taiga* includes the northernmost forests that extend in a broad zone just below the tundra, across Europe, Asia, and North America. These forests contain evergreen coniferous trees, such as pine, spruce, fir, and hemlock. Some of the animals are moose, black bears, wolves, rodents, and birds.

3. *Temperate deciduous forests* contain trees that shed their leaves during the cold winters and grow them back during the warm summers. They occur in the eastern United States, England, and central Europe. Characteristic trees are maple, oak, elm, beech, and birch. Deer, pumas, squirrels, and foxes are some of the animals.

4. *Tropical rain forests* occur in regions of high temperature and ample rainfall. The warm, humid climate of the tropics encourages a rich abundance of plant life. The physical conditions are relatively constant throughout the year. There are many climbing vines and numerous insects and birds.

5. *Grasslands* have less rainfall than the deciduous forests. Because of the recurrence of droughts, trees are generally unable to develop in these areas. Most of the plants are of the grassy type, and serve as food for the grazing animals.

6. *Deserts* have the driest conditions of all. The days are hot and the nights cold. The cactus and other plants are adapted to conserve water. Rodents and snakes may be common.

Effects of Latitude and Altitude

The land biomes take their characteristics largely from their latitude, or their distance from the equator. The further away from the equator—or the greater the latitude—the colder the climate. Altitude has a similar effect. The higher up one goes on a mountain, the greater the change in temperature. The vegetation also changes accordingly. The change is similar to that in traveling north. In the Tropical Zone, the tops of high mountains have conditions resembling those of the Arctic Zone. The similarity between latitudinal and altitudinal life zones can be shown as follows:

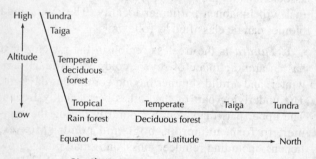

Similarity Between Latitudinal
and Altitudinal Life Zones

Aquatic Biomes

More than 70 percent of the earth's surface is covered by water. Variations in temperature are not extreme. Water absorbs large quantities of heat. It absorbs this heat slowly, and loses it slowly. Much of the solar heat that reaches the earth is absorbed by water, thus keeping the temperature of the earth relatively low. Aquatic areas have the largest and most stable ecosystems on earth. The living things are affected by such physical factors as the amount of dissolved oxygen, the temperature, and the amount of dissolved minerals.

1. *Marine biome.* The oceans of the world make up a huge, continuous body of water. They absorb and hold large quantities of solar heat and regulate the earth's temperature. They contain a relatively constant supply of salts, which originated in the land masses. The greatest amount of food production on earth takes place in the oceans, along the edges of the land masses.

The coastal waters are fairly clear, and photosynthesis takes place where light can penetrate. Since water absorbs much light energy, photosynthesis takes place near the surface. The deeper regions are too dark for photosynthesis.

Near the surface, the basic food is the *plankton*, consisting of microscopic forms of life such as diatoms and other algae, and protozoa. Plankton is the basis of the food chains of the ocean, in which many diverse types of animals are involved, including fish, lobsters, clams, and whales. At the lower depths of the ocean, the food consists of a "rain" of dead organic matter that comes down from above. At the bottom, bacteria and scavengers break down the complex organic molecules into simpler molecules, thereby releasing the mineral content. Through the circulation of the water, these dissolved nutrients are brought to the surface, where they are used over again for food.

2. *Freshwater biomes.* These include lakes, rivers, ponds, and swamps. Life in fresh water is affected by factors similar to those in marine environments, namely, available oxygen, temperature, transparency of the water, depth, and salt concentration. The low concentration of salts in fresh water, compared to the higher concentration in the living cells, causes a diffusion gradient, or difference, to exist. The semipermeable cell membrane controls the passage of water and salts by active transport. Excess water in protozoa is eliminated by specialized structures, the contractile vacuoles. Homeostatic mechanisms are present also in other freshwater inhabitants, such as trout, frogs, and many types of invertebrates.

Human Effects on Ecological Balance

Human beings have not always realized how they affect other living things. The following examples show how human activities have interfered with the ecological balance:

1. A farmer may dislike hawks because they occasionally make off with a chicken. However, hawks are useful to the farmer because they hunt and kill mice, rats, and rabbits, which destroy his crops. If hawks were eliminated, mice, rats, and rabbits would increase in such large numbers that they would ruin crops. Hawks are thus natural enemies of these small mammals;

the numbers of hawks are kept fairly constant by the available food supply.

2. On the Kaibab Plateau of Arizona, pumas, wolves, and coyotes prey on the deer. One year, a drive was made to kill off these enemies of the deer. The result was that the deer multiplied in such large numbers that they did not have enough food. Many of them died of starvation, and others chewed the bark off the trees, killing them. When the trees died, the birds and small mammals lost shelter and food. It was predicted that, if enough trees were killed, the soil would not be held together by their roots, and erosion would take place. The following year, the natural enemies of the deer were allowed to increase in numbers, in order to restore the balance of nature.

3. A few pairs of European rabbits were released in Australia in 1862. This was the first time these mammals had been brought to Australia. Within a few decades there were millions of rabbits. Without natural enemies to keep them in check, they destroyed grasslands and crops and became a national menace. In 1950, a deadly virus disease of the rabbits, *myxomatosis,* was introduced by scientists to control the rabbit scourge.

4. The island of Jamaica was troubled with rats in the sugar-cane fields. To control them, various animals were introduced, including the mongoose. Unfortunately, the mongooses, which kill rats, also killed birds, snakes, and lizards. Without these natural enemies, the cane beetles multiplied, and now do as much damage as the rats once did.

5. In 1916, the first Japanese beetles were found in New Jersey. They are not considered a menace in Japan, where they are held in check by natural enemies. But in this country they spread rapidly and soon were responsible for the loss of millions of dollars' worth of fruit crops, lawns, and flowers. Efforts are now being made by the Department of Agriculture to discover insect, bacterial, and fungus enemies of the beetles.

6. Humans have brought about other insect problems by upsetting the balance of nature. Elm trees are now rapidly being destroyed by the Dutch elm disease, caused by a fungus introduced from Europe in the 1920s and spread by the elm bark beetle. The gypsy moth was brought to Massachusetts in 1869 by a man engaged in research on silkworms; through his carelessness, the insects escaped, and became a serious pest of trees. The European corn borer entered this country before 1917, in broomcorn being imported from Italy for use in broom factories in New England. It now causes losses amounting to hundreds of millions of bushels of corn annually.

7. In the Tonopas area of the Colorado Rockies, the sheep raisers were convinced that their sheep were being menaced by coyotes. Although this predator actually preys on sheep only occasionally, and lives mostly on smaller animals such as rabbits and gophers, its numbers were drastically reduced by the use of poison. When this happened, the rabbits and gophers multiplied in such numbers that they ruined thousands of acres of pasture land. With this lesson in ecology, the cattle growers have now formed the Tonopas Grassland Protective Association, with a campaign to protect the coyotes!

Section Review

Select the correct choice to complete each of the following statements:

1. Plankton is composed of   (A) microscopic life   (B) fish   (C) whales   (D) sharks   (E) humans

2. Green plants are the basis of life for (A) herbivorous animals   (B) carnivorous animals   (C) omnivorous animals   (D) saprophytes   (E) all of the above

3. The study of living things in relation to each other and their environment is known as (A) conservation   (B) eugenics   (C) endocrinology   (D) ecology   (E) evolution

4. A cactus is an example of a   (A) hydrophyte   (B) xerophyte   (C) mesophyte   (D) saprophyte   (E) pteridophyte

5. All the members of a species inhabiting a given location comprise a(n)   (A) community   (B) population   (C) ecosystem   (D) biome   (E) biosphere

6. The next-to-last successful members of a maple-beech climax are   (A) cattails   (B) poplar and pine trees   (C) oak and hickory trees   (D) sumac trees   (E) willow trees

7. All of the following are examples of climax communities *except*   (A) cypress-white cedar   (B) beech-magnolia   (C) tall prairie grass   (D) water lilies   (E) scrub oak

8. Rabbits became a menace in Australia because (A) they transmitted rabbit fever to the kangaroos   (B) their holes in the ground were a hazard to live-

stock   (C) in the absence of natural enemies, they multiplied tremendously   (D) they caused myxomatosis among newborn infants   (E) they removed valuable nutrients from the soil

9. Japanese beetles are a worse pest in the United States than in Japan because here   (A) the soil is better   (B) there are fewer earthquakes   (C) there are few natural enemies   (D) there are more plants   (E) there is a longer summer

10. The gypsy moth is a serious pest of   (A) ferns   (B) trees   (C) cattle   (D) horses   (E) butterflies

11. All of the following are physical factors of the environment *except*   (A) light   (B) temperature   (C) water supply   (D) predators   (E) pH

12. The transfer of materials between organisms and the environment is known as a   (A) niche   (B) cycle   (C) synthesis   (D) free gas   (E) solution

13. All of the following are examples of consumers *except*   (A) carnivores   (B) herbivores   (C) saprophytes   (D) green plants   (E) parasites

14. An example of a parasite is   (A) a barnacle on a whale   (B) nitrogen-fixing bacteria in clover roots   (C) a remora on a shark   (D) a vulture   (E) a disease germ

15. The correct sequence in a food chain is
   (A) mice → green plants → snake → hawk
   (B) mice → snake → green plants → hawk
   (C) green plants → mice → snake → hawk
   (D) green plants → hawk → snake → mice
   (E) hawk → green plants → snake → mice

16. In a pyramid of energy, the greatest amount of energy is present in the level represented by   (A) producers   (B) first-order consumers   (C) second-order consumers   (D) third-order consumers   (E) decomposers

17. An example of a pioneer organisms is a   (A) pond lily   (B) cattail   (C) lichen   (D) fern   (E) wild carrot

18. The northernmost biome is   (A) grassland   (B) desert   (C) taiga   (D) tropical rain forest   (E) tundra

19. At the equator, a taiga type of growth   (A) is impossible   (B) occurs only in winter   (C) occurs in deep alleys   (D) occurs in the mountain heights   (E) occurs only in deserts

20. Compared to land masses, the temperature of the marine biome   (A) is below freezing   (B) is more stable   (C) is less stable   (D) absorbs less solar radiation   (E) has greater extremes

Answer Key

1-A	5-B	9-C	13-D	17-C
2-E	6-C	10-B	14-E	18-E
3-D	7-D	11-D	15-C	19-D
4-B	8-C	12-B	16-A	20-B

Answers Explained

1. (A) Algae make their own food; protozoa and minute invertebrates feed upon the algae. Together, they make up the floating mass of microscopic life called plankton.

2. (E) Green plants make their own food. Herbivorous animals eat the plants. Carnivorous animals eat the herbivores. Omnivorous animals eat both plant and animal foods. Saprophytes live on both plant and animal foods.

3. (D) Ecology is the science that deals with the interrelationships between living things and their environment. Living things do not live in isolation; they are involved with each other and their surroundings.

4. (B) A cactus lives in dry conditions. A xerophyte, it has a deep root system, much water-storage tissue, a thick epidermis, and modified leaves, or spines.

5. (B) All the members of a species living in a given area comprise a population. Example: sunfish in a lake.

6. (C) As the amount of shade cast by oak and hickory trees increases, their seedlings have difficulty surviving and maples and beeches succeed them, then becoming the dominant climax community.

7. (D) As water lilies and other plants in a pond die, their remains build up to form a semisolid base at the edge of the pond. Marsh plants then begin to grow there. As the amount of soil increases, a new community develops to take the place of the pond.

8. (C) Rabbits were introduced to Australia for the first time in 1862. The rabbits had no natural enemies to keep them in check, and soon there were millions of them.

9. (C) In 1916, the first Japanese beetles were found in the United States. They were not considered a menace in Japan, where they are held in check by natural enemies, but in this country they spread rapidly and soon were responsible for the loss of millions of dollars' worth of fruit crops, lawns, and flowers.

10. (B) The gypsy moth was originally brought to Massachusetts in 1869 for research on silkworms. It escaped from the laboratory and its caterpillars became a threat to trees by eating their leaves.

11. (D) Abiotic factors are nonliving physical factors that affect living things. Predators are animals that kill other animals.

12. (B) Important elements such as nitrogen, carbon, oxygen, hydrogen, and phosphorus are constantly cycled between organisms and their environment. These elements pass from inorganic sources to organic forms and back again to inorganic form.

13. (D) Green plants are producers, since they carry on photosynthesis and manufacture food.

14. (E) Disease germs are parasites, because they live within the body of an organism and benefit at its expense.

15. (C) Green plants serve as producers and make food. Mice are first-order consumers, living on plant food. A snake, a second-order consumer, eats mice. A hawk is a third-order consumer that eats the snake.

16. (A) Since there is a loss of energy at each feeding level, it can be seen that the producers, at the bottom of the pyramid of energy, contain the greatest amount of energy.

17. (C) Lichen first appears on bare rock and is the first living thing to populate the area. A pioneer organism, it helps build up soil by breaking down the surface of the rock and releasing its minerals. Mosses then take root, and their remains continue to build up the soil.

18. (E) Tundra is the treeless region in the far North. The underlying part of the soil is permanently frozen. During the summer, the upper part thaws out temporarily, and mosses, lichens, and small plants grow actively.

19. (D) The higher one goes on a mountain, the greater the drop in temperature. The vegetation changes accordingly. At the equator, the higher altitudes of the mountains support a taiga type of growth, with coniferous trees, such as pine, spruce, fir, and hemlock.

20. (B) Water absorbs large quantities of heat. It absorbs this heat slowly and loses it slowly. As a result, variations in temperature are not extreme.

12.2 PROTECTING OUR ENVIRONMENT

On June 1, 1992, an international Earth Summit was convened for the first time by the United Nations. It was held in Rio de Janeiro, Brazil, for the purpose of considering ways of protecting the earth's environment. The delegates of the 143 countries that attended, including President George Bush of the United States, reviewed the damage done to the earth's ecosystems, and discussed measures that should be taken to reduce the growing global environmental crisis. This section discusses some of the threats to our environment that exist today and the measures that are being taken to conserve our natural resources.

Threats to the Environment

Living things live in a delicately balanced equilibrium with each other. When this balance is upset, animals and plants are seriously affected, and many of them die. Human activities have been the major factor in upsetting the balance of nature.

Greenhouse Effect and Global Warming

The amount of carbon dioxide in the atmosphere has been found to be increasing. Much of it comes from the burning of fossil fuels (coal, oil, and natural gas) by factories, homes, and automobiles. Since trees use carbon dioxide in photosynthesis, the cutting down of the world's forests over the years is preventing some of this CO_2 from being absorbed. The actual effects of this increase are not fully known.

Carbon dioxide, like the glass of a greenhouse, allows visible sunlight to pass through to the earth. As the earth warms up, it gives off infrared rays. These are absorbed by CO_2 instead of being emitted into space. It is believed that this "greenhouse effect," as it is called, will eventually cause the earth's atmosphere to warm up. Other heat-trapping emissions that also contribute to the greenhouse effect are methane, chlorofluorocarbons, black particles of diesel and coal soot, and compounds that create smog in the ozone.

One of the possible effects of such an increase in atmospheric temperature was the discovery in the summer of 2000 that, for the first time in history, the thick ice cap at the North Pole had melted, leaving an ice-free patch of ocean about a mile wide. Further melting of polar ice could eventually lead to rising sea levels and the consequent flooding of heavily populated areas along the coast. Other effects of global warming might be the spread of desert areas, the reduction of food crop production, and a warmer climate.

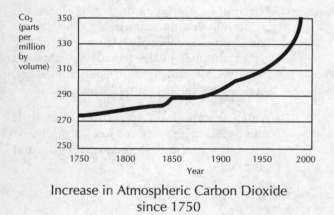

Increase in Atmospheric Carbon Dioxide
since 1750

In recognition of the need to reduce the dangers of global warming, more than 150 countries met in Japan in 1997 to discuss steps to protect the world's environment. The Kyoto Protocol that was adopted legally binds 38 industrialized countries to reduce their greenhouse emissions for the next 15 years below the levels that existed in 1990. The United States must reduce its emissions by 7 percent, the European Union of 15 nations by 8 percent, and Japan by 6 percent. Less dependence on fossil fuels will require development of alternate sources of energy, such as the applications of solar energy, windmill power, and the improved use of fuel cells.

As possible evidence of global warming, climatologists at the Goddard Institute of Space Studies in New York reported that the earth was warmer in 1990 than in any year since 1880, when such records first began to be kept. Some scientists say, however, that the question to be answered is whether this warming trend is due to the greenhouse effect or to natural variations in temperature. Recently, scientists have discovered that adding a small amount of iron to the chilly ocean waters near Antarctica stimulated a large bloom in the growth of the algae there. This led to an increase in photosynthesis, which uses up carbon dioxide in the air. Whether this is a way of dealing with the problem of rising levels of carbon dioxide remains to be seen.

The Endangered Ozone Layer

The ozone layer of the atmosphere, reaching up 10–30 miles, contains a type of oxygen whose molecule is denoted as O_3. It contains three atoms of oxygen, compared to the two atoms in the more familiar molecules of oxygen that we breathe, O_2. The ozone layer protects life on earth by absorbing most of the powerful ultraviolet radiation coming from the sun. Now it appears that chemicals called *chlorofluorocarbons (CFCs)*, commonly used in air conditioners, refrigerators, and aerosol sprays, are acting on the ozone layer, with the possibility of eroding it. In 1990, the world's industrialized countries met in London and agreed to eliminate the production of CFCs by the end of the century.

Early in the 1980s, scientists discovered a large hole in the ozone layer over Antarctica. In the following years, it was determined that the ozone layer had also begun to erode in the area above the North Pole, and then over the United States.

The Environmental Protection Agency has calculated that, if the ozone depletion continues, more than 12 million Americans will develop skin cancer in the next 50 years because of greater exposure to a particular band of ultraviolet radiation called *UV-B*.

One deleterious influence of the ozone thinning is a reduction in the production of plankton. Plankton forms the base of the food chain in the oceans, and is adversely affected by the deeper penetration of ultraviolet rays.

Recently, scientists developed a small, inexpensive device, called a DNA dosimeter, that measures the impact of ultraviolet-B radiation on DNA. It was found that UV-B radiation causes cyclobutane rings to form between adjacent nucleotides in DNA strands whenever genetic damage occurs. By counting the number of rings present, scientists can estimate the radiation dosage that inflicted that damage.

Acid Rain

This is rainfall that can be as acid as vinegar. It is formed when gases of nitrogen oxide and sulfur dioxide are given off into the atmosphere as by-products of fuel combustion by automobiles, homes, factories, and power plants. As the fumes are carried into the air by wind currents, they combine with water vapor molecules and are transformed into microscopic drops of nitric acid and sulfuric acid. When it rains or snows, the precipitation returns the acids to earth, sometimes thousands of miles from their origin. Lakes and streams have become so acidified that the populations of trout, salmon, and other fish are being destroyed. There is also concern about the possible effects of acid rain on soil minerals and nitrogen-fixing bacteria. In 1990, the United States and Canada agreed on joint steps to reduce the threat of acid rain.

Loss of Tropical Rain Forests

Environmentalists are deeply concerned by the rapid destruction of tropical rain forests now taking place in the Amazon area of Brazil, Southeast Asia, and Central America for purposes of lumbering and land clearing for agriculture. At the present time, these forests are disappearing at the rate of 50–100 acres a minute, and will be gone within a century.

Experts maintain that there are many values to these huge rain forests. They regulate the global climate of our entire planet. Through photosynthesis, these forests absorb great quantities of carbon dioxide, which would otherwise contribute to the greenhouse effect. With the burning of the forests to clear the land for farming and cattle raising, more CO_2 is actually being added to the atmosphere. The forests are also valuable because they act as "lungs" for the planet by giving off large quantities of oxygen into the air. In addition, they contribute perhaps 50 percent of the Amazon rainfall through the transpiration of moisture from their leaves into the air.

Moreover, countless thousands of species of plants, animals, and microbes that inhabit the rain forests may become extinct, many before they are discovered, named, and described. One quarter of the prescription drugs now used in the United States originated in rain-forest plants, with many others awaiting development.

Two other effects of continued rain forest destruction are also predicted. First, the Amazon temperature would rise by 4.5°F, affecting the worldwide climate. Second, since tropical forests have infertile soil, with most of the nutrients contained in the vegetation, rather than in the soil, any crops planted there would exhaust the soil after a year or two.

Air Pollution

The quality of the air we breathe is affected by smoke, dust, automobile exhausts, and gaseous wastes from factories. Pollutants such as carbon monoxide, nitrogen dioxide, sulfur dioxide and hydrocarbons irritate the lungs and pose a serious threat, especially to the health of older people and others who are prone to asthma, bronchitis, emphysema, and heart diseases. The oxidation of air pollutants produces the foglike haze over large cities known as *smog*. In serious incidents that occurred in London, England, and Donora, Pennsylvania, air pollution resulted in deaths and widespread illness. It is now common practice for the weather report on the radio to state whether the air quality is expected to be satisfactory for the day. During the Persian Gulf war, massive air pollution was caused when Iraq set fire to 500 oil wells in Kuwait.

Pesticides

The widespread use of chemicals to destroy insects, weeds, fungi, and other pests has resulted in heavy loss of fish and wildlife. The chemicals become especially harmful as they are passed on in food chains, where they become concentrated in the bodies of the various types of animals. These chemicals have also been found to be harmful to humans, causing cancer and contaminating water supplies. *DDT*, which seemed to be a miracle when first used toward the end of World War II, saved millions of people from typhus, malaria, or starvation. However, it was soon found to have harmful effects on fish, birds, and other wildlife, as well as humans. In birds, it interfered with calcium metabolism, causing eggs to become brittle and break, and resulting in the death of the young before hatching. The use of DDT was prohibited in 1972, partly as a result of Rachel Carson's powerful book *Silent Spring*.

A recent advance has been the use of biological controls. One method is to use a pest's natural predator; thus, a small wasp is imported from Australia to control the longhorn borer beetle that infects eucalyptus trees in California. Another approach is to use bacteria against insects. In 1961, *Bacillus thuringiensis* (Bt) was approved for use against the caterpillar of the gypsy moth; it acts by destroying the digestive cells, causing death by starvation. More recently, new strains of the bacteria have been successfully used by potato farmers against the Colorado potato beetle. The new technique of genetic engineering (see Section 10.4) is being tested to protect corn by enabling it to grow its own pesticide; this is accomplished by inoculating the seed with algae that have been equipped with a bacillus gene whose product is toxic to the corn-borer beetle, a major pest of corn.

Toxic Wastes

The chemical wastes from industrial factories flow into our rivers, along with sewage, pesticides, and excess fertilizers from farms. In addition, these hazardous wastes can filter through the soil into the water table, contaminating the groundwater that supplies the wells in the area.

Some pollutants found in hazardous wastes are known to cause cancer in animals. *PCB* (polychlorinated biphenyl) is a product of the electrical industry; *trichloroethylene* is a degreasing agent used in dry cleaning and metal working. *Dioxins* exist as contaminants in herbicides such as 2,4,5-T (phenoxacid) and Agent Orange, used to defoliate tree cover and destroy enemy food crops in Vietnam (1965–71). Dioxins are now suspected as a possible cause of cancer and other ailments in Vietnam veterans.

In one well-publicized case, the residents of the Love Canal area of Niagara Falls, New York, were driven from their homes, from 1978 to 1980, because of the accumulation of toxic chemical wastes that had been dumped there over a period of years by a plastics manufacturing company. The threat to their well-being came to light after there had been an unusual number of illnesses, miscarriages, and birth defects among the local people. A scientific study showed that the residents were risking cancer and chromosomal damage by remaining in the area.

An unusual approach to controlling toxic wastes is to utilize the activity of bacteria in breaking these wastes down to harmless products by a process called *biomediation*. Thus, in the case of oil spills, bacteria have been applied and have reduced the oil to an emulsion of fatty acids. When the oil was gone, the bacteria died.

Dozens of new companies have been organized to develop other uses for bacteria, with some promising results. Bacteria are being used to absorb some of the sulfur in coal, making it cleaner before it is burned. Other bacteria convert coal into methane, an energy-rich gas. Both methods could help reduce the risk of acid rain. Still other bacteria are being used to break down the structure of PCB. Restaurants are buying bacterial products that consume grease. Some bacteria convert spilled chemicals in warehouses and in underground water sources to safe compounds.

Excessive Eutrophication

This happens when a lake receives large amounts of nitrates and phosphates from industrial wastes, sewage, and farm fertilizers. The addition of these nutrient minerals encourages the explosive growth of the algae known as bloom. As they spread rapidly, the algae use up the oxygen supply of the water, killing off fish and other forms of animal life. Dead animal and plant life accumulates, generating unpleasant odors. The use of laundry detergents containing phosphates is being prohibited in many states in an effort to protect their freshwater lakes, including the Great Lakes.

Pollution in Nature

The "red tide" is a name given to a huge outbreak of microorganisms that turns the ocean a bright red color. Fish, clams, and mussels may be killed in massive numbers, and people who eat seafood affected by the red tide may die of paralytic poisoning. The cause of the red tide is a dinoflagellate, a microscopic red alga that moves by means of two flagella. It is classified as being either the *Gonyaulax* or the *Gymnodium* type.

When blooms, or massive growths, of either type of dinoflagellate erupt, a large amount of a powerful nerve toxin is produced. Mussels and clams concentrate this poison within them and become dangerous, if eaten. The cause of a bloom is unknown. It is suspected that it may be triggered by an upwelling, or upward current, of ocean water containing high concentrations of nutrients such as nitrates and phosphates.

Lead Poisoning

Lead poisoning is believed to have been a major reason for the decline of the Roman Empire. Wine was stored in lead pots and became heavily contaminated with lead. Lead plumbing and cookware were also significant sources of lead at the time.

Lead poisoning is known to cause paralysis, brain damage, loss of kidney function, and visual disturbances. In children, in whom the nervous system is still developing, even slight exposure to lead can cause learning deficits, behavioral disorders, growth retardation, and hearing loss. Even small amounts of lead in the body can interfere with neurotransmitter action and the conversion of vitamin D into its active form. Children are especially vulnerable to lead poisoning since they tend to put their fingers into their mouths; fingers can be contaminated by peeling paint chips and by earth that contains lead dust.

There are several sources of lead in the environment. Paint is the primary cause of lead exposure for children. Sandblasting of bridges and buildings before repainting produces a coating of contaminated dust in the adjacent area. Ceramic dinnerware may be another source if the glaze deteriorates. Drinking water may contain traces of lead arising from faucet fixtures and solder joints; for this reason it is advisable to let water run for a minute until cold, before using.

There has been a significant decrease in the amount of atmospheric lead emitted by automobiles since the mandated use of unleaded gasoline in 1978. A new law in New York State requires that pediatricians screen all children 1–2 years old for lead. In addition, doctors are instructed to advise pregnant women on the steps they should take to avoid lead poisoning.

A number of cities, including New York and Atlanta, now add the chemical calcium orthophosphate, a relative of baking soda, to the water supply to prevent lead in pipes from seeping into the water. The chemical bonds with metal pipes and fixtures, adding a thin coating to the inside of the pipes and sealing in the lead.

In recognition of the menace of lead poisoning, the federal government recently reduced the acceptable amount of lead in drinking water from 50 parts per billion to 15 parts per billion.

Mercury Pollution

Mercury is accumulating in fish living in thousands of lakes across the United States and Canada, poisoning wildlife and threatening human health. The principal source of contamination appears to be rain contianing traces of mercury given off by coal-burning power plants, municipal incinerators, and smelters.

Mercury is a liquid that becomes a gas at mild temperatures, making it difficult to control. Since mercury is used to make batteries, electric switches, and many other products, incineration of such discarded items releases mercury into the atmosphere. Here, it may linger in its gaseous form for up to 2 years. When it falls to earth in rain, it contaminates lakes and rivers, and is passed up the food chain from plankton to minnows to large fish to birds of prey and large mammals, including human beings.

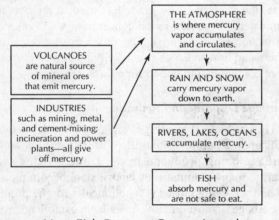

How Fish Become Contaminated with Mercury

If taken into the body, even in tiny amounts, mercury can eventually lead to brain damage or loss of muscle control. About 50 years ago, more than 100 people developed tremors, fell into a coma, and died in Minamata, Japan. The fish caught in Minamata Bay were found to have an average mercury level of eleven thousandths of a gram for every kilogram. This is comparable to one speck of pepper on a 36-ounce steak.

The Food and Drug Administration considers a single meal of fish containing one thousandth of a gram of mercury per kilogram to be dangerous, and prohibits commercial sales of fish with higher amounts.

Historically, in the 19th century, the poisonous effects of mercury were known in the hat industry. Fur that was used in the manufacture of felt hats was first dipped into mercuric nitrate solution. Many of the workers absorbed so much mercury through the skin that they became known for slurred spech and uncontrollable trembling.

Nuclear Waste

The waste material from nuclear reactors is dangerously radioactive. It may contain various isotopes that decay at greatly different rates of speed: iodine-131, which can cause human thyroid cancer, has a half-life of just 8 days: cesium-137 has a half-life of 30 years and can cause many types of cancer. Half-life is the time needed for one-half of a radioactive element to disintegrate.

The problem facing countries that produce nuclear wastes is where to dispose of them safely, so that they do not contaminate groundwater or give off radiations that are harmful to living things, including humans. The United States disposes of such wastes in several ways, including storing them deep in unused mines, and sealing them in concrete containers that are then sunk in the ocean.

There is concern, however, that if leakage occurs in these containers sunk at sea, bottom seasonal currents could carry radiations to new areas where they might be dangerous. In the late 1950s the International Atomic Energy Commission banned ocean dumping of high-level radioactive wastes. It has been learned, however, that the former Soviet Union violated the ban by dumping 2.5 million curies of radioactive wastes, including 18 nuclear reactors from submarines and an icebreaker, in the frigid waters of the Arctic Circle. This amounts to twice the amount previously dumped by all other nations. It also raises questions of possible violation by other countries, as well. Strict international cooperation will be needed in the future to avoid such problems of a global nature.

An unusual method for cleaning up nuclear waste sites contaminated with uranium is under study. The U.S. Geological Survey has found that a rod-shaped bacterium, known as GS-15, has the ability to transform the soluble form of uranium in waste water into a solid form that can be easily filtered out. This offers a possible mechanism for cleaning up, at a very rapid rate, environments contaminated with soluble uranium, leaving purified water behind.

Oil Spills

Oil is transported across the oceans in huge ships known as tankers, which may carry over a million barrels at a time. If one of these tankers is damaged by storms or wrecked by rocks along the coast, the oil is released into the sea, with dire consequences to the environment.

One of the most devastating wrecks was that of the huge *Exxon Valdez*, carrying 125 million barrels of oil, off the coast of Alaska in 1989. Spilled oil covered the water and the shores of Prince William Sound, killing enormous numbers of fish and marine mammals. Oil soaked into the feathers of sea birds, preventing them from flying and poisoning them in a few hours. The coastline was covered with oil, contaminating the habitat of the wildlife.

In a massive cleanup program, bacteria that digest oil were successfully introduced into the oil-covered waters. Through a process of *bioremediation*, in which fertilizers containing nitrogen and phosphorus were added, the bacteria multiplied and fed on the hydrocarbons in the oil, converting them to harmless carbon dioxide and water.

This process was supplemented by the natural wave action and extended washing techniques. After more than 3 years of intensive work, the oil pollution was finally declared to be cleaned up, at a cost of $2.5 billion.

Other oil spills around the world from time to time have had similar adverse environmental effects. During the Persian Gulf War in 1991, the deliberate release by Iraq of oil into the gulf waters was extremely damaging to aquatic life there.

"Sick Buildings"

A relatively recent problem has appeared in newly constructed, air-conditioned office buildings having airtight windows that do not open. Many of the office workers in these buildings have complained of a variety of illnesses, including headaches, nausea, asthma, chest tightness, scratchy throats, irritated eyes, and fatigue.

The cause of these problems appears to be the prevalence of yeasts, molds, and bacteria in the ducts of the air-conditioning system. When the air-conditioning is turned on, these microscopic organisms are circulated throughout the building, causing the health-threatening symptoms. Additional causes may be formaldehyde, asbestos, and other organic compounds used in insulation and building construction. To counteract these "sick building" health problems, the Environmental Protection Agency has recommended increasing the ventilation with fresh outside air.

The complaints caused by "sick buildings" are reminiscent of the mysterious Legionnaire's disease, which in 1976 caused 29 members of the American Legion attending a convention in a Philadelphia hotel to die suddenly from a form of pneumonia. Public health scientists succeeded in tracing the cause to a strain of bacteria, which they named *Legionella pneumophila*, that apparently was being circulated throughout the hotel's air-conditioning and water systems.

Radon

A "natural" environmental problem involves hazards arising from radon in as many as 8 million homes throughout the United States. Radon is a colorless, odorless gas that occurs naturally from the decay of uranium deposits found in many kinds of rock. It can seep into a home through the basement by means of cracks in the foundation. If ventilation in the house is inadequate, the concentration of radon can build up. Over a long period of exposure, it can cause disease, especially cancer of the lungs.

A rock formation known as the Reading Prong, located in parts of Pennsylvania, New Jersey, and New York, has been found to have the highest radon levels. Residents in an area suspected of radon contamination are advised to use a radon detector, which consists of a charcoal canister that is left in the home for 3–7 days and then sent to a laboratory for analysis. Methods of getting rid of the gas in the home include the installation of fans to improve air circulation and of exhaust pipes to draw the radon outdoors.

Soil that has been contaminated by radium-processing plants is another source of radon pollution. In one such case, the state of New Jersey has tried to remove contaminated soil so that radon concentration will not build up in homes in the area. The problem that remains is where and how to deposit such soil and how to prevent it from contaminating underground water sources.

Garbage Disposal and Recycling

The garbage that we get rid of every day usually ends up in a dump, or landfill. With 195 million tons of garbage being produced in the United States every year, in many communities landfills are becoming used up and are being closed. To help deal with the problem of garbage disposal, many states have established recycling laws requiring that newspapers, aluminum cans, bottles, and objects made of plastic be separated from trash and recycled.

Besides reducing the load on landfills, recycling helps protect our environment and conserve our resources. Here are five examples:

1. According to the World Marine Foundation, the production of one ton of paper from discarded waste paper uses "half the energy, half the water, results in 74% less air pollution and 35% less water pollution,

saves 17 pulp trees, reduces solid waste going to land-fills, and creates five times more jobs—compared to the production of one ton of paper from wood pulp."

2. Sixty-two percent of all aluminum cans are being recycled. Ninety-five percent less energy is required to produce new aluminum from old cans than to produce it from raw materials. Also, it has been estimated that, when you toss away an aluminum beverage can, you waste as much potential energy as if you had filled the same can half full of gasoline and poured the fuel on the ground.

3. By using recycled glass containers, manufacturers can reduce the temperatures of their furnaces, saving energy and prolonging the life of their glass-making equipment. In addition, the energy saved from each recy-cled glass bottle could light a 100-watt bulb for 4 hours.

4. A new type of plastic grocery bag now being used by supermarkets is degradable; within a few days of exposure to sunlight, it will start breaking down into nontoxic, environmentally safe dust. The recycling of plastic is a growing field. Plastic soda bottles, salad-dressing bottles, and some other food containers can be reprocessed into new soda bottles and into polyester fibers used for carpets and furniture. Milk and water jugs, as well as laundry detergent bottles, can be recy-cled into new containers and housewares.

Recycling Symbol (enlarged) on the
Bottom of a Plastic Soda Bottle

5. Iron and steel have been recycled for many years, in the form of old toys, automobiles, refrigerators, and food cans. This has resulted in a saving of over 100 bil-lion pounds in 1 year, making iron and steel "America's most recycled materials."

Problems Posed by Dams

The construction of hydroelectric dams has resulted in many benefits. They can generate enormous amounts of "clean" electric power with none of the pollutants causing acid rain that are produced by coal-fired power plants, and none of the radioactive wastes produced by nuclear reactors. For example, the Grand Coulee Dam in the state of Washington generates more than 6,000 megawatts of electric power, as much as six nuclear power plants.

In addition, dams can store large amounts of water in huge, man-made lakes. This allows them to absorb floodwaters, so that downstream communities do not face destructive floods after heavy rains. Dams also allow irrigation in arid areas. In addition, these big bodies of water provide recreational lakes in areas where there are none.

Despite these advantages, however, there are some drawbacks to dams. Fish, such as salmon and shad, that live in the ocean find it difficult to swim up rivers to lay their eggs in freshwater spawning grounds, because of the presence of dams. For example, there are eight dams along the Columbia River, where thousands of sockeye salmon used to migrate up a 900-mile stretch to spawn. In a recent year, only four salmon were able to complete the trip. Now the federal government has declared the sockeye salmon to be an endangered species.

After the large Aswan Dam was built on the Nile River in Egypt, the farmers in the Nile Delta found they could no longer depend on annual flooding for the river's nutrient-rich silt, which used to fertilize their crops. They were forced to apply their own fertilizers artificially, at considerable cost. Another adverse effect of Aswan was the demise of sardine fishing in the east-ern Mediterranean because the river no longer flooded the sea with its nutrient-rich content.

A problem posed by all dams is that they tend to fill up with silt, especially in highly eroded terrain. For example, the Tarbela Dam in Pakistan, which was built at a cost of $1.4 billion, is expected to lose its usable storage capacity in only about 20 years.

Moreover, because dams create lakes that inundate huge areas of land, they displace not only people but also many species of native plant and animal life. For this reason, Canada's proposed James Bay project, which may be the world's largest single producer of hydroelectric power, is being opposed by many envi-ronmental organizations.

The Population Explosion

In 1986, the earth's population reached 5 billion. The trend of growth over the years since the dawn of civilization shows how rapidly humans have multiplied in recent times. It is estimated that in A.D. 1000, after a million years of human existence, there were possibly 275 million people on the earth. By 1830, the figure reached 1 billion. It took only 100 years for the second billion to be added. The third billion came in 30 years. Fifteen years later, the figure of 4 billion was reached, and in a little more than 10 years another billion was added.

At this rate, the U.S. Census Bureau has predicted that by the year 2010, the world's population will have grown to 6.2 billion. The vast majority of the growth will occur in the poorest nations. Because the rate of population growth is increasing so rapidly, it is referred to as a population explosion. The title of a book by Dr. Paul Ehrlich, the *Population Bomb*, dramatizes the danger inherent in this explosive growth.

Concern over this trend is based on the limitations of the food supply. The famous economist Robert Malthus pointed out at the end of the 18th century that the food supply increases at an arithmetic ratio, or gradually, while the population increases geometrically, that is, it doubles itself. He stated that populations are kept in check by disease, famine, and war. There are many regions on earth today, such as Somalia and Sudan, where people live close to a starvation level.

Ways of controlling the "population explosion" are being studied seriously to cope with the future problems. Birth control methods, including "the pill" and the intrauterine device, are being publicized as desirable approaches. In Great Britain, a group of distinguished scientists issued a "Blueprint for Survival" early in 1972, recommending that the nation's population be reduced from 55 million to 30 million in the next 150 years. They also urged that other measures be adopted as well, such as the recycling of resources and the use of agricultural and industrial techniques that do not threaten the stability of the environment.

In 1980, there seemed to be a slight drop in the world population growth from 1.8 to 1.7 percent. An example of this encouraging trend was seen in China, which has a population of about a billion people; it was attempting an educational campaign to limit families to one child per couple. However, the Global 2000 Report to the President has indicated that the continuing high birth rate in the underdeveloped countries of Asia, Africa, and Latin America is sure to extend the population explosion into the 21st century.

Extinction of Animals

Because of the spread of civilization, the balance of nature has been so upset by humans that over 100 different kinds of animals have become extinct. Only a little more than a century ago, passenger pigeons were so numerous that Audubon, the famous naturalist and painter, described their flight as darkening the sky. Uncontrolled hunting gradually led to their destruction, and in 1914 the last passenger pigeon died. Other animals that have become extinct during the past century are the heath hen, Carolina parakeet, sea mink, and Merriam elk. The ivory-billed woodpecker, North America's largest woodpecker, has not been seen since 1952 and may also be extinct.

A bird that became extinct as far back as 1693 was the *dodo*. Now known only in museums, this bird was about 3 feet tall and could not fly. It made little effort to escape when sailors hunted it on its native island of Mauritius in the Indian Ocean. The introduction of pigs, which ate its eggs, led to its extinction on the island.

Animals threatened with extinction   In this country, tremendous herds of buffaloes, or American bisons, once dominated the Great Plains. As the result of uncontrolled hunting and the advancing front of civilization, countless numbers were killed. At the beginning of this century, only a few hundred remained alive. Conservation laws were passed to protect them, and at the present time the survivors are slowly increasing in number.

Also facing extinction are the California condor, America's largest bird, having a wing span of 10 feet, which now totals about 60; and the trumpeter swan, our largest water bird, which was recently down to about 75 birds, and is now slowly beginning to increase in numbers. The best-known mammal facing extinction is the grizzly bear; it is estimated that fewer than 800 grizzlies remain. Other endangered mammals are the sea otter, sea cow, elk, Florida panther, and Key deer. Among rapidly disappearing fish are the lake sturgeon, Great Lakes whitefish, and lake trout.

Protection of endangered species   The extent to which we have lately become aware of the need to protect species threatened with extinction was vividly demonstrated in the "snail darter vs. Tellico Dam" situation. The snail darter, a small fish about 3 inches long, was first discovered in 1973 in the Little Tennessee River; this is the only place where the fish is found. It was threatened by extinction if the nearby, nearly completed Tellico Dam was opened, and a lake formed over the river. Under the Endangered Species Act passed by Congress that year, the snail darter was declared an endangered species.

Various environmental groups argued for the protection of a species that would otherwise be wiped out forever. On the other hand, there was pressure against holding up an expensive dam because of a mere little fish. The issue finally came before the Supreme Court, which ruled in 1978 against the completion of the dam as a threat to the snail darter's habitat. There was much more public controversy before the problem was settled in 1979, when the snail darter was removed to two other, similar rivers nearby, and the $116 million dam was finally completed.

Three species threatened with extinction have received special attention:

1. *Peregrine falcon.*   In 1970, the peregrine falcon had practically died off in the United States from the

use of pesticides and insecticides. At Cornell University, captive peregrines were raised under protection and then released into the wild. Some of these birds were even seen to set their nests high in skyscrapers of New York City and Philadelphia. By 1980, four peregrine chicks were born in the wild, bringing new hope for the survival of the species. Thus, for the first time, an almost-extinct bird population was reproduced from captive stock that had been released back to nature.

2. *Whooping crane.* Also on the verge of extinction is the whooping crane, a large white bird that numbered only 14 in 1938. There are now about 200 individuals. In an attempt to increase this tiny population, Florida biologists are planning to place whooping crane eggs in nests of the Florida sandhill crane. It is hoped that the whooping crane chicks will follow the example of their foster parents and stay in a protected year-round environment, rather than risking the rigors of long migrations that might reduce their numbers.

3. *Bengal tiger.* The Bengal tiger in India, which may have numbered 40,000 in 1900, was down to 1,827 in 1973, and in danger of extinction. The government outlawed hunting of the animal and set up protected preserves to encourage its survival. By 1979, the tiger population had risen to 3,015.

Earth Day

Earth Day was first observed on April 22, 1970, in this country in an organized effort to protect our planet from the problems brought on by threats to the environment. By 1990, 140 countries around the world participated in Earth Day in a growing movement to solve the threats of acid rain, toxic wastes, the greenhouse effect, the destruction of the ozone layer, the loss of tropical rain forests, and the extinction of living species. Organizations such as the federal Environmental Protection Agency, the Sierra Club, and the Environmental Defense Fund plan Earth Day activities in which schools, clubs, corporations, and individuals can participate. Think about what *you* can do for the environment next April 22!

Section Review

Select the correct choice to complete each of the following statements or to answer the question:

1. All of the following statements about DDT are true *except* (A) DDT is harmless to humans (B) DDT is harmful to fish (C) DDT causes bird eggs to become brittle and break (D) After World War II, DDT saved many lives from typhus and malaria (E) DDT use has been prohibited in recent years

2. Radon is a colorless gas that can (A) stimulate the growth of green plants (B) seep into a home from the basement (C) make FM radio reception clearer (D) clean up a landfill quickly (E) be recycled

3. The explosive growth of algae in a lake that receives large amounts of nitrates and phosphates is known as (A) chlorination (B) eutrophication (C) dialysis (D) agglutination (E) alternation of generations

4. Smog is an example of (A) a type of snail (B) a cross between a salamander and a frog (C) a vitamin deficiency (D) soil deficiency (E) air pollution

5. The greenhouse effect is caused by (A) an increase in the number of greenhouses in Greenland (B) a decrease in the number of greenhouses in Greenland (C) an increase in the amount of carbon dioxide in the atmosphere (D) a decrease in the amount of carbon dioxide in the atmosphere (E) an increase in the amount of oxygen in the atmosphere

6. Scientists believe that global warming is caused by (A) atom bomb testing (B) the spread of warfare in equatorial regions (C) acid rain (D) the greenhouse effect (E) the evaporation of mercury

7. Which of the following statements is (are) true of acid rain? (A) It can be as acid as vinegar. (B) It originates in gases of nitrogen oxide and sulfur dioxide given off into the atmosphere as by-products of fuel combustion. (C) It forms from microscopic drops of nitric acid and sulfuric acid in the atmosphere. (D) Lakes have become so acidified that fish are being destroyed. (E) All of these.

8. All of the following statements about tropical rain forests are true *except* (A) they regulate the global climate of our entire planet (B) through photosynthesis, they give off great quantities of carbon dixoide (C) they contribute perhaps 50 percent of the rainfall in the Amazon area (D) they contain countless thousands of species of plants and animals (E) their soil is comparatively infertile

9. The number of atoms of oxygen in a molecule of ozone is (A) 1 (B) 2 (C) 3 (D) 4 (E) none

10. In the year A.D. 1000, there were about 275 million people on the earth. By 1986, the figure had climbed to (A) 1 billion (B) 2 billion (C) 3 billion (D) 4 billion (E) 5 billion

11. Of the following animals, the one that is not extinct, but is in danger of becoming so, is the (A) passenger pigeon (B) dodo (C) California condor (D) heath hen (E) sea mink

12. One way of cleaning up the environment after an oil spill is through the use of (A) bacteria (B) algae (C) fungi (D) viruses (E) a bacteriophage

13. Two harmful sources of lead are (A) paint and drinking water (B) paint and milk (C) paint and decaffeinated coffee (D) paint and regular coffee (E) none of these

14. A food that may be contaminated by mercury is (A) cheese (B) melon (C) chicken (D) fish (E) turkey

15. The environmental problem posed by nuclear wastes is that they (A) contain large amounts of PCB (B) are contained in acid rain (C) are causing the ozone layer to erode (D) may be radioactive for many years (E) take up too much space in neighborhood landfills

Answer Key

1-A	4-E	7-E	10-E	13-A
2-B	5-C	8-B	11-C	14-D
3-B	6-D	9-C	12-A	15-D

Answers Explained

1. (A) Although DDT protected millions of people from typhus and malaria after World War II, it was subsequently found to be harmful to fish, birds, other wildlife, and humans as well. Its use is now prohibited.

2. (B) Radon occurs naturally from the decay of uranium deposits found in many kinds of rock. It can seep into a home through cracks in the foundation and cause disease.

3. (B) The addition to a lake of nitrates and phosphates found in industrial wastes, sewage, and farm fertilizers encourages such rapid growth of algae that they use up the oxygen supply of the water, killing off fish and other aquatic animals. One measure taken against this eutrophication has been to prohibit the use of laundry detergents containing phosphates.

4. (E) The oxidation of air pollutants such as carbon monoxide, nitrogen dioxide, and sulfur dioxide can result in the foglike haze over cities known as smog. This air pollution can be dangerous to people with asthma, bronchitis, emphysema, or heart disease.

5. (C) The amount of carbon dioxide in the atmosphere has been found to be increasing. Like the glass of a greenhouse, CO_2 allows sunlight to pass through to the earth. As the earth is warmed by the sunlight, it gives off infrared rays that are absorbed by the CO_2 in the atmosphere, instead of being given off into space. This greenhouse effect is believed to result in a warming trend on the earth called global warming, which can adversely affect living conditions.

6. (D) As explained in the preceding answer, the greenhouse effect is believed to cause global warming.

7. (E) When it rains or snows, the microscopic drops of acid are brought to earth, sometimes thousands of miles from their origin. All of the statements are true.

8. (B) During photosynthesis, the plants take in, not give off, great quantities of carbon dioxide. Tropical forests have infertile soil, with most of the nutrients contained in the vegetation, rather than in the soil, so any crops planted, if the forests were cut down, would exhaust the soil in a year or two.

9. (C) Ozone contains a type of oxygen whose molecule is designated as O_3.

10. (E) The term "population explosion" has been applied to the rapid increase in the human population.

11. (C) The California condor, America's largest bird, having a wing span of 10 feet, has been reduced in number to less than 60 at the present time.

12. (A) Through the process of bioremediation, bacteria, supplemented by fertilizers, digest the hydrocarbons in spilled oil to harmless carbon dioxide and water.

13. (A) Peeling chips of paint are dangerous sources of lead for small children who nibble on them. Drinking water may contain traces of lead arising from faucet fixtures and soldered pipe joints; therefore, water should be allowed to run for about a minute before drinking.

14. (D) When it rains, mercury vapor in the atmosphere falls to earth and collects in lakes and rivers. Fish ingest the mercury and store it in their bodies, thereby imposing a dangerous risk on humans who catch and eat them.

15. (D) Exposure to the radiations of nuclear wastes can cause serious health problems, including cancer.

12.3 CONSERVATION OF OUR RESOURCES

America is called a land of plenty. Our country enjoys a high standard of living, there is an excess of food, and we enjoy the benefits of many luxuries. Yet, as discussed in Section 12.2, we are in danger of losing our valuable natural resources because of waste and poor planning. We have found that our use of pesticides and insecticides has backfired in the form of contaminated crops and water supplies. The dangerous effects of these chemicals were well depicted in the book *Silent Spring,* by Rachel Carson as early as 1962.

In addition, the threat of pollution is affecting our use of the air and the natural waters. Combustion in industrial plants and automobile engines is adding poisonous wastes to the air that we breathe. Smog is a serious condition affecting the health of large city dwellers. The water supply is polluted by sewage, industrial wastes, silt, fertilizers, and salt water. The Great Lakes, which once had pure, fresh water, are in danger of becoming mere sewage basins in which few living things can survive.

We are warned that we must practice *conservation,* the planned use and preservation of our natural resources, especially (1) forests, (2) soil, and (3) wildlife.

Forests

The destruction of our forests has been taking place ever since the first settlers arrived. There have been three major causes of this loss:

1. *Excessive cutting.* The early settlers cut down trees in order to clear the land for crops and pastures. As our country grew, the original virgin forests were cut down to the point where many of them disappeared, especially in the East. Since the forests were so plentiful at first, and grew down to the water's edge, little thought was given to their value. Later on, when lumbering became well established, poor timber practices consisted of cutting down one forest and then moving on to another. There was no plan for replanting trees for the future.

2. *Forest fires.* Every year, more than 200,000 fires burn over 23 million acres of our forests. The damage to human lives, wildlife, soil, watersheds, recreation areas, and future growth of trees is incalculable. The cost of the bill amounts to about $60 million annually. The worst forest fire in American history was the Peshtigo fire in Wisconsin in 1871; 1,280,000 acres were burned, 1,500 people were killed, and towns and settlements were destroyed. Forest fires are chiefly caused by human carelessness, involving smoking, campfires, and small fires to clear land that get out of control. Lightning and railroad train sparks also are causes.

3. *Insects and diseases.* Certain insects are serious enemies of trees. Some of them bore into the bark, destroy the living cambium cells, and stop the flow of sap; examples are the spruce beetle, pine bark beetle, and elm bark beetle. Others, such as the tussock moth,

spruce budworm, and gypsy moth, eat the leaves, stopping the manufacture of food and causing the trees to starve. In a 6-year period beginning in 1942, 4 billion feet of magnificent spruce forests were destroyed by the spruce beetle in the Rocky Mountains of Colorado. These and other insects destroy far more trees than do forest fires.

Certain deadly fungus diseases also destroy trees; among them is the Dutch elm disease fungus, which is carried by elm bark beetles as they bore into the bark. The white pine blister rust destroys the valuable white pine tree; this fungus also lives on two alternative hosts, currants and gooseberries. The chestnut blight has wiped out the American chestnut tree from Canada to the Gulf states. Heart rot fungus attacks the heartwood of oak, ash, and most types of evergreen trees.

Value of Forests

Forests supply us with many useful products. The paper of this book, as well as the paper used in newspapers, magazines, cardboard boxes, and writing paper, originates in wood pulp. The structure of various parts of buildings requires immense amounts of lumber. Furniture, telephone poles, turpentine, rayon, cellophane, railroad ties, acetone, and methyl alcohol are but a few of the products derived from wood.

Forests serve as a home for wildlife, including such fur bearers as the raccoon, fox, opossum, and mink; game animals such as rabbits, squirrels, deer, and ruffed grouse; birds such as thrushes, warblers, woodpeckers, hawks. When the woodland is destroyed, these creatures disappear.

The woods are also used for recreation by many Americans, who make 21 million visits a year to the national parks alone. They come to picnic, fish, hunt, ski, swim, hike, ride, look, and sit. The first national park to be established was Yellowstone Park in 1872.

An essential value of forests is their role in preventing erosion and floods. The leaves and branches intercept rain, breaking the effect of its fall on the soil below. The organic material of the soil (humus), consisting of decayed plant and animal matter, acts as a spongy layer and soaks up the water, thereby preventing it from running downhill and from washing away the topsoil. Thus, the forests also act as a watershed, retaining water and yielding it slowly for run-off streams.

Forest-Conservation Measures

The national forests were established in 1891. There are now national forests in 30 states. Together with state forests, they serve as demonstration areas of good forestry practices; they also protect watersheds and wildlife, yield valuable forest products, and provide places for recreation. The 230 million acres of national forests are administered by the U.S. Forest Service, a branch of the Department of Agriculture.

One of the recent projects of the Forest Service is to support the nationwide Champion Tree Project, which seeks to identify the largest trees in the country. A question it would like to answer is: Have these giant trees survived because they have superior genes for insect or drought resistance? Since 1996, genetic clones of some of these giant trees have been planted in different climates of the country in an effort to answer this question.

Selective cutting Forests are now looked upon as a growing crop. Selective cutting is practiced, by which mature trees, as well as dying or diseased trees, are cut down. This process gives young, growing trees an opportunity to mature and produce seeds. After older trees are cut down, they are replaced by young trees, which are carefully spaced and cared for. As the trees are felled, they are prevented from damaging other trees. Like any other crop, trees must be harvested or they go to waste. Although conflicts between conservationists and lumberers sometimes arise, many lumberers are actively engaged in forest conservation to protect the trees we now have, and to ensure a continuous growth of more trees for the future. As a result, we are today finally growing more wood than we are using.

Reforestation This is another practice that is restoring woodlands. To date, over 8 billion trees and shrubs have been grown in nurseries and have been planted in needed areas. In addition, seeds have been planted directly in the woodlands.

Proper forest management Forest rangers are trained to manage and protect forests. They patrol the forests from fire towers that are distributed throughout the woods, and from airplanes and helicopters. At the first sign of a fire they hurry into action, using various methods to put out the blaze. One of the first steps is to make a fire line, or barrier, down to the soil, all around the fire, to keep it from spreading. Then all the burning material within the fire area is extinguished. Among the most important weapons of fire-fighting are educational campaigns to make the public aware of the need for exercising great care with matches, cigarettes, and campfires.

Insect and fungus control A continuous program of research is carried on to control the insect and fungus menace to forests. Airplane spraying of dieldrin is a good method of insect control, most effective against insects that eat leaves. Genetics has supplied a new tool against forest diseases; experiments are being conducted to breed trees that are resistant to fungus infections. Thus, the American chestnut has been hybridized with the Chinese chestnut to produce a promising resistant timber tree.

Pheromones, the sex chemical compounds given off into the air by insects, are being used as a substitute for insecticides. Some pheromones given off by female moths can attract males many miles away. Scientists are now extracting these substances from insects and using them to bait traps, where the insects can be collected and killed.

There are several advantages in using pheromones over insecticides. These compounds are harmless to other living things and are specific for a particular type of insect. Thus, they can control a certain pest without disrupting populations of desirable insects and without harming any other form of animal life.

Soil Erosion

Such early books as *Our Plundered Planet* (1948), by Fairfield Osborn, and *Deserts on the March* (1950), by Paul B. Sears, emphasized the threat to our national survival in the critical problem of soil erosion. When the country was first settled, there was an average of about 9 inches of topsoil; since then, about one third of the topsoil has been lost. It takes about 500 years for 1 inch of topsoil to form under ideal conditions, by the slow decay of organic material and rock. This valuable resource can be lost in only a few years.

Erosion is the wearing away of the topsoil by the action of water and wind. *Gully erosion,* which is common on slopes, is caused by the cutting action of running water in small channels that become ever wider. *Sheet erosion* results when a sloping area is so full of water that the topsoil is lifted up in a sheet and floated away. *Wind erosion* results when the wind carries off dry topsoil that is not held together by the roots of plants.

Causes of Erosion

Erosion has been brought about by several factors:

1. *Forest destruction.* The roots of trees hold soil together. The soil acts as a sponge and absorbs much water during a rainfall. When the forests were cut down, the soil was not held together, and was washed away by the action of falling raindrops and running water. With the loss of topsoil, floods resulted, wash-

ing away still more topsoil and doing considerable damage.

2. *Cultivation of soil.* As farmers plowed up the soil and planted such crops as wheat and corn, the binding action of roots on the soil was weakened, leading to water and wind erosion. In addition, the mineral content of the soil was depleted when the same crop was planted year after year. The Dust Bowl in the 1930s had dust storms in which the valuable topsoil was blown away from countless farms in the Southwest.

3. *Overgrazing.* Poor grazing practices on fields and in forests led to the loss of grass and young trees, opening the way to eventual soil loss.

4. *Desertification.* With the increasing destruction of vegetation and the consequent erosion of the land, there has been a growing spread of desert areas, or desertification. In North Africa, the Sahara Desert has been creeping northward toward the Mediterranean Sea. This area, once the granary of the Roman Empire, now has to import food for its expanding population. About 1,200 years ago, Mesopotamia was one of the most fertile areas of the world. Today, it is largely desert. The same threat of environmental destruction is menacing other regions in all of the continents.

Soil Conservation Measures

A number of practices help to overcome the loss of valuable topsoil:

1. *Forest conservation.* The protection of forests, including reforestation, is an important step in conserving soil.

2. *Windbreaks and shelterbelts.* Trees have been planted around farms as a *windbreak* to reduce the eroding action of the wind. *Shelterbelt* rows of trees have also been planted in strips across the prairies to reduce the blowing action of the wind and to reduce excessive evaporation.

3. *Cover crops.* During the winter, fields of corn or other cultivated clean-till crops are planted with rye grass, so that the soil will be held together and improved. Perennial lespedeza and kudzu plants have converted severely eroded land in many parts of the South to excellent grazing areas in which the soil particles are bound together. The Department of Agriculture demonstrated that in one area of Texas on which corn was grown more than 20 tons of soil an acre were lost, along with 13.5 percent of the rainfall; when Bermuda grass was grown, the soil loss was only 0.02 ton an acre, and the water loss only 0.05 percent of the precipitation. In the state of Washington, the Soil Conservation Service planted a cover crop to convert a river that

caused floods in the spring and ceased to flow during late summer into a continuously flowing river that no longer floods and no longer carries away vast quantities of productive soil.

4. *Contour plowing.* Fields are plowed along the contour of sloping land, instead of up and down the slopes. The furrows hold back the water after a rain, and allow it to soak into the ground instead of running downhill and carving away the soil.

5. *Strip cropping.* Cultivated crops such as cotton or corn are planted on slopes in alternate strips with grasses and legumes. Any soil that is washed away from the cultivated rows is held by the cover crop below, and is not lost.

6. *Terracing.* On steep slopes, terraces or steps of soil and rocks are built across the hill and planted with soil-binding plants. This method holds the water back and prevents the loss of soil.

7. *Crop rotation.* Land that is planted with the same crop year after year (corn, for example) loses its mineral value and its capacity to hold water and soil; the yield of the crop is also reduced. By means of a 3-year rotation of corn, oats, and clover, the yield of corn is increased, while the loss of soil is reduced by the cover of oats and clover. In addition, the nitrogen content of the soil is increased by the action of nitrogen-fixing bacteria in the nodules of the clover. Other legume crops besides clover that are useful are alfalfa, lespedeza, and vetch.

8. *Soil improvement.* As plants grow, they remove valuable minerals from the soil. The farmer restores them by adding chemical fertilizers containing important compounds of nitrogen, phosphorus, magnesium, potassium, and other elements. When soils become too acid, he adds lime. He also spreads animal manure to increase the organic and mineral content of the soil. He further improves the soil by plowing into it the remains of his crops, such as cotton, wheat, or corn, and by plowing up fields of legumes and grasses; they are known as *green manures.* The resulting increased humus not only provides a greater mineral content, but also increases the water-holding capacity of the soil.

Also, when the surface is covered with leaves and other plant material, the splashing action of falling raindrops in causing the breaking up of the soil is reduced.

Water Conservation

The Earth's Water

Fresh water is essential for life to exist on earth. Humans may be able to go without food for nearly a month; without water, however, they will not be able to survive for more than a week.

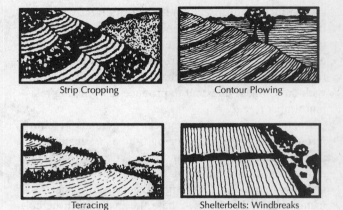

Methods of Soil Conservation

Although 97.2 percent of the earth's water is found in the oceans, it is too salty to use. The remaining 2.8 percent is fresh water, but most of it (2.38 percent) is frozen in the icecaps and glaciers. The remainder (less than one-half of one percent of fresh water) is found in lakes, streams, and underground sources. This small amount of available fresh water is therefore precious in making life on land possible for plants, animals, and humans.

The Water (Hydrologic) Cycle

The earth's water is continuously being interchanged between its surface and the atmosphere through evaporation and precipitation. This natural circulation is known as the water cycle, or *hydrologic cycle.* Water evaporates from the ocean, lakes, rivers, and land. It then collects in the atmosphere in the form of water vapor.

The water vapor forms clouds and returns to the earth's surface during precipitation in the form of rain, snow, or sleet. Much of this precipitation falls directly

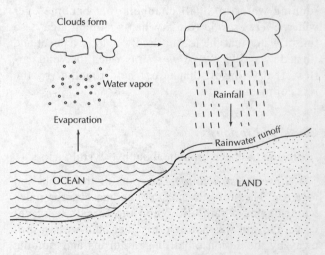

The Water Cycle

into the oceans, which cover 71 percent of the earth's surface. The remainder ends up on the land surface, where some eventually collects in rivers that flow into the sea.

Water for Human Use

An average American uses as much as 80 gallons of water daily. In most cases, water is easily available by merely turning on the faucet. Does anyone ever stop to think about where this endless flow of water comes from? Probably not. Cities depend on reservoirs and purification systems to provide a safe water supply. New York City obtains 80 percent of its water from the Catskill Mountains, which are almost 100 miles away. The city depends on laws and health department procedures to keep this *watershed* area safe. A watershed is a mountainous area that serves as a source of water for lakes, rivers, and streams.

Many communities and individual homes depend on wells that pump up ground water. Such wells are adequate, provided the supply of ground water continues to exist. In areas of the world that are close to the sea, plans are under consideration to extract salt from ocean water. At the present time, such a *desalination* process requires too much energy and is too expensive to be practical. However, many cruise ships do obtain fresh water for their passengers by desalination.

Protecting the Water Supply

During drought, it often becomes necessary to limit water use. Restrictions prohibit the watering of lawns, washing of cars, and the use of fountains in parks and monuments. Residents are also encouraged to conserve water by using low-volume toilets and low-flow shower heads.

A serious threat to fresh water is the discharge of untreated industrial and municipal wastes into lakes and rivers. To reduce this threat, Congress passed the Clean Water Act in 1972, which has successfully reduced pollution. Another source of pollution includes bacteria which can cause diseases such as typhoid and cholera. Although laws may help protect our water supply, the individual person can also play a role daily in water conservation by observing the slogan, "Waste not, want not."

Wetlands

Wetlands is a collective term that refers to areas like swamps, marshes, and bogs. These have a natural supply of water that keeps them wet for a good part of the year. Their importance has long been misunderstood. Only recently have ecologists come to recognize that they are among the most valuable ecosystems in the world.

Countless shorebirds—herons, egrets, gulls, and pelicans—depend on wetlands for feeding, living, and nesting habitats. Many species of fish as well as clams, oysters, and crabs inhabit wetlands. Microorganisms conduct important phases of the nitrogen, carbon, phosphorus, and sulfur cycles, breaking down harmful pollutants into simple forms which, in turn, can be utilized by other organisms for food.

Wetlands also serve to filter and cleanse the nation's water and help to retain floodwaters. By keeping our waterways clean and safe, they help protect the country's water supply.

Under the terms of the Clean Water Act, builders of homes and factories in wetland areas must first obtain permits before construction to ensure that the natural environment will not be adversely affected. The importance of wetlands was nationally recognized when the month of May was designated as National Wetlands Month by the Environmental Protection Agency.

Our Vanishing Wildlife

With the spread of civilization, the once-abundant wildlife of our country has been threatened, including mammals, birds, fish, and flowers and shrubs.

Factors Leading to Wildlife Destruction

Specific factors leading to the destruction of animals and plants have been the eradication of breeding and feeding grounds such as forests, prairies, and swamps; uncontrolled hunting and fishing; and stream pollution by sewage and industrial wastes.

Wildlife Conservation Measures

Conservation measures have included the following:

1. *Wildlife refuges.* *Sanctuaries* have been set aside in which wildlife is protected. From time to time, when there is a surplus of animals, open seasons are permitted for controlled hunting and fishing; agencies such as the U.S. Fish and Wildlife Service set up the regulations that help maintain a balance in the animal populations. Windbreaks and stripcropped areas also provide refuge for small mammals and birds. In addition, small ponds are built by farmers in soil-conservation districts, which are stocked with fish and which serve to attract ducks; about 87,000 of these ponds are built annually.

2. *Hunting and fishing limitations.* Game laws regulate the season when hunting and fishing may be carried on. There are limits as to the types of animals that may be hunted, the number that may be killed, the hunting of females, and the size and number of fish caught. Recently, a limited number of 400 hunters were picked at a public drawing from 8,000 applicants for the first open season since 1936 on moose in New Brunswick, Canada; at that time, the number of moose had been so depleted that they had to be fully protected until their numbers were found to be on the increase again.

3. *Bird laws.* Birds are protected by a number of state and federal laws. Songbirds, which destroy countless insects, are fully protected. Ducks and geese may be hunted only during a restricted open season. Migrating birds are protected by treaties with Canada and Mexico.

4. *Fish protection.* Such fish as trout, bass, salmon, and perch are raised in fish hatcheries. When the young fish are large enough, they are released in streams and ponds, thereby helping to replenish the supply. Erosion control also protects fish by preventing the water from becoming polluted by silt, which covers their food and interferes with spawning. The salmon is assisted in its migration upstream by the construction of *fish ladders* along dams.

In colonial days, the waters of the Hudson River in New York State were so pure that they were teeming with fish. Then, as industrial factories were built along the river, the water became polluted and most fish died off. Beginning with the Clean Water Act passed by Congress in 1972, succeeding laws prohibited the dumping of harmful chemicals such as PCBs and dioxin into the river. As a result, the water became cleaner, and fish are now returning in increasing numbers.

5. *Wild flower protection.* As forests, swamps, and fields were destroyed by the advancing front of civilization, many attractive wild flowers were threatened. Some of them have been picked to the point of extinction. State laws now prohibit the picking of such attractive flowering plants as dogwood, lady's slipper, trailing arbutus, Jack-in-the-pulpit, and Dutchman's breeches. The value of preserving even obscure plants is illustrated in the rare plant, Rosy Periwinkle. A compound, *vincristine*, is derived from it to treat childhood leukemia.

Section Review

Select the correct choice to complete each of the following statements:

1. All of the following are major causes of forest destruction *except* (A) excessive cutting (B) forest fires (C) conservation (D) insects (E) fungus diseases

2. Two insects that are enemies of trees are the (A) elm bark beetle and *Drosophila* fly (B) *Drosophila* fly and honeybee (C) honeybee and gypsy moth (D) elm bark beetle and monarch butterfly (E) elm bark beetle and gypsy moth

3. A tree that has been practically wiped out by disease is the (A) American chestnut (B) Norway maple (C) sycamore (D) weeping willow (E) staghorn sumac

4. Decayed organic matter in the soil is known as (A) aqueous humor (B) humus (C) hilum (D) hydra (E) hyphae

5. Trees prevent floods because (A) their roots soak up excess water (B) their leaves store excess water (C) their xylem ducts store excess water (D) their roots bind the soil (E) they carry on excess transpiration

6. Forests are now looked upon as a (A) source of fires (B) growing crop (C) home for harmful insects (D) waste of valuable space (E) menace to wildlife

7. All of the following are useful in controlling forest fires *except* (A) fire lines (B) smoke jumpers (C) smokescreens (D) fire breaks (E) backfires

8. When the country was first settled, the average thickness of the topsoil layer was about (A) 6 inches (B) 9 inches (C) 12 inches (D) 6 feet (E) 9 feet

9. Soil erosion is caused by (A) shelterbelts (B) windbreaks (C) reforestation (D) overgrazing (E) spraying with insecticides

10. A useful cover crop is (A) Bermuda grass (B) wheat (C) corn (D) cotton (E) rice

11. Terracing is practiced to (A) hold water back (B) enrich the soil (C) increase the nitrogen content of soil (D) add green manure to the soil (E) keep rodents away

12. All of the following are examples of legumes *except* (A) clover (B) alfalfa (C) vetch (D) oats (E) lespedeza

13. All of the following are examples of good soil conservation measures *except* (A) contour plowing (B) cultivation of soil (C) cover crops (D) strip cropping (E) crop rotation

14. Songbirds are useful because they destroy (A) field mice (B) moles (C) snakes (D) earthworms (E) insects

15. A protected wildflower is the (A) daisy (B) dandelion (C) clover (D) flowering dogwood (E) aster

Answer Key

1-C	4-B	7-C	10-A	13-B
2-E	5-D	8-B	11-A	14-E
3-A	6-B	9-D	12-D	15-D

Answers Explained

1. (C) Conservation is the planned use and protection of such natural resources as forests, soil, and wildlife.

2. (E) The elm bark beetle bores into the bark of the elm tree, destroys the living cambium cells, and stops the flow of sap. The caterpillar of the gypsy moth eats the leaves of trees, depriving them of their source of food.

3. (A) At one time, chestnut trees were common. Now the chestnut blight has wiped out practically all of the American chestnut trees from Canada to the Gulf states.

4. (B) Humus, the organic material of the soil, consists of decayed plant and animal matter.

5. (D) The roots of trees grow into the soil and attach to the soil particles, helping to hold the soil together. When forests are cut down, the soil is not held together and is washed away by the action of running water. With the loss of topsoil, which normally acts as a spongy layer, the water runs off in large amounts, causing floods.

6. (B) Selective cutting is the practice in which mature trees as well as dying or diseased trees are cut down. This process gives young, growing trees an opportunity to mature and produce seeds. After older trees are cut down, they are replaced by young trees that are carefully spaced and cared for.

7. (C) A fire line is made down to the soil all around the fire to keep it from spreading. Smoke jumpers are parachute fire fighters dropped by airplane, while fires are still small. Fire breaks are fire lanes built in the forest to divide it into small blocks. A backfire is started within a fire line to travel backward toward an oncoming fire and burn out the potential fuel.

8. (B) When the country was first settled, there was an average of about 9 inches of topsoil; since then about one third of the topsoil has been lost.

9. (D) Poor grazing practices on fields lead to the loss of grass, opening the way to eventual soil loss.

10. (A) When only corn or other cultivated crops are grown, valuable topsoil is lost because plowing weakens the binding actions of roots on soil. Bermuda grass binds the soil and controls the problem of erosion.

11. (A) On steep slopes, terraces, or steps of soil and rock, are built across the hill and planted with soil-binding plants. This method holds the water back and prevents the loss of soil.

12. (D) Legume plants increase the nitrogen content of the soil through the action of nitrogen-fixing bacteria in the nodules on their roots. Oat is not a legume.

13. (B) When soil is cultivated, it is plowed and planted with crops such as wheat and corn. The binding action of roots on soil is weakened, leading to erosion. In addition, the mineral content is depleted by planting the same crop year after year.

14. (E) Songbirds eat countless numbers of insects as food. For this reason, they are protected by state and federal laws.

15. (D) State laws have been passed to prevent the picking of attractive flowering plants that might otherwise become extinct. Examples: flowering dogwood, lady's slipper, trailing arbutus, Jack-in-the-pulpit, and Dutchman's breeches.

THE VARIETY OF LIVING THINGS ON EARTH

CHAPTER

13

13.1 HOW LIVING THINGS ARE CLASSIFIED

Biology is the study of living things. Altogether, there are probably at least 2 million different kinds of living things on earth. They live in a great variety of places: in the steaming tropics, in the frozen North, in the temperate zone; high on mountains, deep in the ocean, in deserts, in marshes, at the seashore, in crowded cities. To understand and study these living things, scientists have arranged them in groups. Aristotle first tried to classify organisms over 2,000 years ago. The present system of classification was devised by the Swedish biologist Carolus Linnaeus (1707–1778).

Binomial Classification

Linnaeus used two Latin or Greek names for each plant and animal in his system of binomial nomenclature, a *genus* and a *species* name. Thus a human being is known as *Homo sapiens,* while the dog is *Canis familiaris,* and the house cat is *Felis domestica.* A species is considered to be a closely related group of organisms that are similar in structure and can interbreed. A genus consists of one or more species that show many similarities. Thus, the house cat species, *Felis domestica,* is grouped in the same genus, *Felis,* as

such other species of the cat family as *Felis leo,* the lion, and *Felis tigris,* the tiger. The genus name is spelled with a capital letter, while the species name begins with a small letter; both names are printed in italic type.

Modern Views of Classification

Taxonomy

This division of biology deals with the classification of living things based on similarities in their structure, development, and evolutionary history. There is not universal agreement among present-day biologists on a single system of classification.

Three Domains

In 1997, a new classification system was proposed, based on the discovery 20 years earlier of unusual microbes living in the superheated waters of undersea volcanoes. These microbes have since been classified as *Archaea*—one of three newly named domains, or divisions of living things. The other two domains are Prokarya (prokaryotes) and Eukarya (eukaryotes). The complete genome of one of the Archaea, *Methanoccus janaschii*, has already been deciphered and shows some genes that are similar to those of prokaryotes and some to eukaryotes; most of the genes however, are completely different from both.

Five Kingdoms

Before the discovery of the Archaea, the commonly accepted classification system was based on five kingdoms: Monera, Protista, Fungi, Plants, and Animals. Each kingdom is divided into lower groups as follows:

Kingdom — phylum — subphylum — class — order — family — genus — species — variety

The smaller the grouping, the more similar the types of individuals will be. Those that are placed in the same genus, for example, will be comparable in structure and development; they undoubtedly are descended from a common ancestor and have evolved rather recently, along somewhat different lines from each other.

As an example of the classification scheme described above, the house cat would be classified as follows:

Kingdom:	Animal
Phylum:	Chordata (animals with a notochord)
Subphylum:	Vertebrates (animals with a backbone of vertebrae)
Class:	Mammalia (warm-blooded animals having hair and producing milk)
Order:	Carnivora (meat-eating animals with long canine teeth)
Family:	Felidae (cat family, with sharp, curved claws)
Genus:	*Felis* (closely related members such as cat, lion, tiger)
Species	*domestica* (the house cat)
Variety	specific types of house cat, such as the Persian cat, Siamese cat, Manx cat

CLASSIFICATION OF MONERA

Monera are prokaryotes (primitive cells) and are classified into phyla of bacteria and blue-green algae.

Phylum	Characteristics	Examples	
1. Bacteria	Smallest cells known; have a cell wall; may form spores; no nuclear membrane; some have flagella	Bacilli, cocci, spirilla	
2. Blue-green algae	Cells larger than bacteria; contain chlorophyll, but not in chloroplasts; carry on photosynthesis	*Nostoc, Oscillatoria*	

CLASSIFICATION OF PROTISTA

Protists are classified into phyla of protozoa, algae, and slime molds.

Phylum	Characteristics	Examples	
1. Protozoa	Animal-like; single-celled; some form colonies, mostly free-moving	Ameba, paramecium, euglena, malaria plasmodium	
2. Algae	Contain chlorophyll; may be unicellular or in colonies	Spirogyra; diatoms; green, brown, red algae	
3. Slime molds	Naked mass of protoplasm; many nuclei; no internal cell membranes; spores form flagellate cells, which fuse into ameboid form	*Physarum*	

CLASSIFICATION OF FUNGI

Phylum	Characteristics	Examples	
Fungi	Appear to be plantlike; lack chlorophyll; spores	Yeast, molds, mushrooms	

CLASSIFICATION OF ANIMALS

Animals are classified according to the following table, which includes most of the invertebrate phyla and the vertebrate subphylum.

Phylum	Characteristics	Examples	
1. Porifera	Two-layered organisms with pores; attached	Sponges	
2. Coelenterata	Two layers of cells; hollow digestive cavity, with tentacles at opening	Hydra, jellyfish, coral, sea anemone	
3. Platyhelminthes (Flatworms)	Three layers of cells; mostly parasitic	Tapworm, planaria, liver fluke	
4. Nematoda (Roundworms)	Threadlike; digestive system includes an anus; many parasitic	Hookworm; *Ascaris*; trichina worm	
5. Annelida (Segmented worms)	Long, cylindrical body with ringed segments; nervous system; circulatory system	Earthworm, leech	
6. Rotifera	Many-celled; beating crown of cilia at head end; digestive tube	Rotifer	
7. Echinodermata (Spiny-skinned)	Radial symmetry; spiny exoskeleton; marine dwelling	Starfish, sea urchin, sea cucumber	
8. Mollusca	Soft bodies, mostly protected by a shell	Clam, snail, octopus	
9. Arthropoda	Jointed legs; exoskeleton; segmented body		
Class 1. Crustacea	Gills for breathing; two pairs of antennae; jointed legs with claws	Crab, lobster	
Class 2. Insecta	Six legs; three body parts (head, thorax, abdomen); one pair of antennae	Butterfly, bee, grasshopper	
Class 3. Arachnida	Eight legs; two body parts; no antennae	Spider, scorpion, tick	
Class 4. Chilopoda	One pair of legs per segment	Centipede	

CLASSIFICATION OF ANIMALS *(Continued)*

Phylum	Characteristics	Examples	
Class 5. Diplopoda	Two pairs of legs per segment	Millipede	
10. Chordata	Possess a notochord, supporting rod structure during life history; dorsal nerve cord; gill slits	Subphyla: vertebrates, amphioxus, tunicates, acorn worms	
SUBPHYLUM Vertebrate	Backbone enclosing spinal cord		
Class 1. Pisces (Fish)	Most have bony skeleton; gills; scales; two-chambered heart	Salmon, trout, cod	
Class 2. Amphibia	Live in water in early stages, with gills; develop lungs; thin, moist skin; three-chambered heart	Frog, toad, salamander	
Class 3. Reptilia	Lungs; dry scales; eggs with a horny covering; cold-blooded; mostly three-chambered heart	Snake, lizard, turtle	
Class 4. Aves	Feathers; wings; warm-blooded; four-chambered heart; eggs with shell	Sparrow, chicken, ostrich	
Class 5. Mammalia	Warm-blooded; hair; diaphragm; young born alive, fed on milk	Human, dog, bat, whale	

CLASSIFICATION OF PLANTS

Plants are green and multicellular, and generally live on land. They are classified into two phyla.

Phylum	Characteristics	Examples	
1. Bryophyta	Simple stems; rootlike and leaflike structures; alternation of generations; produce spores		
Class 1. Hepaticae (Liverworts)	Flat, leaflike structure; simple rootlike structures	*Marchantia*	
Class 2. Musci (Mosses)	Usually erect with stem; capsule contains spores	*Sphagnum*	
2. Tracheophyta	Have a vascular system; true roots, stems, leaves		

CLASSIFICATION OF PLANTS *(Continued)*

Phylum	Characteristics	Examples
SUBPHYLUM 1. Lycopsida	Creeping stems	Club moss, ground pine
SUBPHYLUM 2. Sphenopsida	Scalelike leaves; spores in conelike structures	Horsetail, scouring rush
SUBPHYLUM 3. Pteropsida	Broad-leaved	
Class 1. Filicineae (True ferns)	Spores on leaves or special structures	Royal fern, Boston fern
Class 2. Gymnospermae	Naked seeds, often in a cone; needlelike leaves, mostly evergreen	Pine, hemlock, spruce
Class 3. Angiospermae	Seeds enclosed in fruit or nut; broad leaves, not evergreen	
Subclass 1. Dicotyledons	Net-veined leaves; seeds with two cotyledons	Maple, rose, bean
Subclass 2. Monocotyledons	Parallel-veined leaves; seed with one cotyledon	Corn, lily, grass

Section Review

Select the correct choice to complete each of the following statements:

1. Linnaeus originated the system of classification which employs   (A) bilateral symmetry   (B) binomial nomenclature   (C) continuity of germ-plasm   (D) conservation of energy   (E) valence of elements

2. The division of biology that deals with classification is   (A) cytology   (B) histology   (C) botany   (D) morphology   (E) taxonomy

3. Of the following, the most closely related group of organisms is known as a(n)   (A) genus   (B) species   (C) order   (D) family   (E) class

4. The number of protist phyla is   (A) 1   (B) 3   (C) 5   (D) 7   (E) 9

5. Classification is based on all of the following *except*   (A) structure   (B) development   (C) evolutionary history   (D) common ancestry   (E) size

6. Starting with the largest group, the correct sequence of the following groups: 1-class; 2-family; 3-phylum; 4-order; 5-genus; 6-species, is   (A) 3–4–1–6–5–2   (B) 3–2–4–1–5–6   (C) 3–1–2–4–5–6   (D) 3–1–4–2–5–6   (E) 3–4–1–5–6–2

7. The Latin words of the name given to a human being, *Homo sapiens*, include the   (A) genus and species   (B) genus and family   (C) family and order   (D) order and class   (E) genus and class

8. Of the following, the animal not included in the same genus as the house cat is the   (A) lion   (B) tiger   (C) dog   (D) leopard   (E) Siamese cat

9. Two invertebrate phyla are (A) protozoa and thallophyta (B) protozoa and mammalia (C) mollusks and arthropoda (D) protozoa and pteridophyta (E) coelenterates and bryophytes

10. Monera include (A) brown algae and bacteria (B) blue-green algae and bacteria (C) brown algae and protozoa (D) blue-green algae and protozoa (E) bacteria and protozoa

Answer Key

1-B	3-B	5-E	7-A	9-C
2-E	4-B	6-D	8-C	10-B

Answers Explained

1. **(B)** Linnaeus used two Latin or Greek names for each plant or animal, a genus and a species name. This system is known as binomial nomenclature.

2. **(E)** Taxonomy is the division of biology that deals with the classification of living things, based on similarities in their structure, development, and evolutionary history.

3. **(B)** A species is a closely related group of organisms that are similar in structure and can interbreed.

4. **(B)** The protist phyla are protozoa, algae, and slime molds.

5. **(E)** Size is not a factor in classifying organisms. Structure, development, evolutionary history, and ancestry are determining factors.

6. **(D)** Of the groups named, the phylum is the largest group of organisms. It is divided into classes, which, in turn, are divided into orders. Then in turn come families, genera, and species.

7. **(A)** According to Linnaeus's system of binomial nomenclature, a human is classified into the genus *Homo* and the species *sapiens*.

8. **(C)** The genus of the cat, lion, tiger, and leopard is *Felis*.

9. **(C)** Invertebrates are animals without a backbone.

10. **(B)** Monera are prokaryotes and are classified into phyla of bacteria and blue-green algae.

13.2   THE MONERA KINGDOM

In the most recent system of classification, bacteria and blue-green algae are grouped into the Monera kingdom. They have a primitive form of cell structure and lack a definite nucleus with a nuclear membrane. They also lack organelles commonly found in other cells, such as mitochondria, endoplasmic reticulum, Golgi complex, and chloroplasts. Their DNA is contained in a single chromosome. Such primitive cells are known as *prokaryotes*.

By contrast, the cells of the other four kingdoms (Protista, Fungi, Animal, Plant) are known are *eukaryotes*. Their DNA is contained in paired chromosomes located within a nucleus that has a nuclear membrane. They also contain such other organelles as mitochondria, endoplasmic reticulum, Golgi complex, and, in the case of green plants and algae, chloroplasts.

The cell walls of organisms classified in the Monera kingdom are different from those found in eukaryotes. They never contain cellulose or chitin. Instead, they are built of large molecules called peptidoglycans.

Phylum 1 Bacteria

Bacteria are the smallest cells known, varying from 1 to 3 micrometers (µm) in size, compared to about 10 µm for most other cells. (There are 1,000 µm in 1 millimeter.) Some bacteria are the cause of serious diseases in humans. Others are useful allies in the decay of organic substances in the soil, and in the dairy industry. Bacteria are so important in human affairs that they are considered in greater detail in Chapter 8.

Phylum 2 Blue-green algae

These are extremely simple in structure and are considered by some biologists to be essentially bacteria, or cyanobacteria. They are able to carry on photosynthesis, but their chlorophyll is not contained in chloroplasts. Instead, long chlorophyll-containing membranes are scattered through the outer region of the cell. The cells of blue-green algae also contain a blue pigment, phycocyanin, which is always present, and occasionally a red pigment, phycoerythrin. The colors of blue-green algae may vary from blue-green to green, to yellow, to red. In fact, the Red Sea derives its name from blue-green algae of the red type.

Blue-green algae live in a wide diversity of habitats, ranging from snow in the Antarctic to hot springs where the temperature may reach 85°C or more. Some are the algae members of lichens; many are able to carry on nitrogen fixation; others live in fresh water, in the soil, or in the ocean. Some are anaerobic, obtaining their energy from the oxidation of hydrogen sulfide. They reproduce asexually.

One type, *Oscillatoria*, consists of long filaments that slowly swing to and fro in the water. Another, *Nostoc*, contains numerous filaments surrounded by gelatinous sheaths that may sometimes form balls almost 60 centimeters in diameter.

Section Review

Select the correct choice to complete each of the following statements:

1. The smallest known cells are (A) protozoa (B) blue-green algae (C) algae (D) fungi (E) bacteria

2. Monera contain none of the following structures *except* (A) mitochondria (B) a nucleus (C) DNA (D) a nuclear membrane (E) chloroplasts

3. Eukaryotes are different from prokaryotes because eukaryotes contain (A) cytoplasm (B) DNA (C) cell walls (D) endoplasmic reticulum (E) chlorophyll

4. The kingdom that contains prokaryotes is (A) Monera (B) Protista (C) Fungi (D) Plants (E) all of these

5. Eukaryotes may contain (A) a Golgi complex (B) an endoplasmic reticulum (C) chloroplasts (D) a nuclear membrane (E) all of these

6. The chlorophyll of blue-green algae is contained in (A) chloroplasts (B) chlorophyll-containing membranes (C) DNA (D) RNA (E) phycocyanin

7. Monera include (A) algae and bacteria (B) protozoa and bacteria (C) blue-green algae and bacteria (D) blue-green algae and protozoa (E) algae and protozoa

8. Organisms that do not have a nucleus are classified in the kingdom (A) Monera (B) Protista (C) Fungi (D) Animals (E) Plants

9. DNA is found in (A) all Monera (B) all Protista (C) all Fungi (D) all Plants (E) all of these

10. The color of blue-green algae may be (A) blue-green (B) green (C) yellow (D) red (E) all of these

Answer Key

1-E	3-D	5-E	7-C	9-E
2-C	4-A	6-B	8-A	10-E

Answers Explained

1. (E) Bacteria average only 1–3 micrometers in size.

2. (C) Monera have DNA contained in a single chromosome.

3. (D) Eukaryotes contain organelles within their cells, including endoplasmic reticulum.

4. (A) Monera have a primitive form of cell structure and are considered to be prokaryotes because their cells do not contain organelles.

5. (E) Eukaryotes have cells that contain organelles.

6. (B) Blue-green algae are extremely simple in structure. Their chlorophyll is not contained in chloroplasts, but in long chlorophyll-containing membranes scattered throughout the outer region of the cell.

7. (C) The Monera kingdom consists of two phyla, bacteria and blue-green algae.

8. (A) The members of the Monera kingdom are prokaryotes; their cells do not contain a nucleus.

9. (E) In Monera, DNA is found in a single chromosome. In all the other kingdoms, DNA is located in paired chromosomes.

10. (E) The cells of blue-green algae may contain the pigments chlorophyll, phycocyanin, and occasionally, phycoerythrin. Their colors may vary from blue-green to green to yellow to red.

13.3 **THE PROTISTA KINGDOM**

Originally, living things were classified into either the Animal kingdom or the Plant kingdom. However, recent investigations have shown that certain groups of living things do not fit easily into such a classification. Some types of protozoa and algae appear to have characteristics of both kingdoms. Consequently, there has been general agreement among most biologists that the Protista kingdom should be set up.

The protists include small living things whose classification is not clear cut. They are believed to have appeared early in evolutionary history. They are composed essentially of single cells, or colonies of cells.

In general, their cells differ from those of higher animals and plants. They are such an uncertain group that some of them (protozoa) may be considered by some biologists to be animals, while others (algae) may be considered to be plants. Some types (euglenoids) are included in either kingdom because they have characteristics of both—they possess chlorophyll, and move about by flagella.

Phylum 1 **Protozoa**

This phylum includes about 15,000 different species. Most of them live in fresh water, some live in the ocean, and some are parasitic.

Class 1 **Sarcodina** The best-known member of this class is the *ameba*. It consists of a little mass of protoplasm without any particular shape. It moves by forming projections or *pseudopodia* (false feet) and flowing into them. As it moves along, it surrounds and engulfs its food. Some members of this class form a skeleton of lime or silica. The chalk cliffs of Dover were once under water, and are composed of the remains of countless numbers of such shell-forming protozoa.

Class 2 **Flagellata** These protozoa have long, hairlike projections, *flagella,* that beat in whiplike fashion, and by which they move.

The *euglena* is unusual in having chloroplasts, a characteristic of plants. It also has a red eyespot that is sensitive to sunlight. The *trypanosome* causes African sleeping sickness and is spread by the bloodsucking tsetse fly. Another type of flagellate lives in the intestines of termites and digests the wood that these insects eat. Without them, the termites would not be able to make use of the wood and would starve.

Class 3 **Infusoria** The *paramecium* is undoubtedly the best-known member of this group. It moves by means of *cilia* and has two kinds of nuclei. The cilia are used both to move and to create currents in the region of the mouth for feeding. As food enters the body, it forms into a *food vacuole,* where digestion occurs. Two *contractile vacuoles* at either end eliminate excess water. Another ciliate, the *vorticella,* has a row of cilia at one end of its bell-shaped body and a sensitive stalk at the other end which contracts when it is disturbed.

Class 4 **Sporozoa** These are parasitic protozoa, such as the malaria *plasmodium,* which form spores. In this process, the nucleus divides many times; a little cytoplasm collects around each nucleus, and a number of different spores are produced and spread. The plasmodium is carried by the female *Anopheles* mosquito. When it enters the human body, it invades and destroys red blood corpuscles causing malaria.

Phylum 2 **Algae**

Algae are simple protists that contain chlorophyll, and so are capable of making their own food. About 20,000 different algae have been classified. A well-known type of green algae is the *spirogyra,* which is found in freshwater ponds. It consists of single cells arranged in a long, silky thread or filament; the chloroplast is present as a spiral ribbon. Another example is the *pleurococcus,* commonly found growing on tree trunks and mistakenly called moss. Other common types include single-celled *desmids,* and *diatoms.* The latter have a hard, sculptured wall and are responsible for thick deposits of diatomaceous earth that have accumulated over the ages; this earth now is used for toothpaste, polishes, and filters.

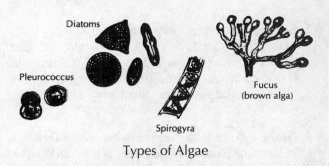

Diatoms

Pleurococcus

Spirogyra

Fucus (brown alga)

Types of Algae

In addition to the green chlorophyll, algae contain various other colored pigments. Brown algae and red

Ameba   Euglena   Vorticella

algae include most of the seaweeds. The rocks of the seashore are covered with deposits of brown algae. *Kelps* are brown algae that may reach a length of over 100 feet. Agar-agar, a gelatinous substance used for growing bacteria, is obtained from red algae growing in the Pacific Ocean.

Algae make up a good part of plankton, the floating mass of microscopic food in the sea, upon which small animals and, in turn, the larger fish of the oceans depend for their existence.

Phylum 3 Slime Molds

This phylum consists of organisms that have characteristics of plants and animals. They consist of a naked mass of protoplasm called a *plasmodium,* which

spreads slowly over decaying logs, somewhat like a huge ameba. It may achieve a size of up to 1 foot in diameter. Its streaming cytoplasm contains many nuclei. Its mass is not subdivided into cells.

At times, the plasmodium begins to resemble a fungus. It forms spore cases, or *sporangia*, on short stalks. Spores are released, and in some cases germinate into flagellate cells. These cells then fuse to form an ameboid, multinucleate mass, which becomes a new plasmodium.

The common slime mold *Physarum* may be found in the woods, growing in a moist area on a fallen log or decaying tree stump, as a glistening, bright yellow plasmodium. It can be cultured in the laboratory on moist toweling, or mimeograph paper, in a covered battery jar. It can be fed simply by sprinkling oatmeal powder over it.

Section Review

Select the correct choice to complete each of the following statements:

1. Protists are   (A) more complex than plants but simpler than animals   (B) more complex than animals but simpler than plants   (C) more complex than both plants and animals   (D) simpler than plant and animals   (E) too complex to classify

2. Protists are classified separately from plants and animals because   (A) they appeared later in evolutionary history   (B) their classification is not clearcut   (C) they are all multicelled   (D) Linnaeus did not know of their existence   (E) they are harmful to humans

3. Some protists   (A) have characteristics of both plants and animals   (B) are invisible   (C) are plants at one time and then turn into animals   (D) are animals at one time and then turn into plants   (E) originated in other planets

4. Of the following, the organism that is *not* included in the same phylum as the others is the   (A) ameba   (B) vorticella   (C) paramecium   (D) *Physarum*   (E) trypanosome

5. One-celled green protists are known as   (A) bacteria   (B) algae   (C) fungi   (D) rusts   (E) slime molds

6. The malaria plasmodium is a(n)   (A) parasitic protozoan   (B) autotrophic protozoan   (C) example of Monera   (D) parasitic fungus   (E) parasitic mosquito

7. All of the following are algae *except*   (A) the pleurococcus   (B) diatoms   (C) the ameba   (D) the spirogyra   (E) kelps

8. Plankton is made up largely of   (A) bacteria   (B) bacilli   (C) algae   (D) cocci   (E) slime molds

9. All of the following are examples of protists *except*   (A) algae   (B) slime molds   (C) protozoa   (D) the euglena   (E) blue-green algae

10. Organisms that consist of a naked mass of protoplasm are   (A) slime molds   (B) pleurococci   (C) kelps   (D) diatoms   (E) plankton

Answer Key

1-D	3-A	5-B	7-C	9-E
2-B	4-D	6-A	8-C	10-A

Answers Explained

1. (D) The protists include single cells or colonies of cells. They may be protozoa, algae, or slime molds, and are simpler than plants and animals.

2. (B) The cells of protists are different from the cells of animals and plants. Some of them, protozoa, may be considered animals; others, algae, may be considered plants; and some, flagellates, have characteristics of both plants and animals.

3. (A) Flagellates, such as the euglena, possess chlorophyll, like plants, and move about, like animals.

4. (D) *Physarum* is not a member of the protozoa phylum. It is a slime mold.

5. (B) Algae are simple protists that contain chlorophyll and so are capable of making their own food.

6. (A) The malaria plasmodium is a parasitic protozoan that is spread by the bite of the *Anopheles* mosquito. When it enters the body, it invades and destroys red blood cells.

7. (C) Ameba is a protozoan that consists of a little mass of protoplasm without any particular shape.

8. (C) Plankton is the floating mass of microscopic algae and protozoa in the sea upon which small animals feed.

9. (E) Blue-green algae are classified as Monera because their cells do not have well-defined organelles.

10. (A) Slime molds consist of a naked mass of protoplasm called a plasmodium that spreads slowly over decaying logs, somewhat like a huge ameba.

13.4   THE FUNGI KINGDOM

Although fungi appear to be plantlike, recent studies of their genes that produce ribosomal RNA indicate that, in their evolutionary development, fungi are more closely related to animals than to plants. Lacking chlorophyll, their nutrition is heterotrophic.

The basic structure of a fungus is a threadlike filament called a *hypha*. A hypha is multinucleate; that is, it has many nuclei distributed throughout the cytoplasm without intervening cell membranes or cell walls. A mass of hyphae is known as a *mycelium*. Fungi reproduce asexually by spores, as well as by gametes involved in sexual reproduction. Their cell walls usually contain chitin, although some have cellulose.

Many are *parasitic*, attaching themselves to other living things as a source of food; they injure and sometimes kill their host. Others, such as mushrooms, cheese molds, and yeast, are *saprophytes*, obtaining their food from nonliving organic matter. Still others are *decomposers*—organisms of decay—that break down dead animal and plant remains.

About 65,000 species of fungi have been identified and arranged in several classes. *Yeast* plants are single-celled. They are used in the baking of bread and cakes to make the dough rise. They also give beer its characteristic flavor. These activities result from the *fermentation* of sugar, giving rise to the products carbon dioxide and alcohol.

Another class includes larger fungi, the *molds*. Some of them are useful in the making of Roquefort cheese and in the production of antibiotics (e.g., *Penicillium notatum*). Others attack foods such as bread and fruits, spoiling them. A few are parasitic and cause athlete's foot and ringworm. Molds reproduce by spores and form a mass of threadlike structures (*mycelium*) as they grow.

Mushrooms, *smuts*, and *rusts* constitute another group of fungi. Wheat and corn crops are seriously affected by rust and smut parasites. Mushrooms obtain their food from organic materials in the soil. The undersurface of a mushroom cap contains sections or gills where spores are produced. Poisonous mushrooms, or toadstools, may be recognized by the ring or collar below the cap. Some mushrooms are edible, but usually only an expert can tell whether or not a mushroom is poisonous.

Lichens really consist of two organisms, a fungus and a one-celled alga, living together in a helpful partnership. The alga makes food, while the fungus furnishes moisture and minerals. This mutually helpful relationship is known as *mutualism*. Lichens are found on the surface of rocks and logs. An important example of a lichen is reindeer moss, an important food for reindeer, which is not a moss at all.

Section Review

Select the correct choice to complete each of the following statements.

1. All of the following are examples of fungi *except* (A) yeast (B) a mushroom (C) smuts (D) rusts (E) a vorticella

2. The lichen is an example of (A) an alga and a fungus (B) a protozoan (C) bacteria (D) an alga and a slime mold (E) an alga and yeast

3. Fungi may be all of the following *except* (A) heterotrophic (B) autotrophic (C) parasitic (D) saprophytic (E) decomposers

4. Fungi do not contain (A) cell walls (B) hyphae (C) chlorophyll (D) mycelium (E) spores

5. A saprophyte obtains its food from (A) nonliving organic matter (B) nonliving inorganic matter (C) autrotrophic nutrition (D) parasitic nutrition (E) photosynthesis

6. Yeast is useful in the baking industry because it carries on (A) aerobic respiration (B) fermentation (C) photosynthesis (D) the Krebs cycle (E) symbiosis

7. Some fungi are harmful because (A) they are parasitic (B) they cause athlete's foot (C) they are poisonous mushrooms (D) they attach themselves to and obtain nourishment from a host (E) all of these

8. Some fungi are beneficial because (A) they produce antibiotics (B) they cause athlete's foot (C) they are rusts (D) they are smuts (E) all of these

9. All of the following are true of mushrooms *except* (A) some are edible (B) some are poisonous (C) they produce spores (D) some produce chlorophyll (E) they live on organic materials in the soil

10. Ringworm is caused by a (A) protozoan (B) bacterium (C) slime mold (D) fungus (E) roundworm

Answer Key

1-E	3-B	5-A	7-E	9-D
2-A	4-C	6-B	8-A	10-D

Answers Explained

1. (E) Vorticella is a bell-shaped protozoan on a slender stalk. The stalk contracts rapidly when it is touched.

2. (A) A lichen really consists of two organisms, a fungus and a one-celled alga, living together in a helpful relationship called mutualism. The alga makes food, while the fungus furnishes moisture and minerals.

3. (B) Fungi are simple, plantlike organisms that do not contain chlorophyll and therefore cannot make their own food. An autotrophic organism can make its own food.

4. (C) Fungi do not contain chlorophyll and therefore are heterotrophic.

5. (A) Saprophytes, such as mushrooms, cheese molds, and yeast, obtain their food from nonliving organic matter.

6. (B) Yeast carries on anaerobic respiration, or fermentation, giving off the products carbon dioxide and alcohol.

7. (E) Some fungi obtain their nourishment from a host and are parasitic—for example, athlete's foot fungus. Poisonous mushrooms are harmful if eaten.

8. (A) Fungi such as *Penicillium notatum* are useful because they produce antibiotics, such as penicillin. An antibiotic is a chemical produced by an organism, such as a fungus, that kills germs.

9. (D) Mushrooms do not contain chlorophyll. They obtain their nourishment from organic materials in the soil.

10. (D) Ringworm is a fungus disease of the skin in which there are circular infected patches on the skin.

13.5 THE ANIMAL KINGDOM

The study of animals is known as *zoology*. About a million different species of animals have been named.

Phylum 1 Porifera

Sponges, or "pore animals," are simple, many-celled animals that do not move and that are attached to rocks or sticks. Water enters through many pores into the hollow central cavity, and eventually leaves through a large opening (excurrent pore) at the top. The current of water is created by the beating of flagella located in *collar cells* lining the cavities of the sponge. As the water streams by, these cells engulf particles of food and digest them. The continuous current of water passing through a sponge also supplies oxygen to the cells and removes carbon dioxide and nitrogenous wastes.

Most sponges live in the ocean, although a few species are found in fresh water. Commercial sponges are collected in the warm, shallow waters near Florida and the Bahamas, and in the Mediterranean Sea. When alive, they resemble slimy pieces of raw liver. They are prepared for commercial use through decomposition of the soft protoplasmic material by bacteria. The final product, the bath sponge, consists of the fibrous skeleton. Other members of the phylum are the glass sponge, Venus flower basket, and the freshwater sponge, *Spongilia*.

Phylum 2 Coelenterata

This group includes the jellyfish, hydra, sea anemone, and coral. These animals have a central cavity, and consist of two layers of cells: an outer *ectoderm* and an inner *endoderm*. The only body opening, the mouth, is surrounded by a ring of tentacles containing stinging cells (*nematocysts*). When small animals are paralyzed by these cells, they are carried to the mouth by the tentacles. Digestion is carried on by the endodermal cells. Indigestible remains are eliminated through the mouth. There are also primitive muscle, nerve, and gland cells in the body of these animals.

Jellyfish

Jellyfish vary in size, from some that are microscopic to others that grow as large as a human. They are made of 99 percent water, and move by a pulsating motion. They show radial symmetry. When they come into contact with a person swimming in the water, their stinging cells may cause a severe reaction. The *hydra* is a freshwater coelenterate that looks like a piece of string, less than 1 inch long, with frayed ends where the tentacles are. When disturbed, it contracts to the size of a pinhead. It feeds on water fleas and small worms. *Corals* are colonial forms that take lime out of the water and secrete limestone on the outside of their ectoderm layer. Accumulations of this coral skeleton in tropical waters form coral reefs and even whole islands.

Phylum 3 Platyhelminthes (*Flatworms*)

Beginning with the flatworms, all the higher animals have a third layer, the *mesoderm,* between the ectoderm and the endoderm. This layer produces the muscles and the reproductive organs. Definite organs also begin to form, each composed of several types of tissues.

Planaria are small flatworms about 1/2 inch long that live in fresh water under stones. They have the beginning of a nervous system, with a concentration of nerve tissue in the head comprising a simple brain. They also have a digestive system, and both male and female reproductive organs.

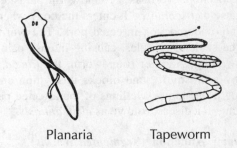

Planaria   Tapeworm

The *tapeworm* is a long, flat, ribbonlike animal made up of sections, which lives as an adult in the human intestines and may reach a length of more than 20 feet. It attaches itself to the wall of the intestine by means of suckers and hooks on its head. It does not have a digestive system, but absorbs digested food from its host directly into its body. The mature sections drop off when full of eggs and are passed out of the body with solid wastes. Another host, a cow, may take these eggs into its body as it grazes, and may also become infected. Thus, the beef tapeworm lives in two hosts.

Another parasitic flatworm, the *blood fluke,* is a serious menace in the Orient, where it infects millions of people during rice-planting time. The worm bores through the skin of their legs as they stand barefooted in the shallow water of a rice field. Eventually it settles in the blood vessels of the intestines, causing body pains, severe dysentery, anemia, and general weakness.

A snail is the intermediate host of this fluke. Another parasite, the Chinese liver fluke, has two other hosts besides humans—a fish and a snail.

Phylum 4 Nematoda (*Roundworms*)

The roundworms, or threadworms, have a digestive canal with a mouth opening at one end and an anus at the other. Many are parasitic on plants as well as on animals. The *hookworm* causes a serious health problem in the South, where it lives in the soil and bores its way through the soles of people walking barefoot. It travels through the bloodstream, heart, and lungs before finally boring into the intestines, where it attaches itself and feeds on blood. Victims show anemia, listlessness, and even a retardation of mental and physical development.

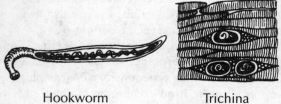

Hookworm   Trichina
(encysted in human muscle)

The *trichina* worm is a roundworm parasite that is the cause of *trichinosis*. It enters the body when a person eats undercooked infected pork. The worms bore into the skeletal muscles, causing intense pain, fever, and swelling. Another roundworm, the *filaria*, is carried by a mosquito, and blocks the lymph channels, causing tremendous swellings of the affected parts and resulting in a disease known as *elephantiasis*.

Phylum 5 Annelida (*Segmented Worms*)

The third group of worms are called segmented because their body structure is arranged in rings or segments. The digestive tube runs the length of the worm's body, and constitutes a tube within a tube. The muscles are well developed for wriggling. The circulatory system has blood vessels with muscular walls for contracting and driving the blood along. The nervous system is well developed, including a brain, a long nerve cord, and nerve centers or *ganglia* in each segment.

Earthworm   Sandworm   Leech

The *earthworm* is useful because of its habit of turning over the soil and making it porous. It does this by swallowing earth, which passes through its digestive canal, where the organic material is digested and absorbed. The *sandworm* (*Nereis*) is a marine relative of the earthworm. It has two pairs of eyes, and a pair of extensions or simple paddles on each segment for swimming. The *leech* has a sucker at each end of its body by which it attaches itself to an animal, and then sucks its blood. While the leach is feeding, it secretes the chemical *hirudin,* which prevents the blood from clotting.

Phylum 6 Rotifers

Rotifers are microscopic animals, of little economic importance, that are somewhat more complicated than the flatworms in some ways, and less so in others. They are grouped in the phylum *Rotifera*, meaning "wheel-bearers," and take their name from the beating crown of cilia at the head end. Although they are no larger than the one-celled paramecium, rotifers are many-celled. They posses a digestive tube with a mouth and an anus. The whole body can be folded together in telescopic fashion when they contract. They reproduce sexually and are commonly found in pond water.

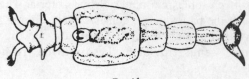

Rotifer

Phylum 7 Echinodermata

The body plan of these animals is different from that of most animals in that it shows radial symmetry. There is a central disc from which five or more arms radiate, like the spokes of a wheel. The *starfish* is a well-known example. Its mouth is located on the undersurface of the disk, and it has the characteristic spiny skin of the phylum. Locomotion is achieved by means of a water vascular system that has canals going into each of the arms.

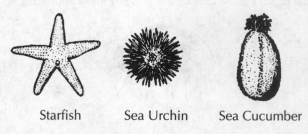

Starfish   Sea Urchin   Sea Cucumber

The starfish is quite harmful to the oyster industry. It feeds by slowly forcing the shell of an oyster open with its many small tube feet. Then it turns the lower part of its stomach inside out, extends it through its mouth, and envelops the soft body of the oyster, digesting it. Oyster fishermen used to hack starfish to pieces in order to get rid of them. Starfish have unusual powers of regeneration, however, and can re-form a new organism from a piece containing only one arm and a small part of the disk.

Other members of this phylum are *sea urchins,* which have so many spines that they resemble pin cushions; *sea cucumbers,* which are elongated with finely branched arms around the mouth; and *sea lilies,* deep-sea forms that have a stalk with branched arms at one end.

Phylum 8 Mollusca

The mollusks are soft-bodied animals, many of which have a shell. They have a muscular foot, and a heavy fold of tissue, the mantle, which secretes the shell. There are three major types: the snail, having a coiled shell; the oyster and clam, with two shells or valves—hence the name bivalve; and the octopus and squid, with a thin, horny shell on the inside, and a foot containing tentacles. About 80,000 different species of mollusks have been described. Most of them live in salt water, but some are found in fresh water and on land. Clams, oysters, and some of the other mollusks make up our large supply of invertebrate food.

Snail   Oyster   Clam

Snails are generally scavengers scraping fragments of food as they glide along slowly on their foot. Some types are destructive to garden plants. *Clams, oysters,* and *mussels* feed by drawing in currents of water between the slightly open shells and straining the food particles. When disturbed, they close their shells tightly by means of two large muscles at either end. People who eat "scallops" are really eating such muscles. If an irritating substance such as a grain of sand becomes located between the mantle and the shell of an oyster, a secretion is formed around it to produce a pearl. Buttons are made from the shells of clams.

The *squid* and *octopus* have suckers on their tentacles for seizing their prey. When attacked, they contract their ink sac, giving off a cloud of inky material. They can dart backward with considerable speed by expelling a jet of water. They have well-developed eyes, which, like the human eye, are built on the same principle as a camera.

Phylum 9 Arthropoda

This is the largest phylum of the invertebrates. Its members have three or more pairs of jointed legs, an outer skeleton (exoskeleton) made of *chitin,* and segmented bodies. They are arranged in five classes, including crustaceans (e.g., crabs and lobsters), insects, spiders, centipedes, and millipedes.

***Class 1* Crustacea** Examples are the crab, lobster, crayfish, barnacle, and shrimp. The head and thorax are fused together and are distinct from the abdomen. These animals have two pairs of antennae or feelers. They breathe by means of gills. Their jointed legs may include large pincers for holding and crushing, as in the lobster. In addition to five pairs of jointed legs for walking, they also have additional appendages for swimming and eating. Because they have a rigid exoskeleton, they grow by molting, or shedding their skin.

Lobster

***Class 2* Insecta** There are more species of insects than of all the other animals on earth put together. They are considered humans' closest competitor because of their success in living practically everywhere; the damage they do to crops, clothing, and buildings; the diseases they transmit; and their enormous powers of reproduction. On the other hand, they are helpful in pollinating plants.

Insects have three pairs of legs, one pair of antennae, and a body with three distinct parts—head, thorax, and abdomen. They have compound eyes with many lenses, as well as simple eyes. Some of them, such as butterflies and bees, have four wings; flies and mosquitoes have two wings; fleas have no wings. Some (butterflies and mosquitoes) have four stages in their life history, or complete metamorphosis, including egg, larva, pupa, and adult. Others (grasshoppers and plant lice) have three stages or incomplete metamorphosis, including egg, nymph, and adult. Some (bees, wasps, ants) live in social communities with a high degree of organization. Serious diseases are transmitted by the *Anopheles* mosquito (malaria); *Aedes* mosquito (yellow fever); body louse (typhus); flea (bubonic plague); and tsetse fly (African sleeping sickness).

***Class 3* Arachnida** Spiders have four pairs of legs and two main body parts—a fused head and thorax, and an abdomen. They have no antennae. Many of them have silk glands for spinning webs. Other members of this class are ticks, scorpions, and horseshoe crabs.

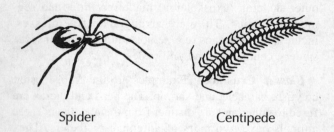

Spider   Centipede

***Class 4* Chilopoda** The centipedes ("hundred leggers") have one pair of legs on each segment. They can move rather rapidly. One pair of appendages contains poison claws. Centipedes are carnivorous, and live on earthworms and insects.

***Class 5* Diplopoda** Millipedes ("thousand leggers") have two pairs of legs on each segment. Despite the large number of legs, they move slowly. They are plant-eating scavengers.

Millipede

Phylum 10 Chordata

Chordates are animals that have a notochord, which is a cartilagelike supporting structure along the back, in some stage of their life history. Above the notochord is a hollow nerve cord. Chordates also develop gill slits. There are three small subphyla, including animals known as the lancelet (amphioxus), sea squirt (tunicate), and acorn worm. The fourth subphylum is the large group of vertebrates, animals that develop a backbone to replace the notochord.

Subphylum Vertebrata

Vertebrates are animals with a backbone or spinal column composed of sections called vertebrae, and enclosing a spinal cord. The brain is contained in a bony skull, or *cranium*. The vertebrates are classified into five important classes:

***Class 1* Pisces** Fish are cold-blooded animals that live in either salt or fresh water. They breathe by means of gills; as water passes over the gills dissolved oxygen is taken in by the many tiny blood vessels located there and carbon dioxide is given off. They have a two-chambered heart, consisting of an auricle and a ventricle. Fish move by means of paired fins,

which are like the front and hind limbs of higher animals, and a tail. The streamlined shape of the body permits them to glide through the water with a minimum of resistance. Fish possess an air bladder, which helps them rise, descend, or remain stationary. The surface of the body of most fish is covered with scales that are made slippery by mucous glands in the skin. The nostrils are used for smell.

Most fish lay eggs that are fertilized outside the body. Codfish may produce several million eggs in 1 year. Migratory ocean fish such as the salmon travel long distances to breed in the fresh waters of rivers. By contrast, eels leave inland waters to breed in the ocean. Lungfish are unusual fish that bury themselves in the mud during the dry season and breathe with their air bladder. Bony fish are classified as the subclass Osteichthyes. Another group of fish, which have a skeleton made of cartilage, include sharks and rays. They are classified in the subclass Chondrichthyes.

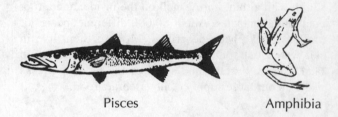

Pisces   Amphibia

***Class 2* Amphibia** The amphibians, such as frogs, toads, and salamanders, are cold-blooded and have two stages in their life history. They develop from eggs laid in the water and fertilized externally. During the *tadpole* stage, they breathe by means of gills. They later form legs and lungs, and live on land. This completes their *metamorphosis,* or life changes. The skin is thin and moist, permitting the absorption of some oxygen through it. The heart has three chambers: two auricles and one ventricle. The red corpuscles are oval cells with nuclei. Food, consisting of insects and worms, is caught by means of the sticky tongue, which is flipped out from its place of attachment at the front part of the mouth. Toads, with a rough skin, do not cause warts, as some people believe, and can remain in dry places for a longer period of time than frogs or salamanders.

***Class 3* Reptilia** Examples are the snake, lizard, turtle, and alligator. These animals are cold-blooded and breathe by means of lungs. Their bodies are generally covered with dry scales. Fertilization is internal, but development of the young from eggs generally takes place externally. The heart is three-chambered, becoming four-chambered among the alligators. Dinosaurs are extinct reptiles that dominated the earth millions of years ago. Although snakes have no legs, they can move fairly rapidly by wavelike contractions of muscles attached to the numerous ribs and scales. Most snakes are useful, judging by the many insects and mice they catch. The few harmful ones, such as the rattlesnake and the copperhead, have a pair of hollow fangs in their upper jaw through which poison is injected when they bite. Prompt first-aid treatment and antivenin are needed for snakebite.

Reptilia   Aves   Mammalia

Class 4 **Aves** Birds are best-known for their feathers and wings. They are warm-blooded, and have a four-chambered heart. They lay eggs that are covered with a shell, and that they incubate with their bodies until the young are born. Their beaks are horny and do not contain teeth. Parts of their skeleton are hollow, thus assisting in flight. Bird migration, which sometimes involves a flight of thousands of miles, is still an unexplained phenomenon.

The ostrich and the penguin are birds that cannot fly. Birds of prey like the hawk and owl have strong curved beaks. Ducks have a flattened beak for straining food from water. Woodpeckers have powerful beaks for drilling into trees for insects. Songbirds have slender beaks for catching insects. Birds constitute one of our strong defenses against the insect menace.

Class 5 **Mammalia** Mammals are the most highly developed animals. They are warm-blooded and have a four-chambered heart. They breathe by means of lungs, and have a diaphragm that divides the body into two distinct cavities: the thorax or chest, and the abdomen. Mammals are covered with hair. The young are born alive after developing in a special structure, the uterus, and are fed with milk secreted by mammary glands.

These characteristics apply to all mammals, including the human, dog, whale, bat, and elephant. The duckbill platypus and the anteater, however, are exceptions; they lay eggs that hatch outside the mother's body. The mother then nurses the young on her milk. The pouched mammals, such as the kangaroo and the opossum, give birth to premature young that they place in a pouch to complete their development and nourishment. Sea-living mammals such as the whale, porpoise, and seal have limbs, modified as flippers. The blue whale, the largest animal on earth, grows more than 100 feet long and weighs 150 tons. Other mammals differ in their tooth structure: grinding teeth—hoofed mammals such as cow, deer, giraffe; large canine teeth—bear, dog, lion; long front incisor teeth for gnawing—rodents such as rat, beaver, squirrel.

The primates, the mammals with the most highly developed brain, walk more or less erect. Their limbs have a hand with an opposable thumb for grasping. Examples are the human, chimpanzee, gorilla, and monkeys. The human, with the most highly developed brain of all, stands at the top of the Animal kingdom and is classified as *Homo sapiens*—man, the wise. All human beings today are members of this one genus and species.

Section Review

Select the correct choice to complete each of the following statements:

1. Simple animals that contain pores leading to a hollow central cavity are classified as (A) hydras (B) coelenterates (C) mollusks (D) sponges (E) echinoderms

2. Two warm-blooded groups of vertebrates are (A) mammals and sharks (B) birds and amphibians (C) mammals and birds (D) amphibians and reptiles (E) mammals and amphibians

3. The science dealing with the study of animals is known as (A) physiology (B) physiography (C) psychology (D) zoology (E) botany

4. An animal that is *not* an invertebrate is the (A) jellyfish (B) starfish (C) goldfish (D) sponge (E) plasmodium

5. Hydra and coral are classified as members of the (A) Coelenterata (B) Porifera (C) Planaria (D) Infusoria (E) Echinodermata

6. Three phyla that include worms are (A) Annelida, Nematoda, and Flagellata (B) Annelida, Sarcodina, and Platyhelminthes (C) Annelida, Nematoda, and Platyhelminthes (D) Annelida, Flagellata, and Sarcodina (E) Annelida, Flagellata, and Platyhelminthes

7. All of the following are parasitic worms *except* the (A) filaria (B) leech (C) liver fluke (D) hookworm (E) ringworm

8. Animals without backbones are classified as (A) invertebrates (B) vertebrates (C) protozoa (D) protists (E) rotifers

9. A phylum that is characterized by radial symmetry is (A) Chordata (B) Echinodermata (C) Pisces (D) Aves (E) Arthropoda

10. The animal that is *not* included in the same phylum as the others is the (A) starfish (B) sea urchin (C) sea lion (D) sea cucumber (E) sea lily

11. The octopus, clam, and snail are alike in (A) having a backbone (B) having jointed legs (C) being classified as mollusks (D) being classified as vertebrates (E) having a three-chambered heart

12. Animals having a chitinous exoskeleton and jointed legs are classified as (A) turtles (B) echinoderms (C) coelenterates (D) arthropods (E) chordates

13. Of the following, the closest relative of the lobster is the (A) grasshopper (B) squid (C) sea horse (D) oyster (E) eel

14. Spiders may be distinguished from insects because they have (A) two antennae (B) jointed legs (C) eight legs (D) a separate head, thorax, and abdomen (E) biting mouth parts

15. Animals that possess a notochord are classified as (A) social insects (B) crustacea (C) chilopoda (D) notobrates (E) chordates

16. All of the following are classes of vertebrates *except* (A) Pisces (B) Amphibia (C) Reptilia (D) Arthropoda (E) Mammalia

17. An animal with a four-chambered heart is the (A) lungfish (B) whale (C) codfish (D) eel (E) shark

18. An animal that has a tadpole stage in its life history is the (A) toad (B) turtle (C) tortoise (D) tiger moth (E) lizard

19. Although the duckbill platypus lays eggs, it is classified as a (A) bird (B) reptile (C) rodent (D) mammal (E) primate

20. Of the following, the closest relative of the bat is the (A) robin (B) seal (C) crow (D) penguin (E) ostrich

21. Two classes of vertebrates that always breathe by means of lungs are (A) birds and amphibians (B) birds and insects (C) mammals and amphibians (D) mammals and insects (E) birds and mammals

22. The characteristic that makes a kangaroo a mammal is (A) internal fertilization (B) external development of its young (C) presence of a backbone (D) production of milk (E) presence of lungs

23. All of the following are primates *except* the (A) chimpanzee (B) gorilla (C) giraffe (D) human (E) monkey

24. Animals that have a pupa stage in their development are (A) protozoa (B) insects (C) crustaceans (D) segmented worms (E) frogs

25. The most numerous group of animals is the (A) amphibians (B) protozoa (C) worms (D) birds (E) insects

Answer Key

1-D	6-C	11-C	16-D	21-E
2-C	7-E	12-D	17-B	22-D
3-D	8-A	13-A	18-A	23-C
4-C	9-B	14-C	19-D	24-B
5-A	10-C	15-E	20-B	25-E

Answers Explained

1. (D) Sponges are simple, many-celled animals that do not move, are attached to rocks or sticks, and contain pores leading to a hollow central cavity.

2. (C) Fish, amphibians, and reptiles are cold-blooded animals whose body temperature is the same as that of their environment. Warm-blooded animals (birds and mammals) have a constant temperature.

3. (D) The study of animals is known as zoology. The study of plants is called botany.

4. (C) The goldfish has a backbone and is classified as a vertebrate. Invertebrates do not have a backbone.

5. (A) Coelenterates have a central body cavity and consist of two layers of cells, an outer ectoderm and an inner endoderm.

6. (C) Annelida are segmented worms, such as the earthworm. Nematoda are roundworms, such as the hookworm. Platyhelminthes are flatworms, such as the tapeworm.

7. (E) Ringworm is a fungus disease in which there are circular infected patches on the skin.

8. (A) Invertebrates are animals without a backbone, such as sponges, worms, starfish, mollusks, and arthropods.

9. (B) The body plan of Echinodermata is different from that of animals in the other phyla in that it shows radial symmetry. There is a central disc from which five or more arms radiate.

10. (C) The sea lion is a mammal that lives in the sea and is like a seal. The other animals mentioned are members of the Echinodermata.

11. (C) Mollusks such as the snail, clam, and octopus are soft-bodied animals, many of which have shells. They have a muscular foot and a heavy fold of tissue, the mantle, which secretes the shell.

12. (D) Arthropoda have three or more pairs of jointed legs, segmented bodies, and an exoskeleton made of chitin. Examples: crustaceans, insects, spiders, centipedes, and millipedes.

13. (A) The grasshopper, like the lobster, has a segmented body, jointed legs, and a chitinous exoskeleton. Both are arthropods.

14. (C) Spiders are distinguished from insects by having eight legs instead of six, and two main body parts (a fused head and thorax, and an abdomen) instead of three (head, thorax, abdomen).

15. (E) Chordates are animals that have a notochord, which is a cartilagelike supporting structure along the back, in some stage of their life history.

16. (D) The Arthropoda phylum consists of invertebrates that have jointed legs, segmented bodies, and a chitinous exoskeleton. Examples: crustaceans, insects, spiders, centipedes, and millipedes.

17. (B) The whale is a mammal; it has a four-chambered heart and is warm-blooded.

18. (A) Toads are amphibians. Like frogs and salamanders, they develop from fertilized eggs into tadpoles that live in the water. Then, by the process of metamorphosis, they develop lungs and legs and begin to live on land.

19. (D) The platypus is a primitive mammal that lays eggs which hatch outside the mother's body; it then nurses the young.

20. (B) The bat and the seal are mammals. They give birth to live young and feed them on milk when they are born. They are also warm-blooded and have a four-chambered heart.

21. (E) Birds and mammals are warm-blooded animals, have a four-chambered heart, and breathe through lungs.

22. (D) Besides the production of milk, the kangaroo has the following characteristics of mammals: it is warm-blooded, has a four-chambered heart, and is covered with hair.

23. (C) Primates are mammals with the most highly developed brain; they walk more or less erect, and have an opposable thumb for grasping.

24. (B) Insects go through a life history that includes metamorphosis from the egg to the adult. Some have complete metamorphosis, including egg, larva, pupa, and adult stages.

25. (E) There are more species of insects than of all the other animals put together.

13.6 THE PLANT KINGDOM

Botany is the study of plants. About a third of a million species have been named and classified.

Phylum 1 Bryophyta

Bryophytes are small green land plants that include the mosses and liverworts. They have simple stems with leaflike structures, without flowers or true roots. Their life history includes an alternation of generations, with both a spore stage (*sporophyte*) and a sex-cell stage (*gametophyte*).

Class 1 Hepaticae The *liverworts,* such as *Marchantia,* have a small, leaflike structure that lies flat on the ground, with simple rootlike projections. There are separate structures for producing sperm and egg cells. After fertilization, spores are produced, each of which may give rise to a new plant. Liverworts are found in moist, shaded places. About 4,000 species have been identified.

Class 2 Musci The *mosses* are somewhat better known than the liverworts, and a little more advanced in structure. They have a more erect appearance, with many leaflike structures on a simple stem and with larger, rootlike structures. They have a separate spore case at the end of a stalk.

Sphagnum is a moss that accumulates in swamps or bogs to form *peat*. Peat moss is used by gardeners to improve the soil. Peat eventually becomes partly carbonized and can be used as fuel. About 15,000 species of moss have been classified.

Phylum 2 Tracheophyta

The tracheophytes are the large group of vascular plants, that is, plants with well-defined conducting systems. The sporophyte stage is prominent, with a tiny gametophyte stage. The plants have well-developed roots, stems, and leaves.

Subphylum 1 Lycopsida

This subphylum includes the club-moss and ground pine. In former geologic eras, these plants were as large as trees. Now they remain as low plants that grow close to the ground in the woods. They produce spores in such large numbers that at times they seem to be covered with a yellow dust. Lycopod powder consists of these spores. About 600 club mosses have been identified.

Subphylum 2 Sphenosida

The *horsetails* are relatives of the ferns that grow in sandy places and along railroad tracks, and contain large amounts of minerals. For this reason, they were used to scour pots before the days of scouring powders. Instead of true leaves, horsetails have whorls of scalelike leaves along their stem. Spores are produced in a conelike structure at the top of the stem.

Subphylum 3 Pteropsida

This group includes broad-leaved tracheophytes such as ferns, gymnosperms, and angiosperms.

Class 1 **Filicineae** This class contains the true *ferns*, such as the royal fern, Boston fern, and Christmas fern. They generally have an underground woody stem that sends up leaves. The undersurface of the leaves may be seen to contain tiny brown dots; these are the *sori*, the spore cases that produce spores. Other ferns produce spores on separate stems. Most ferns grow in wooded areas. About 4,400 ferns have been identified.

The sporophyte stage is well developed, while the gametophyte is reduced in size to less than $1/2$ inch. About 250 million years ago ferns were very prominent and grew to the size of trees. Over the ages, compres-

sion and decay of masses of these ancient ferns formed our present coal deposits.

Classes 2 and 3 **Seed Plants** Seed-producing plants are the most advanced, numbering about 175,000 species. *Annuals* (wheat, bean, marigold) live for one year. *Biennials* (carrot, foxglove) complete their life cycle in two years. *Perennials* (phlox, aster, dandelion) live on year after year. Some are dormant in the winter and then reappear in the spring. Seed plants are classified into two classes, based on their manner of forming seeds.

Class 2 **Gymnospermae** Pines, spruce, hemlock, and other gymnosperms form "naked seeds" that are usually produced on the scales of a cone, and drop out when ripe. Many gymnosperms have leaves that are modified as needles and are evergreen. The giant sequoia redwood trees, which are over 2,000 years old, are also members of this class.

Class 3 **Angiospermae** The angiosperms include flowering plants (rose, violet), trees with broad leaves (oak, maple), and plants used as crops (corn, bean). They all have covered seeds that develop in the part of a flower called the *ovary*. Most of them are *deciduous*, losing their leaves in the fall. Botanists distinguish between two subclasses of angiosperms:

Subclass 1 **Dicotyledons** The dicots have seeds that contain, in addition to the tiny embryo plant, two parts or cotyledons, such as occur in the bean or peanut. Their leaves have veins arranged in a network. They include many families of flowering plants, such as the legumes (peas, beans, clover), composites (daisy, dandelion, aster), rose (roses, strawberry, apple) and mustard (cabbage, cauliflower). The flower parts are arranged in groups of four or five. The vascular bundles in the stem are arranged in a circular ring.

| Corn | Lily | Bean | Maple |
| (one cotyledon) | (parallel veining) | (two cotyledons) | (net veining) |

Monocots   Dicots

Subclass 2 **Monocotyledons** The monocots have seeds with one cotyledon. Their leaves have parallel veins. Examples: the grains (corn, wheat, rice and barley), grasses, onion, pineapple, orchid. The flower parts are arranged in groups of three. The vascular bundles are scattered throughout the stem.

Section Review

Select the correct choice to complete each of the following statements.

1. The branch of biology that deals with the study of plants is   (A) paleontology   (B) genetics   (C) zoology   (D) botany   (E) endocrinology

2. All of the following plants have a life history that includes an alternation of generations *except*   (A) the liverwort   (B) sphagnum   (C) moss   (D) the horsetail   (E) spirogyra

3. Plants possessing leaves, stems, and roots, but no seeds, are classified as   (A) Musci   (B) Hepaticae   (C) Filicinae   (D) Bryophta   (E) angiosperms

4. The most highly developed plants are classified as   (A) saprophytes   (B) angiosperms   (C) bryophytes   (D) vertebrates   (E) lycopods

5. An example of a biennial plant is   (A) wheat   (B) bean   (C) cotton   (D) marigold   (E) carrot

6. An example of a gymnosperm is the   (A) jimson weed   (B) pine   (C) oak   (D) maple   (E) daisy

7. All of the following are deciduous plants *except* the   (A) spruce   (B) oak   (C) aster   (D) willow   (E) poplar

8. The corn plant   (A) is a monocot   (B) has leaves with veins arranged in a network   (C) is a gymnosperm   (D) has an embryo with several cotyledons   (E) is a dicot

9. Plants that have covered seeds are classified as   (A) gymnosperms   (B) angiosperms   (C) horsetails   (D) ground pines   (E) *Marchantia*

10. The closest relatives of mosses are   (A) smuts   (B) legumes   (C) algae   (D) liverworts   (E) molds

Answer Key

1-D	3-C	5-E	7-A	9-B
2-E	4-B	6-B	8-A	10-D

Answers Explained

1. (D) Botany is the study of plants. Zoology is the study of animals.

2. (E) Spirogyra is an alga that reproduces by either fission or conjugation.

3. (C) The class Filicineae consists of the true ferns, such as the royal fern. They usually contain spores on the underside of their leaves.

4. (B) Angiosperms have covered seeds that develop in a part of the flower called the ovary. Example: maple.

5. (E) A biennial plant like a carrot completes its life cycle in 2 years.

6. (B) A gymnosperm forms naked seeds that are usually produced on the scales of a cone and drop out when ripe.

7. (A) Deciduous plants lose their leaves in the fall; the spruce has needles and is an evergreen.

8. (A) A monocot such as corn has seeds with only one cotyledon. Its leaves have parallel veins.

9. (B) Angiosperms have seeds that develop within the ovary of a flower. Example: violet.

10. (D) Both mosses and liverworts are classified in the phylum Bryophyta. They have simple stems with leaflike structures, without true roots or flowers. Their life history includes an alternation of generation.

PRACTICE TESTS

PART
IV

Answer Sheet: Practice Test 1

1. Ⓐ Ⓑ Ⓒ Ⓓ Ⓔ
2. Ⓐ Ⓑ Ⓒ Ⓓ Ⓔ
3. Ⓐ Ⓑ Ⓒ Ⓓ Ⓔ
4. Ⓐ Ⓑ Ⓒ Ⓓ Ⓔ
5. Ⓐ Ⓑ Ⓒ Ⓓ Ⓔ
6. Ⓐ Ⓑ Ⓒ Ⓓ Ⓔ
7. Ⓐ Ⓑ Ⓒ Ⓓ Ⓔ
8. Ⓐ Ⓑ Ⓒ Ⓓ Ⓔ
9. Ⓐ Ⓑ Ⓒ Ⓓ Ⓔ
10. Ⓐ Ⓑ Ⓒ Ⓓ Ⓔ
11. Ⓐ Ⓑ Ⓒ Ⓓ Ⓔ
12. Ⓐ Ⓑ Ⓒ Ⓓ Ⓔ
13. Ⓐ Ⓑ Ⓒ Ⓓ Ⓔ
14. Ⓐ Ⓑ Ⓒ Ⓓ Ⓔ
15. Ⓐ Ⓑ Ⓒ Ⓓ Ⓔ
16. Ⓐ Ⓑ Ⓒ Ⓓ Ⓔ
17. Ⓐ Ⓑ Ⓒ Ⓓ Ⓔ
18. Ⓐ Ⓑ Ⓒ Ⓓ Ⓔ
19. Ⓐ Ⓑ Ⓒ Ⓓ Ⓔ
20. Ⓐ Ⓑ Ⓒ Ⓓ Ⓔ
21. Ⓐ Ⓑ Ⓒ Ⓓ Ⓔ
22. Ⓐ Ⓑ Ⓒ Ⓓ Ⓔ
23. Ⓐ Ⓑ Ⓒ Ⓓ Ⓔ
24. Ⓐ Ⓑ Ⓒ Ⓓ Ⓔ
25. Ⓐ Ⓑ Ⓒ Ⓓ Ⓔ
26. Ⓐ Ⓑ Ⓒ Ⓓ Ⓔ
27. Ⓐ Ⓑ Ⓒ Ⓓ Ⓔ

28. Ⓐ Ⓑ Ⓒ Ⓓ Ⓔ
29. Ⓐ Ⓑ Ⓒ Ⓓ Ⓔ
30. Ⓐ Ⓑ Ⓒ Ⓓ Ⓔ
31. Ⓐ Ⓑ Ⓒ Ⓓ Ⓔ
32. Ⓐ Ⓑ Ⓒ Ⓓ Ⓔ
33. Ⓐ Ⓑ Ⓒ Ⓓ Ⓔ
34. Ⓐ Ⓑ Ⓒ Ⓓ Ⓔ
35. Ⓐ Ⓑ Ⓒ Ⓓ Ⓔ
36. Ⓐ Ⓑ Ⓒ Ⓓ Ⓔ
37. Ⓐ Ⓑ Ⓒ Ⓓ Ⓔ
38. Ⓐ Ⓑ Ⓒ Ⓓ Ⓔ
39. Ⓐ Ⓑ Ⓒ Ⓓ Ⓔ
40. Ⓐ Ⓑ Ⓒ Ⓓ Ⓔ
41. Ⓐ Ⓑ Ⓒ Ⓓ Ⓔ
42. Ⓐ Ⓑ Ⓒ Ⓓ Ⓔ
43. Ⓐ Ⓑ Ⓒ Ⓓ Ⓔ
44. Ⓐ Ⓑ Ⓒ Ⓓ Ⓔ
45. Ⓐ Ⓑ Ⓒ Ⓓ Ⓔ
46. Ⓐ Ⓑ Ⓒ Ⓓ Ⓔ
47. Ⓐ Ⓑ Ⓒ Ⓓ Ⓔ
48. Ⓐ Ⓑ Ⓒ Ⓓ Ⓔ
49. Ⓐ Ⓑ Ⓒ Ⓓ Ⓔ
50. Ⓐ Ⓑ Ⓒ Ⓓ Ⓔ
51. Ⓐ Ⓑ Ⓒ Ⓓ Ⓔ
52. Ⓐ Ⓑ Ⓒ Ⓓ Ⓔ
53. Ⓐ Ⓑ Ⓒ Ⓓ Ⓔ
54. Ⓐ Ⓑ Ⓒ Ⓓ Ⓔ

55. Ⓐ Ⓑ Ⓒ Ⓓ Ⓔ
56. Ⓐ Ⓑ Ⓒ Ⓓ Ⓔ
57. Ⓐ Ⓑ Ⓒ Ⓓ Ⓔ
58. Ⓐ Ⓑ Ⓒ Ⓓ Ⓔ
59. Ⓐ Ⓑ Ⓒ Ⓓ Ⓔ
60. Ⓐ Ⓑ Ⓒ Ⓓ Ⓔ
61. Ⓐ Ⓑ Ⓒ Ⓓ Ⓔ
62. Ⓐ Ⓑ Ⓒ Ⓓ Ⓔ
63. Ⓐ Ⓑ Ⓒ Ⓓ Ⓔ
64. Ⓐ Ⓑ Ⓒ Ⓓ Ⓔ
65. Ⓐ Ⓑ Ⓒ Ⓓ Ⓔ
66. Ⓐ Ⓑ Ⓒ Ⓓ Ⓔ
67. Ⓐ Ⓑ Ⓒ Ⓓ Ⓔ
68. Ⓐ Ⓑ Ⓒ Ⓓ Ⓔ
69. Ⓐ Ⓑ Ⓒ Ⓓ Ⓔ
70. Ⓐ Ⓑ Ⓒ Ⓓ Ⓔ
71. Ⓐ Ⓑ Ⓒ Ⓓ Ⓔ
72. Ⓐ Ⓑ Ⓒ Ⓓ Ⓔ
73. Ⓐ Ⓑ Ⓒ Ⓓ Ⓔ
74. Ⓐ Ⓑ Ⓒ Ⓓ Ⓔ
75. Ⓐ Ⓑ Ⓒ Ⓓ Ⓔ
76. Ⓐ Ⓑ Ⓒ Ⓓ Ⓔ
77. Ⓐ Ⓑ Ⓒ Ⓓ Ⓔ
78. Ⓐ Ⓑ Ⓒ Ⓓ Ⓔ
79. Ⓐ Ⓑ Ⓒ Ⓓ Ⓔ
80. Ⓐ Ⓑ Ⓒ Ⓓ Ⓔ

PRACTICE TEST 1

CHAPTER

14

NOTE: The College Board test offered in Biology E/M contains a common core of 60 questions numbered 1–60, covering the general field of biology, and a specialized section of 20 questions numbered 61–80 for Biology E (Ecology) or questions 81–100 for Biology M (Molecular). On the day of the examination, you may take ONE of the Biology E/M tests, either E or M, by marking the appropriate grid. Practice Test 1 of this book is designed to prepare you for the Biology E Test. All these Practice Tests have been created to present different core questions 1–60 in order to give you more experience in test preparation.

Practice Test 1: Biology E

Part A (Core Questions 1–60)

Directions: Each of the questions or incomplete statements below is followed by five suggested answers or completions. Choose the one that is best in each case and then blacken the corresponding space on the answer sheet.

1. Over time, the balance of nature has been upset most seriously by

   (A) the effects of lightning in forests
   (B) the spread of the English sparrow
   (C) human activities
   (D) spring floods
   (E) the spread of the gypsy moth

2. When two glucose ($C_6H_{12}O_6$) molecules are combined to form a molecule of maltose ($C_{12}H_{22}O_{11}$), the formula of the maltose molecule is not $C_{12}H_{24}O_{12}$ because

   (A) hydrolysis takes place
   (B) dehydration synthesis takes place
   (C) transpiration takes place

   (D) polypeptides are formed
   (E) water is added

3. Of the following terms, the one that includes all the others is

   (A) oxidation
   (B) respiration
   (C) excretion
   (D) metabolism
   (E) digestion

4. Which of the following groups of ecological terms is in correct order, from simplest to most complex?

   (A) Organism, population, community, ecosystem, biosphere
   (B) Organism, population, community, biosphere, ecosystem
   (C) Population, organism, community, ecosystem, biosphere
   (D) Biosphere, ecosystem, organism, community, population
   (E) Ecosystem, biosphere, organism, community, population

Questions 5–7 refer to the following diagram of the human digestive system.

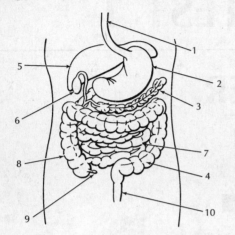

5. Glycogen is stored by

(A) 4
(B) 8
(C) 3
(D) 5
(E) 2

6. Which of the following produces hydrochloric acid?

(A) 5
(B) 7
(C) 2
(D) 4
(E) 8

7. An enzyme that starts protein digestion is produced by

(A) 5
(B) 7
(C) 3
(D) 8
(E) 2

8. Viruses do not fit into our concept of the cell because they

(A) lack cytoplasm
(B) cannot multiply
(C) contain DNA
(D) are extremely small
(E) contain enzymes in their chlorophyll

9. Of the following organic compounds, the one that represents a protein is

(A) $C_{12}H_{22}O_{11}$
(B) $C_6H_{12}O_6$

(C) $C_{17}H_{35}COOH$
(D) $(C_6H_{10}O_5)_n$
(E) $C_{708}H_{1130}O_{224}N_{180}S_4$

10. Near the equator, at 5°S latitude, the top of a 2,200-meter mountain would have a biome that most closely resembled a

(A) tropical rain forest
(B) desert
(C) tundra
(D) coniferous forest
(E) taiga

11. Rivers become muddy because of

(A) salmon migration upstream
(B) the use of check-dams to reduce erosion
(C) the building of dams by beavers
(D) strip-cropping, which loosens the soil
(E) erosion resulting from forest destruction

12. That evolution has occurred is now generally accepted as a doctrine rather than a theory because

(A) Congress passed a law to that effect in 1958
(B) Darwin received a posthumous Nobel Prize for his book on natural selection
(C) it is now certain that the earth is at least four billion years old
(D) there is ample evidence that species have changed over the ages
(E) uranium-lead studies have traced the history of the horse

13. Ciliated epithelial cells in the nasal passages are useful because they

(A) keep out dust and bacteria
(B) provide the sense of smell
(C) reduce the breathing rate when the air is impure
(D) reduce the humidity of inhaled air
(E) fan the air and cool it, especially in summer

14. Carbohydrates are organic compounds containing carbon, hydrogen, and oxygen, in which the hydrogen and oxygen occur in the same ratio as in water. Of the following compounds, the carbohydrate is

(A) stearin, $C_{57}H_{110}O_6$
(B) thiamin, $C_{12}H_{18}N_4O_2S$
(C) palmatin, $C_{51}H_{98}O_6$
(D) cellulose, $(C_6H_{10}O_5)_n$
(E) riboflavin, $C_{17}H_{20}N_4O_6$

15. Although the duckbill platypus lays eggs, it is classified as a

    (A) bird
    (B) reptile
    (C) rodent
    (D) mammal
    (E) primate

16. An independent organism is discovered that does not contain a nucleus. In all likelihood, it would be classified in the kingdom

    (A) Monera
    (B) Protista
    (C) Fungi
    (D) Animal
    (E) Plant

17. Why do legume plants enrich the soil?

    (A) They remove selenium, a deadly poison.
    (B) They encourage the breeding of earthworms.
    (C) Their roots contain nitrogen-fixing bacteria.
    (D) Their root systems penetrate 2 meters below the surface.
    (E) The underground stems of potatoes are considered to be tubers.

18. Messenger RNA is important in protein synthesis because it

    (A) contains the 20 essential amino acids
    (B) carries the code from DNA to the nucleus
    (C) carries the code from DNA to the ribosomes
    (D) is transmitted to the nucleotides
    (E) contains the pyrimidine base thymine

19. In which of the following life processes is ATP (adenosine triphosphate) produced?

    I. Photosynthesis
    II. Aerobic respiration
    III. Anaerobic respiration

    (A) I only
    (B) II only
    (C) I and II only
    (D) II and III only
    (E) I, II, and III

20. The ATP → ADP relationship may be summed up as

    (A) energy-rich → charged
    (B) energy-poor → discharged
    (C) energy rich → discharged

    (D) energy-poor → charged
    (E) discharged → charged

21. A man who is normal for color vision marries a normal heterozygous woman. What is the chance of their son being color blind?

    (A) 0%
    (B) 25%
    (C) 50%
    (D) 75%
    (E) 100%

22. If John's father has type A blood and his mother has type O blood, John's blood type will most likely be

    (A) A or B
    (B) A or O
    (C) A or AB
    (D) B or O
    (E) AB or O

23. In some of Mendel's experiments, three quarters of the offspring showed the dominant trait. Which of the following is most likely to be true about the parents?

    (A) Both were recessive.
    (B) Both were heterozygous.
    (C) Both were dominant.
    (D) One was heterozygous; the other, homozygous dominant.
    (E) One was recessive; the other, homozygous dominant.

24. In ecological succession, since lichens grow on bare rock, they are considered to be

    (A) primary consumers
    (B) pioneer organisms
    (C) climax organisms
    (D) producers
    (E) decomposers

25. A forest fire may be a cause of flooding because

    (A) the roots of trees absorb excess water
    (B) the roots of the dead trees no longer bind the topsoil, which would absorb water
    (C) the leaves are not able to carry on transpiration fast enough
    (D) the xylem loses its power of capillarity
    (E) phloem cells are not able to store excess water

26. At times, hyenas feed on the remains of animals killed by other animals. At other times, hyenas themselves kill animals for food. Therefore, hyenas may best be described as

(A) scavengers and herbivores
(B) scavengers and parasites
(C) scavengers and predators
(D) herbivores and predators
(E) herbivores and parasites

27. Which of the following statements is most true of producer organisms?

(A) They are eaten by carnivores.
(B) They are eaten by scavengers.
(C) They are parasitic.
(D) They contain chlorophyll.
(E) None of these.

Questions 28–31 refer to the following food chain:

Maple tree → aphid → ladybird beetle →
frog → garter snake → hawk.

28. Which organism in this food chain can transform light energy to chemical energy?

(A) Frog
(B) Ladybird beetle
(C) Aphid
(D) Garter snake
(E) Maple tree

29. At which stage in the food chain will the smallest number of organisms occur?

(A) Frog
(B) Aphid
(C) Hawk
(D) Garter snake
(E) Ladybird beetle

30. Which organism in the food chain is herbivorous?

(A) Maple tree
(B) Frog
(C) Ladybird beetle
(D) Aphid
(E) Garter snake

31. Which organism in the food chain is considered to be a primary consumer?

(A) Maple tree
(B) Ladybird beetle
(C) Frog

(D) Garter snake
(E) Aphid

32. In which of the following organic compounds is a COOH (carboxyl) group found?

 I. Carbohydrate
 II. Lipid
III. Protein

(A) I only
(B) II only
(C) I and III only
(D) II and III only
(E) I, II, and III

33. The cytoplasm of a paramecium has a higher concentration of mineral salts than its external water environment. This higher concentration is due to the action of

(A) DNA
(B) messenger RNA
(C) ATP
(D) chlorosis
(E) meiosis

34. Chloroplasts are found in which of the following kingdoms?

 I. Monera
 II. Protista
III. Plant

(A) I only
(B) II only
(C) III only
(D) I and III only
(E) II and III only

35. After watching the behavior of earthworms in the ground, a biologist suggested that the penetration of air into the soil promotes root development of plants. He then set up the following experiment:

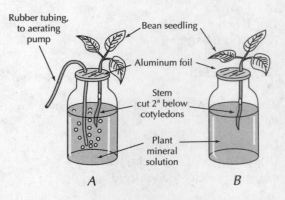

Rubber tubing, to aerating pump

Bean seedling

Aluminum foil

Stem cut 2" below cotyledons

Plant mineral solution

A   *B*

The important data to be recorded in the experiment will come from observation of the increase in

(A) leaf size
(B) stem size
(C) number of leaves
(D) number of roots
(E) number of buds

36. A green plant cell contains all of the following structures EXCEPT

(A) DNA
(B) genes
(C) a cell wall
(D) a centriole
(E) a nucleus

37. If the magnification of a microscope is 440 and the high-power objective is marked 44X, the ocular is marked

(A) 1X
(B) 5X
(C) 10X
(D) 100X
(E) 1000X

38. The presence of cyclic AMP in a cell

(A) helps produce the action of a particular hormone
(B) causes the cell to die
(C) activates the microtubules
(D) prevents the endoplasmic reticulum from synthesizing protein
(E) causes the cytoplasm to gel

39. The innermost chamber of the respiratory system into which air can be drawn is the

(A) bronchiole
(B) bronchus
(C) air sac
(D) bronchial tube
(E) sinus

40. Which of the following animals develops a placenta?

(A) A goldfish
(B) A tropical fish
(C) A lizard
(D) A salamander
(E) A cow

41. In aerobic respiration, the final hydrogen acceptor is

(A) chlorophyll
(B) carbon dioxide
(C) water
(D) ATP
(E) molecular oxygen

42. The splints in a horse's foot are the remains of

(A) toes
(B) teeth
(C) flippers
(D) a coccyx
(E) gill slits

43. Antibodies are chemicals that are

(A) nonspecific
(B) produced by the body in response to antigen
(C) synthesized from carbohydrates
(D) synthesized from glycogen
(E) transported by red blood cells

44. Hormones are distributed through the body by

(A) blood plasma
(B) ducts
(C) lacteals
(D) endocrines
(E) enzymes

45. Why are insects and spiders classified as arthropods?

(A) They are carnivorous.
(B) They have jointed legs.
(C) They have a backbone in the embryonic stages.
(D) They are land-dwellers.
(E) Their eyes are simple.

46. The chloroplasts in an *Elodea* green-plant cell move because they

(A) are equipped with cilia
(B) have flagella
(C) are carried around by the streaming protoplasm
(D) form pseudopodia
(E) carry on plasmolysis

47. In the gene pool of a given population of rabbits, 80 percent of all the gametes carry the dominant allele for gray coat and 20 percent carry the recessive allele for white coat. From this information, it can be predicted that the percentage of rabbits which will be homozygous for gray coat will be

   (A) 20%
   (B) 40%
   (C) 64%
   (D) 80%
   (E) 96%

Directions: Each set of lettered choices below refers to the numbered statements immediately following it. Choose the one lettered choice that best fits each statement and then blacken the corresponding space on the answer sheet. A choice may be used once, more than once, or not at all in each set.

Questions 48–50

   (A) Natural selection
   (B) Passive immunity
   (C) Heterotroph hypothesis
   (D) Vestigial structure
   (E) Use and disuse

48. Strains of mosquitoes have appeared that are not affected by certain insecticides.

49. A steelworker develops large shoulder muscles.

50. Amino acids are formed from molecules acted on by various forms of energy.

Questions 51–53

   (A) Habit
   (B) Instinct
   (C) Tropism
   (D) Reflex
   (E) Conditioned response

51. How a person's eye reacts when a cinder blows in it

52. How a secretary takes shorthand notes

53. How a horse responds to the command "Whoa"

Questions 54–56

   (A) Blending
   (B) Test cross
   (C) Dominance
   (D) Independent assortment
   (E) Linkage

54. Some people have blond hair and blue eyes, while others have brown hair and blue eyes.

55. When tall yellow-seeded dihybrid pea plants are cross-pollinated, short yellow-seeded offspring appear.

56. Geneticists find that *Drosophila* flies with black bodies always seem to have short wings.

Questions 57–58

   (A) Respiration
   (B) Photosynthesis
   (C) Protein synthesis
   (D) Assimilation
   (E) Diffusion

57. How oxygen is used by a green-plant cell

58. How carbon dioxide is used by a green-plant cell

Questions 59–60

   A student studied a drop of pond water with the low power of a compound microscope and made the following exact drawing of an organism she observed:

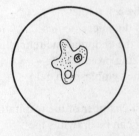

59. The diameter of the microscope field of vision is 1.8 millimeters. This is equivalent to

   (A) 1.8 micrometers
   (B) 18 micrometers
   (C) 180 micrometers
   (D) 1,800 micrometers
   (E) 18,000 micrometers

60. What is the approximate length of the organism?

   (A) 6 micrometers
   (B) 60 micrometers
   (C) 600 micrometers
   (D) 1.8 micrometers
   (E) 180 micrometers

Part B (Choice E Questions 61–80)

Directions: Each of the following group of questions concerns a laboratory or experimental situation. In each case, first study the description of the situation. Then select the one best answer to each question following it and blacken the corresponding space on the answer sheet.

Questions 61–62

A group of 24 frogs was separated into two equal groups. Group *A* was placed in an environment in which the temperature was a constant 2°C. Group *B* was placed in a similar environment, except that the temperature was kept at a constant 18°C. Both groups were given equal amounts of food at the start of the experiment and every 24 hours thereafter. Immediately before each daily feeding, the excess food from the previous feeding was removed and measured. Thus, it was possible to determine the daily consumption of food by each group. Each day, the frogs in each group were checked for heartbeat and breathing rate. At the end of the experiment, the accompanying bar graphs were prepared.

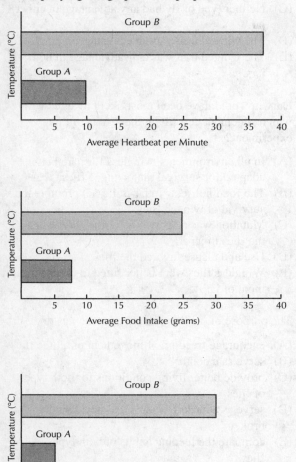

61. What was the primary question that the scientist was probably studying in this experiment?

(A) How do living things adapt to a change in temperature?
(B) In what ways do frogs adapt to a change in temperature?
(C) How much oxygen does a frog need when the temperature changes?
(D) How much blood is circulated through the heart of a frog when the temperature changes?
(E) How much food does a frog need?

62. Under natural conditions in a pond environment, the frogs in Group *A* would most likely react to the lower temperature by

(A) reproducing
(B) hibernating
(C) hopping around more actively to keep warm
(D) seeking more insect prey
(E) increasing their kick reflex

Questions 63–65

Pea seedlings were grown for 7 days after they germinated from seeds. Seedling #1 was germinated and grown in complete darkness for the entire 7 days. Seedling #2 was germinated and grown in continuous white light for the entire 7 days. Seedling #3 was grown and germinated in continuous dim, red light for the entire 7 days. After 7 days, the seelings were examined. They appeared as shown in the accompanying drawings.

#1: Darkness   #2: Continuous White Light   #3: Continuous Dim, Red Light

63. The pea seedling grown in darkness

(A) was very short
(B) was dark green
(C) was white
(D) carried on photosynthesis
(E) had green chloroplasts

64. The pea seedling grown in continuous white light

(A) was shorter than the seedling grown in darkness
(B) was longer than the seedling grown in darkness
(C) had fewer leaflets than the seedling grown in darkness
(D) had leaves with less stored food starch than the seedling grown in darkness
(E) developed without chloroplasts

65. The seedling grown in continuous, dim red light

(A) was shorter than the seedling grown in continuous darkness
(B) was shorter than the seedling grown in continuous white light
(C) had more leaflets than the seedling grown in continuous white light
(D) had more leaflets than the seedling grown in continuous darkness
(E) was stunted in appearance

Questions 66–70

Fruit-fly food nutrient was placed at the bottom of two jars, *A* and *B*, that were plugged with absorbent cotton. A strip of sticking flypaper was suspended in *A*. Equal numbers of fruit flies with wings and wingless fruit flies were introduced into both containers. After 7 days, only the wingless flies were alive in *A*, while both types of flies were still alive in *B*.

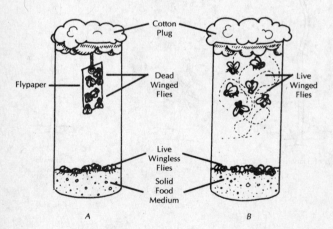

66. All of the following statements about this experiment are true EXCEPT

(A) the wingless flies readily survived in jar *A*
(B) the wingless flies readily survived in jar *B*
(C) since winged flies can both fly and walk, they were able to survive in both jars
(D) the winged flies became stuck on the flypaper
(E) the wingless flies did not become stuck on the flypaper

67. At the end of the experiment, when the flies were counted in both jars, it was found that there were

(A) equal numbers of live flies in both jars
(B) equal numbers of dead flies in both jars
(C) more live flies in jar *B*
(D) more dead flies in jar *B*
(E) more live flies in jar *A*

68. What is the most likely conclusion to be drawn from this experiment?

(A) Winglessness was an advantage in jar *A*.
(B) Winglessness was an advantage in jar *B*.
(C) Neither type of fly had any advantage in either jar.
(D) The winged trait was an advantage in jar *B*.
(E) The winged trait was a disadvantage in both jars.

69. Darwin would have been most likely to agree with which statement describing the results of this experiment?

(A) In all environments, wingless flies are better adapted for survival than winged flies.
(B) The food nutrients in jar *B* did not promote the survival of winged flies.
(C) Mutation was responsible for the survival of all the flies in jar *B*.
(D) Use and disuse favored the flies in jar *B*.
(E) Wingless flies were better fitted to the environment of jar *A*.

70. The purpose of jar *B* in this experiment was to

(A) encourage freedom of movement of all the flies
(B) serve as a control
(C) provide better living conditions for both types of flies
(D) serve as a backup if jar *A* was accidentally broken
(E) compare the feeding habits of both types of flies

Questions 71–73

The 24-hour clocks shown below illustrate the effects of different periods of light and dark on a chrysanthemum and a black-eyed Susan. In I, the long night produces flowering in chrysanthemum only. In II, the short night produces flowering only in black-eyed Susan.

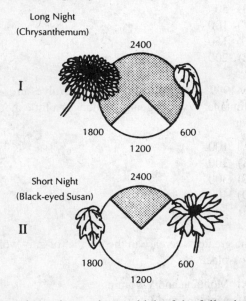

71. From these observations, which of the following is a reasonable conclusion?

(A) Chrysanthemums bloom in the summer, when nights are short.
(B) Chrysanthemums bloom in the fall, when nights are long.
(C) Chrysanthemums and black-eyed Susans bloom at the same time.
(D) Black-eyed Susans bloom when the nights are long.
(E) It is not possible to predict under what conditions black-eyed Susans will bloom.

72. To encourage chrysanthemum flowering in July, when the days have more than 12 hours of sunlight, an experimenter would

(A) place the plant in a dark room after 11 hours of daylight each afternoon, and then move it outdoors again each morning
(B) place the plant in a dark room at 11 P.M. each night, and then return it outdoors again early each morning
(C) place a transparent plastic cover over the plant every night
(D) place a transparent cover over the plant every morning
(E) cover the plant with an opaque box every night

73. An experimenter desires to get a black-eyed Susan to produce flowers in the late fall, when the nights are long. Which course of action should he take?

(A) Grow the plant in a hydroponic solution.
(B) Apply a fertilizer that encourages flowering.
(C) Cover the plant with a dark cloth every morning.
(D) Expose the plant to artificial illumination for 6 hours daily, beginning at sunrise.
(E) Expose the plant to artificial illumination for 6 hours daily, beginning at sunset.

Questions 74–76

The zebra mussel, a type of small striped mollusk that originated in Europe and Asia, has found its way to the Hudson River where it has had an impact on the ecology of the river. One problem is that the mussels use up the dissolved oxygen in the water, limiting the ability of desirable fish and other species to survive. Graph A shows the density of the zebra mussel population during the late summer from 1990 to 1999. Graph B shows the relative percentage of oxygen in the water at Kingston, N.Y. before and after the increase in the zebra mussel population.

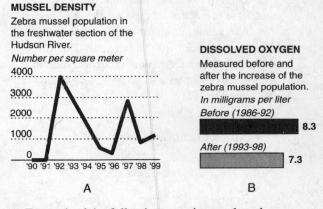

For each of the following questions, select the correct answer and mark your answer sheet accordingly.

74. According to Graph A, the zebra mussel density was at its height in

(A) 1991
(B) 1992
(C) 1993
(D) 1996
(E) 1997

75. According to Graph B, the percentage of dissolved oxygen in the water

   (A) was unaffected by the presence of zebra mussels
   (B) was highest between 1986–1992
   (C) was highest between 1993–1998
   (D) was lowest in 1992
   (E) was highest in 1993

76. All of the following are practical conclusions to be derived from the studies of the zebra mussel EXCEPT

   (A) find a way to reduce the population of the zebra mussel.
   (B) introduce a natural enemy of the zebra mussel.
   (C) study the ecology of the zebra mussel where it thrives.
   (D) study the concentrations of oxygen before and after the mussel appeared.
   (E) increase the food supply the mussel depends on.

77–80. The gray wolf, *Canis lupus*, which was on the verge of extinction in the Rocky Mountain area, has begun to extend its range in recent years under the protection of the Endangered Species Act. As a result, its status is being downgraded from "endangered" to "threatened." The following graph shows its population distribution in three states (Montana, Wyoming, and Idaho) beginning in 1980.

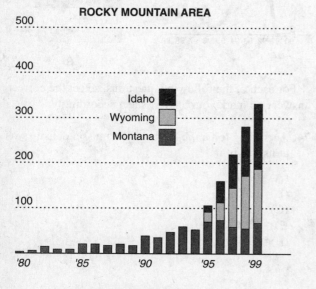

ROCKY MOUNTAIN AREA

The Gray Wolf Returns

For each of the following questions, select the correct answer and mark your answer sheet accordingly.

77. Between the years 1980 and 1994, the number of gray wolves in Wyoming was

   (A) 0
   (B) 50
   (C) 100
   (D) 150
   (E) 200

78. By 1999, the total number of gray wolves in Montana, Wyoming, and Idaho was a little more than

   (A) 100
   (B) 200
   (C) 300
   (D) 400
   (E) 500

79. The greatest increase in the number of gray wolves took place in

   (A) Montana and Wyoming
   (B) Montana and Idaho
   (C) Idaho and Wyoming
   (D) 1995
   (E) 1998

80. Since 1980, the number of gray wolves in Montana

   (A) has increased to less than 100
   (B) has increased to more than 100
   (C) has increased to more than that in Wyoming
   (D) has increased to more than that in Idaho
   (E) has decreased in number

Answer Key: Practice Test 1

1. (C)	17. (C)	33. (C)	49. (E)	65. (D)
2. (B)	18. (C)	34. (E)	50. (C)	66. (C)
3. (D)	19. (E)	35. (D)	51. (D)	67. (C)
4. (A)	20. (C)	36. (D)	52. (A)	68. (A)
5. (D)	21. (C)	37. (C)	53. (E)	69. (E)
6. (C)	22. (B)	38. (A)	54. (D)	70. (B)
7. (E)	23. (B)	39. (C)	55. (D)	71. (B)
8. (A)	24. (B)	40. (E)	56. (E)	72. (A)
9. (E)	25. (B)	41. (E)	57. (A)	73. (E)
10. (C)	26. (C)	42. (A)	58. (B)	74. (B)
11. (E)	27. (D)	43. (B)	59. (D)	75. (B)
12. (D)	28. (E)	44. (A)	60. (C)	76. (E)
13. (A)	29. (C)	45. (B)	61. (B)	77. (A)
14. (D)	30. (D)	46. (C)	62. (B)	78. (C)
15. (D)	31. (E)	47. (C)	63. (C)	79. (C)
16. (A)	32. (D)	48. (A)	64. (A)	80. (A)

Self-Evaluation Chart: Your Road to More Knowledge and Improved Scores

Your Raw Score

Using the Answer Key at the end of the test, place a ✔ next to each correct answer and an ✘ next to each incorrect answer.

A) Number of correct (✔) answers   _____

B) Number of incorrect (✘) answers   _____

   Raw Score (A – B)   _____

Your College Board Score

Turn to the Biology E/M Conversion Table in Chapter 1 and determine your equivalent

   College Board Score   _____

Improving Your Score

1. In column *A* below, list the numbers of the questions that you did not answer correctly.

2. Turn to the Answers Explained section, and for each question number listed in column *A* write in column *B* the key word or phrase that best summarizes the main topic or point of the answer.

3. Look up the topic in the Index and review the material.

4. Go back to the test and try to answer again each of the questions you answered incorrectly the first time. Write your new answers in column *C*.

5. Compare the answers in column *C* with the Answer Key.

6. Calculate your revised Raw Score:
   Number of correct answers (A above + number of column *C* correct answers)   _____

   Revised Equivalent College Board Score   _____

A. Incorrectly answered questions	B. Main point(s) of the answer	C. Answers to questions in column A
_____	_____	_____
_____	_____	_____
_____	_____	_____
_____	_____	_____
_____	_____	_____
_____	_____	_____
_____	_____	_____
_____	_____	_____
_____	_____	_____
_____	_____	_____
_____	_____	_____
_____	_____	_____
_____	_____	_____
_____	_____	_____
_____	_____	_____
_____	_____	_____
_____	_____	_____

Answers Explained: Practice Test 1

1. (C) The spread of human civilization has interfered with wildlife, forests, and soil.

2. (B) In dehydration synthesis, large organic molecules are built up from smaller building blocks, with the release of water. When two glucose molecules are linked to form maltose, a hydrogen atom (H) is removed from one glucose and a hydroxyl group (OH) is removed from the other, to form a molecule of water.

3. (D) Metabolism refers to all of the life activities.

4. (A) An organism is either a single-celled or a many-celled living thing. Large numbers of organisms of one species make up a population. A community consists of all the plant and animal populations interacting in a given environment. An ecosystem is a self-sustaining living community in relation to the physical environment. The biosphere is that portion of the earth in which ecosystems operate, and includes the biologically inhabited soil, water, and air.

5. (D) Liver

6. (C) Stomach

7. (E) Stomach

8. (A) Viruses contain a central core of nucleic acid (DNA or RNA) surrounded by a protein covering, but no cytoplasm.

9. (E) A protein contains not only carbon (C), hydrogen (H), and oxygen (O), but also nitrogen (N) and sometimes other elements such as sulfur (S) and phosphorus (P).

10. (C) At high altitudes, the climate is as cold as it is in the North.

11. (E) When a forest is destroyed, the trees can no longer bind the soil with their roots. Rains loosen the soil and carry it down to the rivers, which become muddy in appearance.

12. (D) Evidence for evolution is taken from many fields, including paleontology, comparative anatomy, embryology, vestigial structures, physiology, geographic distribution, classification, and heredity.

13. (A) The cilia beat dust and bacteria outward, thus keeping them out of the respiratory passages.

14. (D) To have the same ratio as water, there must be twice as much hydrogen as oxygen; in cellulose, $(C_6H_{10}O_5)_n$.

15. (D) The duckbill platypus is a rare mammal that lays eggs. It nurses its young on milk, has hair, and is warm-blooded.

16. (A) Members of the Monera kingdom, such as bacteria and blue-green algae, do not have nuclei in their cells.

17. (C) Nitrogen-fixing bacteria are located in the nodules of legume plants. They convert nitrogen, which cannot be used directly by plants, into nitrates, which can be used.

18. (C) Messenger RNA is formed as a copy of part of DNA. It separates from DNA and moves out of the nucleus to the ribosomes, where it supplies the code for the formation of a specific protein.

19. (E) ATP is produced during all three processes. The greatest amount is produced during aerobic respiration.

20. (C) In the thermodynamics of the cell, ATP may be considered as the energy-rich, or "charged," form of the energy carrier, and ADP as the energy-poor, or "discharged," form.

21. (C) When XY and XX are crossed, the possible results are:

	X	Y
X	XX	XY
X	XX	XY

Male: 50% normal; 50% color-blind
Female: 100% normal (50% heterozygous)

22. (B) Red blood cells may contain either A or B proteins for either type A or type B blood. These proteins are determined by genes A and B. Both of these genes are dominant to gene O, which does not cause either protein to be produced, and results in blood type O. When the genes A and B are both present, they produce both proteins, resulting in blood type AB. The following diagram shows that John's father, with type A blood, can have either AA or AO genes, while his mother, with O blood, can have only OO genes.

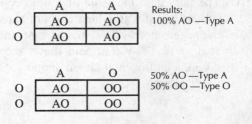

	A	A
O	AO	AO
O	AO	AO

Results: 100% AO —Type A

	A	O
O	AO	OO
O	AO	OO

50% AO —Type A
50% OO —Type O

23. (B) When hybrids are crossed, the genes segregate into the gametes, and then recombine in the fertilized eggs in a ratio of 25% homozygous dominant, 50% heterozygous dominant, and 25% recessive. The total percentage of dominant offspring is therefore 75%, or three quarters of all.

24. (B) Lichens are considered pioneer organisms because they first populate a given area. They may appear on bare rock. They build up soil as they break down the surface of the rock, releasing its minerals. Moss spores may then germinate on this very thin layer of soil and build it up. After a time, other plants appear on the accumulating soil and a succession of communities is on its way.

25. (B) The roots of trees hold soil together. The soil acts as a sponge and absorbs much water. When a forest fire destroys the trees, the soil is not held together and is washed away, very often leading to flooding.

26. (C) As a scavenger, the hyena will loiter in the vicinity of an animal killed by a lion, and after the lion leaves, will hurry over to feast on the remains. At other times, as a predator, it will hunt and kill an antelope or other small animal.

27. (D) Producer organisms are green plants that contain chlorophyll and synthesize organic compounds by photosynthesis.

28. (E) The maple tree has chlorophyll, which can use the energy of light to make organic compounds containing chemical energy.

29. (C) At each level of the food chain there is a loss of energy. Each member uses up some of the energy in its life processes. Therefore, the number of individuals at the top of the pyramid of energy is lower than the numbers at the levels further down.

30. (D) Since the aphid feeds on the leaves of the maple tree, it is considered to be herbivorous.

31. (E) Organisms such as aphids that feed directly on green plants are known as primary consumers.

32. (D) The fatty acid molecule of a lipid contains a COOH group with a long chain of carbon and hydrogen atoms. The structure of an amino acid molecule of a protein shows a carboxyl group, a hydrogen atom, an amino group, and a side group.

33. (C) In active transport, the cell uses the energy of ATP to move molecules across the cell membrane from an external region of low concentration to the inner region of high concentration. This movement is against the concentration gradient.

34. (E) Although the blue-green algae of the Monera kingdom contain chlorophyll, it is not present in chloroplasts, as it is in the algae of the Protista kingdom and in green plants.

35. (D) The biologist was interested in determining whether aeration promotes root development. He would find the answer by determining whether there was an increase in the number of roots in the experimental plant.

36. (D) Centrioles are present outside the nucleus in animal cells.

37. (C) The magnification of the ocular (10) is multiplied by the magnification of the high-power objective (44), to give a total magnification of 440.

38. (A) Cyclic AMP is called the second messenger because it is produced when a hormone (first messenger) activates a receptor on the surface of the cell. This leads to the formation of cyclic AMP from ATP; it then causes the cell to perform its specialized function.

39. (C) After entering the nasal passages, air passes through the trachea into the bronchi, the bronchial tubes, and finally into the air sacs, where its oxygen enters the blood and where it picks up carbon dioxide given off by the blood.

40. (E) All mammals, with the exception of the duck-bill platypus, the anteater, and pouched mammals, develop a placenta from which food and oxygen diffuse from the bloodstream of the mother into the bloodstream of the embryo.

41. (E) In the Krebs cycle of aerobic respiration, hydrogen atoms are passed from one carrier to another, releasing energy along the way. The final acceptor is oxygen, which combines with the hydrogen to form water.

42. (A) The ancestor of the horse, *Eohippus*, had four toes on its forelegs. With the passage of millions of years, one toe became enlarged, and two others were reduced to small bones or splints high up on the horse's foot. The fourth toe disappeared altogether.

43. (B) When foreign substances (antigens) such as germs enter the body, antibodies are produced in the blood tissue to counteract their effect.

44. (A) Hormones are produced by the ductless glands and are circulated in the blood plasma.

45. (B) Insects have six jointed legs; spiders have eight legs; they also have an exoskeleton.

46. (C) The streaming nature of protoplasm is illustrated in *Elodea* cells.

47. (C) Rabbits that are homozygous for gray coat have two dominant alleles, *GG*. One of these alleles came from an egg, and the other from a sperm. Since 80% of the eggs and 80% of the sperm carried *G*, when they were united, the percentage of the resulting zygotes having *GG* alleles would be 64 ($0.80 \times 0.80 = 0.64$).

48. (A) By natural selection, mosquitoes that were unaffected by certain insecticides survived, and produced a new variety of mosquitoes.

49. (E) Exercise (use) causes the steelworker's shoulder muscles to increase in size.

50. (C) According to the heterotroph hypothesis, amino acids were bonded together in the "hot, thin soup" of primitive oceans from molecules originating in the gases of the atmosphere, under the influence of such forms of energy as solar radiation, cosmic rays, lightning, and the earth's radioactivity.

51. (D) A reflex action occurs; tears are stimulated to flow, and the eye closes.

52. (A) By habit formation, the secretary has learned to take notes automatically and quickly.

53. (E) The horse has learned to associate the sound of the command with a pull on the reins, and has developed a conditioned response to stop at the command.

54. (D) Each characteristic is inherited independently of the other.

55. (D) Each characteristic is inherited independently of the other.

56. (E) The genes for both characteristics are located on the same chromosome, and so they are inherited together.

57. (A) Green plants, like animals, carry on respiration. They release energy from food nutrients in the presence of oxygen, and give off carbon dioxide and water as wastes.

58. (B) Carbon dioxide, which is normally a waste product resulting from respiration, is taken in by green plants in the light during photosynthesis to make carbohydrates.

59. (D) There are 1,000 micrometers in a millimeter. In 1.8 millimeters, there are 1,800 micrometers ($1.8 \times 1,000$).

60. (C) If the length of the organism is estimated to be one third of the field of vision, its approximate length is $^1/_3$ of 1.8 millimeters (1,800 micrometers), or 600 micrometers ($1,800 \div 3$).

61. (B) Since other factors were kept constant, the experiment obviously deals with reactions to different temperatures. The scientist was specifically examining the adaptation of the frog.

62. (B) Frogs react to the lower temperature of advancing winter by hibernating. As their metabolic activity begins to slow down, they bury themselves in the mud and remain inactive during the cold weather.

63. (C) Plants that are germinated from seeds in the dark are white because chlorophyll is formed only in the presence of light.

64. (A) Plants that are grown in darkness are spindly and long. In the presence of light, plants develop normally, and by comparison are shorter, with green leaves, and can be seen in the accompanying illustration.

65. (D) The illustration shows the presence of leaflets on seeding #3. Photosynthesis proceeds most actively in the red and violet regions of the spectrum, probably encouraging the development of leaflets on this seedling.

66. (C) Since the winged fruit flies had the ability to fly, they were able to land on the sticky flypaper in jar *A*. Once there, they were stuck and could not reach the food. They were not able to survive, whereas winged flies had no problem surviving in jar *B*.

67. (C) Jar *B* had no sticky flypaper, so the winged flies, as well as the wingless flies, were able to survive there.

68. (A) Since the winged flies became stuck on the flypaper in jar *A* and did not survive, the wingless condition was an advantage. The wingless flies, which did not get stuck on the flypaper, had no problem in reaching the food.

69. (E) Darwin's theory included the idea of the survival of the fittest. In jar *A* the wingless flies were better fitted and survived because they did not become stuck on the flypaper.

70. (B) The control in an investigation represents the normal condition, allowing it to be compared with the experimental procedure. This experiment involved the use of flypaper in jar *A*.

71. (B) This is an example of photoperiodism. Chrysanthemums require a long period of darkness in order to flower. In the fall, when the hours of daylight are less than 12–13, such short-day flowers will bloom.

72. (A) These actions would reduce the exposure of the plant to light, and increase the length of the dark period. Since a chrysanthemum is a long-night plant, it would be stimulated to flower in July.

73. (E) This action would lengthen the exposure of the plant to light and reduce its exposure to darkness. Since a black-eyed Susan is a short-night plant, it would be stimulated to flower.

74. (B) The graph shows that in 1990 and 1991 there were no mussels in the river. Then, in 1992, there was a sudden increase in density amounting to 4,000 per square meter. After that, it declined, reaching a low level in 1996 before it increased to almost 3,000 in 1997. After that it declined again.

75. (B) The graph shows that the percentage of dissolved oxygen was at its highest between 1986 and 1992, reaching a level of 8.3 milligrams per liter. In the years after that the percentage of dissolved oxygen declined, and between 1993–1998, it was at a low point of 7.3 milligrams per liter.

76. (E) The problem is that there has not been a way to reduce the population of the zebra mussel. A natural enemy has not been found that would control it. Studies of the concentration of oxygen have been made of the river water and the mussel has been found to reduce the supply, limiting the growth of fish.

77. (A) According to the part of the bar graph representing Wyoming, there were no gray wolves until 1995, when they first appeared. After that they steadily increased in numbers until 1999.

78. (C) The length of the bar graph representing the total number of gray wolves in Montana, Wyoming, and Idaho shows that by 1999 there were more than 300 of them. Their numbers increased steadily since the year 1995.

79. (C) The bar graph for 1999 shows that the number of gray wolves in Idaho and Wyoming increased considerably. In the same period, the number of gray wolves in Montana increased slightly.

80. (A) Gray wolves began to appear in Montana as early as 1980, but their numbers never reached 100, as the bar graph shows.

Answer Sheet: Practice Test 2

1. (A) (B) (C) (D) (E)
2. (A) (B) (C) (D) (E)
3. (A) (B) (C) (D) (E)
4. (A) (B) (C) (D) (E)
5. (A) (B) (C) (D) (E)
6. (A) (B) (C) (D) (E)
7. (A) (B) (C) (D) (E)
8. (A) (B) (C) (D) (E)
9. (A) (B) (C) (D) (E)
10. (A) (B) (C) (D) (E)
11. (A) (B) (C) (D) (E)
12. (A) (B) (C) (D) (E)
13. (A) (B) (C) (D) (E)
14. (A) (B) (C) (D) (E)
15. (A) (B) (C) (D) (E)
16. (A) (B) (C) (D) (E)
17. (A) (B) (C) (D) (E)
18. (A) (B) (C) (D) (E)
19. (A) (B) (C) (D) (E)
20. (A) (B) (C) (D) (E)
21. (A) (B) (C) (D) (E)
22. (A) (B) (C) (D) (E)
23. (A) (B) (C) (D) (E)
24. (A) (B) (C) (D) (E)
25. (A) (B) (C) (D) (E)
26. (A) (B) (C) (D) (E)
27. (A) (B) (C) (D) (E)

28. (A) (B) (C) (D) (E)
29. (A) (B) (C) (D) (E)
30. (A) (B) (C) (D) (E)
31. (A) (B) (C) (D) (E)
32. (A) (B) (C) (D) (E)
33. (A) (B) (C) (D) (E)
34. (A) (B) (C) (D) (E)
35. (A) (B) (C) (D) (E)
36. (A) (B) (C) (D) (E)
37. (A) (B) (C) (D) (E)
38. (A) (B) (C) (D) (E)
39. (A) (B) (C) (D) (E)
40. (A) (B) (C) (D) (E)
41. (A) (B) (C) (D) (E)
42. (A) (B) (C) (D) (E)
43. (A) (B) (C) (D) (E)
44. (A) (B) (C) (D) (E)
45. (A) (B) (C) (D) (E)
46. (A) (B) (C) (D) (E)
47. (A) (B) (C) (D) (E)
48. (A) (B) (C) (D) (E)
49. (A) (B) (C) (D) (E)
50. (A) (B) (C) (D) (E)
51. (A) (B) (C) (D) (E)
52. (A) (B) (C) (D) (E)
53. (A) (B) (C) (D) (E)
54. (A) (B) (C) (D) (E)

55. (A) (B) (C) (D) (E)
56. (A) (B) (C) (D) (E)
57. (A) (B) (C) (D) (E)
58. (A) (B) (C) (D) (E)
59. (A) (B) (C) (D) (E)
60. (A) (B) (C) (D) (E)
61. (A) (B) (C) (D) (E)
62. (A) (B) (C) (D) (E)
63. (A) (B) (C) (D) (E)
64. (A) (B) (C) (D) (E)
65. (A) (B) (C) (D) (E)
66. (A) (B) (C) (D) (E)
67. (A) (B) (C) (D) (E)
68. (A) (B) (C) (D) (E)
69. (A) (B) (C) (D) (E)
70. (A) (B) (C) (D) (E)
71. (A) (B) (C) (D) (E)
72. (A) (B) (C) (D) (E)
73. (A) (B) (C) (D) (E)
74. (A) (B) (C) (D) (E)
75. (A) (B) (C) (D) (E)
76. (A) (B) (C) (D) (E)
77. (A) (B) (C) (D) (E)
78. (A) (B) (C) (D) (E)
79. (A) (B) (C) (D) (E)
80. (A) (B) (C) (D) (E)

PRACTICE TEST 2

CHAPTER

15

NOTE: The College Board test offered in Biology E/M contains a common core of 60 questions numbered 1–60, covering the general field of biology and a specialized section of 20 questions numbered 61–80 for Biology E (Ecology) or questions 81–100 for Biology M (Molecular). On the day of the examination, you may take ONE of the Biology E/M tests, either E or M, by marking the appropriate grid. Practice Test 2 of this book is designed to prepare you for the Biology E test. All of these Practice Tests have been created to present different core questions 1–60 in order to give you more experience in test preparation.

Practice Test 2: Biology E

Part A (Core Questions 1–60)

Directions: Each of the questions or incomplete statements below is followed by five suggested answers or completions. Choose the one that is best in each case and then blacken the corresponding space on the answer sheet.

1. Which of the following does this structural formula represent?

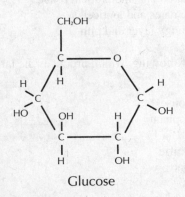

Glucose

(A) An atom
(B) An ion
(C) A molecule
(D) An element
(E) A mixture

2. Which one of the following statements about photosynthesis and bioluminescence is correct?

(A) In photosynthesis chemical reactions produce light; in bioluminescence light initiates chemical reactions.
(B) Both photosynthesis and bioluminescence produce light from chemical reactions.
(C) Light promotes chemical reactions in both photosynthesis and bioluminescence.
(D) Both photosynthesis and bioluminescence are inhibited by light.
(E) In photosynthesis light initiates chemical reactions; in bioluminescence chemical reactions produce light.

3. The breaking apart of the platelets in the human bloodstream leads to the

   (A) production of Rh negative antigens
   (B) formation of a clot
   (C) formation of antibodies
   (D) clumping of Type A blood
   (E) deamination of amino acids

4. A characteristic of organisms that excrete uric acid as their main nitrogenous waste is that they

   (A) usually live in salt water
   (B) usually live in fresh water
   (C) are usually land dwellers
   (D) can carry on only anaerobic respiration
   (E) cannot metabolize proteins

5. Of the following chemical substances, the only one related to the nervous system is

   (A) gibberellic acid
   (B) acetylcholine
   (C) insulin
   (D) deoxyribonucleic acid
   (E) opsonin

6. From what part of a plant does the seed develop?

   (A) Hilum
   (B) Pollen tube
   (C) Anther
   (D) Oviduct
   (E) Ovule

7. Red corpuscles are to hemoglobin as chloroplasts are to

   (A) guard cells
   (B) palisade cells
   (C) chlorophyll
   (D) photosynthesis
   (E) cytoplasm

8. Foods have to be digested before they can be used by the body because

   (A) the stomach is the center of all digestion
   (B) the villi digest only nutrients they can absorb
   (C) only insoluble materials can pass through membranes
   (D) only soluble materials can pass through membranes
   (E) assimilation always takes place before digestion

9. Why are weeds harmful?

   (A) Their seeds serve as food for birds.
   (B) They interfere with the growth of useful plants.
   (C) They prevent the cross-pollination of useful plants.
   (D) They poison the soil.
   (E) Their roots do not bind the soil.

10. The arm of a human, the wing of a bat, and the flipper of a whale have the same basic structure because

   (A) they are used for the same purpose
   (B) these animals had a common ancestor
   (C) these animals have identical genes
   (D) these animals all have backbones
   (E) these animals are descended from each other

11. The bean is classified as a dicot because

   (A) its leaves have parallel veins
   (B) its flower parts are in groups of three
   (C) when it germinates, the plumule remains underground
   (D) it has two cotyledons in its seed
   (E) its seed has the diploid number of chromosomes

12. When a nerve impulse is initiated at a receptor, which pathway does an impulse follow to stimulate an effector in the finger?

   (A) interneuron → motor neuron → sensory neuron
   (B) interneuron → sensory neuron → motor neuron
   (C) motor neuron → sensory neuron → interneuron
   (D) sensory neuron → motor neuron → interneuron
   (E) sensory neuron → interneuron → motor neuron

13. The leaves, stem, and roots of a corn plant all contain

   (A) xylem and phloem tissue
   (B) palisade cells and xylem tissue
   (C) guard cells and phloem tissue
   (D) stomates and lenticels
   (E) spongy layer and pith

14. Of the following, which produces the largest ovum?

   (A) Mouse
   (B) Elephant
   (C) Whale
   (D) Giraffe
   (E) Sparrow

15. A color-blind man married a normal woman who is heterozygous for color vision. What is the probability that their two daughters will be color blind?

(A) 0   (B) $\frac{1}{8}$ (C) $\frac{1}{4}$ (D) $\frac{1}{2}$ (E) $\frac{3}{4}$

16. The basal metabolism of a normal adult man was determined to proceed at such a rate as to liberate 39 calories an hour for each square meter of skin surface. Which of the following would be most likely to represent the basal metabolism rate of a patient with myxedema?

(A) 39 calories per hour
(B) 24 calories per hour
(C) 50 calories per hour
(D) 75 calories per hour
(E) 100 calories per hour

17. In a large litter of guinea pigs, three quarters of the offspring were black (black is dominant). The genotypes of the parents were most likely

(A) $BB \times bb$
(B) $Bb \times bb$
(C) $Bb \times Bb$
(D) $Bb \times BB$
(E) $BB \times BB$

18. In one of his experiments, Gregor Mendel crossed tall pea plants. He found that in the next generation there were both tall and short plants. The genetic makeup of the original pea plants was most likely

(A) homozygous tall
(B) heterozygous tall
(C) heterozygous short
(D) homozygous short
(E) recessive short

19. By which of the following can movement of materials across animal cell membranes be accomplished?

I. Active transport
II. Diffusion
III. Pinocytosis

(A) I only
(B) II only
(C) III only
(D) I and II only
(E) I, II and III

20. During early development, the embryo of a chicken and the embryo of a pig both share many similarities, including gill slits, tails, and a two-chambered heart. The similarity of these embryos suggests that chickens and pigs most probably

(A) have a common ancestry
(B) use gills for breathing in the early amniotic fluid
(C) carry on anaerobic respiration as embryos
(D) have gill slits for breathing in an emergency as adults
(E) have a two-chambered heart as adults

21. A popular supposition about the extinction of dinosaurs at the end of the Mesozoic Era postulates that an asteroid smashed into earth, causing such catastrophic environmental changes that the dinosaurs died off in a relatively short time, thus changing the course of evolution. This concept is an example of

(A) the theory of gradualism
(B) the theory of punctuated equilibrium
(C) the heterotroph hypothesis
(D) geographic isolation
(E) the Hardy-Weinberg Principle

22. A lipid molecule is composed of glycerol and fatty acid molecules in a ratio of

(A) 1:1
(B) 1:2
(C) 1:3
(D) 1:4
(E) 1:5

23. Trees grow in which of the following biomes?

I. Tundra
II. Taiga
III. Temperate deciduous forest

(A) I only
(B) II only
(C) III only
(D) I and III only
(E) II and III only

24. Different strata of rock in an undisturbed region are found to contain two different fossils: A and B. Fossil A is located in the layer of rock above the layer with fossil B. Which of the following statements is most likely true?

(A) Fossil B is older than fossil A.
(B) Fossil A is older than fossil B.
(C) Fossil A is that of an organism that evolved from fossil B.
(D) Fossil B is that of an organism that evolved from fossil A.
(E) Fossils A and B are closely related and evolved from a common ancestor.

25. Of the following, the one that contains the largest number of different types of cells is

(A) mucous membrane
(B) smooth muscle
(C) nerve
(D) small intestine
(E) blood

26. Humans are similar in structure to apes such as the chimpanzee and the orangutan. A possible explanation for this similarity is that

(A) humans are descended from apes
(B) apes are descended from humans
(C) humans are descended from Neanderthals who came from apes
(D) humans and apes had a common ancestor
(E) early humans and apes lived in caves

27. Animals that have jointed appendages include which of the following?

I. Spider
II. Crab
III. Snail

(A) I only
(B) II only
(C) III only
(D) I and II only
(E) I, II and III

28. The blood type known as the universal donor is

(A) A
(B) B
(C) AB
(D) O
(E) Rh factor

29. All of the following are functions of the skin EXCEPT

(A) protection
(B) sensation
(C) excretion
(D) manufacture of vitamin D
(E) exhalation

30. In what way are chloroplasts and mitochondria similar?

(A) They utilize the rays of the sun.
(B) They contain structures known as grana.
(C) They are sites for ATP synthesis.
(D) They occur in animal cells.
(E) They function only during daylight hours.

31. A nerve cell that transmits impulses from a sense organ to the nervous system is known as a(n)

(A) sensory neuron
(B) motor neuron
(C) interneuron
(D) plexus
(E) ganglion

32. The weakness in Darwin's theory of how evolution occurs was his inability to explain the

(A) mechanisms that produce variations
(B) reasons for overproduction
(C) role played by natural selection
(D) adaptations of living organisms for survival
(E) inheritance of acquired characteristics

33. Which of the following represents the sequence of bases on the messenger RNA formed from a DNA base sequence of AAC-ATC?

(A) AAG-ATG
(B) AAC-ATC
(C) UUG-UAG
(D) TTG-TAG
(E) UUT-TUT

34. An increase in the diameter of an oak tree is caused chiefly by the activity of the

(A) vascular ducts
(B) bark
(C) lenticels
(D) cambium
(E) annual rings

35. What effect does the hydrolytic action of enzymes have on organic molecules?

    (A) They are converted to more complex forms.
    (B) Their hydrogen-ion concentration is increased.
    (C) Their hydrogen-ion concentration is decreased.
    (D) They become chemically inactive.
    (E) They become smaller.

36. The marrow of bones in humans provides

    (A) cartilage for the ends of the bones
    (B) white fibrous connective tissue
    (C) structural support for the skeletal system
    (D) a source of new blood cells
    (E) connections between the tendons and ligaments

37. What biological relationship is illustrated by this well-known jingle of the 18th-century satirical writer Jonathan Swift?

Big fleas have little fleas
Upon their backs to bite 'em,
And little fleas have lesser fleas
And so, ad infinitum

    (A) Life in a niche
    (B) Abiotic factors
    (C) Food chain
    (D) Food guide pyramid
    (E) Population in a community

38. Which of the following explains why carnivorous animals depend on green plants for their food?

    (A) Carnivores need to achieve a balanced diet.
    (B) Only herbivorous animals have grinding teeth.
    (C) Only green plants can make food.
    (D) Carnivores need protection against vitamin deficiency diseases.
    (E) Only omnivores are both carnivores and herbivores.

39. Although the island of Madagascar is separated from Africa only by a narrow strait, many plants and animals common on the mainland are unknown on the island. What principle does this fact illustrate?

    (A) Incomplete dominance
    (B) Independent assortment
    (C) Evolutionary equilibrium
    (D) Evolution in isolated populations
    (E) Ecological succession

40. Which of the following is indicated by the information on the accompanying graph?

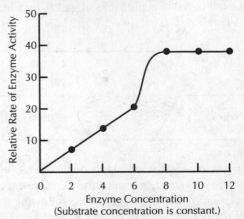

(Substrate concentration is constant.)

    (A) The rate of enzyme action is directly dependent upon the substrate concentration.
    (B) The rate of enzyme action increases constantly with an increase in enzyme concentration.
    (C) The rate of enzyme action becomes stabilized when a certain enzyme concentration is reached.
    (D) Enzyme concentration has no effect upon the rate of enzyme action.
    (E) When the substrate concentration is increased, the enzyme concentration is decreased.

41. When Mendel crossed pea plants that were hybrid for smooth seed form (smooth is dominant; wrinkled is recessive), he obtained 1,850 wrinkled seeds out of a total of 7,324. Which of the following would be most likely to represent the number of smooth seeds?

    (A) 1,850
    (B) 3,700
    (C) 5,474
    (D) 7,324
    (E) 9,174

42. Animals fed vitamin B_{12} show increased growth. Pure vitamin B_{12} is extracted from waste materials left in vats in which antibiotics were made. Animals fed on a diet that includes wastes from the antibiotic vats grow faster than those fed only pure vitamin B_{12}. Which of the following is the most probable explanation?

    (A) Waste from the antibiotic vats contains a growth promoter other than vitamin B_{12}.
    (B) Pure vitamin B_{12} is not a growth promoter at all.
    (C) Vitamin B_{12} is a good growth promoter if it is in impure form.
    (D) The waste material in the vats contains vitamin B_{12} that the process does not extract.
    (E) Antibiotics are better growth promoters than vitamin B_{12}.

43. A population originally inhabiting the entire area shown on the accompanying diagram has become separated into two populations, *A* and *B*, by a barrier (water).

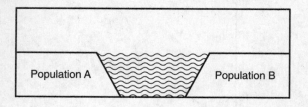

If the environment inhabited by population *A* undergoes severe changes and the environment of population *B* does not, which of the following will most likely be true about the rate of evolution of population *A*?

(A) It will be consistently slower than that of population *B*.
(B) It will be consistently faster than that of population *B*.
(C) It will be the same as that of population *B*.
(D) It will be slower at first and then faster than that of population *B*.
(E) It will depend on the rate of evolution of population *B*.

44. A climax community in North America can exist for a long period of time if it

(A) is host to many pioneer organisms
(B) alters the local climate
(C) is in equilibrium with the environment
(D) contains a variety of mosses and ferns
(E) contains a mixture of maples, pines, and tall grasses

45. Plankton is composed of

(A) microscopic life
(B) platelets
(C) plasmids
(D) fish
(E) whales

46. Competition among members of a prairie dog population in a given area would probably increase as a result of an increase in the

(A) rate of reproduction of their predators
(B) prairie dog reproduction rate
(C) spreads of rabies among the prairie dogs
(D) number of prairie dogs killed by cars on the road
(E) number of secondary consumers

47. An example of a parasite is

(A) nitrogen-fixing bacteria on the roots of alfalfa
(B) a barnacle growing on the back of a whale
(C) a peregrine hawk
(D) a remora fish attached to a shark
(E) the protozoan that causes malaria

48. The thin layer of life surface of the earth, including the biologically inhabited soil, water, and air is known as the

(A) biogeography
(B) biome
(C) biosphere
(D) biomass
(E) biotin

Questions 49–51 refer to the following diagram of a paramecium.

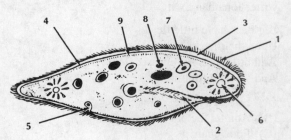

49. Food is digested by

(A) 2
(B) 4
(C) 5
(D) 6
(E) 7

50. Which of the following contains the chromosomes?

(A) 3
(B) 5
(C) 6
(D) 7
(E) 8

51. Which of the following permits diffusion of dissolved gases?

(A) 3
(B) 4
(C) 5
(D) 6
(E) 8

Directions: Each set of lettered choices below refers to the numbered statements immediately following it. Choose the one lettered choice that best fits each statement and then blacken the corresponding space on the answer sheet. A choice may be used once, more than once, or not at all in each set.

Questions 52–54

    (A) Left atrium
    (B) Left ventricle
    (C) Right atrium
    (D) Right ventricle
    (E) Aorta

52. Receives blood from the lungs

53. Sends blood to the lungs

54. Receives blood from the head

Questions 55–57

    (A) Test cross
    (B) Mutation
    (C) Sex linkage
    (D) Independent assortment
    (E) Segregation

55. How hemophilia is inherited

56. How a gene affecting pigmentation is changed to produce an albino

57. How dark-haired parents with brown eyes have a child with blond hair and brown eyes

Questions 58–60

    (A) Phototaxis
    (B) Geotropism
    (C) Thigmotropism
    (D) Hydrotropism
    (E) Chemotaxis

58. How a plant grows after it has been blown over by the wind

59. How a paramecium reacts to salt

60. How a euglena moves in a partly shaded pool

Part B (Choice E Questions 61–80)

Questions 61–63

In an ecological study of vegetation in a cultivated garden bed that had been neglected for eight weeks, a record was made of the number of weeds that were found in an area measuring a square foot. Twenty-five random trials were made and the findings summarized in Table I. By multiplying the totals by 4, the percentage occurrence in 100 trial areas is obtained. The species were then arranged in five groups according to frequency of occurrence, as shown in Table II.

TABLE I

Species	Total Occurrence in 25 Trials
Poa	25
Stellaria	25
Capsella	24
Senecio	10
Lamium	1
Veronica	2

TABLE II

Group	Percentage Occurrence
I	0–20
II	21–40
III	41–60
IV	61–80
V	81–100

61. The number of species present in Group II is

    (A) 0
    (B) 1
    (C) 2
    (D) 3
    (E) 4

62. The least common species classified in Group I is

    (A) *Stellaria*
    (B) *Capsella*
    (C) *Lamium*
    (D) *Senecio*
    (E) *Veronica*

63. From this study it may be concluded that the two most common weeds to be found in this entire neglected garden bed were

(A) *Senecio* and *Capsella*
(B) *Senecio* and *Lamium*
(C) *Poa* and *Veronica*
(D) *Poa* and *Lamium*
(E) *Poa* and *Stellaria*

Questions 64–66

A scientist grew separate laboratory cultures of two closely related species of paramecium, *Paramecium caudatum* and *Paramecium aurelia*. He then mixed the two species together and studied the effect. His results are summarized in the accompanying graphs, where the solid curves represent the population growths of each species alone, and the dotted curves show the population growths when the species were combined.

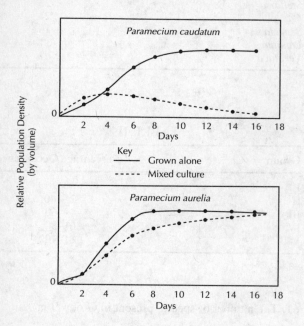

64. When the two species were cultured separately,

(A) their population growths followed a similar pattern
(B) *P. caudatum* reached maximum population growth earlier than *P. aurelia*
(C) *P. aurelia* did not reach maximum population growth

(D) *P. caudatum* reached a higher population density than *P. aurelia*
(E) *P. aurelia* reached the same population density as *P. caudatum*

65. When the two species were cultured together,

(A) *P. caudatum* increased its growth until the 8th day
(B) *P. caudatum* increased more rapidly than when grown alone
(C) *P. caudatum* failed to survive
(D) *P. aurelia* grew at a faster rate than when grown alone
(E) *P. aurelia* failed to survive

66. The change in population growth when the two species were cultured together was most likely due to

(A) commensalism
(B) mutalism
(C) inability of *P. aurelia* to survive
(D) competition
(E) parasitism by *P. caudatum*

Questions 67–71

There are two forms of the peppered moth (*Biston betularia*), one dark in color and one light. Scientists observed that in the industrial area of Manchester, England, the originally prominent light form was replaced by the dark form between the years 1848 and 1895. At first, there were only light forms; later, the dark form comprised 98 percent of the total population.

A scientist explained this evolutionary change as follows: The moths rest on tree trunks during the day, and through their protective coloration avoid being seen and eaten by insectivorous birds. In the earlier years, before 1848, any dark forms were conspicuous on the light-colored tree trunks and were easily found by birds. With the coming of many factories after 1848, tree trunks became blackened by the soot given off in chimney smoke. Then the dark forms of moths resembled the background more closely, while the light forms stood out and were easily seen, and eaten, by the birds.

Use the graphs shown on the following page to answer questions 67–69.

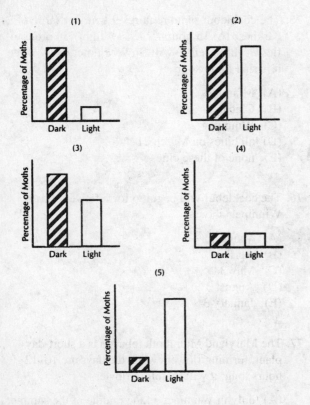

67. Which graph represents the original populations of the peppered moth?

(A) 1
(B) 2
(C) 3
(D) 4
(E) 5

68. Which graph represents the moth population after the coming of the factories?

(A) 1
(B) 2
(C) 3
(D) 4
(E) 5

69. The rural area of Dorset, England, has no factories, and the tree trunks are light in color. In a scientific study, equal numbers of the dark and light forms of the moth were released into the area. Which graph represents the percentages of the surviving moths?

(A) 1
(B) 2
(C) 3
(D) 4
(E) 5

70. Which of the following offers the best explanation for the change in moth color after 1848?

(A) Inheritance of acquired characteristics
(B) Gene mutations caused by the soot
(C) Natural selection of favorable variations
(D) Lamarck's theory of evolution
(E) Ingestion of the soot particles

71. What question would scientists need to answer in order to determine whether the two forms, light and dark, of the moth have become different species?

(A) Are the two forms the same size?
(B) Do the dark moths fly more frequently during the day than the light moths?
(C) Do light moths fly more frequently during the day than the dark ones?
(D) Can the two forms interbreed?
(E) Do the two forms feed on different food?

Questions 72–74

A project was conducted to determine the factors involved in frog hibernation. The effects of reduced temperature on the breathing rate were studied. Graph I was prepared to show the results. Graph II shows the results when the experiment was repeated. During the first experiment, the frog attempted to hibernate at 4.4°C. During the second experiment it attempted to hibernate at 12.8°, 11.5°, 10.0°, 6.7°, 5.0°, 2.2°, and 1.1°. The attempts to hibernate included these activities: The frog closed its eyes and expelled air from its lungs; it attempted to dig in at the bottom of the jar as though there were mud there.

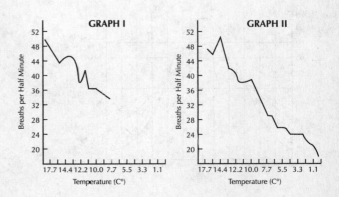

72. From this project, it can be concluded that

(A) frogs hibernate when the temperature is reduced
(B) as the temperature is lowered, the heartbeat is reduced
(C) frog hibernation is not related to temperature
(D) the breathing rate is reduced as the temperature is lowered
(E) frogs stop breathing when they attempt to hibernate

73. The experiment was done a second time for all of the following reasons EXCEPT to

(A) improve on the techniques of the first experiment
(B) check on the first set of results
(C) find the rate of respiration from 8.8° to 6.1°C
(D) locate the brain center that controls hibernation
(E) see whether hibernation could be induced artificially

74. At 15°C, the breathing rate

(A) was higher than at 16.6° in Graph I
(B) was lower than 16.6° in Graph II
(C) was at its highest point in Graph II
(D) was at its lowest point in Graph I
(E) showed an increase in Graph I

Questions 75–77

It has been found that the flowering rate of plants is related to the length of daylight. This phenomenon is known as photoperiodism. Some plants are short-day plants; they flower only when they are exposed to short periods of light. Other plants are long-day plants; they produce flowers only when they are exposed to long periods of light.

The graph below shows the annual change in the length of day throughout the year at four latitudes: Miami (26°), San Francisco (37°), Chicago (42°), and Winnipeg (50°), and also the hours of darkness required for three different plants to flower.

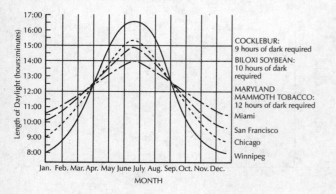

75. The cocklebur plant requires 9 hours or more of darkness (or 15 hours or less of light), in order to flower. It therefore will flower immediately when it is ripe to do so in

(A) Miami
(B) Chicago
(C) Winnipeg
(D) all cities on March 21
(E) none of these cities

76. The cocklebur will begin to form flower buds in Winnipeg about

(A) March 3
(B) April 3
(C) May 3
(D) August 3
(E) January 3

77. The Maryland Mammoth tobacco is a short-day plant, forming flowers when the days are 10–12 hours long; it will therefore flower

(A) only in Winnipeg in the middle of the summer
(B) in all the cities except Winnipeg in the middle of the summer
(C) very early in the summer in all the cities
(D) in the middle of the summer in all the cities
(E) very late in the summer in all the cities

Questions 78–80

A farmer collected 203 pellets of indigestible remains dropped by barn owls on his farm. He took the pellets to a nearby museum for analysis of the type of food eaten by the owls. The following results were obtained:

429	meadow mice	96	short-tailed shrews
4	lemming mice	1	squirrel
1	pine mouse	5	cottontail rabbits
12	white-footed deer mice	23	unidentified mice
18	jumping mice	5	small birds
21	star-nosed moles	1	Brewer's mole

78. The most common prey of the barn owls were

(A) white-footed deer mice
(B) jumping mice
(C) pine mice
(D) meadow mice
(E) lemming mice

79. Which of the following was the only type of nonmammalian prey?

(A) Squirrels
(B) Moles
(C) Birds
(D) Rabbits
(E) Shrews

80. On the basis of these findings, the best course of action for the farmer to take would be to

(A) kill all owls as a menace to his chickens
(B) feed poisoned pellets to owls to get rid of them
(C) offer special protection to barn owls
(D) scare owls away with white mice
(E) inform his neighbors that he now has proof that owls are a great menace to small birds

Answer Key: Practice Test 2

1. (C)		17. (C)		33. (C)		49. (E)		65. (C)
2. (E)		18. (B)		34. (D)		50. (E)		66. (D)
3. (B)		19. (D)		35. (E)		51. (B)		67. (E)
4. (C)		20. (A)		36. (D)		52. (A)		68. (A)
5. (B)		21. (B)		37. (C)		53. (D)		69. (E)
6. (E)		22. (C)		38. (C)		54. (C)		70. (C)
7. (C)		23. (E)		39. (D)		55. (C)		71. (D)
8. (D)		24. (A)		40. (C)		56. (B)		72. (D)
9. (B)		25. (D)		41. (C)		57. (D)		73. (D)
10. (B)		26. (D)		42. (A)		58. (B)		74. (C)
11. (D)		27. (D)		43. (B)		59. (E)		75. (A)
12. (E)		28. (D)		44. (C)		60. (A)		76. (D)
13. (A)		29. (E)		45. (A)		61. (B)		77. (E)
14. (E)		30. (C)		46. (B)		62. (C)		78. (D)
15. (D)		31. (A)		47. (E)		63. (E)		79. (C)
16. (B)		32. (A)		48. (C)		64. (A)		80. (C)

Self-Evaluation Chart: Your Road to More Knowledge and Improved Scores

Your Raw Score

Using the Answer Key at the end of the test, place a ✔ next to each correct answer and an ✘ next to each incorrect answer.
A) Number of correct (✔) answers _____
B) Number of incorrect (✘) answers _____
   Raw Score (A – B) _____

Your College Board Score

Turn to the Biology E/M Conversion Table in Chapter 1 and determine your equivalent
   College Board Score _____

Improving Your Score

1. In column *A* below, list the numbers of the questions that you did not answer correctly.

2. Turn to the Answers Explained section, and for each question number listed in column *A* write in column *B* the key word or phrase that best summarizes the main topic or point of the answer.
3. Look up the topic in the Index and review the material.
4. Go back to the test and try to answer again each of the questions you answered incorrectly the first time. Write your new answers in column *C*.
5. Compare the answers in column *C* with the Answer Key.
6. Calculate your revised Raw Score:
   Number of correct answers (A above + number of column *C* correct answers) _____

   Revised Equivalent College Board Score _____

A. Incorrectly answered questions	B. Main point(s) of the answer	C. Answers to questions in column A

Answers Explained: Practice Test 2

1. (C) The diagram shows the structural formula of the glucose molecule. A molecule is the smallest part of a substance that still has the properties of the substance.

2. (E) In photosynthesis, light strikes the chlorophyll of a leaf, raising it to a higher, or an excited, level. This is the beginning of photosynthesis. High electrons are passed along to form ATP and to change the coenzyme NADP to NADPH. Water is split and O_2 gas is given off. In the Calvin cycle, CO_2 is converted through a series of intermediate compounds into PGAL, which is rearranged to form the end products of photosynthesis: sucrose and starch. In bioluminescence, chemical reactions result in light being produced in certain organisms, such as the firefly, deep-sea fish, mushrooms, and certain algae.

3. (B) When a wound occurs, the platelets break up and start a series of reactions. In the process, one of the dissolved proteins in the plasma, fibrinogen, becomes insoluble and turns into threads of fibrin. The red blood cells become entangled in this network, causing a clot to form.

4. (C) In the land-dwelling grasshopper, nitrogenous wastes are largely excreted as uric acid. Since this substance is very insoluble, little water is used to carry it in solution. Instead, it is removed in solid crystal form by the Malpighian tubules, which open into the digestive system. The wastes are then eliminated through the anus along with undigested food materials. This method of excretion serves as a water-conservation mechanism for certain land-dwelling organisms.

5. (B) Acetylcholine is produced along the length of a neuron and plays a part in the transmission of a nervous impulse. It is also formed in a synapse.

6. (E) When double fertilization occurs within the ovule, one part develops into an embryo, and the rest becomes the food and covering. The matured ovule becomes the seed.

7. (C) Red blood cells receive their color from the hemoglobin in them; chloroplasts are green because they contain chlorophyll.

8. (D) During digestion, the nutrients of food are digested to simpler, soluble form, and can then diffuse into the bloodstream to be carried to the cells of the body.

9. (B) Weeds crowd out more desirable plants, and keep them from growing properly.

10. (B) These animals have changed since the time of their common ancestor, but they still retain enough of the same genes for their forelimbs to be similar in structure.

11. (D) Bean leaves also have a network of veins.

12. (E) In a simple reflex, the sensation is received at a receptor in the skin, and the impulse travels along a sensory neuron to the spinal cord. It then crosses a synapse and enters an interneuron. It passes out across another synapse and enters a motor neuron where it travels along to stimulate an effector, a muscle, to contract and pull the finger back from the sensation, such as a flame.

13. (A) The roots of the plant contain xylem and phloem tissue which extend up through the stem into the leaves. Xylem transports water and minerals up to the stem and leaves. Phloem transports manufactured food from the leaves to other parts of the plant.

14. (E) All the animals are mammals, with microscopic ova having practically no stored food, except the sparrow, which has relatively large eggs (ova) containing a large amount of stored food.

15. (D) The possible results are:

	X	Y
X	XX	XY
X	XX	XY

Female: 50% normal; heterozygous 50% color-blind
Male: 50% normal; 50% color-blind

16. (B) In myxedema, there is an underproduction of thyroxin, resulting in a reduced basal metabolism rate.

17. (C) When hybrids are crossed, the results are:

	B	b
B	BB	Bb
b	Bb	bb

25% homozygous black ⎫ 75%
50% heterozygous black ⎬ black
25% white

18. (B) In pea plants, tall is dominant to short. Tall pea plants that produce short offspring must carry a recessive gene, and so are heterozygous tall. This can be shown as follows:

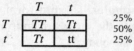

25% *TT*—homozygous tall
50% *Tt*—heterozygous tall
25% *tt*—short

19. (D) During active transport, the cell uses energy to move soluble materials across cell membranes, from a region of low concentration to a region of high concentration. By diffusion, molecules of soluble substances pass through cell membranes from a region of high concentration to a region of low concentration; this is passive transport since the cell does not use energy. In pinocytosis, relatively large particles, such as proteins, are engulfed by pockets of the cell membrane.

20. (A) The similarity of these embryos indicates that these animals are most probably descended from a common ancestor and that they still have some similar genes in their early development. The gill slits will disappear, as will the tail in the chicken embryo; the two-chambered heart will develop into four chambers in both animals.

21. (B) The theory of punctuated equilibrium proposed by Eldridge and Gould holds that some species remain unchanged for long periods. Then there are short periods during which new species are rapidly formed. It differs from Darwin's theory, which is based on the concept of gradualism, in which organisms change gradually during the ages.

22. (C) A lipid molecule consists of a glycerol molecule bonded to three fatty acid molecules.

23. (E) Tundra is the treeless region in the Arctic Zone, where the underlying part of the soil is permanently frozen.

24. (A) Strata of rock were originally formed from layers of sediment that were deposited under water and became cemented together by pressure over the ages. Thus, sand turned into sandstone, and shale was formed from deposits of clay. The lower strata were laid down first, so any fossils found in them would be older than fossils above them that became embedded in the upper strata later on.

25. (D) The small intestine is an organ composed of various types of tissues, including all those mentioned in the other choices, that work together.

26. (D) Studies of the DNA of humans and apes have shown that the sequence of nucleotides is very similar in both species. Other similarities in structure and biochemical composition have led scientists to speculate that humans and apes are derived from a common ancestor.

27. (D) The spider and the crab are members of the Arthropoda phylum, whose characteristics include jointed appendages, an exoskeleton of chitin, and segmented bodies. The snail is classified as a mollusk; it is a soft-bodied animal, with a muscular foot and a mantle that secretes the shell.

28. (D) Blood of the O type may be safely given in a transfusion to people who have other blood groups.

29. (E) The skin does not exhale air; this is done only through the lungs.

30. (C) Chloroplasts contain chlorophyll which becomes activated by light energy. High energy electrons of the excited chlorophyll are passed along by a series of coenzymes to split water and to form ATP. This takes place during photosynthesis in which sucrose and starch are end products. Mitochondria are "the powerhouses of the cell," in which the energy stored in glucose is converted, during aerobic respiration, to molecules of ATP, with carbon dioxide and water as the end products.

31. (A) A sensory neuron is also known as an afferent neuron.

32. (A) Darwin's theory of natural selection was published after many years of careful observation and study. He came to his conclusions after he had accumulated much evidence to support them, mostly based on the observations he himself had made. He was frank to admit that there were aspects of evolution he could not explain because he did not have enough data for a full understanding, for example, how variations were caused, and how life originated on earth.

33. (C) RNA contains a uracil (U) base instead of a thymine (T) base, which lines up with adenine (A). Cytosine (C) lines up with guanine (G). The adenine (A) of RNA lines up with the thymine (T) of DNA.

34. (D) Cambium consists of actively dividing cells that differentiate into xylem, phloem, and additional cambium cells, adding to the diameter of the trunk.

35. (E) A complex chain of organic molecules may be broken down during the process of hydrolysis. Water is added while enzymes break apart the bond between the molecules, and give rise to individual smaller molecules. Example: The disaccharide, maltose, is hydrolyzed by the enzyme maltase into the simpler molecules of glucose.

36. (D) Red blood cells and white blood cells are manufactured in the marrow of bones. Red blood cells form special round disc cells that lose their nuclei before they enter the bloodstream; they also contain hemoglobin, which gives them their red color.

37. (C) A food chain represents the different links along which food passes from one organism to another. The various links can be identified as: *producers* (green plants) which make food; *primary consumers* (herbivores) which eat green plants; *second- and third-order* consumers (carnivores) which prey on herbivores; and *decomposers*, which break down the wastes and dead bodies of producers and consumers.

38. (C) Without green plants there would be no food. Carnivorous animals eat herbivorous animals, which depend on green plants for their food.

39. (D) Because of the water barrier, the plants and animals on the mainland were prevented from migrating to the islands and interbreeding. Therefore, changes that took place in one area were not transmitted to the other area. Consequently, over a long period of time, these changes became accentuated, making the mainland and island species quite different from each other. This illustrates the principle that evolution in isolated populations proceeds in different directions.

40. (C) The graph shows that the relative rate of enzyme activity rises as the enzyme concentration increases until the concentration is about 7. After that point, the relative rate of enzyme activity levels off at about 35, and remains the same.

41. (C) When hybrids are crossed, the results are in a ratio of 3 dominant : 1 recessive. There would therefore be about three times as many smooth seeds as wrinkled seeds ($1,850 \times 3 = 5,550$, or approximately 5,474).

42. (A) Waste from antibiotic vats contains many substances. One of them, vitamin B_{12}, is known to increase the growth of animals. Evidently there must be an additional growth promoter other than vitamin B_{12} in the waste.

43. (B) If there is a drastic change in the environment of population *A*, many organisms that are not adapted to the new conditions will die out. The few that are adapted will survive and will pass their genes on to the next generation. By natural selection over a period of time, organisms that are fit for the new environment will continue to survive, increasing the gene frequency for the favorable characteristics. This may lead to evolutionary changes in population *A*.

44. (C) In a climax community, each species reproduces at a rate that maintains itself at the same number from year to year. There is an equilibrium between the living things in it and it may continue indefinitely. Examples of this include maple-beech forest in northeastern United States, pine forests in New Jersey, and tall grasses on the prairies.

45. (A) Plankton is the floating mass of microscopic life composed of algae that make their own food by photosynthesis, as well as protozoa and minute invertebrates that feed on the algae.

46. (B) As the prairie dog reproduction rate increases, there will be more animals depending on the available food in the given area. This will inevitably increase the competition among them for the food supply.

47. (E) Plasmodium, the protozoan that enters the body through the bite of the Anopheles mosquito, invades and destroys the red blood corpuscles of its host. It is a deadly parasite.

48. (C) The biosphere includes all the living things dwelling at or near the earth's surface. It is also thought of as including a system of relationships between the living things and the materials and energy surrounding them.

49. (E) Food vacuole

50. (E) Micronucleus

51. (B) Cell membrane

52. (A) The pulmonary veins bring oxygenated blood from the lungs to the left atrium of the heart.

53. (D) When the right ventricle contracts, it sends blood into the pulmonary arteries, which carry it to the lungs.

54. (C) The right atrium receives blood from the body circulation through the superior vena cava (from the head and upper part of the body) and the inferior vena cava (from the lower part of the body).

55. (C) Hemophilia is inherited as a sex-linked recessive trait.

56. (B) The change in a normal gene for pigment to the albino condition is inherited and is known as a mutation.

57. (D) Hair color and eye color are traits that are inherited independently of each other.

58. (B) The stem grows upward; this is negative geotropism.

59. (E) The paramecium moves away from salt, illustrating negative chemotaxis.

60. (A) The euglena moves toward the lighted side, illustrating positive phototaxis.

61. (B) Only *Senecio* is present. Ten multiplied by 4 gives a percentage occurrence of 40% for Group II.

62. (C) Only one example of *Lamium* was found in 25 trials.

63. (E) *Poa* and *Stellaria* were found to occur in every one of the 25 trials, making them the most numerous.

64. (A) The population curve for each species was not affected by the other species and resembled a population curve under normal conditions. This type of curve is often referred to as an S-shaped growth curve: at first, at low densities, it shows an accelerating rate of growth; then growth decelerates as the population density approaches the carrying capacity of the environment; finally, there is no further increase in density and the population continues in a steady state.

65. (C) The graph shows that *P. caudatum* eventually died out when mixed with *P. aurelia*.

66. (D) When the species were grown together, the population growth rate of *P. aurelia* was slower than normal, while *P. caudatum* died out. Both species were apparently adversely affected by competition for food and space.

67. (E) In the earlier years before the development of factories, the tree trunks were light in color. The protective coloration of the light-colored moths allowed them to escape the notice of insectivorous birds and to multiply freely. On the other hand, dark moths were conspicuous and more easily found by the birds. Graph 5 shows a far greater number of light-colored moths.

68. (A) Once the factories were established, the soot given off by the chimney smoke became deposited on the tree trunks, blackening them. The dark moths resembled this background and escaped the attention of the birds. The light moths, however, were conspicuous and were readily eaten by the birds. Graph 1 shows a predominance of dark moths.

69. (E) The light moths blended into the background of the light-colored tree trunks and escaped detection by the birds. The dark moths stood out against this background and were readily seen and eaten by the birds. Graph 5 shows the relative percentages of the survivors.

70. (C) The favorable variation was the color of the moths that resembled the color of the tree trunks, which had become dark from soot. These moths could not be readily seen by the birds. As a consequence, they escaped detection and survived. This natural selection did not favor the light-colored moths, which stood out against the color of the tree trunks, and they died out in larger numbers.

71. (D) A species is a closely related group of organisms that are similar in structure and can interbreed. If the two forms could not interbreed, they would be considered different species.

72. (D) The graphs show that the number of breaths per minute decreased with the reduction in temperature.

73. (D) All of the items except (D) represent good scientific procedure.

74. (C) At 15°C, the breathing rate in Graph II was 50 breaths per minute. In Graph I, at 16.6°C, it was only 44. In Graph II, it was 46 at 16.6°C.

75. (A) Since the day length in Miami is less than 14 hours throughout the summer, the cocklebur plant has more than 9 hours of darkness daily, and so can flower as soon as it is ripe to do so. In other parts of the country it will begin to flower when the days become short enough for this amount of darkness to set in daily.

76. (D) In early August, the day length in Winnipeg becomes reduced to 15 hours, and then proceeds to get shorter. This exposes the cocklebur plants to 9 or more hours of darkness, and they begin to form flower buds.

77. (E) Toward the end of September, all of the cities begin to have 12 hours or less of daylight. The tobacco plant therefore begins to produce flowers after that date.

78. (D) There were 429 pellets of meadow mice.

79. (C) Birds are not mammals; all the other choices are mammals.

80. (C) The owl is a very useful part of the farm environment, since it destroys animals that eat the farmer's grain and his crops. The evidence indicates that owls rarely go after birds, but live mainly on rodents and other small mammals.

Answer Sheet: Practice Test 3

1. Ⓐ Ⓑ Ⓒ Ⓓ Ⓔ
2. Ⓐ Ⓑ Ⓒ Ⓓ Ⓔ
3. Ⓐ Ⓑ Ⓒ Ⓓ Ⓔ
4. Ⓐ Ⓑ Ⓒ Ⓓ Ⓔ
5. Ⓐ Ⓑ Ⓒ Ⓓ Ⓔ
6. Ⓐ Ⓑ Ⓒ Ⓓ Ⓔ
7. Ⓐ Ⓑ Ⓒ Ⓓ Ⓔ
8. Ⓐ Ⓑ Ⓒ Ⓓ Ⓔ
9. Ⓐ Ⓑ Ⓒ Ⓓ Ⓔ
10. Ⓐ Ⓑ Ⓒ Ⓓ Ⓔ
11. Ⓐ Ⓑ Ⓒ Ⓓ Ⓔ
12. Ⓐ Ⓑ Ⓒ Ⓓ Ⓔ
13. Ⓐ Ⓑ Ⓒ Ⓓ Ⓔ
14. Ⓐ Ⓑ Ⓒ Ⓓ Ⓔ
15. Ⓐ Ⓑ Ⓒ Ⓓ Ⓔ
16. Ⓐ Ⓑ Ⓒ Ⓓ Ⓔ
17. Ⓐ Ⓑ Ⓒ Ⓓ Ⓔ
18. Ⓐ Ⓑ Ⓒ Ⓓ Ⓔ
19. Ⓐ Ⓑ Ⓒ Ⓓ Ⓔ
20. Ⓐ Ⓑ Ⓒ Ⓓ Ⓔ
21. Ⓐ Ⓑ Ⓒ Ⓓ Ⓔ
22. Ⓐ Ⓑ Ⓒ Ⓓ Ⓔ
23. Ⓐ Ⓑ Ⓒ Ⓓ Ⓔ
24. Ⓐ Ⓑ Ⓒ Ⓓ Ⓔ
25. Ⓐ Ⓑ Ⓒ Ⓓ Ⓔ
26. Ⓐ Ⓑ Ⓒ Ⓓ Ⓔ
27. Ⓐ Ⓑ Ⓒ Ⓓ Ⓔ

28. Ⓐ Ⓑ Ⓒ Ⓓ Ⓔ
29. Ⓐ Ⓑ Ⓒ Ⓓ Ⓔ
30. Ⓐ Ⓑ Ⓒ Ⓓ Ⓔ
31. Ⓐ Ⓑ Ⓒ Ⓓ Ⓔ
32. Ⓐ Ⓑ Ⓒ Ⓓ Ⓔ
33. Ⓐ Ⓑ Ⓒ Ⓓ Ⓔ
34. Ⓐ Ⓑ Ⓒ Ⓓ Ⓔ
35. Ⓐ Ⓑ Ⓒ Ⓓ Ⓔ
36. Ⓐ Ⓑ Ⓒ Ⓓ Ⓔ
37. Ⓐ Ⓑ Ⓒ Ⓓ Ⓔ
38. Ⓐ Ⓑ Ⓒ Ⓓ Ⓔ
39. Ⓐ Ⓑ Ⓒ Ⓓ Ⓔ
40. Ⓐ Ⓑ Ⓒ Ⓓ Ⓔ
41. Ⓐ Ⓑ Ⓒ Ⓓ Ⓔ
42. Ⓐ Ⓑ Ⓒ Ⓓ Ⓔ
43. Ⓐ Ⓑ Ⓒ Ⓓ Ⓔ
44. Ⓐ Ⓑ Ⓒ Ⓓ Ⓔ
45. Ⓐ Ⓑ Ⓒ Ⓓ Ⓔ
46. Ⓐ Ⓑ Ⓒ Ⓓ Ⓔ
47. Ⓐ Ⓑ Ⓒ Ⓓ Ⓔ
48. Ⓐ Ⓑ Ⓒ Ⓓ Ⓔ
49. Ⓐ Ⓑ Ⓒ Ⓓ Ⓔ
50. Ⓐ Ⓑ Ⓒ Ⓓ Ⓔ
51. Ⓐ Ⓑ Ⓒ Ⓓ Ⓔ
52. Ⓐ Ⓑ Ⓒ Ⓓ Ⓔ
53. Ⓐ Ⓑ Ⓒ Ⓓ Ⓔ
54. Ⓐ Ⓑ Ⓒ Ⓓ Ⓔ

55. Ⓐ Ⓑ Ⓒ Ⓓ Ⓔ
56. Ⓐ Ⓑ Ⓒ Ⓓ Ⓔ
57. Ⓐ Ⓑ Ⓒ Ⓓ Ⓔ
58. Ⓐ Ⓑ Ⓒ Ⓓ Ⓔ
59. Ⓐ Ⓑ Ⓒ Ⓓ Ⓔ
60. Ⓐ Ⓑ Ⓒ Ⓓ Ⓔ
61. Ⓐ Ⓑ Ⓒ Ⓓ Ⓔ
62. Ⓐ Ⓑ Ⓒ Ⓓ Ⓔ
63. Ⓐ Ⓑ Ⓒ Ⓓ Ⓔ
64. Ⓐ Ⓑ Ⓒ Ⓓ Ⓔ
65. Ⓐ Ⓑ Ⓒ Ⓓ Ⓔ
66. Ⓐ Ⓑ Ⓒ Ⓓ Ⓔ
67. Ⓐ Ⓑ Ⓒ Ⓓ Ⓔ
68. Ⓐ Ⓑ Ⓒ Ⓓ Ⓔ
69. Ⓐ Ⓑ Ⓒ Ⓓ Ⓔ
70. Ⓐ Ⓑ Ⓒ Ⓓ Ⓔ
71. Ⓐ Ⓑ Ⓒ Ⓓ Ⓔ
72. Ⓐ Ⓑ Ⓒ Ⓓ Ⓔ
73. Ⓐ Ⓑ Ⓒ Ⓓ Ⓔ
74. Ⓐ Ⓑ Ⓒ Ⓓ Ⓔ
75. Ⓐ Ⓑ Ⓒ Ⓓ Ⓔ
76. Ⓐ Ⓑ Ⓒ Ⓓ Ⓔ
77. Ⓐ Ⓑ Ⓒ Ⓓ Ⓔ
78. Ⓐ Ⓑ Ⓒ Ⓓ Ⓔ
79. Ⓐ Ⓑ Ⓒ Ⓓ Ⓔ
80. Ⓐ Ⓑ Ⓒ Ⓓ Ⓔ

PRACTICE TEST 3

CHAPTER

16

NOTE: The College Board test offered in Biology E/M contains a common core of 60 questions numbered 1–60, covering the general field of biology and a specialized section of 20 questions numbered 61–80 for Biology E (Ecology) or questions 81–100 for Biology M (Molecular). On the day of the examination, you may take ONE of the Biology E/M tests, either E or M, by marking the appropriate grid. Practice Test 3 of this book is designed to prepare you for the Biology M test. All of these Practice Tests have been created to present different core questions 1–60 in order to give you more experience in test preparation.

Practice Test 3: Biology M

Part A (Core Questions 1–60)

Directions: Each of the questions or incomplete statements below is followed by five suggested answers or completions. Choose the one that is best in each case and then blacken the corresponding space on the answer sheet.

1. Lower forms of life such as protozoa, sponges, and jellyfish have continued to exist since earliest geologic times. Which of the following statements would Darwin have considered to be the best explanation for the continued existence of lower life forms to modern times?

(A) These organisms are still in the process of becoming extinct.
(B) There is a fairly direct connection in evolution between the number of new forms that appear and the number of old ones that disappear.
(C) Some simpler forms of life will survive no matter what environmental changes occur.
(D) Some simpler organisms have a broad adaptation to the conditions of life.
(E) Simpler organisms continue to exist because they are able to inherit acquired characteristics.

2. In what part of the cell are proteins synthesized?

(A) Ribosome
(B) Nucleus
(C) Centriole
(D) Mitochondrion
(E) Golgi complex

3. The process of fertilization in both monocotyledons and mammals

(A) requires an external watery environment
(B) normally results in an embryo that requires yolk for food
(C) normally results in the production of monoploid offspring
(D) restores the monoploid number of chromosomes
(E) occurs within female reproductive organs

4. Which of the following procedures may be applied to help determine whether a child will be born having Down syndrome?

(A) Electron microscope examination of sperm and egg cells
(B) Weekly analysis of the mother's urine sample
(C) Amniocentesis
(D) Cloning
(E) X-ray diffraction of deoxyribose molecules

5. Which of the following represents the sequence of bases on the messenger RNA formed from a DNA sequence of CGG–ACT?

(A) CGG–TGA
(B) UGG–TGA
(C) GCC–UGA
(D) CGC–ACT
(E) GGG–TGA

6. A synapse is

(A) part of an axon
(B) part of a neuron
(C) a gap between neurons
(D) a chemical impulse
(E) an electrical impulse

7. The adrenal gland is located next to the

(A) liver
(B) brain
(C) bladder
(D) kidney
(E) neck

8. The Mesozoic Era is called the Age of Reptiles because at that time

(A) reptiles first appeared
(B) reptiles reached their greatest development
(C) the dinosaurs exterminated all other types of reptiles
(D) reptiles successfully resisted humans
(E) the dinosaurs became extinct

9. Darwin found that the finches on the various Galapagos Islands were somewhat different from each other because of

(A) different food
(B) geographic isolation
(C) different natural enemies
(D) different climates
(E) the inheritance of acquired characteristics

10. In diabetes, excess sugar is found in the urine because

(A) the basal metabolism is decreased
(B) the kidneys are in a state of hypertension
(C) the cells cannot utilize glucose
(D) an excess of insulin is produced
(E) insufficient thyroxin is formed

11. A paramecium is able to avoid obstacles in its path because its

(A) sense organs receive responses
(B) contractile vacuoles burst
(C) flagellum moves it away
(D) protoplasm has the property of irritability
(E) conditioned responses are inborn

12. Why can't a grasshopper be drowned by keeping its head under water?

(A) It breathes through openings in its abdomen.
(B) The spiracles in its head filter out oxygen.
(C) It can hold its breath for hours.
(D) It is cold-blooded.
(E) The maxillae and mandibles serve as an airtight valve.

13. Chestnut trees are becoming extinct in the eastern part of the United States because

(A) the chestnut crop has been overpicked
(B) the Japanese beetle eats the leaves and roots
(C) the tent caterpillar defoliates the trees
(D) the trees are susceptible to the white pine blister rust
(E) a fungus disease, the blight, is killing the trees

14. What is the reason that a blood clot in the coronary artery may be fatal?

(A) It leads to atherosclerosis.
(B) Varicose veins invariably result as a by-product.
(C) The portal circulation is interfered with.
(D) The flow of blood to the heart muscle is blocked.
(E) The clot leads to coagulation of lymph.

15. The fact that, under certain conditions, stomatal openings become smaller enables plants to avoid excessive loss of

    (A) carbon dioxide
    (B) oxygen
    (C) water
    (D) essential plant hormones
    (E) chlorophyll

16. Down syndrome is a human defect caused by the presence of an extra chromosome; the affected individual has 47 chromosomes in the body cells instead of 46. Which of the following is the cause of this condition?

    (A) Linkage
    (B) Nonlinkage
    (C) Nondisjunction
    (D) Sex linkage
    (E) Chromosomal replication

17. To the species involved, a major advantage of sexual reproduction over asexual reproduction is that sexual reproduction results in

    (A) greater number of offspring
    (B) greater variety of offspring
    (C) greater size of offspring
    (D) more rapid development of offspring
    (E) less vulnerability of offspring

18. The phase-contrast microscope has aided our understanding of the cell by permitting

    (A) detailed observations of unstained living cells
    (B) magnification of cells up to 100,000 times their normal size
    (C) observation of objects a centimeter or more in thickness
    (D) chemical analysis of parts of the cell
    (E) ultraviolet-light penetration of cells

19. A man who is normal for blood clotting ability marries a normal heterozygous woman who came from a family having a history of hemophilia. The chance of their daughters being normal heterozygous is

    (A) 25%
    (B) 0%
    (C) 100%
    (D) 50%
    (E) 75%

20. A brown male guinea pig was crossed with two different black females as shown in the following diagram:

Brown Male × Black Female (A)	Brown Male × Black Female (B)
Offspring:	*Offspring:*
9 Black	15 Black
8 Brown	0 Brown

Which of the following statements is most likely true?

    (A) All three parents were heterozygous.
    (B) All three parents were homozygous.
    (C) Female (A) is heterozygous.
    (D) All offspring of female (B) are homozygous.
    (E) All black guinea pigs are heterozygous.

21. Lactic acid, alcohol, and 2 ATPs are produced as a result of which of the following processes?

    I. Aerobic respiration
    II. Anaerobic respiration
    III. Photosynthesis

    (A) I only
    (B) II only
    (C) I and III only
    (D) II and III only
    (E) I, II, and III

22. All of the following are parasitic worms EXCEPT

    (A) leech
    (B) liver fluke
    (C) hookworm
    (D) tapeworm
    (E) ringworm

23. Which of the following kingdoms includes single-celled organisms?

    I. Monera
    II. Protista
    III. Fungi

    (A) I only
    (B) II only
    (C) I and III only
    (D) II and III only
    (E) I, II, and III

24. A measurement of 0.6 millimeter is equal to

    (A) 6 micrometers
    (B) 60 micrometers
    (C) 600 micrometers
    (D) 6,000 micrometers
    (E) 60,000 micrometers

25. The process of protein synthesis involves which of the following cellular organelles?

    I. Mitochondrion
    II. Ribosome
    III. Nucleus

    (A) I only
    (B) II only
    (C) III only
    (D) I and III
    (E) II and III

26. A drop of anti-A blood-typing serum was placed on the left side of a slide, and a drop of anti-B serum on the right side. Then a drop of blood was mixed into the serum on the left, and another drop into the serum on the right. If the blood was type A, the red blood cells would be clumped on

    (A) the right side only
    (B) the left side only
    (C) both sides
    (D) neither side
    (E) the type AB side

27. Enzyme synthesis in a living cell is most likely to occur in the

    (A) centriole
    (B) centrosome
    (C) chromosome
    (D) ribosome
    (E) food vacuole

28. Which of the following is true of the organic molecule depicted here?

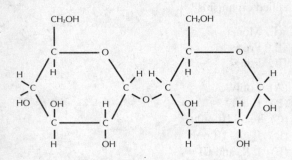

    (A) It is hydrolyzed to form glycerol.
    (B) It represents the building block of a type of protein.
    (C) It represents a building block utilized in the synthesis of a DNA molecule.
    (D) It represents a polysaccharide.
    (E) It is acted upon by the enzyme maltase.

29. In a nephron of the kidney, reabsorption of glucose generally takes place by the process of

    (A) osmosis
    (B) diffusion
    (C) active transport
    (D) passive transport
    (E) cyclosis

30. Which of the following plants is a member of the legume family?

    (A) Queen Anne's lace
    (B) Clover
    (C) Wheat
    (D) Buckwheat
    (E) Corn

31. The expression "Rh negative" refers to

    (A) immunity to rheumatic fever
    (B) susceptibility to rheumatic fever
    (C) a blood type
    (D) a tropism
    (E) a rootlike structure of a minus strain of mold

32. Digestive enzymes that hydrolyze molecules of fat into fatty acid and glycerol molecules are known as

    (A) lactases
    (B) lipases
    (C) leukocytes
    (D) lipids
    (E) lacteals

33. Which of the following conditions is NOT essential in a self-sustained ecosystem?

    (A) A means to permit the cycling of carbon between living organisms and their environment
    (B) A means to permit the cycling of water between living organisms and their environment
    (C) A constant supply of energy
    (D) A living system capable of incorporating energy into organic compounds
    (E) Equal numbers of plants and animals

34. The Hardy-Weinberg principle is a mathematical formulation that explains why gene frequencies can remain constant from generation to generation, under certain conditions. The fact that gene frequencies in a given population remain constant would be an indication that

   (A) evolution within the population would not take place
   (B) evolution would take place, but at a very slow rate
   (C) evolution would take place but at a very rapid rate
   (D) dominant characteristics would increase in the population and would eventually replace the recessive ones
   (E) recessive characteristics would increase in the population and would eventually replace the dominant ones

35. It is generally believed by scientists that chlorophyll and hemoglobin have a common origin. What is the best reason for believing this to be true?

   (A) All organisms contain either chlorophyll or hemoglobin.
   (B) Heterotroph aggregates have the same basic structure as chlorophyll and hemoglobin.
   (C) Chlorophyll can be changed into hemoglobin.
   (D) Both substances perform the same biological function.
   (E) The two substances are similar in chemical structure.

36. The concentration of carbon dioxide in the atmosphere remains relatively constant at about 0.04 percent as a result of established equilibrium between the processes of

   (A) assimilation and excretion
   (B) oxidation and photosynthesis
   (C) photosynthesis and assimilation
   (D) photosynthesis and reproduction
   (E) respiration and reproduction

37. Which of the following could be expected to cause competition among the chipmunk population in a certain area to increase?

   (A) A temporary increase in the chipmunk reproduction rate
   (B) An epidemic of rabies among the chipmunks
   (C) An increase in the number of chipmunks killed on the highways
   (D) An increase in the number of hawks that prey on chipmunks
   (E) An increase in secondary consumers

38. Which of the following human disorders is caused by a dietary lack?

   (A) Hemophilia
   (B) Phenylketonuria (PKU)
   (C) Pellagra
   (D) Sickle-cell anemia
   (E) Tay-Sachs disease

39. The methods of agriculture used by humans have created serious problems with insects, chiefly because these methods

   (A) increase soil erosion
   (B) provide concentrated areas of food for insects
   (C) increase the effectiveness of insecticides over a long period of time
   (D) Make it possible to grow crops in former desert areas
   (E) encourage insect resistance to their natural enemies

40. Under natural conditions large quantities of organic matter decay after each year's plant growth has been completed. Which of the following occurs as a result?

   (A) Many plants are deprived of adequate food supplies.
   (B) Many animals are deprived of adequate food supplies.
   (C) Soils soon become exhausted if not fertilized.
   (D) Soils maintain their fertility.
   (E) Decomposers are placed at a disadvantage in the food chain they are part of.

41. Hawks are useful to the farmer because they

   (A) eat only sickly chickens
   (B) scare crows away from cornfields
   (C) destroy mice
   (D) eat harmful insects
   (E) keep other birds from eating useful crops

42. After a surgical operation, which one of the following substances produced in the body is directly responsible for the rejection of a transplanted organ such as the kidney?

   (A) Amylase
   (B) Antibodies
   (C) Anticodons
   (D) Antigens
   (E) Asters

43. In the accompanying pedigree of a family, brown eye color (*B*) is indicated as ○, and blue eye color (*b*) as ●. From this chart it can be determined that

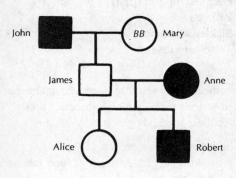

(A) Robert has the same genotype as his grandmother, Mary.
(B) Robert has the same genotype as his father, James.
(C) Alice has the same genotype as her grandfather, John.
(D) Alice has the same genotype as her father, James.
(E) Alice has the same genotype as her grandmother, Mary.

44. The complete oxidation of one molecule of glucose can be summarized in the following two equations:

(a) 1 glucose + 2 ATP + enzymes → 2 pyruvic acid + 4 ATP
(b) 2 pyruvic acid + O_2 + enzymes → CO_2 + H_2O + 34 ATP

What is the net gain of ATP molecules?

(A) 2
(B) 4
(C) 34
(D) 36
(E) 38

45. Dead trees partially buried under soil and stones may have termites living there and eating the wood. Within the digestive canal of the termites, the wood is digested by protozoa that live there. The abiotic factors in this habitat are the

(A) soil and stones
(B) termites and soil
(C) termites and protozoa
(D) protozoa and soil
(E) soil and tree

46. When crossing-over occurs during meiosis, the result is that

(A) the chromatids of nonhomologous chromosomes twist about each other
(B) the chromatids thicken and stay apart
(C) the chromatids fail to sort independently, creating an abnormal chromosome number
(D) genes are rearranged, increasing the variability of the next generation
(E) genes are replicated exactly to produce offspring identical to the parents

47. Cytochrome *c* is a protein that is present in mitochondria of cells. It acts as an electron carrier during cellular respiration. It has been determined that it is made up of a chain of 104 amino acids. When the sequence of these amino acids was compared for different organisms, it was found that only 1 molecule was different in monkeys and humans. By comparison, the cytochrome *c* of other organisms differ from humans as follows: rabbit-9; duck-11; turtle-15; tuna fish-21; yeast-45; bread mold-48. On the basis of this information, all of the following conclusions are correct EXCEPT for

(A) The greater the number of amino acid differences between two organisms, the more distant their evolutionary relationship.
(B) Organisms with few differences in their cytochrome *c* amino acids have a closer evolutionary relationship.
(C) Cold-blooded animals are more distant from humans in evolutionary relationships than warm-blooded animals.
(D) Humans are descended from monkeys.
(E) Cytochrome *c* molecules of humans and monkeys are very similar.

48. Most of the food and oxygen in the environment is produced by the action of

(A) heterotrophic organisms
(B) autotrophic organisms
(C) saprophytic bacteria
(D) aerobic bacteria
(E) archaebacteria

49. The most northern biome on earth is known as

(A) taiga
(B) grassland
(C) desert
(D) temperate deciduous forests
(E) tundra

50. Which one of the following concepts is NOT associated with the work of Gregor Mendel?

    (A) Dominant traits
    (B) Recessive traits
    (C) Structure of the gene
    (D) Segregation
    (E) Independent assortment

51. During experiments dealing with inheritance in plants, paper bags are placed over flowers to

    (A) keep them from blowing away
    (B) keep them warm
    (C) protect them from excessive sunlight
    (D) keep stray pollen away
    (E) prevent self-pollination

52. The part of the nervous system that controls the contractions of cardiac muscles is the

    (A) autonomic nervous system
    (B) somatic nervous system
    (C) cerebellum
    (D) double ventral nerve cord
    (E) cerebrum

53. The action of platelets in the blood breaking apart leads to the

    (A) dilation of arterioles
    (B) formation of red blood cells
    (C) formation of a clot
    (D) production of hemoglobin
    (E) production of oxyhemoglobin

54. Which one of the following structures is present in a human nephron?

    (A) Urethra
    (B) Ureter
    (C) Flame cell
    (D) Bowman's capsule
    (E) Malpighian tubule

Questions 55–57 refer to the following diagram of flower fertilization.

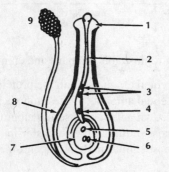

55. Which structure will develop into the fruit?

    (A) 1
    (B) 5
    (C) 7
    (D) 8
    (E) 9

56. After fertilization, the embryo of the seed is formed by

    (A) 3
    (B) 4
    (C) 5
    (D) 7
    (E) 8

57. After fertilization, the food of the seed is formed by

    (A) 4
    (B) 5
    (C) 6
    (D) 7
    (E) 8

Questions 58–60

Directions: Each set of lettered choices below refers to the numbered statements immediately following it. Select the one lettered choice that best fits each statement and then blacken the corresponding space on the answer sheet. A choice may be used once, more than once, or not at all in each set.

    (A) Adrenal gland
    (B) Thyroid gland
    (C) Pancreas
    (D) Pituitary gland
    (E) Small intestine

58. A man is very thin, very active, and has bulging eyes.

59. A boy who is being chased by a bully never ran so fast.

60. The fat lady of the circus weighs 300 pounds.

Part B (Choice M Questions 61–80)

61. Scientists estimate that the earth was formed about 4.6 billion years ago. A 12-month calendar used to represent the earth's history would show the following equivalent dates:

January 1 — The earth forms
April 30 — Earliest life appears
November 29 — First land plants
December 18 — Reptiles dominant
December 27 — Mammals abundant
December 31 — About 10 P.M. — Humans appear

On the basis of this calendar, which one of the following statements is correct?

(A) The heterotroph hypothesis went into effect on January 2
(B) Molecules of DNA appeared on January 3
(C) Archaebacteria appeared around hot vents on January 4
(D) Mammals became abundant when dinosaurs died out after December 18
(E) Fossils of Neanderthals were formed in caves on December 27

62. The correct order of molecules involved in protein synthesis is

(A) messenger RNA → transfer RNA → DNA → polypeptide
(B) DNA → messenger RNA → polypeptide → transfer RNA
(C) transfer RNA → polypeptide → DNA → messenger RNA
(D) DNA → messenger RNA → transfer RNA → polypeptide
(E) messenger RNA → DNA → transfer RNA → polypeptide

Questions 63–65

The molecular structure of the sugars of DNA and RNA can be represented as follows:

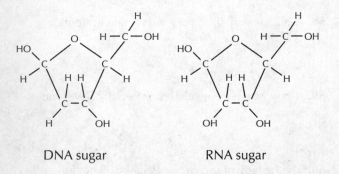

DNA sugar   RNA sugar

63. The chief difference in the sugar molecules of DNA and RNA is that

(A) one atom of oxygen is missing from the RNA sugar
(B) one atom of oxygen is missing from the DNA sugar
(C) a ribosome is missing in the RNA sugar
(D) a ribosome is missing in the DNA sugar
(E) a ribosome should be present in both sugars

64. The sugar of RNA is called

(A) glucose
(B) sucrase
(C) sucrose
(D) ribose
(E) dextrose

65. All of the following nitrogenous bases can be attached to the sugar of DNA EXCEPT

(A) adenine
(B) cytosine
(C) guanine
(D) thymine
(E) uracil

Questions 66–68

Scientists have made a biochemical comparison of five types of proteins found in both humans and chimpanzees. Their findings are summarized in the following table:

Protein	Amino Acid Differences	Number of Amino Acids
Hemoglobin	1	579
Myoglobin	1	153
Cytochrome *c*	0	104
Serum albumin	6*	580
Transferrin	8*	647

approximate

66. According to this table, humans and chimpanzees

(A) have equal amounts of all the proteins
(B) have equal amounts of the amino acids
(C) have equal amounts of serum albumin
(D) are most similar in their transferrin
(E) are most similar in their cytochrome *c*

67. The largest protein in the list is

(A) hemoglobin
(B) myoglobin
(C) cytochrome *c*
(D) serum albumin
(E) transferrin

68. The results for cytochrome *c* lead to the conclusion that

(A) neither humans nor chimpanzees need it in their respiratory chain
(B) both humans and chimpanzees differ in their amino acids for this protein
(C) humans and chimpanzees are closely related in their evolutionary origin
(D) fossils for humans and chimpanzees have identical DNA
(E) humans and chimpanzees have identical hemoglobin

Questions 69–71

The following drawing of the Fluid Mosaic model of the cell membrane depicts the approximate location of the cell's molecular activity involved in the sodium-potassium pump.

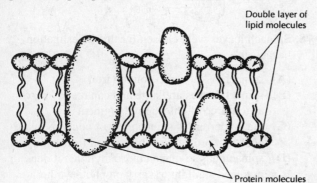

69. All of the following statements about the sodium-potassium pump are true EXCEPT

(A) It requires ATP as the energy source.
(B) It is an example of active transport.
(C) It results in a higher concentration of potassium ions within the cell.
(D) It results in a lower concentration of sodium ions outside the cell.
(E) It results in a lower concentration of sodium ions within the cell.

70. The sodium-potassium pump is located in

(A) a protein molecule of the cell membrane
(B) a lipid molecule of the cell membrane
(C) an ATP molecule of the cell membrane

(D) an ADP molecule of the cell membrane
(E) the cytoplasm itself

71. The action of the sodium-potassium pump results in

(A) a movement of ions against the concentration gradient
(B) a movement of ions in a direction equal to the concentration gradient
(C) a movement of ions in the opposite direction of the concentration gradient
(D) the exchange of equal numbers of sodium and potassium ions
(E) the addition of phosphate groups to the ATP molecule

Questions 72–74

The molecular composition of a biochemical reaction is represented below:

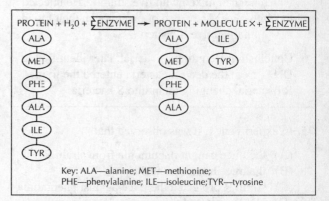

72. Molecule X can best be described as a

(A) disaccharide
(B) monosaccharide
(C) polysaccharide
(D) dipeptide
(E) diploploda

73. The bond that exists between methionine and phenylalanine is known as

(A) a phosphate bond
(B) a weak hydrogen bond
(C) a peptide bond
(D) an ionic bond
(E) a nucleotide bond

74. The reaction shown in the diagram is known as

(A) hydrolysis
(B) dehydration synthesis
(C) glycolysis
(D) photolysis
(E) carbon fixation

Questions 75–77

Early evidence that DNA is the genetic material of life was derived from experiments with pneumococcus bacteria. A particular strain (S) causes pneumonia in mice. Another strain (R) does not give them pneumonia. A scientist conducted the following experiments:

1. He injected strain S into mice.

2. He injected strain R into other mice.

3. He killed strain S with heat and injected it into mice; they lived.

4. He injected a mixture of dead strain S and living strain R into mice. They died of pneumonia. Blood samples now showed the presence of live S bacteria.

5. He grew strain S on a culture medium, killed it with heat, ground it up, and prepared an extract which he then mixed with live strain R. He injected this into mice. They died of pneumonia; their blood now contained live S bacteria.

6. Conclusion: Hereditary material, later identified as DNA, from the dead S bacteria entered the living R bacteria, changing them into S bacteria.

75. In experiment 1, it was observed that

  (A) the mice caught pneumonia from strain R
  (B) the mice remained healthy
  (C) strain R protected the mice against pneumonia
  (D) DNA protected the mice against pneumonia
  (E) the mice died of pneumonia

76. In experiment 5:

  (A) By evolution, the DNA of strain R became changed and no longer caused pneumonia.
  (B) Strain S extract contained living pneumococcus bacteria.
  (C) The mice caught pneumonia from strain R extract.
  (D) Strain S extract plus strain R caused pneumonia.
  (E) Although strain S was killed by heat, it still caused pneumonia without strain R.

77. The change in the heredity of strain R caused by DNA of strain S is known as

  (A) nucleic acid
  (B) transformation
  (C) transduction
  (D) transmission
  (E) extraction

Questions 78–80

A committee of students studying photosynthesis decided to separate the pigments contained in chlorophyll. They prepared an extract of chlorophyll from spinach leaves and placed a drop of this chlorophyll solution near the bottom of a strip of absorbent filter paper. When the drop dried, they added another drop to make it more concentrated. They repeated this step three more times.

After the last drop dried, they carefully suspended only the tip of the paper in a test tube containing a small amount of a solution made up of a mixture of petroleum ether and acetone. The spot of chlorophyll extract was now just above the level of the solution. As the solution was absorbed by the filter paper, it moved up, dissolving and depositing various pigments as shown in the accompanying diagram.

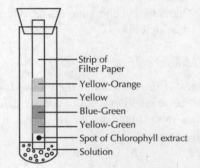

Strip of Filter Paper
Yellow-Orange
Yellow
Blue-Green
Yellow-Green
Spot of Chlorophyll extract
Solution

78. Spinach leaves were chosen for this investigation because

  (A) spinach is known to contain iron
  (B) spinach leaves are large and can readily serve as a source of chlorophyll pigments
  (C) the epidermis of spinach leaves contains guard cells that protect the chlorophyll from injury
  (D) spinach has so much chlorophyll that it does not need sunlight to carry on photosynthesis
  (E) when leaves are removed from a spinach plant, they cease to carry on transpiration

79. On the basis of this investigation, which of the following conclusions would be most valid?

  (A) The same results would be obtained with any other green plant.
  (B) The same results would be obtained only with large leaves of trees.
  (C) The same results would be obtained only with leaves of garden vegetables.
  (D) The same results would be obtained with leaves grown in the dark.
  (E) The same results could not be predicted with leaves of other plants.

80. The technique used to separate pigments is known as

(A) microanalysis
(B) X-ray diffraction
(C) chromatography
(D) spectography
(E) homeostasis

Answer Key: Practice Test 3

1. **(D)**	17. **(B)**	33. **(E)**	49. **(E)**	65. **(E)**
2. **(A)**	18. **(A)**	34. **(A)**	50. **(C)**	66. **(E)**
3. **(E)**	19. **(D)**	35. **(E)**	51. **(D)**	67. **(E)**
4. **(C)**	20. **(C)**	36. **(B)**	52 **(A)**	68. **(C)**
5. **(C)**	21. **(B)**	37. **(A)**	53. **(C)**	69. **(D)**
6. **(C)**	22. **(E)**	38. **(C)**	54. **(D)**	70. **(A)**
7. **(D)**	23. **(E)**	39. **(B)**	55. **(D)**	71. **(A)**
8. **(B)**	24. **(C)**	40. **(D)**	56. **(C)**	72. **(D)**
9. **(B)**	25. **(E)**	41. **(C)**	57. **(C)**	73. **(C)**
10. **(C)**	26. **(B)**	42. **(B)**	58. **(B)**	74. **(A)**
11. **(D)**	27. **(D)**	43. **(D)**	59. **(A)**	75. **(E)**
12. **(A)**	28. **(E)**	44. **(D)**	60. **(B)**	76. **(D)**
13. **(E)**	29. **(C)**	45. **(A)**	61. **(D)**	77. **(B)**
14. **(D)**	30. **(B)**	46. **(D)**	62. **(D)**	78. **(B)**
15. **(C)**	31. **(C)**	47. **(D)**	63. **(B)**	79. **(E)**
16. **(C)**	32. **(B)**	48. **(B)**	64. **(D)**	80. **(C)**

Self-Evaluation Chart: Your Road to More Knowledge and Improved Scores

Your Raw Score

Using the Answer Key at the end of the test, place a ✔ next to each correct answer and an ✘ next to each incorrect answer.

A) Number of correct (✔) answers   _____

B) Number of incorrect (✘) answers   _____

   Raw Score (A − B)   _____

Your College Board Score

Turn to the Biology E/M Conversion Table in Chapter 1 and determine your equivalent

   College Board Score   _____

Improving Your Score

1. In column *A* below, list the numbers of the questions that you did not answer correctly.
2. Turn to the Answers Explained section, and for each question number listed in column *A* write in column *B* the key word or phrase that best summarizes the main topic or point of the answer.
3. Look up the topic in the Index and review the material.
4. Go back to the test and try to answer again each of the questions you answered incorrectly the first time. Write your new answers in column *C*.
5. Compare the answers in column *C* with the Answer Key.
6. Calculate your revised Raw Score:
   Number of correct answers (A above + number of column *C* correct answers)   _____

   Revised Raw Score (number of correct answers − ¼ number of incorrect answers)   _____

   Revised Equivalent College Board Score   _____

A. Incorrectly answered questions	B. Main point(s) of the answer	C. Answers to questions in column A
_____	_____	_____
_____	_____	_____
_____	_____	_____
_____	_____	_____
_____	_____	_____
_____	_____	_____
_____	_____	_____
_____	_____	_____
_____	_____	_____
_____	_____	_____
_____	_____	_____
_____	_____	_____
_____	_____	_____
_____	_____	_____
_____	_____	_____
_____	_____	_____
_____	_____	_____

Answers Explained: Practice Test 3

1. (D) Darwin stated that the organism that possessed favorable variations would be most fit in the struggle for existence, and would survive. The lower forms of life are broadly adapted to the conditions surrounding them, and have continued to survive.

2. (A) A ribosome is a tiny granule located on the endoplasmic reticulum of the cytoplasm. It is the place where amino acids are bonded together to synthesize proteins.

3. (E) Monocotyledons are flowering plants in which fertilization occurs internally within the ovule after pollination. In mammals, fertilization occurs internally within the oviduct, following the introduction of sperm during mating.

4. (C) In amniocentesis, a doctor removes a small sample of amniotic fluid from the uterus of an expectant mother. This liquid surrounds the developing fetus and contains cells that are normally shed by it. The chromosomes are studied using a microscope. The presence of an extra chromosome in chromosome pair 21 may lead to Down syndrome, in which there is mental and physical impairment.

5. (C) RNA differs from DNA in having the base uracil (U) instead of thymine (T). In RNA, cytosine (C) bonds with guanine (G), and adenine (A) bonds with uracil (U) instead of with thymine (T).

6. (C) Neurons are not directly connected with each other. Impulses pass from the end brush of the axon of one neuron to the dendrites of another across the area of a synapse.

7. (D) The adrenal gland is located at the top of the kidneys.

8. (B) The dinosaurs were giant reptiles that dominated the earth during the Mesozoic Era.

9. (B) As the result of being isolated, any changes that occurred among the finches of one island remained concentrated there. These birds did not interbreed with those from other islands, and remained distinct from them.

10. (C) Insulin permits the cells to use glucose for energy. In diabetes, where there is a shortage of insulin, the glucose remains unused and is excreted as waste.

11. (D) A paramecium moves forward until it bumps into an object. Then it reverses the movement of its cilia, backs up, and moves forward again at a different angle. In this way, it can avoid obstacles in its way. At other times, it simply reverses its movement and goes off in a completely different direction.

12. (A) The grasshopper breathes by a system of tubes that connect with the openings, or spiracles, in its abdomen. Air is pumped in and out of the tubes by the contractions and expansions of the abdomen.

13. (E) The chestnut blight is a parasitic fungus that began destroying chestnut trees at the beginning of this century. By now there are practically no chestnut trees left in the eastern United States.

14. (D) The coronary arteries supply the cells of the heart with nourishment. A clot blocks the flow of blood through them and causes a heart attack. This condition is known as coronary thrombosis.

15. (C) Stomates are the tiny openings in the epidermis of a leaf. The size of the stomatal openings is controlled by pairs of guard cells. Water is constantly given off through these openings during transpiration. During conditions of dryness, the guard cells reduce the size of the openings, thereby avoiding the excessive loss of water.

16. (C) In nondisjunction, a pair of homologous chromosomes fails to separate during meiosis. This results in gametes having one chromosome more or less than the monoploid number. After fertilization, the new individual has more (as in Down syndrome) or less than the normal $2n$ number of chromosomes.

17. (B) In sexual reproduction, many unlike gametes are involved, each having its own genetic makeup. When gametes are brought together during fertilization, there will be many different combinations of genes, resulting in a great deal of variety in the offspring.

18. (A) The phase microscope uses different arrangements of light waves that are out of phase with each other, resulting in variations in light intensity. This makes possible the examination of normal living cells without the necessity of staining them.

19. (D) The possible genetic combinations would be:

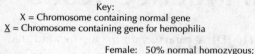

Key:
X = Chromosome containing normal gene
X̲ = Chromosome containing gene for hemophilia

Female: 50% normal homozygous;
         50% normal heterozygous
Male: 50% normal;
         50% hemophiliac

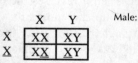

	X	Y
X	XX	XY
X̲	XX̲	X̲Y

20. (C) According to the diagram, black is dominant to brown in guinea pigs. Brown guinea pig is represented by bb. Black female (A) must be heterozygous Bb. The results of this cross may be represented as follows.

	B	b
b	Bb	bb
b	Bb	bb

Possible offspring
50% Bb – black (heterozygous)
50% bb – brown

Female (B) must be homozygous black, BB. The results of this cross may be represented as follows.

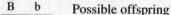

	B	B
b	Bb	Bb
b	Bb	Bb

Possible offspring
100% Bb – black (heterozygous)

The correct answer is that female (A) is heterozygous, Bb.

21. (B) Anaerobic respiration, also referred to as fermentation, takes place in the absence of oxygen. The only yield of energy is 2 ATPs. There is still a great deal of energy stored in lactic acid and alcohol molecules, where the C-C and C-H bonds remain intact.

22. (E) Ringworm is a parasitic fungus infection of the skin, not a worm.

23. (E) Bacteria are single-celled members of the Monera kingdom; paramecia are members of the Protista kingdom; yeasts are members of the Fungi kingdom.

24. (C) One millimeter contains 1,000 micrometers. Therefore, 0.6 mm contains 600 micrometers $(1,000 \times 0.6)$.

25. (E) RNA is produced in the nucleus and carries the genetic code for the formation of proteins to the ribosomes, where amino acids are assembled into proteins.

26. (B) Anti-A blood-typing serum contains an agglutinin that makes the red blood cells of type A blood clump together, or agglutinate. Anti-B serum contains an agglutinin that makes the red blood cells of type B blood clump. The blood mixed with each serum in this case was type A, since it clumped the red blood cells on the left side of the slide, which contained the anti-A serum.

27. (D) The ribosome is a tiny granule found in the endoplasmic reticulum of the cell. It contains most of the RNA of the cell, and is the site of protein synthesis, including enzyme formation.

28. (E) The structural formula of the disaccharide maltose is shown. It is a polymer of two glucose units. Under the influence of the enzyme maltase, it is hydrolyzed and converted into glucose.

29. (C) The reabsorption of glucose and other materials takes place by active transport. Energy is released by the mitochondria of the nephron to move the glucose molecules through the membrane back into the bloodstream. This movement is opposite to that which would normally occur through diffusion.

30. (B) Other members of the legume family are alfalfa, peanut, bean, and pea. These plants have nodules on their roots containing nitrogen-fixing bacteria. Their fruit is usually in the form of a pod.

31. (C) About 15% of the population has the Rh-negative blood type.

32. (B) Lipases are digestive enzymes that break down the relatively large molecules of fat into smaller molecules of fatty acid and glycerol.

33. (E) An ecosystem is a community of plants and animals that are self-sustaining in relation to their physical environment. In order to be self-sustaining, the community requires autotrophic organisms as its basis. These make their own food, using the constant source of energy available in the sun. The various cycles, such as carbon, water, and nitrogen, help to make the important elements available for the use of the organisms. As a result, matter and energy are constantly being exchanged between the community and the physical environment. Equal numbers of plants and animals are not significant in these relationships.

34. (A) According to the Hardy-Weinberg principle, gene frequencies remain constant from generation to generation if mating is at random, and no new factors are introduced, such as mutation, selection, or migration. Under such theoretical conditions, the species would remain predictably the same.

35. (E) Hemoglobin is closely related, chemically, to chlorophyll. However, the hemoglobin molecule contains iron, while the chlorophyll molecule contains magnesium. Both substances belong to a class of pigments known as porphyrins. According to the heterotroph hypothesis, primitive protein molecules may have developed, leading to porphyrin molecules that served as a common origin for chlorophyll and hemoglobin.

36. (B) During oxidation, living things take in oxygen and give off carbon dioxide and water as wastes. During photosynthesis, carbon dioxide and water are taken in by green plants and oxygen is given off as a waste. As a result, the concentrations of carbon dioxide and oxygen in the atmosphere remain relatively constant.

37. (A) A temporary increase in the chipmunk reproduction rate would cause an increase in the number of chipmunks seeking food in the given area. This would result in increased competition among them for the available food and other necessities of life.

38. (C) Pellagra is a nutritional disorder caused by lack of the vitamin niacin.

39. (B) Insects are the greatest enemies of humans. Although other forms of life have had difficulty maintaining themselves with the spread of civilization, insects have continued to multiply and spread. Humans have promoted this growth of insects by planting crops in large fields, thereby providing the insects with concentrated areas of food. The potato beetle and the corn borer, for example, do not have far to go for their food, once they settle in the right field.

40. (D) The organic material in plants is attacked by bacteria of decay and by fungi. The protein compounds are digested by them and broken down first into amino acids, and then into simpler nitrates and other compounds. Thus, the fertility of soils is maintained.

41. (C) The hawk is a carnivorous bird that is useful because it destroys more rodents than other birds. Mice are harmful because they eat grain.

42. (B) The transplanted organ from another person may act as an antigen and stimulate antibodies to attack it, causing the organ to be rejected by the recipient's body. Specifically, T cell antibodies are involved. A doctor may prescribe the drug cyclosporin to suppress the immune reaction of these antibodies.

43. (D) James is heterozygous for brown eyes, since he received a gene for blue eyes from his father, John, and a gene for brown eyes from his mother, Mary. James's daughter Alice is also heterozygous for brown eyes, having received a gene for brown eyes from him, and a gene for blue eyes from her mother, Anne.

44. (D) According to the first of the two equations, glucose is broken down by enzymes to yield 4 molecules of ATP. However, in order for some of the processes to occur, the energy from 2 ATP was used up. As a result, the total gain of energy during this phase is only 2 ATP molecules. The next equation shows the breakdown of pyruvic acid as yielding 34 ATP. The net gain of both equations is thus 2 + 34 = 36 ATP.

45. (A) Abiotic factors are the nonliving influences in the environment. In addition to soil and stones, these influences could also include light, temperature, water supply, oxygen supply, minerals, and pH.

46. (D) During synapsis, in the first part of meiosis, the chromatids of homologous chromosomes may twist about each other and then separate, exchanging adjacent parts. As a result, sets of genes cross over from one chromosome to another, producing new combinations of characteristics.

47. (D) Although they have similarities in their cytochrome *c* molecules, humans are not descended from monkeys. Scientists believe both are descended from a common ancestor. Since then, both species have developed differently. But there is much similarity between the two, as can be seen in their cytochrome *c* amino acids. In addition the sequence of nucleotides in the DNA molecule is similar and their proteins are almost identical. And when their skeletons are studied, the bones of humans and apes are very similar.

48. (B) Green plants that contain chlorophyll are considered to be autotrophic organisms because they can make their own food by photosynthesis. In the process, they give off oxygen into the atmosphere.

49. (E) Tundra is the treeless region in the most northern biome of the earth. The underlying part of the ground is permanently frozen. During the brief summer, the upper part thaws temporarily, and mosses and small plants grow abundantly.

50. (C) Gregor Mendel's paper on his experiments dealing with inheritance in pea plants was first published in 1866, long before the structure of the gene in the DNA molecule was identified in the middle of the 20th century.

51. (D) By covering the flowers with paper bags, stray pollen is prevented from introducing factors that are not part of the experiment.

52. (A) The autonomic nervous system consists of masses of ganglia located on either side of the backbone and plexuses, which control several involuntary actions including the beating of the heart. The heart is made up of cardiac muscle.

53. (C) The breaking up of platelets starts a series of reactions that ends in the formation of a clot. At the beginning, the enzyme thromboplastin is given off, which causes prothrombin to change into thrombin. This acts on fibrinogen to produce threads of fibrin. Red blood cells become entangled in the resulting network, and a clot is formed.

54. (D) Blood vessels in the kidney branch into a network of capillaries called the glomerulus, which is surrounded by a thin-walled cup, known as Bowman's capsule, that extends into a long nephron tubule.

55. (D) Ovary

56. (C) Egg nucleus

57. (C) Endosperm nuclei

58. (B) The man undoubtedly is suffering from an overactive thyroid gland. As a result of the extra amount of thyroxin, his metabolism is carried on at a faster rate.

59. (A) During times of emotional stress, such as the boy is experiencing, a greater amount of adrenaline is secreted by the adrenal glands; this causes an increased amount of glucose and oxygen to be brought to the muscles, supplying them with additional energy.

60. (B) In obesity, the thyroid gland is underactive, resulting in a lower metabolism rate. Food is not entirely oxidized, and the excess is stored as fat.

61. (D) Mammals became abundant during the present Cenozoic era after the dinosaurs, which had predominated during the earlier Mesozoic era, had died out.

62. (D) A part of a DNA molecule, which is a gene, serves as a template, or pattern, for the synthesis of a messenger RNA molecule. This mRNA now carries the genetic code, which consists of triplets of nucleotides called codons. The new mRNA leaves the nucleus, enters the cytoplasm, and becomes located in a ribosome. Here, small transfer RNA molecules (tRNA) which picked up amino acid molecules in the cytoplasm, line up with mRNA. The tRNA has a group of three molecules called an anticodon, which fits the appropriate codon on the mRNA. The tRNA separates when amino acid molecules are linked together to form a polypeptide.

63. (B) DNA represents deoxyribonucleic acid, indicating that its five-carbon sugar is deoxyribose. The *de* prefix means that it is missing an oxygen atom. By comparison, RNA has a five-carbon ribose sugar that contains an additional oxygen atom.

64. (D) As explained in the previous answer, the RNA molecule contains a five-carbon ribose sugar. This is different from the deoxyribose sugar of DNA which has one oxygen atom less.

65. (E) RNA contains the same bases as DNA, namely adenine, cytosine, and guanine, except for uracil which takes the place of thiamine.

66. (E) This table shows that there are no amino acid differences in cytochrome *c* of humans and chimpanzees, thus making it the most similar of the proteins in the list.

67. (E) The table shows that there are 647 amino acids in transferrin, the largest number of any of the proteins listed.

68. (C) Biochemical tests of protein similarities between humans and chimpanzees indicate that chimpanzees are the closest relatives to humans. The study of cytochrome *c*, a protein that functions in cellular respiration, shows that it is identical for the two species and that there are no amino acid differences in this protein between them. This is part of a large body of evidence indicating that humans and chimpanzees probably descended from a common ancestor during their evolutionary history.

69. (D) The action of the sodium-potassium pump results in the passage of sodium ions out of the cytoplasm into the exterior of the cell, where there is a higher concentration of sodium ions. This is against the concentration gradient and requires energy, which is supplied by ATP. This is an example of active transport.

70. (A) According to the Fluid Mosaic Model, the cell membrane consists of two layers of lipid molecules in which proteins are embedded, both on the surface and in the interior of the membrane. During the action of the sodium-potassium pump, a protein molecule becomes activated by ATP. It transports sodium ions out of the cell and potassium ions into the cell.

71. (A) As a result of the action of the sodium-potassium pump, sodium ions are moved out of the cell to the exterior, where there is a higher concentration of sodium ions. Also, potassium ions are moved from the exterior of the cell into the cytoplasm, resulting in a higher concentration of potassium ions. In both cases, ions are moved from a region of low concentration to a region of higher concentration. This is against the concentration gradient and requires the energy of ATP.

72. (D) A dipeptide consists of two molecules of amino acids, such as ILE and TYR, that combine by forming a bond between the amino group ($-NH_2$) of one and the acid group ($-COOH$) of the other.

73. (C) When two amino acids are combined they form a bond between the amino group of one and the acid group of the other. The amino group loses an atom of hydrogen, while the acid group loses an atom of hydrogen and an atom of oxygen. The hydrogen atoms combine with the oxygen atom to form a molecule of water, H_2O. A peptide bond connects the two amino acids and the water splits off.

74. (A) During the chemical process of hydrolysis, a large protein molecule is split into smaller protein molecules by adding water. Digestive enzymes speed up hydrolysis.

75. (E) In experiment (1), mice were injected with strain S pneumococcus bacteria. This strain of bacteria caused pneumonia, and the mice died of the disease.

76. (D) In experiment (5), strain S bacteria were grown on a culture medium. They were then killed by heat and ground up. An extract of the dead bacteria was prepared and mixed with live strain R which is normally harmless. However, in this case, the mice died of pneumonia. Under the microscope, their blood was seen to contain strain S bacteria.

77. (B) Transformation is the transfer of DNA from dead bacteria into living bacteria, as was done in experiments (4) and (5). The result was that the heredity of harmless strain R bacteria became changed into harmful strain S bacteria.

78. (B) Spinach leaves are readily available, are relatively large, and are a ready source of chlorophyll.

79. (E) One should not generalize about the results of a single investigation and try to apply the same conclusion to other plants. It is necessary to repeat the experiment with other plants before drawing any conclusions.

80. (C) Much information about the various steps of photosynthesis has been obtained through the application of refined techniques of chromatography. In paper chromatography, the various pigments are separated out at different levels as they are absorbed at different rates on a strip of absorbent filter paper.

Answer Sheet: Practice Test 4

1. (A) (B) (C) (D) (E)
2. (A) (B) (C) (D) (E)
3. (A) (B) (C) (D) (E)
4. (A) (B) (C) (D) (E)
5. (A) (B) (C) (D) (E)
6. (A) (B) (C) (D) (E)
7. (A) (B) (C) (D) (E)
8. (A) (B) (C) (D) (E)
9. (A) (B) (C) (D) (E)
10. (A) (B) (C) (D) (E)
11. (A) (B) (C) (D) (E)
12. (A) (B) (C) (D) (E)
13. (A) (B) (C) (D) (E)
14. (A) (B) (C) (D) (E)
15. (A) (B) (C) (D) (E)
16. (A) (B) (C) (D) (E)
17. (A) (B) (C) (D) (E)
18. (A) (B) (C) (D) (E)
19. (A) (B) (C) (D) (E)
20. (A) (B) (C) (D) (E)
21. (A) (B) (C) (D) (E)
22. (A) (B) (C) (D) (E)
23. (A) (B) (C) (D) (E)
24. (A) (B) (C) (D) (E)
25. (A) (B) (C) (D) (E)
26. (A) (B) (C) (D) (E)
27. (A) (B) (C) (D) (E)

28. (A) (B) (C) (D) (E)
29. (A) (B) (C) (D) (E)
30. (A) (B) (C) (D) (E)
31. (A) (B) (C) (D) (E)
32. (A) (B) (C) (D) (E)
33. (A) (B) (C) (D) (E)
34. (A) (B) (C) (D) (E)
35. (A) (B) (C) (D) (E)
36. (A) (B) (C) (D) (E)
37. (A) (B) (C) (D) (E)
38. (A) (B) (C) (D) (E)
39. (A) (B) (C) (D) (E)
40. (A) (B) (C) (D) (E)
41. (A) (B) (C) (D) (E)
42. (A) (B) (C) (D) (E)
43. (A) (B) (C) (D) (E)
44. (A) (B) (C) (D) (E)
45. (A) (B) (C) (D) (E)
46. (A) (B) (C) (D) (E)
47. (A) (B) (C) (D) (E)
48. (A) (B) (C) (D) (E)
49. (A) (B) (C) (D) (E)
50. (A) (B) (C) (D) (E)
51. (A) (B) (C) (D) (E)
52. (A) (B) (C) (D) (E)
53. (A) (B) (C) (D) (E)
54. (A) (B) (C) (D) (E)

55. (A) (B) (C) (D) (E)
56. (A) (B) (C) (D) (E)
57. (A) (B) (C) (D) (E)
58. (A) (B) (C) (D) (E)
59. (A) (B) (C) (D) (E)
60. (A) (B) (C) (D) (E)
61. (A) (B) (C) (D) (E)
62. (A) (B) (C) (D) (E)
63. (A) (B) (C) (D) (E)
64. (A) (B) (C) (D) (E)
65. (A) (B) (C) (D) (E)
66. (A) (B) (C) (D) (E)
67. (A) (B) (C) (D) (E)
68. (A) (B) (C) (D) (E)
69. (A) (B) (C) (D) (E)
70. (A) (B) (C) (D) (E)
71. (A) (B) (C) (D) (E)
72. (A) (B) (C) (D) (E)
73. (A) (B) (C) (D) (E)
74. (A) (B) (C) (D) (E)
75. (A) (B) (C) (D) (E)
76. (A) (B) (C) (D) (E)
77. (A) (B) (C) (D) (E)
78. (A) (B) (C) (D) (E)
79. (A) (B) (C) (D) (E)
80. (A) (B) (C) (D) (E)

PRACTICE TEST 4

CHAPTER

17

NOTE: The College Board test offered in Biology E/M contains a common core of 60 questions numbered 1–60, covering the general field of biology, and a specialized section of 20 questions numbered 61–80 for Biology E (Ecology) or questions 81–100 for Biology M (Molecular). On the day of the examination, you may take ONE of the Biology E/M tests, either E or M, by marking the appropriate grid. Practice Test 4 of this book is designed to prepare you for the Biology M test. All these Practice Tests have been created to present different core questions 1–60 in order to give you more experience in test preparation.

Practice Test 4: Biology M

Part A (Core Questions 1–60)

Directions: Each of the questions or incomplete statements below is followed by five suggested answers or completions. Choose the one that is best in each case and then blacken the corresponding space on the answer sheet.

1. Niacin will most likely be found in foods containing vitamin

   (A) A
   (B) B₁
   (C) C
   (D) D
   (E) E

2. The housecat is classified as *Felis domestica*. The name "Felix the Cat" is derived from which part of the scientific name?

   (A) The name of the phylum
   (B) The name of the class
   (C) The name of the order
   (D) The name of the genus
   (E) The name of the species

3. The process that most directly enables a root hair cell to absorb minerals by active transport, and enables a muscle cell to contract, is

   (A) circulation
   (B) ingestion
   (C) digestion
   (D) excretion
   (E) respiration

4. Which of the following is an exception to the cell theory?

   (A) A virus
   (B) A protozoan
   (C) A bacterium
   (D) A human
   (E) A rosebush

5. A semipermeable membrane is placed in the middle of a U-tube. This membrane is completely permeable to water molecules but not to sucrose molecules. A 20 percent sucrose solution is placed in side *A* of the U-tube and equal amount of distilled water in side *B*.

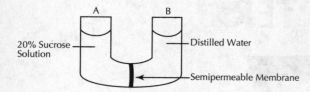

After 3 hours, what change will be evident?

(A) The liquid level will be higher in side *B* than in side *A*.
(B) The liquid level will be higher in side *A* than in side *B*.
(C) The liquid level will be the same in both sides.
(D) The concentration of sucrose will be the same in both sides.
(E) The concentration of sucrose will be higher in side *B*.

6. A person who has tested positive for the virus known as HIV is likely to develop

(A) Lyme disease
(B) trichinosis
(C) AIDS
(D) malaria
(E) rickets

7. Which of the following represent(s) a carbohydrate?

I. $C_{12}H_{22}O_{11}$
II. $C_{17}H_{35}COOH$
III. $(C_6H_{10}O_5)_n$

(A) I only
(B) II only
(C) III only
(D) I and II only
(E) I and III only

8. The cells of a *Drosophilia* fly have been shown to contain 8 chromosomes. The number of chromosomes in the fertilized egg is

(A) 2
(B) 4
(C) 8
(D) 12
(E) 16

9. A paramecium contains all of the following EXCEPT

(A) cilia
(B) a contractile vacuole
(C) a food vacuole
(D) an anal spot
(E) a chloroplast

10. The following diagram represents a fluid mosaic model of a cell membrane:

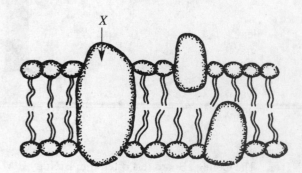

The structure labeled *X* represents a

(A) protein
(B) carbohydrate
(C) lipid
(D) fatty acid
(E) glycerol

11. Which process is represented in the following equation:

$$C_6H_{12}O_6 + 6O_2 \xrightarrow[\text{enzymes}]{\text{numerous}} 6CO_2 + 6H_2O?$$

(A) Aerobic respiration
(B) Anaerobic respiration
(C) Fermentation
(D) Phosphorylation
(E) Photosynthesis

12. Chromatin is to chromosomes as DNA is to

(A) maturation
(B) daughter cells
(C) genes
(D) mitosis
(E) meiosis

13. The fact that ameba subjected to a very bright light will move away from it shows that

(A) its protoplasm has the property of irritability
(B) the ameba is not able to make its own food
(C) movement occurs only when conditions are favorable
(D) transport is an inherited characteristic
(E) nucleotides are not formed in very bright light

14. The tiny openings in the epidermis of a leaf are known as

(A) guard cells
(B) stomates
(C) lenticels
(D) grana
(E) veins

15. The evaporation of water from leaves is called

(A) translocation
(B) transformation
(C) translation
(D) transpiration
(E) transcription

16. An organelle that is visible only with the aid of an electron microscope is the

(A) nucleus
(B) chloroplast
(C) cell wall
(D) ribosome
(E) vacuole

17. A biochemist analyzed an unknown organic compound and determined its chemical makeup. The accompanying chart summarizes the results of the biochemist's analysis

Element	Number of Atoms per Molecule
C	16
H	32
O	2
S	0
N	0
P	0

On the basis of this chart, which class of organic compounds was the biochemist studying?

(A) Carbohydrates
(B) Lipids
(C) Proteins
(D) Nucleic acids
(E) Amino acids

18. On the basis of their patterns of nutrition, most animals are classified as

(A) decomposers
(B) autotrophic
(C) heterotrophic
(D) photosynthetic
(E) anaerobic

19. The U.S. Department of Agriculture has issued the following Food Guide Pyramid, containing recommendations for daily food choices:

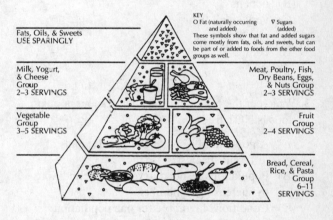

Health authorities believe that, if more Americans heeded these food guidelines, there would be a decrease in the number of cases of

(A) Alzheimer's disease and hepatitis
(B) acromegaly and pellagra
(C) heart disease and cancer
(D) pernicious anemia and genital herpes
(E) malaria and ringworm

20. What are the dominant species in a taiga biome?

(A) Sagebrush and cactus
(B) Maple and oak
(C) Deciduous trees
(D) Spruce and pine
(E) Moss and lichen

21. In the following food chain, what is the role of the rabbit?

     Cabbage plant → rabbit → fox

     (A) Saprophyte
     (B) Consumer
     (C) Producer
     (D) Decomposer
     (E) Carnivore

22. The basis of life in the ocean food chain is

     (A) heterotroph aggregates
     (B) lichens
     (C) very small fish
     (D) plankton
     (E) coral

23. Which of the following statements about populations is NOT true?

     (A) A population is made up of individuals of different species.
     (B) Populations respond to favorable conditions by increasing in number.
     (C) Conditions favorable for one population may be unfavorable for another.
     (D) Individuals of a population interact with each other.
     (E) Individuals of a population can interbreed.

24. All the living things present in 30 cubic centimeters of soil make up a

     (A) community
     (B) niche
     (C) species
     (D) biome
     (E) biosphere

25. The lobster, spider, and wasp are classified in the same phylum. Which of the following sets of characteristics do they have in common?

     (A) Segmented body and six appendages
     (B) Jointed appendages and a digestive system with a single opening
     (C) Jointed appendages and a chitinous exoskeleton
     (D) A dorsal nerve cord and a chitinous exoskeleton
     (E) A dorsal nerve cord and jointed appendages

26. In pea plants, the allele for colored flower is dominant over the allele for white flower. In an experiment, plants with colored flowers are cross-pollinated with plants having white flowers. The cross results in 271 colored flowers and 273 white flowers. On the basis of these results, the genotype of the parental plant with colored flowers can be correctly described as

     (A) homozygous
     (B) heterozygous
     (C) incompletely dominant
     (D) pure dominant
     (E) pure recessive

27. The hemoglobin of a person with sickle-cell anemia differs from normal molecules of hemoglobin by one

     (A) monosaccharide, fructose
     (B) disaccharide, sucrose
     (C) fatty acid, glutamic acid
     (D) lipid, oleic acid
     (E) amino acid, valine

28. The phenomenon known as crossing-over occurs during

     (A) mitosis
     (B) meiosis
     (C) fertilization
     (D) geographic distribution
     (E) active transport

29. If a pair of heterozygous black guinea pigs are mated, and there are four offspring in the litter, their appearance may be

     (A) all black
     (B) 3 black, 1 white
     (C) 2 black, 2 white
     (D) 1 black, 3 white
     (E) any of the above

30. The organs of excretion in human beings include which of the following?

       I. Lungs
      II. Kidneys
     III. Skin

     (A) I only
     (B) II only
     (C) I and II only
     (D) II and III only
     (E) I, II, and III

31. An increase in the diameter of an oak tree is caused chiefly by the activity of the

    (A) vascular ducts
    (B) bark
    (C) lenticels
    (D) cambium
    (E) annual rings

32. Which organ of the human digestive system produces hydrochloric acid?

    (A) Spleen
    (B) Gall bladder
    (C) Stomach
    (D) Small intestine
    (E) Large intestine

33. If the heart of a frog that has just been killed is suspended in a beaker of Ringer's solution (which includes 0.45 percent common salt, NaCl), the heart will continue to beat at the same rate as that of a live frog. If the heart is then immersed in a 0.6 percent water solution of common salt, it will beat more slowly and weakly. If it is then replaced in Ringer's solution, it will recover its normal rate.

    The best interpretation of this demonstration is that

    (A) salt is harmful to the frog's tissues
    (B) there is not enough salt in the 0.6 percent water solution
    (C) Ringer's solution is chemically identical to vertebrate plasma
    (D) Ringer's solution provides a chemical environment similar to that of a frog's heart
    (E) the heart is too complicated an organ to be able to beat outside the frog's body

34. A substance named calvacin, obtained by extraction from a species of giant puffball mushroom, has been used to prevent the growth of certain types of tumors. The most probable immediate use of this substance would be as a

    (A) supplement to the diet of cancer patients
    (B) supplement to the use of radioisotopes in treating cancer patients
    (C) basis for new cancer experiments
    (D) new method of treating cancer
    (E) new method of destroying unwanted growths in both plants and animals

35. Special structures for the absorption of digested food in the small intestine are called

    (A) alveoli
    (B) tubules
    (C) mucous glands
    (D) villi
    (E) peristalsis

36. A person who is exercising breathes more rapidly because

    (A) more carbon dioxide is given off, which stimulates the medulla
    (B) the lungs use up more energy
    (C) the calorie production goes down
    (D) glycogen in the liver is converted to glucose
    (E) glycogen in the muscles is converted to glucose

37. The body often has extra strength in an emergency because

    (A) the storage of glucose is stimulated
    (B) the activity of the digestive tract is stimulated
    (C) the production of pancreatic juice is stimulated
    (D) the rate of peristalsis is stimulated
    (E) the secretion of adrenaline is stimulated

38. Urine leaves the human body through the

    (A) ureter
    (B) urethra
    (C) glomerulus
    (D) epiglottis
    (E) alveoli

39. A combination of phosphate, ribose, and a nitrogen base is characteristic of a

    (A) carbohydrate
    (B) lipid
    (C) protein
    (D) nucleotide
    (E) fatty acid

40. A substrate is the substance

    (A) from which an enzyme is formed
    (B) by which an enzyme is deactivated
    (C) on which an enzyme acts
    (D) with which an enzyme forms bonds
    (E) by which an enzyme is phosphorylated

41. The oldest known dinosaur fossil, *Eoraptor,* was found in Argentina embedded in rocks near the Andes mountain. Paleontologists estimate that this early dinosaur lived about 225 million years ago. This fossil was probably formed in

(A) igneous rock
(B) a glacier
(C) sedimentary rock
(D) amber
(E) molten lava that cooled

42. Earth Day on April 22 is observed around the world in more than 140 countries in an organized effort to protect our planet from problems brought on by threats to the environment. Among these threats is

(A) contour plowing
(B) the pyramid of energy
(C) pheromone secretion by insects
(D) the nitrogen cycle
(E) destruction of tropical rain forests

43. In the accompanying pedigree of a family, brown eye color is indicated as ◯, and blue eye color as ●.

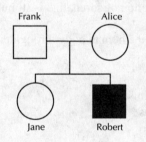

From this chart it can be determined that

(A) Frank is homozygous and Alice is heterozygous for brown eyes
(B) Robert is heterozygous for blue eyes
(C) Frank and Alice are both homozygous for brown eyes
(D) Frank and Alice are both heterozygous for brown eyes
(E) Frank is heterozygous and Alice is homozygous for brown eyes

44. The primary function of NAD in the Krebs cycle is to serve as

(A) an oxygen acceptor
(B) an oxygen donor
(C) a source of phosphate ions
(D) a hydrogen ion acceptor
(E) a source of glycolysis

45. Which of the following result(s) in an inherited change?

I. Gene mutation
II. Chromosomal aberration
III. Mineral deficiency

(A) I only
(B) II only
(C) III only
(D) I and II only
(E) I and III only

46. The belief that, because of a need, animals acquired the ability to move from an aquatic environment onto land is most closely associated with a theory proposed by

(A) Darwin
(B) De Vries
(C) Weissmann
(D) Lamarck
(E) Mendel

47. The Human Genome Project was planned by scientists to map the positions of all the genes in their proper sequence on the chromosomes of a cell. What is expected to be one of the benefits of this undertaking?

(A) It will prove the reliability of Mendel's Law of Dominance.
(B) It will offer a more precise understanding of the defective genes that cause inherited disease.
(C) It will help compare the chromosomes of humans and *Drosophila* flies.
(D) It will prove or disprove Lamarck's theory of evolution.
(E) It will show why the gene pool of a population remains stable.

48. The most likely explanation for the presence of useless hipbones in the whale is that

(A) the whale is descended from the coelocanth
(B) the whale is descended from an ancestor that used hipbones
(C) all vertebrates have hipbones
(D) the comparative anatomy of the whale is like that of any other water animal
(E) all vertebrates have four limbs

49. The bat is classified in the same class as a chimpanzee for all of the following reasons EXCEPT

(A) it has hair
(B) it suckles
(C) it forms eggs with the diploid number of chromosomes
(D) its young are developed internally
(E) it is warm-blooded

Questions 50–52 refer to the following diagram of the human heart.

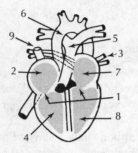

50. Chamber that receives blood from the lungs

(A) 1
(B) 3
(C) 5
(D) 7
(E) 9

51. Chamber that receives blood from the lower and upper parts of the body

(A) 1
(B) 2
(C) 3
(D) 4
(E) 7

52. Artery carrying deoxygenated blood

(A) 1
(B) 5
(C) 6
(D) 7
(E) 9

Directions: Each set of lettered choices below refers to the numbered statements immediately following it. Choose the one lettered choice that best fits each statement and then blacken the corresponding space on the answer sheet. A choice may be used once, more than once, or not at all in each set.

Questions 53–55

(A) Ciliated epithelium
(B) Diaphragm
(C) Pancreas
(D) Spinal cord
(E) Esophagus

53. Helps to remove foreign particles in the trachea

54. Produces both enzymes and hormones

55. Is part of the alimentary canal that does not secrete enzymes

Questions 56–58

(A) Greenhouse effect
(B) Depletion of the ozone layer
(C) Toxic waste
(D) Acid rain
(E) Excessive eutrophication

56. May result in global warming

57. Results in an increase in ultraviolet radiation

58. Is caused by an increase in the carbon dioxide and methane contents of the atmosphere

Questions 59–60

(A) Mitochondria
(B) Ribosomes
(C) Centrosomes
(D) Endoplasmatic reticulum
(E) Lysosomes

59. Contains enzymes associated with cell respiration

60. Contain(s) digestive enzymes that break down worn-out organelles within the cell

Part B (Choice M Questions 61–80)

61. The activity of an enzyme at various temperatures is summarized in the accompanying graph.

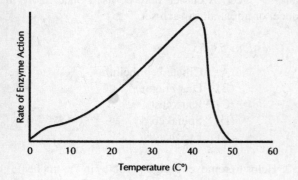

A valid conclusion to be drawn from the graph is that

(A) at a temperature of 30°, the activity of the enzyme is twice as low as it is at 20°
(B) at a temperature of 10°, the activity of the enzyme is twice as low as at 30°
(C) the activity of the enzyme reaches a maximum at 40°
(D) the activity of the enzyme reaches a maximum at 50°
(E) at 60°, the rate of activity is at a maximum

62. Glycolysis occurs during

 I. Anaerobic respiration
 II. Aerobic respiration
 III. Photosynthesis

(A) Only I
(B) Only II
(C) Only I and II
(D) Only I and III
(E) Only III

63. The following diagram represents part of the process of cellular respiration:

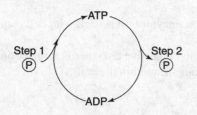

Energy is released and made available for metabolic activities at

(A) only step 1
(B) only step 2
(C) both step 1 and step 2
(D) neither step 1 nor step 2
(E) P

64. Anticodon is to tRNA as codon is to

(A) codeine
(B) sickle cell anemia
(C) dominant gene
(D) mRNA
(E) ribosome

65. A lipid is composed of

(A) three fatty acids and one glycerol molecule
(B) three glycerol molecules and one fatty acid
(C) three glycerol molecules and one hydrocarbon chain
(D) a hydrocarbon chain and a carboxyl group
(E) carbon, hydrogen, oxygen, and one sulfur group

66. All of the following statements about photolysis are true EXCEPT

(A) It takes place during photosynthesis.
(B) It is part of the Krebs cycle.
(C) It results in the splitting of water into hydrogen ions and oxygen.
(D) It is part of the so-called light reaction.
(E) It is centered in chlorophyll.

67. The method by which a virus attacks a host includes all of the following steps EXCEPT

(A) It attaches its protein coat to the host cell.
(B) It injects its nucleic acid into the host cell.
(C) Its nucleic acid replicates using the nucleic acid of the acid host cell.
(D) It reproduces by spore formation.
(E) Its replicated units burst out of the host cell.

Questions 68–70 refer to two stages of a metabolic process shown below:

Stage 1
   $X + 2 \text{ ATP} \rightarrow$ pyruvic acid $+ 4 \text{ ATP}$

Stage 2
   2 pyruvic acid + oxygen $\rightarrow CO_2 + H_2O + 34 \text{ ATP}$

---

68. The metabolic process is

(A) dehydration synthesis
(B) photosynthesis
(C) fermentation
(D) aerobic respiration
(E) anaerobic respiration

69. Which raw material, represented by the letter X, is needed for the stage 1 reaction to occur?

(A) chlorophyll
(B) nitrogen
(C) PGAL
(D) $C_6H_{12}O_6$
(E) sunlight

70. What is the net gain in ATP from the two stages in this metabolic process?

(A) 40
(B) 36
(C) 34
(D) 30
(E) 38

Questions 71–73.
Base your answers on the following two chemical reactions.

---
1)   maltose + water $\xrightarrow{A}$ glucose + B

2)   fat + water $\xrightarrow{C}$ glycerol + fatty acids

---

71. Letter A most likely represents

(A) a hormone
(B) a neurotransmitter
(C) an organic catalyst
(D) an amino group
(E) a carboxyl group

72. Letter B represents a

(A) deoxyribose sugar
(B) ribose sugar
(C) dipeptide molecule
(D) polymer
(E) monosaccharide

73. Compound C is most likely

(A) protease
(B) sucrase
(C) maltase
(D) amylase
(E) lipase

Questions 74–78.
The mitochondria of cells have been found to contain cytochrome *c*, an enzyme involved in the process of respiration. The amino acid sequence of cytochrome *c* was compared for different species. The differences in these sequences are shown in the adjacent chart. The numbers indicate how many amino acids in the cytochrome *c* content of a certain species differ from those of the other species.

	Human	Monkey	Horse	Rabbit	Duck	Turtle	Tuna fish	Moth	Yeast	Bread mold
Human	—									
Monkey	1	—								
Horse	12	11	—							
Rabbit	9	8	4	—						
Duck	11	10	8	10	—					
Turtle	15	14	11	9	7	—				
Tuna fish	21	21	19	17	17	18	—			
Moth	31	30	29	26	27	28	32	—		
Yeast	45	45	46	45	46	49	47	47	—	
Bread mold	48	47	46	46	46	49	48	47	41	—

Differences in the number of amino acid sequences of cytochrome *c* in various species

For each of the following questions, select the correct answer and mark your answer sheet accordingly.

74. Based on the amino acid differences in cytochrome *c* which of the following pairs of organisms are most closely related to each other?

(A) Rabbit and monkey
(B) Monkey and duck
(C) Turtle and tuna fish
(D) Rabbit and duck
(E) Bread mold and yeast

75. Which of the following organisms listed below is LEAST closely related to humans?

(A) Tuna fish
(B) Turtle
(C) Moth
(D) Yeast
(E) Bread mold

76. According to the chart, the DNA codes for cytochrome *c* in monkeys is most similar to that in

(A) a rabbit
(B) a horse
(C) a human
(D) tuna fish
(E) yeast

77. The information reported in this chart demonstrates that cytochrome *c* is present in

(A) arthropods and insects only
(B) prokaryotes and eukaryotes only
(C) animals and fungi
(D) mammals only
(E) viruses and bacteria

78. Which of the following best explains why there are relatively few differences in the cytochrome *c* of the various organisms in the chart?

(A) DNA replication is universal.
(B) All organisms undergo protein synthesis.
(C) Members of these groups have the same proteins.
(D) Mitochondria are of recent evolutionary development.
(E) Respiratory reactions are similar in eukaryotes.

Questions 79–80.
Base your answers on the following chart:

Messenger RNA Codes for Certain Amino Acids

mRNA CODE	AMINO ACID
C-G-A	Arginine
C-C-A	Leucine
A-A-A	Lysine
U-U-U	Phenylaline
G-U-U	Valine

79. Which base sequence of a DNA molecule transcribes to an mRNA codon that will result in the amino acid arginine being lined up into a protein?

(A) G-C-T
(B) C-G-A
(C) G-C-U
(D) C-G-U
(E) G-C-A

80. Which amino acid will be brought to a ribosome by a transfer RNA molecule containing the triplet code A-A-A?

(A) lysine
(B) valine
(C) leucine
(D) phenylalanine
(E) none of these

Answer Key: Practice Test 4

1. (B)	17. (B)	33. (D)	49. (C)	65. (A)
2. (D)	18. (C)	34. (C)	50. (D)	66. (B)
3. (E)	19. (C)	35. (D)	51. (B)	67. (D)
4. (A)	20. (D)	36. (A)	52. (B)	68. (D)
5. (B)	21. (B)	37. (E)	53. (A)	69. (D)
6. (C)	22. (D)	38. (B)	54. (C)	70. (B)
7. (E)	23. (A)	39. (D)	55. (E)	71. (C)
8. (C)	24. (A)	40. (C)	56. (A)	72. (E)
9. (E)	25. (C)	41. (C)	57. (B)	73. (E)
10. (A)	26. (B)	42. (E)	58. (A)	74. (D)
11. (A)	27. (E)	43. (D)	59. (A)	75. (E)
12. (C)	28. (B)	44. (D)	60. (E)	76. (C)
13. (A)	29. (E)	45. (D)	61. (C)	77. (C)
14. (B)	30. (E)	46. (D)	62. (C)	78. (E)
15. (D)	31. (D)	47. (B)	63. (B)	79. (A)
16. (D)	32. (C)	48. (B)	64. (D)	80. (D)

Self-Evaluation Chart: Your Road to More Knowledge and Improved Scores

Your Raw Score

Using the Answer Key at the end of the test, place a ✔ next to each correct answer and an ✗ next to each incorrect answer.

A) Number of correct (✔) answers   _____

B) Number of incorrect (✗) answers   _____

Raw Score (A – B)   _____

Your College Board Score

Turn to the Biology E/M Conversion Table in Chapter 1 and determine your equivalent

College Board Score   _____

Improving Your Score

1. In column *A* below, list the numbers of the questions that you did not answer correctly.

2. Turn to the Answers Explained section, and for each question number listed in column *A* write in column *B* the key word or phrase that best summarizes the main topic or point of the answer.

3. Look up the topic in the Index and review the material.

4. Go back to the test and try to answer again each of the questions you answered incorrectly the first time. Write your new answers in column *C*.

5. Compare the answers in column *C* with the Answer Key.

6. Calculate your revised Raw Score:
   Number of correct answers (A above + number of column *C* correct answers)   _____

Revised Equivalent College Board Score   _____

A. Incorrectly answered questions	B. Main point(s) of the answer	C. Answers to questions in column A
_____	_____	_____
_____	_____	_____
_____	_____	_____
_____	_____	_____
_____	_____	_____
_____	_____	_____
_____	_____	_____
_____	_____	_____
_____	_____	_____
_____	_____	_____
_____	_____	_____
_____	_____	_____
_____	_____	_____
_____	_____	_____
_____	_____	_____
_____	_____	_____

Answers Explained: Practice Test 4

1. (B) Niacin is a member of the vitamin B complex, which also includes vitamin B_1 (thiamine).

2. (D) The name "Felix the Cat" is derived from the genus name, *Felis*. The species name is *domestica*. The cat is classified in the order Carnivora, the class Mammalia, and the phylum Chordata.

3. (E) Active transport makes possible the passage of minerals through the cell membrane of a root hair from the soil, where they are in low concentration. This movement of minerals is against the concentration gradient and therefore requires the expenditure of energy, which is made available by the process of respiration. Also, in order to contract, a muscle cell requires energy, which is derived from respiration.

4. (A) According to the cell theory, living things are made of cells. A virus, however, is not composed of cells. It contains a core of either DNA or RNA and a proteinaceous coat.

5. (B) The solution in side *A* has a 20% concentration of sugar and an 80% concentration of water. There is a flow of water from the region of 100% concentration in side *B* through the semipermeable membrane into side *A*, where there is a lower concentration of water. This incoming water causes the liquid level in side *A* to rise. Since sucrose molecules cannot pass through the membrane, they do not appear in side *B* and do not affect the liquid level there.

6. (C) The presence of HIV, or human immunodeficiency virus, in a person's body leads to the development of AIDS. The virus kills a type of white blood cell called the helper T cell, which would otherwise protect the body from infection. As a result, the person is liable to develop a variety of infections and tumors, ultimately leading to death. Scientists are seeking a cure for AIDS, either through a vaccine or a chemical that will destroy HIV.

7. (E) A carbohydrate contains the elements carbon, hydrogen, and oxygen. Also, hydrogen and oxygen are present in the same ratio as in water, 2:1. I represents a disaccharide such as sucrose; II, a fatty acid, stearic acid; and III, a polysaccharide such as starch. In II the 2:1 ratio of hydrogen and oxygen is not present.

8. (C) The fertilized egg multiplies by the process of mitosis, resulting in all the somatic cells having the same number (in this case, 8) of chromosomes as the fertilized egg.

9. (E) A chloroplast would be present only in a green plant cell.

10. (A) The cell membrane is depicted as consisting of a double lipid layer in which large proteins are embedded.

11. (A) During aerobic respiration, glucose is acted on in a series of steps involving glycolysis, the Krebs cycle, and the electron transport chain. Carbon dioxide and water are released, with the net formation of 36 ATP.

12. (C) During the process of mitosis, the chromatin material of the nucleus becomes more distinguished as separate chromosomes. The arrangement of the nucleotides in DNA leads to the expression of the genes. In other words, a gene is made up of DNA.

13. (A) The protoplasm of an ameba is sensitive to stimuli (e.g., bright light) and responds to them. This property of protoplasm is known as irritability

14. (B) The epidermis of a leaf contains numerous tiny openings called stomates. Each stomate is surrounded by two guard cells that control the size of the opening.

15. (D) Water is transported up to the leaves through the xylem, after which it evaporates through the stomates in the process known as transpiration.

16. (D) The ribosome is a tiny granule found either along the endoplasmic reticulum or in the cytoplasm of a cell. It is too small to be seen with a compound microscope.

17. (B) Although both a lipid and a carbohydrate may contain carbon, hydrogen, and oxygen, the H and O atoms are present in carbohydrates in the same ratio as in water, H_2O. Here, the ratio is 16:1, and the compound is a lipid, palmitic acid.

18. (C) A heterotroph is an organism that cannot make its own food and depends directly or indirectly on green plants as the primary source of food. Animals are therefore heterotrophic as far as nutrition is concerned.

19. (C) The Food Guide Pyramid is based on the concern that the dietary habits of many Americans may lead to such serious disorders as heart disease, cancer, high blood pressure, obesity, stroke, and diabetes. Among the recommendations are to use fats, oils, and sweets sparingly and to eat a larger vareity of starches, fruits and vegetables, and other foods containing fiber.

20. (D) Taiga is the terrestrial biome that includes the northernmost forests of coniferous trees such as spruce and pine, and is characterized by long, severe winters.

21. (B) The cabbage plant serves as the producer in this food chain, since it forms food. The rabbit is a first-order consumer that feeds directly on cabbage, while the fox, a carnivore, is a second-order consumer that preys on the rabbit.

22. (D) The basic food in the marine biome is plankton, which consists of microscopic life such as diatoms and other algae, and protozoa. Plankton is the basis of the food chain of the ocean, with these algae serving as producers.

23. (A) A population is defined as the members of a species that breed together and live together in an area. Examples: leopard frogs in a pond; dandelions in a lawn.

24. (A) A community is defined as the populations of all the species living in a given area, here 30 cubic centimeters of soil. In this case, the community may consist of plants such as grass and clover and animals such as earthworms and nematodes, as well as soil bacteria and fungi.

25. (C) They belong to the phylum Arthropoda. Members of this phylum have three or more pairs of jointed legs, an outer skeleton, or exoskeleton made of chitin, and segmented bodies. They are arranged in five classes, including crustaceans (crabs and lobsters), insects, spiders, centipedes, and millipedes.

26. (B) If the plant with colored flowers is heterozygous (*Cc*) and is cross-pollinated with white flowers (*cc*), the results can be depicted as follows:

	C	c	
c	Cc	cc	50% *Cc* (heterozygous colored)
c	Cc	cc	50% *cc* (white)

27. (E) In sickle-cell anemia, there is a substitution of the amino acid valine for the normal amino acid, glutamic acid, in the hemoglobin molecule. Otherwise the two hemoglobins are identical.

28. (B) At the beginning of meiosis, the homologous chromosomes of the primary sex cell come together in pairs in a process called synapsis. Some time during the pairing, the chromosomes twist about each other and exchange parts. This phenomenon is called crossing-over.

29. (E) Large numbers of offspring are needed to produce the ideal ratios. With a small number, such as four, the actual results will rarely show the expected ratio, and any of the possibilities listed in the answer choices may occur.

30. (E) The lungs excrete carbon dioxide and water. The kidneys excrete urine, which contains about 95 percent water, the nitrogenous waste called urea, and mineral salts, uric acid, and other materials. The skin excretes water and some salts through its sweat glands.

31. (D) Cambium consists of actively dividing cells that differentiate into xylem, phloem, and additional cambium cells, thus adding to the diameter of the tree trunk.

32. (C) The stomach produces gastric juice, which consists of hydrochloric acid, water, and the enzymes pepsin and rennin. These enzymes assist in the digestion of proteins.

33. (D) When the heart of a freshly killed frog is removed from its body and placed in Ringer's solution, it will continue to beat because the solution provides a chemical environment similar to the normal one of the frog's heart. In addition to common salt (NaCl), Ringer's solution contains potassium chloride, sodium bicarbonate, calcium chloride, and water. Varying the composition of the solution alters the chemical environment and affects the heartbeat.

34. (C) Tumors are growths of cells that may be types of cancer. If calvacin prevents the growth of certain types of tumors, it may be useful in controlling cancer. It would therefore be the basis of new experiments to determine whether it can really be useful in this way.

35. (D) Villi are the tiny, fingerlike projections that line the inside of the small intestines. Their walls are one cell thick, permitting the diffusion of digested nutrients through them into the capillaries and lacteals.

36. (A) The medulla controls the rate at which the diaphragm and chest-wall muscles contract and expand in breathing. When, during exercise, the concentration of carbon dioxide increases in the blood passing through the medula, the breathing rate becomes more rapid.

37. (E) Adrenaline is produced by the adrenal glands, which are also known as the "glands of emergency." Among its effects, adrenaline stimulates the liver to convert glycogen to glucose, causes the heart to beat faster so that more glucose is brought to the muscles and brain, and increases the breating rate, providing more oxygen. Under these conditions, the muscles have greater energy, and the brain can think more clearly, in an emergency.

38. (B) After being formed in the kidneys, urine passes through the ureters and collects in the urinary bladder. At intervals, it leaves the bladder and is eliminated through a tube called the urethra.

39. (D) A nucleotide is made up of three types of molecular units: (1) a phosphate; (2) ribose or deoxyribose, a five-carbon sugar; and (3) a nitrogen base, either a purine or a pyrimidine.

40. (C) A substrate is the specific substance acted on by an enzyme. For example, maltose is the substrate acted on by the enzyme maltase. The association between enzyme and substrate is thought to be a close physical one, but does not lead to the formation of bonds between them. The localized region of the enzyme that acts on the substrate is called the active site.

41. (C) When *Eoraptor* died, it was probably covered by sediment such as mud, sand, or clay, carried by some form of flowing water. Under the pressure of the water during the succeeding ages, the material in the sediment became cemented together, forming a type of sedimentary rock—limestone from mud, sandstone from sand, or shale from clay. When the land was uplifted by geologic forces, the rock containing the fossil was exposed.

42. (E) Environmentalists are deeply concerned about the rapid destruction of tropical rain forests in the Amazon area of Brazil, in Southeast Asia, and in Central America for purposes of lumbering and land clearing for agriculture. These huge rain forests are presently disappearing at the rate of 50–100 acres per minute and will be gone within a century. Experts maintain that there are many values to tropical rain forests.

43. (D) Since Robert has blue eyes, each parent must have had a gene for blue eyes. Frank and Alice are therefore heterozygous for brown eyes. Robert is homozygous for blue eyes.

44. (D) During the Krebs cycle of respiration, hydrogen is released and picked up by the coenzyme NAD (nicotin-amide-adenine-dinucleotide). Other hydrogen acceptors are the coenzymes riboflavin and cytochrome. The electrons of the hydrogen atoms are then passed on in the electron transport chain to release energy in small units. The final hydrogen acceptor is oxygen and the combination forms water.

45. (D) A gene mutation results from a change in the DNA molecule, such as a change in the sequence of the nitrogen bases. A chromosome aberration is a change in the number of chromosomes or in their structure. Both of these types of changes are inherited. A mineral deficiency, however, is an environmental factor that is not inherited.

46. (D) Lamarck believed that animals changed by the inheritance of acquired characteristics. He thought that when an animal uses an organ to adapt itself to the environment that the organ would become well developed and would be inherited as such. Also, new organs would arise according to the needs of the animal.

47. (B) The Human Genome Project was considered complete in June 2000. One of the anticipated benefits is a more precise understanding of the defective genes that cause inherited diseases, such as sickle-cell anemia and phenylketonuria.

48. (B) A distant ancestor of the whale had hind legs attached to hipbones. With the passage of time, the legs disappeared and the hipbones became useless. However, genes are still present for the development of these bones. They are now considered vestigial structures.

49. (C) On the basis of its characteristics, the bat is classified in the class Mammalia, along with other mammals such as the chimpanzee, whale, elephant, dog, and human. Like them, its eggs contain the haploid (not diploid) number of chromosomes.

50. (D) Aerated blood from the lungs flows through the pulmonary veins into the left atrium (7).

51. (B) The inferior (lower) vena cava brings blood from the lower part of the body and the superior (upper) vena cava brings blood from the upper part of the body; both large veins empty into the right atrium (2).

52. (B) The pulmonary artery (5) receives blood from the right ventricle and carries it to the lungs, where it is exposed to oxygen. This is the only artery that carries deoxygenated blood.

53. (A) The inner surface of the trachea is lined with ciliated epithelial cells. The cilia of these cells are constantly in motion, beating foreign particles such as dust upward and outward.

54. (C) The pancreas secretes pancreatic juice into the upper end of the small intestine by means of the pancreatic duct. This juice contains three enzymes: trypsin, which digests peptones and proteoses into amino acids; amylopsin, which changes starch into maltose; and lipase, which changes fats into fatty acids and glycerol. The islands of Langerhans in the pancreas secrete the hormones insulin and glucagon.

55. (E) After food has been chewed in the mouth, it passes into the esophagus and is carried by peristaltic action into the stomach. The esophagus is lubricated by mucus, which is secreted by glands in its lining. The esophagus does not secrete digestive enzymes.

56. (A) The greenhouse effect refers to the increasing amounts of carbon dioxide and methane in the atmosphere due to the burning of coal, oil, and natural gas by factories, homes, and automobiles. CO_2, like glass in a greenhouse, allows visible sunlight to pass through to the earth. As the earth warms up, it gives off infrared rays. These are absorbed by the CO_2 in the atmosphere, instead of being given off into space. It is believed that this greenhouse effect will cause the atmosphere of the globe to warm up, with consequent harmful results to living things.

57. (B) The ozone layer of the atmosphere, reaching up 10–30 miles, contains a type of oxygen molecule consisting of three atoms of oxygen, called ozone. The ozone layer protects life on earth by absorbing most of the powerful ultraviolet radiation emanating from the sun.

Chemicals called CFCs (chlorofluorocarbons), commonly used in refrigerators and air conditioners, are suspected of eroding this ozone layer, causing large holes in it. These could increase the danger of more UV radiation reaching the earth, with harmful effects on plants and animals. For example, it could lead to an increase in the incidence of skin cancer.

58. (A) The greenhouse effect is described in the answer to question 56.

59. (A) Mitochondria contain many enzymes and coenzymes that function in a chain of events leading to the release of energy in a cell. Aerobic respiration takes place in the presence of oxygen. During anaerobic respiration, there is a lack of oxygen.

60. (E) Lysosomes are saclike organelles in a cell that contain digestive enzymes. These enzymes break down large organic molecules and worn-out organelles within the cell.

61. (C) Most enzymes function best at about the temperature of the body, 37°C. As the temperature is raised, their activity increases until a maximum is reached at about 40°C, as shown in the graph. Beyond this point, the shape of the enzyme molecules becomes distorted and the enzyme ceases to function.

62. (C) During cellular respiration, one molecule of glucose is broken down into two molecules of pyruvic acid and the production of ATP, which is the molecule that is the source of energy in the cell. This process is known as glycolysis. When it occurs in the absence of oxygen, during anaerobic respiration, only a small part of the energy of glucose is released by glycolysis, with the production of either CO_2 and alcohol, or lactic acid. In the presence of oxygen, during aerobic respiration, all of the energy of glucose is released, leading to the production of CO_2 and H_2O. Photosynthesis deals with the production of sucrose and starch, not its breakdown into energy.

63. (B) The release of the chemical-bond energy of food into energy that can be used for the life activities of the cell is known as respiration. Every cell contains molecules of the nucleotide ATP (adenosine triphosphates) which is the actual source of energy. Each ATP molecule stores energy in its phosphate bonds. When one

of these bonds is broken, energy is released to the cell for its various activities and ATP is converted to ADP (adenosine diphosphate), giving off a phosphate group, P. It is this change of ATP into ADP that supplies the cell with energy.

64. (D) Messenger RNA (mRNA) molecules carry triplets of nucleotides called codons and become located in a ribosome. Here, mRNA lines up with another group of three nucleotides called an anticodon, carried by a transfer RNA (tRNA) molecule. For example, if tRNA has an anticodon containing the sequence UGC, it will fit in with the mRNA codon ACG. The arrangement of the nucleotides on mRNA dictates the order in which the amino acids are lined up and bonded into polypeptides. This process of turning the instructions from mRNA into protein in the ribosome is known as translation.

65. (A) Lipids consist of carbon, hydrogen, and oxygen, but their hydrogen-oxygen ratio is greater than that in carbohydrates, which is 2:1. A typical lipid molecule consists of a glycerol molecule bonded to three fatty acid molecules.

66. (B) The Krebs cycle is part of the process of cellular respiration that takes place in the mitochondria and where energy is produced by the breakdown of glucose. On the other hand, photolysis occurs during photosynthesis, and leads to the production of sucrose and starch from CO_2 and H_2O.

67. (D) A virus consists of a core containing nucleic acid (either DNA or RNA) surrounded by a protein covering. When the virus attacks a host cell, its nucleic acid enters and the empty shell of the covering is left behind. The nucleic acid redirects the metabolism of the host nucleic acid to form additional virus units with protein coverings. These burst out of the host cell to infect other hosts.

68. (D) The metabolic process is aerobic respiration, through which energy is derived from glucose. Many enzymes are involved in the breakdown of glucose, which takes place in a series of steps in the mitochondria. When the whole process takes place in the presence of oxygen, it is referred to as aerobic respiration. Without oxygen, it is anaerobic respiration.

69. (D) The raw material is glucose, $C_6H_{12}O_6$. The first step in respiration is glycolysis, in which the six-carbon compound of glucose is converted to two 3-carbon molecules of pyruvic acid $C_3H_4O_3$. Energy is released in this stage and is stored in 4 molecules of ATP. But in order for some of these processes to take place, the energy from 2 ATP was used up. As a result, the net gain of energy during glycolysis is 2 ATP.

70. (B) In the second stage of respiration, the Krebs cycle, the pyruvic acid molecules are decarboxylated in a series of reactions and CO_2 is removed. Two additional ATP are formed. In the final phase (the electron transport chain), hydrogen is passed along to a number of hydrogen acceptors to release energy in small units. The final hydrogen acceptor is oxygen and the combination forms water. The energy is stored in 32 ATP molecules along the way in this phase. The total number for the process is thus 36 ATP (2 ATP in glycolysis + 2 ATP in the Krebs cycle + 32 in the electron transport chain).

71. (C) Enzymes are considered organic catalysts because they affect the rate of a chemical reaction without being changed. They can be used over and over again. They are protein in nature and specific in their action. In this case, the enzyme is maltase and, in a process called hydrolysis, breaks the disaccharide maltase down into its two glucose subunits.

72. (E) During the hydrolysis of the disaccharide maltose, the enzyme maltase splits the maltose into two monosaccharides of glucose. Water is required in hydrolysis to complete the formation of the glucose molecules. The disaccharide maltose, with the formula $C_{12}H_{22}O_{11}$, is broken down into the two monosaccharides, glucose-$C_6H_{12}O_6$.

73. (E) Lipase is an enzyme that acts on fat to convert it into glycerol and fatty acids. Typically, a lipid molecule (fat) consists of a glycerol molecule bonded to 3 fatty acid molecules. In the presence of water, the lipid molecule undergoes hydrolysis in which the enzyme breaks down the bond between the glycerol and the fatty acid molecules. Protease is an enzyme that acts on protein. Sucrase and maltase are enzymes that act on sucrose and maltose, respectively. Amylase of the saliva in the mouth digests starch into maltose.

74. (E) The chart shows how many amino acids in the cytochrome *c* of a particular species differ from those of the various species listed. The rabbit is identified as having 9 amino acids in cytochrome *c*, while the duck has 11. The difference in the cytochrome *c* content between the rabbit and the duck is thus 2 molecules. The difference between the other pairs of organisms is greater. (D) is therefore the correct answer.

75. (E) The difference for the bread mold amino acid sequence of cytochrome *c* is listed as 48. This is greater than that of any other organism on the list, making it the least closely related to humans. All the other organisms have a smaller number of differences, making them more similar to humans in terms of sharing amino acids in the makeup of cytochrome *c*. Although these organisms are different from humans, they do share, by varying amounts, the structure of the cytochrome *c* amino acids.

76. (C) The structure of cytochrome *c* is determined by DNA which transmits the code to messenger RNA. The order of the nucleotides of messenger RNA was determined by the arrangement of nucleotides on the DNA molecule. Messenger RNA contains triplets of nucleotides called codons, and becomes located in a ribosome. Small transfer RNA pick up specific amino acid molecules in the cytoplasm and line up with messenger RNA. Each transfer RNA molecule has a group of three nucleotides called an anticodon, and its bases fit the appropriate codon with its three bases of messenger RNA. The arrangement of the nucleotides on messenger RNA dictates the order in which the amino acids are lined up to form cytochrome *c*. The chart shows that the amino acids in cytochrome *c* of monkeys is most similar to that of humans, differing by only 1 amino acid molecule.

77. (C) All the organisms listed on the chart are animals, except for the last two which are fungi. Both of these groups of organisms contain cytochrome *c*. There are no prokaryotes, viruses, or bacteria on the chart. While some of the animals are mammals, the duck, turtle, tuna fish, and moth are not. The moth is an insect which is the only arthropod represented, so (A) is not the correct answer.

78. (E) All the organisms on the chart are eukaryotes, that is their cells contain a nucleus, mitochondria, and several other organelles. Cytochrome *c* is a protein found in mitochondria which is active as an electron carrier during cellular respiration. It is composed of a sequence of 104 amino acids. It has been compared for the different organisms listed on the chart. This chain of amino acids differs by only one molecule in monkeys and humans. In bread mold, it differs from humans by 48 molecules. But all the organisms on the chart are similar to each other in depending on cytochrome *c* for its role in respiration.

79. (A) A portion of a DNA molecule serves as a pattern for the synthesis of messenger RNA from free RNA nucleotides in the nucleus. The free RNA nucleotides line up next to the appropriate DNA nucleotides, cytosine (C) with guanine (G), and thymine (T) with adenine (A). The process of transferring the coded information from DNA to the new strand of RNA is called transcription. The newly formed RNA molecule, which is a "reverse copy" of the DNA that produced it, separates from the strand and moves from the nucleus to the cytoplasm. The messenger RNA molecule now carries the genetic code which consists of triplets of nucleotides called codons. The mRNA codon for arginine is C-G-A, so the base sequence of DNA which transcribed it is G-C-T.

80. (D) The transfer RNA anticodon is complementary to the codon of messenger RNA. The complementary triplet code for A-A-A which is carried on the transfer RNA would be U-U-U on the mRNA, which is the amino acid phenylalanine. The process of turning the instruction from mRNA into an amino acid sequence in the ribosome is called translation.

APPENDIX
Glossary of Biological Terms

Acetylcholine   A neurotransmitter secreted by a neuron that plays a part in the transmission of an impulse.

Active site   The specific region of an enzyme that reacts with the substrate.

Active transport   The passage of a substance through the cell membrane against the concentration gradient (that is, from a region of low concentration to a region of higher concentration); the expenditure of enery is required.

Adaptive radiation   The evolution of a species into different species adapted to various environmental niches. Example: the finches on the Galapagos Islands.

Adenine   A nitrogen base that is paired with thymine in DNA, and with uracil in RNA.

Adrenaline (Epinephrine)   A hormone, produced in the medulla of the adrenal gland, that provides quick energy in emergency situations by increasing the breathing, heart, and metabolic rates.

Albino   An individual that lacks pigment.

Allele   One of a pair of genes that occupies a specific location on a pair of chromosomes and determines a particular characteristic. Example: In hybrid tall pea plants, the alleles are *T* and *t*.

Alveolus   A minute air sac in the lungs through which oxygen enters the bloodstream and through which carbon dioxide and water are excreted from the bloodstream.

Amino acid   One of the building blocks of a protein.

Amniocentesis   A method of prenatal testing in which a small amount of amniotic fluid is withdrawn from the uterus of a pregnant woman, and the cells shed by the fetus that are contained in it are analyzed.

Amniotic fluid   The fluid that surrounds the developing embryo in mammals, birds, and reptiles.

Amyotrophic lateral sclerosis (ALS)   *See* **Lou Gehrig's disease.**

Anaerobic   Lacking oxygen. The term may refer to a type of cellular respiration, or to certain bacteria that live in the absence of oxygen.

Anaphase   The stage of mitosis or meiosis in which the homologous chromosomes separate and move to the opposite poles of the dividing cell.

Angiosperms   Higher plants that reproduce by flowers.

Anther   The part of a stamen that produces pollen.

Antibody   A protein produced by the immune system that destroys a particular antigen or foreign substance.

Anticodon   Sequence of three bases in transfer RNA that bind with the complementary three bases, or codon, in messenger RNA, leading to protein formation.

Antigen   A foreign substance that enters the body and stimulates the body to produce antibodies against it.

Aorta   The largest artery of the body; it receives blood directly from the left ventricle and sends it to most of the body.

Atom   The smallest particle of an element, consisting of electrons that surround a nucleus which contains protons and neutrons.

ATP (adenosine triphosphate)   The molecule that serves as the source of energy of a cell.

Atrium (auricle)   The chamber of the heart that receives blood and pumps it into a larger chamber called the ventricle.

Autonomic nervous system   The part of the nervous system that controls internal, involuntary activities such as heartbeat, peristalsis, kidney action, glandular secretions, and the size of the bronchioles and capillaries.

Autosome   One of the chromosomes in a cell that is not a sex chromosome.

Autotroph   An organism capable of making its own food from inorganic raw materials.

Auxin   A type of plant hormone that promotes cell elongation and also produces other effects, such as tropism response and the formation of roots on cuttings.

Axon   The long extension of a neuron that conducts nerve impulses away from the cell body.

Bacillus   A rod-shaped bacterium.

Bacteriophage   A virus that attacks bacteria.

Binomial nomenclature   The system of classification introduced by Linnaeus in which an organism is given a genus and a species name. Example: *Felis domestica,* the house cat.

Biomass   The total mass of all the organisms in a particular area.

Biome   A major ecological grouping of organisms on a broad geographical basis. Example: tundra.

Bioremediation   The action of bacteria in breaking down oil spills into harmless compounds.

Bioterrorism   A new approach to warfare in which germs or other living things are used as weapons or threats of mass destruction.

Blastula   The hollow-ball stage of embryonic development in animals.

Bronchiole   A small tube that branches throughout the lungs and terminates in an alveolus, or air sac.

Bryophyta   A simple phylum of the plant kingdom, including the mosses and the liverworts.

Calorie   A measure of the energy value of food; the amount of heat needed to raise the temperature of one gram of water by one degree Celsius.

Cambium   A part of a plant consisting of rapidly dividing cells present in the fibrovascular bundles and located between the phloem and the xylem.

Capillary   The smallest blood vessel, whose wall is one cell thick and through which materials are exchanged between the blood and the surrounding tissue.

Carbohydrate   An organic compound such as sugar, starch, and cellulose, which is composed of the elements carbon, hydrogen, and oxygen; the latter two are present in a ratio of 2:1.

Carboxyl group   A —COOH group present in fatty acids and amino acids.

Carcinogen   A cancer-causing agent, which may be a chemical substance or radiations from sunlight or X ray.

Carnivore   An animal that eats other animals, either alive or dead. Examples: hawk, jackal.

Carotene   Yellow, orange, or red pigment present in green leaves, where it is masked by chlorophyll; also found in yellow and green leafy vegetables; a precursor to vitamin A.

Cartilage   A specialized type of dense connective tissue in which the cells are contained in a rubbery matrix that is smooth, firm, and flexible, but not as hard as bone; occurs in joints at the ends of bones and in the ears, nose, and windpipe.

Catalyst   A substance that accelerates the rate of a chemical reaction without itself being changed; an enzyme is a biological catalyst.

Cellulose   A complex polysaccharide found in the cell walls of plants.

Centriole   An organelle, found mostly in the cytoplasm of animal cells, that is attached to the spindle during mitosis and meiosis.

Centromere   A special region on a chromosome to which the spindle is attached during mitosis and meiosis.

Cerebellum   The specialized rear part of the brain of vertebrates that controls muscular coordination.

Cerebrum   The main part of the brain of vertebrates that controls the sense organs; in humans, the center of thought and memory.

Chitin   A polysaccharide that comprises the hard exoskeleton of insects and crustaceans; also present in the cell walls of fungi.

Chlorophyll   The green pigment in plants that absorbs light and carries on photosynthesis.

Chloroplast   A plastid in green plants and algae that contains chlorophyll.

Chromatid   One member of a replicated chromosome that is held together to the other member by a centromere.

Chromatin   The genetic material within the nucleus that takes the form of chromosomes during cell division.

Chromatography   A method of separating substances such as leaf pigments by their varying rates of adsorption on a medium such as filter paper.

Chromosome   A rodlike structure, present in the nucleus of a cell, on which the genes are located.

Cleavage   The repeated division of a fertilized egg into cells that give rise to the early embryo of an animal.

Climax community   A final, stable stage reached in the process of ecological succession.

Clone   An organism produced artificially from a parent and having the identical genetic makeup.

Coccus   A spherical bacterium.

Codominance   A form of inheritance in which neither of two parents has a dominant or recessive trait for a characteristic, and whose offspring have a blend of both.

Codon   A combination of three nucleotides in messenger RNA that directs the formation of the various amino acids in building proteins.

Coenzyme   A compound that is not a protein and that plays an important role in the catalytic action of an enzyme. Example: riboflavin.

Colonoscopy   The use, by a doctor, of a thin, lighted instrument to look for evidence of cancer along the inside of the colon.

Commensalism   A form of symbiosis in which one organism benefits from the relationship while the other neither receives benefit nor is harmed. Example: barnacle living on the hide of a whale.

Community   The populations of all the species living in a given area.

Compound   A substance formed from two or more elements that has different properties from either of them.

Conditioned behavior   A type of learned response in which a new response is associated with an original stimulus.

Conjugation   A form of sexual reproduction in which two similar gametes unite.

Contractile vacuole   An organelle found in protozoa that ejects excess fluid from the cell.

Control, experimental   In a laboratory or research procedure, a part of the experiment that represents the original condition and is not exposed to different treatments; the control serves as a basis of comparison of the results.

Corolla   The group of petals in a flower.

Crossing-over   The exchange of adjacent parts between two homologous chromosomes during meiosis.

Cyclic AMP (cAMP, or cyclic adenosine monophosphate)   A compound formed from ATP that acts as a second messenger in bringing about the effects of hormones.

Cytochrome   Pigments containing iron that function in the electron-transfer chain of photosynthesis and cellular respiration.

Cytokinin (Kinetin)   A plant hormone that stimulates cell division.

Cytoplasm   The living material in a cell, outside the nucleus.

Cytosine   A nitrogen base in the structure of DNA and RNA that pairs with guanine.

Deciduous   Referring to a plant that sheds its leaves each autumn.

Decomposers   Bacteria or fungi that break down the bodies of dead plants and animals into simpler compounds that are then recycled and used again by other living organisms.

Dendrite   Many-branched fiber of a neuron that receives impulses across the synapse and directs them toward the cell body.

Dicot (dicotyledon)   A type of angiosperm plant in which the embryo contains two cotyledons in the seed.

Diffusion   The movement of molecules, which are in a constant state of movement, from one place to another.

Diploid (2n)   Pertaining to a cell that has two chromosomes of each type.

Disaccharide   A double sugar, such as sucrose, with the formula $C_{12}H_{22}O_{11}$.

DNA (deoxyribonucleic acid)   A large molecule in the nucleus of a cell that contains the genetic information of the organism. It is composed of the sugar deoxyribose, a phosphate, and the bases adenine, thymine, guanine, and cytosine.

Dominant   Referring to an allele that exerts its full effect, even though a recessive allele may be present.

Ecology   The study of living things in relation to each other and to their environment.

Ecosystem   A community of plants and animals in relation to their physical environment.

Ectoderm   The outer layer of the gastrula, which differentiates into the nervous system and the skin.

Effector   A part of an organism (a muscle or a gland) that produces a response to a stimulus.

Electron   A negatively charged particle located outside the nucleus of an atom.

Endocrine gland   A ductless gland whose hormone secretions diffuse into the blood and are transported to other parts of the body.

Endoderm   The innermost layer of the gastrula, which differentiates into the lining of the digestive and respiratory systems.

Endoplasmic reticulum   An organelle located in the cytoplasm of a cell that consists of membrane-bound channels and functions in transport within the cell.

Endosperm   Cells in the seed of a plant that contain food for the embryo.

Enzyme   A protein that acts as a catalyst.

Epinephrine   *See* **Adrenaline.**

Epithelial tissue   The tissue that covers the external and internal surfaces of the body.

Estrogen   A hormone, secreted by the ovary, that stimulates the development of female secondary sexual characteristics.

Eukaryote   An organism whose cells contain a nucleus, as is the case in all organisms except bacteria and blue-green algae.

Evolution   The genetic change in living things over a long period of time.

Exoskeleton   The outer skeleton, made of hard chitin, that covers the bodies of arthropods.

Fauna   All the animals of an area.

Feedback   The process in which a mechanism is controlled by excessive or limited amounts of its product. Example: A feedback mechanism controls the release of thyroxin from the thyroid gland.

Fermentation   A method of anaerobic respiration in which alcohol and carbon dioxide, or lactic acid, are generated.

Fertilization   The union of a sperm and an egg.

Fetus   The later stage of development of an embryo in the uterus; in humans, from the beginning of the third month until birth.

Flagellum   A long, whiplike appendage used by a one-celled organism for locomotion.

Flora   All the plants of an area.

Follicle   The cavity in the ovary in which an egg develops and is stored until it is released.

Food chain   The sequence of producers, consumers, and decomposers through which food and energy pass; a segment of a food web.

Food web   A sequence of overlapping food chains in which many kinds of producers are eaten by different kinds of consumers.

Fruit   The mature ovary of a plant.

Gamete   The mature sexual reproductive cell: the egg or the sperm.

Gametophyte   The haploid phase of a plant that undergoes alternation of generations and produces gametes.

Ganglion   A structure in the nervous system containing a mass of cell bodies.

Gastrula   The early embryonic stage of development which consists of an outer ectoderm layer, an inner endoderm layer, and a middle mesoderm layer.

Gene   The portion of a DNA molecule that serves as a unit of heredity.

Gene pool   The sum total of all the genes in a population.

Genome   The genes contained in a haploid set of chromosomes of an individual in a particular species.

Genotype   The genetic makeup of an individual for one or more characteristics.

Genus   In the classification of organisms a group of one or more species that show many similarities. Examples: *Drosophila, Canis.*

Geotropism   The response of a plant to gravity in which the roots grow downward and the stem grows upward.

Gibberellin   A plant hormone that causes stem elongation and other effects.

Glomerulus   A part of a nephron consisting of a dense network of capillaries.

Glucose   A simple sugar, $C_6H_{12}O_6$, that plays a key role in cellular metabolism.

Glycogen   A form of animal starch, or polysaccharide, that is the principal form in which carbohydrates are stored in the liver and muscles.

Golgi complex   An organelle in the cytoplasm of a cell, composed of membranous sacs, that concentrates proteins into vesicles that migrate to the cell membrane and are released outside the cell.

Gonads   The reproductive organs: the ovaries and the testes.

Greenhouse effect   The warming of the earth's surface that occurs as heat is trapped in the atmosphere by carbon dioxide, instead of being given off into space.

Guanine   A nitrogen base in the structure of DNA and RNA that pairs with cytosine.

Gymnosperm   A type of seed-producing plant, such as pine and spruce, that forms naked seeds on the scales of a cone; most gymnosperms are evergreen and contain modified leaves called needles.

Half-life   The period of time in which a radioactive element loses half of its radioactivity.

Haploid (*n*)   Pertaining to a cell, specifically a gamete, that has half the chromosome number characteristic of the species.

Hardy-Weinberg principle   A doctrine stating that the gene pool of a population remains constant from generation to generation, provided that the population is large, matings are random and no mutations occur.

Hemoglobin   The red, iron-containing protein in the red blood cell that carries oxygen.

Herbivore   An animal that feeds on plants. Examples: deer, rabbit.

Heterotroph   An organism that cannot make its own food and depends directly or indirectly on green plants as the source of food.

Heterozygous   Having two different alleles of a particular gene, such as *Tt* in hybrid tall pea plants.

Homeostasis   The stable state maintained in an organism by various feedback and control mechanisms.

Homozygous   Having two similar alleles of a particular gene, such as *tt* in short pea plants.

Hormone   A chemical compound that is secreted into the blood by an endocrine gland and that affects activities or other parts of the body.

Huntington's disease   An inherited disease caused by a dominant gene that shows up after the age of 30; the person develops involuntary, jerky motions, mental deterioration, and dementia, and often dies within 15 years.

Hybrid   The offspring of parents that differ in a genetic trait, such as a hybrid tall pea plant produced by crossing homozygous tall and short pea plants; also, the offspring of two different species, such as a mule, from the union of a horse and a donkey.

Hydrolysis   The breaking apart of a compound into simpler components, through the addition of water.

Hypertonic   Referring to a solution with a higher concentration of a dissolved substance, resulting in an intake of water.

Hypha   The filament of a fungus.

Hypotonic   Referring to a solution with a lower concentration of a dissolved substance, resulting in a loss of water.

Independent assortment   A type of inheritance in which genes on nonhomologous chromosomes are inherited independently from other genes.

Insulin   A hormone, secreted by the islands of Langerhans in the pancreas, that helps the cells utilize glucose for energy; it also causes the liver to convert excess glucose into glycogen for storage.

Interphase   The stage of mitosis when chromosome replication takes place, before nuclear division begins.

Invertebrate   An animal without a backbone.

Ion   An electrically charged atom with either a positive or a negative charge.

Isotonic   Referring to a solution with the same concentration of molecules as is present in a cell, resulting in neither a gain nor a loss of water by the cell.

Isotope   An atom that differs from another atom of the same element by having a different number of neutrons in its nucleus. Examples: ^{235}U and ^{238}U, two isotopes of uranium.

Karyotype   A photograph or diagram of the chromosomes of a cell, which have been arranged in homologous pairs according to size.

Kinetin   *See* **Cytokinetin.**

Leucocyte   A white blood cell.

Ligament   A type of connective tissue that connects bones in a joint.

Linkage   The inheritance together of genes located on the same chromosome.

Lipase   An enzyme that digests fats and converts them into fatty acids and glycerol.

Lipid   A type of organic compound that includes fats, oils, waxes, and steroids.

Lou Gehrig's disease (Amyotrophic lateral sclerosis)   An inherited disease that affected the famous baseball player of that name; it causes muscular and nervous degeneration and eventual death.

Lymph   The watery intercellular fluid that surrounds all the cells; it originates in the blood plasma and is returned to it by lymph vessels.

Lysosome   A small organelle, located in the cytoplasm of a cell, that contains enzymes for digesting large organic molecules within the cell.

Malpighian tubules   Small tubes of the excretory system of insects that carry nitrogenous wastes into the hindgut.

Mammal   A warm-blooded vertebrate that has a four-chambered heart, has hair, and gives birth to live young, which it feeds on milk.

Mammogram   An X ray of a woman's breast to determine whether a lump that may be cancerous is present.

Marsupial   A mammal without a placenta that places its immature young in a pouch, where they feed on milk. Examples: kangaroo, opossum.

Meiosis   Division of a diploid cell in the reproductive organs to produce haploid eggs or sperm.

Meristem   Actively dividing tissue of plants, found in the tips of roots and stems and in the cambium.

Mesoderm   The middle layer of the gastrula stage of embryonic development, which gives rise to the muscles, skeleton, circulatory system, excretory system, and reproductive organs.

Messenger RNA (mRNA)   The type of RNA that carries the code of DNA out of the nucleus to the ribosomes for the formation of proteins.

Metabolism   The sum total of all the body's chemical activities, including the breakdown of molecules to release energy (catabolism) and the buildup of complex molecules (anabolism).

Metamorphosis   The change of an immature animal into an adult. Examples: the development of a tadpole into a frog; the transformation of an insect larva into an adult.

Metaphase   The stage in mitosis or meiosis in which the chromosomes are located in the middle of the cell.

Micropyle   The tiny opening in the ovule of a flower through which the pollen tube enters, carrying the sperm nuclei to unite with the egg and endosperm nuclei.

Microtubule   Organelle in the cytoplasm that helps support the structure of a cell; it forms the spindle during cell division.

Mitochondria   Organelles in the cytoplasm in which aerobic respiration takes place, providing energy for the cell.

Mitosis   Nuclear division in which two new cells are formed with the same number of chromosomes as in the parent cell.

Molecule   The smallest unit of a substance, consisting of two or more atoms.

Monera   In the classification of organisms, the kingdom of prokaryotes, including bacteria and blue-green algae.

Monocot (monocotyledon)   A type of angiosperm plant in which the embryo contains one cotyledon in the seed.

Monosaccharide   A simple sugar such as glucose or fructose, with the formula $C_6H_{12}O_6$.

Motor neuron   A nerve cell that carries impulses from the central nervous system to either a muscle or a gland.

Mutation   A change in the genetic material of an organism.

Mutualism   A form of symbiosis in which both organisms benefit from the relationship. Example: nitrogen-fixing bacteria in the root nodules of legume plants.

Mycelium   The mass of hyphae in a fungus.

Natural selection   A concept, basic to Darwin's theory of evolution, to the effect that the organisms best adapted to their environment tend to survive and perpetuate their species.

Nematocyst   In coelenterates, a stinging cell used to paralyze prey.

Nephridium   An excretory tube in invertebrates, such as the earthworm, that leads to the outside.

Nephron   The excretory unit in the kidney of vertebrates, consisting of Bowman's capsule, the renal tubule, and the loop of Henle.

Nerve net   A network of nerve cells extending throughout the body of the hydra and other coelenterates.

Neuron   A nerve cell, specialized for the transmission of impulses from one part of the body to another.

Neurotransmitter   A chemical stimulant, such as acetylcholine, that is secreted by a neuron and causes the transmission of an impulse across a synapse.

Neutron   A particle in the nucleus of an atom that has a neutral charge.

Niche   The specific ecological role of a particular organism in its environment.

Notochord   The cartilage-like supporting rod present along the backs of chordates at some stage of their life history.

Nucleolus   A dense body inside the nucleus of a cell that contains RNA.

Nucleotide   The building block of a nucleic acid, such as DNA or RNA, consisting of a phosphate, a five-carbon sugar, and a nitrogen base.

Nucleus   A large organelle of the cell that contains the chromosomes and one or more nucleoli; also, the central part of an atom, which contains protons and neutrons.

Nutrient   A substance found in food and used by an organism in its metabolism as a source of energy, growth, repair, and maintenance of good health.

Omnivore   An animal that eats both plants and other animals. Examples: rat, human.

Oncogene   A normal gene that may be activated by radiation or a chemical carcinogen to cause cancer.

Oogenesis   The meiotic process by which egg cells are produced with the haploid number of chromosomes.

Organ   A group of tissues working together to perform a specific function in the body of an organism. Examples: liver, kidney.

Organelle   A specialized structure in a cell that performs a specific function. Examples: nucleus, mitochondrion, chloroplast, ribosome.

Organic compound   A chemical compound containing carbon.

Organism   A living thing.

Osmosis   The passage of water molecules, through a semipermeable membrane, from a region of higher concentration to a region of lower concentration.

Ovary   The female reproductive organ that produces egg cells.

Ovule   A structure in the ovary of a flower that contains an egg and develops into a seed after fertilization.

Ovum   A female gamete or sex cell; an egg.

Palisade layer   A group of elongated cells, located under the epidermis of a leaf, that carry on photosynthesis.

Parasitism   A type of symbiosis in which one organism benefits at the expense of its host organism. Example: a tapeworm in a human.

Parthenogenesis   The production of offspring from an unfertilized egg.

Pathogen   A microorganism that causes disease. Example: the malaria germ.

Peptide bond   A chemical bond between amino acids, joining the amino group of one amino acid and the carboxyl group of another.

Peristalsis   The wavelike contractions and expansions of the smooth muscles in the walls of the alimentary canal that move food along during digestion.

Petiole   The thin stalk by which the blade of a leaf is attached to the stem of a plant.

pH   A unit to express the hydrogen-ion concentration, or acidity, of a solution. A pH of 7 is neutral; lower values are increasingly acidic and higher values are increasingly alkaline (basic).

Phenotype   An expressed genetic trait. Example: A pea plant that is heterozygous for tallness (Tt) appears tall.

Pheromone   A chemical substance produced by an animal, such as an insect, that serves to attract individuals of the same species but the opposite sex a long distance away; pheromones may also mark a path to a source of food.

Phloem   Vascular tissue in higher plants through which manufactured food is transported from the leaves to other parts.

Photoperiodism   The flowering response of a plant to the length of day and of night.

Photosynthesis   The process by which green plants and algae use the energy of light to manufacture carbohydrates from carbon dioxide and water; oxygen is given off as a by-product.

Phototropism   The turning response of a plant toward a source of light.

Phylum   In the classification of organisms, a major grouping consisting of one or more classes and belonging to one of the five kingdoms. Examples: Annelida, Echinodermata.

Pinocytosis   The intake of liquids or very small particles by the cell membrane of a cell.

Pistil   The female reproductive organ of a flower, consisting of the ovary, ovules, stigma, and style.

Placenta   The organ in female mammals (except marsupials) through which the developing fetus receives nourishment from the parent and wastes are eliminated; it is composed of tissues from both the mother and the fetus.

Plankton   A floating mass of microscopic organisms at the surface of a body of water.

Plasma   The liquid part of the blood, without its cells and platelets.

Plasmolysis   The shrinkage of the cytoplasm of a cell away from the cell wall when water leaves the cell under the influence of a hypertonic solution.

Polar body   One of three nonfunctional cells produced during meiosis in the ovary, when the egg is being formed.

Pollination   The transfer of pollen from the anther to the stigma of the pistil.

Polypeptide   A chain of amino acids held together by peptide bonds.

Polysaccharide   A complex carbohydrate composed of hundreds of glucose units bonded together. Examples: starch, glycogen, cellulose.

Population   All the members of a species inhabiting a particular location at a given time.

Predator   An organism that eats another organism.

Producer   An organism in a food chain that synthesizes organic compounds. Examples: green plants, algae.

Prokaryote   A simple organism that does not have a nucleus. Examples: bacteria, blue-green algae.

Prophase   The first stage of mitosis or meiosis, in which the chromosomes condense into rodlike structures.

Protein   A complex organic compound consisting of many amino acids.

Protista   In the classification of organisms, the kingdom of simple life forms such as protozoa, brown algae, and slime molds.

Proton   A positively charged particle found in the nucleus of an atom.

Protoplasm   The living material of a cell.

Pseudopod   A temporary cytoplasmic extension of a cell, such as the ameba, that engulfs food particles and is also used for locomotion.

Punctuated equilibrium theory   A theory of evolution that explains the origin of new species as occurring relatively rapidly after a long period of equilibrium.

Pyruvic acid   One of two three-carbon molecules produced from glucose early in the process of cellular respiration.

Receptor   A specialized molecule on a cell's membrane to which an outside substance such as a neurotransmitter or a virus becomes attached; also, an organ of the nervous system, such as the eye or ear, that receives stimuli.

Recessive   Referring to an allele that does not express itself when the dominant allele is present.

Recombinant DNA   A molecule of DNA artificially produced in the laboratory by joining pieces of DNA from different sources.

Reduction   In living reactions, a gain in energy by the addition of hydrogen electrons and the removal of oxygen.

Reflex   An inborn, automatic form of behavior in which a receptor rapidly passes a stimulus through the spinal cord to an effector.

Replication   The exact duplication of a DNA molecule.

Respiration   On the cellular level, the release of energy by the oxidation of glucose; also, the act of breathing, by which O_2 is taken in and CO_2 is given off.

Ribosome   A small organelle in the cytoplasm of a cell that contains RNA and functions in protein synthesis.

RNA (ribonucleic acid)   A single strand of nucleic acid containing a ribose sugar and the nitrogen base uracil; it may take the form of messenger RNA, transfer RNA, or ribosomal RNA, and functions in the production of proteins.

Saprophyte   A heterotrophic organism (nongreen plant, fungus, or bacterium) that lives on dead organic material.

Scion   In grafting, the bud or stem of one plant that is attached to another plant (the stock).

Segregation law   A doctrine stating that, when hybrids are crossed, the alleles separate into different gametes and then recombine during fertilization to produce dominant and recessive offspring in a ratio of 3:1.

Sensory cell   A type of neuron that leads from a receptor to the central nervous system.

Sepals   The outer layer of small green leaves at the base of a flower that collectively comprise the calyx.

Sex chromosomes   The pair of chromosomes that determine the sex of an individual; these chromosomes are designated as XX (female) and XY (male).

Sieve tube   The part of the phloem in a fibrovascular bundle that carries food from the leaves to other parts of a plant.

Sigmoidoscopy   The use, by a doctor, of a thin, lighted instrument to look for evidence of cancer in the rectum and the adjacent portion of the colon.

Species   In the classification of organisms, forms so closely related that they can breed with each other and bear fertile offspring; they descended from a common ancestor with other species that are grouped together with them in a genus.

Sperm   The male gamete, which consists of a head containing the nucleus, and a tail composed of cytoplasm.

Spermatogenesis   The meiotic process by which sperm are produced with the haploid number of chromosomes.

Spindle   Microtubule structures in the cell to which the chromosomes are attached during mitosis and meiosis.

Spore   An asexual reproductive cell in fungi, mosses, and ferns that develops directly into a new organism; also, a special stage of bacteria that can resist unfavorable conditions.

Stamen   The male reproductive organ of a flower; it consists of an anther, which produces pollen, and a filament, which supports the anther.

Stem cells   Cells in the early embryo that have the ability to differentiate into different types of tissue cells, such as those found in neurons, heart, kidney, and liver. Stem cells have been isolated from the placenta after the birth of a baby. Adult stem cells have generated blood cells, brain, and bone tissue, and have also been isolated from fat tissue.

Stigma   The sticky top part of the pistil of a flower, which receives pollen during pollination.

Stock   In grafting, the stem onto which the scion (bud or stem) is attached.

Stomate   A tiny opening in the epidermis of a leaf through which gases are exchanged, and whose opening is controlled by a pair of guard cells.

Style   The stalk of the pistil of a flower that leads from the ovary at one end to the stigma at the other.

Substrate   The substance acted on by an enzyme; also, the base, such as the soil, on which an organism lives.

Succession   The orderly sequence in which communities gradually replace each other in a given area, eventually resulting in a climax community.

Sucrose   A double sugar (disaccharide), $C_{12}H_{22}O_{11}$, such as sucrose or maltose.

Symbiosis   An ecological relationship in which two organisms of different species live intimately together in a close association.

Synapse   The junction between two neurons, across which an impulse passes.

Synapsis   The arrangement of homologous chromosomes in pairs during meiosis.

Taiga   The terrestrial biome that includes northernmost forests of coniferous trees and is characterized by long, severe winters.

Taxis   The movement of an organism toward or away from a stimulus.

Taxonomy   The division of biology that deals with the classification of living things on the basis of similarities in their structure, development, and evolutionary history.

Telophase   The final phase of mitosis or meiosis, during which the chromosomes at each end of the cell collect to form a nucleus and the nuclear membrane forms, just before the cell divides.

Testis   The male sex organ that produces sperm.

Thymine   A nitrogen base that is paired with adenine in DNA.

Thyroxin   A hormone, produced by the thyroid gland, that regulates the rate of metabolism.

Tissue   A group of similar cells that perform a common function.

Toxin   A poison given off by germs that attack an organism.

Trachea   The windpipe; a tube, supported by rings of cartilage, that extends from the pharynx to the bronchi of the lungs in vertebrates.

Transcription   The transfer of genetic information from a DNA molecule to new strands of messenger RNA that pass from the nucleus to the cytoplasm.

Transduction   The transfer of genes by a virus from one type of bacterial cell to another.

Transfer RNA (tRNA)   A type of RNA that transports a specific amino acid molecule from the cytoplasm to the ribosome, where it is assembled into a protein.

Transformation   The transfer of DNA from dead bacteria into living bacteria of another genotype.

Translation   The formation of proteins under the influence of messenger RNA.

Translocation   The movement of materials in plants from one location to another, mostly through phloem; also, the attachment of a chromosomal fragment to another, nonhomologous chromosome.

Transpiration   The release of water vapor from a plant through the stomates of its leaves.

Tropism   The turning or growth of a plant toward or away from a stimulus.

Tundra   The terrestrial biome in the far North, where the underlying part of the soil is permanently frozen.

Turgidity   The condition of plant cells in which they are expanded and full of water.

Uracil   A nitrogen base in the structure of RNA that pairs with adenine; it is replaced by thymine in DNA.

Urea   A nitrogenous waste product of mammals and some other vertebrates, produced when the liver breaks down amino acids in a process called deamination.

Ureter   A duct that carries urine from the kidney to the bladder.

Urethra   A duct that carries urine from the bladder to the exterior of the body.

Uric acid   Insoluble nitrogenous waste produced by insects and other arthropods, reptiles, and birds.

Uterus (Womb)   In mammals, the specialized part of the female reproductive system in which the embryo develops.

Vacuole   A storage sac in the cytoplasm of a cell that is filled with liquid and is surrounded by a membrane.

Ventricle   The chamber of the heart that receives blood when the atrium contracts.

Vertebrate   An animal that has a backbone (spinal column), in which the spinal cord is enclosed.

Villi   Tiny, fingerlike projections on the walls of the small intestines, through which digested food is absorbed.

Virus   A submicroscopic infectious agent, composed of an outer protein coat and a core containing either DNA or RNA, that can reproduce only in living cells.

Vitamin   Any of a group of chemical nutrients that, in small quantities, are essential for normal growth, development, and metabolism.

Womb   *See* **Uterus.**

X chromosome   A sex chromosome. Female mammals have two X chromosomes in each cell; males have one X and one Y chromosome in each cell.

Xerophyte   A plant that lives in dry conditions, such as prevail in a desert. Example: cactus.

Xylem   In plants, woody, vascular tissue that transports water and dissolved minerals up from the roots through the stem to the leaves.

Y chromosome   The male sex chromosome.

Yolk   The stored food, rich in proteins and lipids, in an egg.

Zygote   The single diploid cell resulting from the union of two gametes.

Index

Abiotic factors, 227
Absorption, 37
Acetylcholine, 115, 117, 172
Acid rain, 236–237
Acquired immunity, 140–141
Acquired immunodeficiency syndrome. *See* AIDS
Acromegaly, 105, 152
Actin, 35, 53
Active immunity, 141
Active transport, 36
Adaptive radiation, 220
Addiction, 131–132
Addison's disease, 105, 153
Adenine arabinoside, 145
Adenoids, 94
Adenosine diphosphate, 48
Adenosine triphosphate, 48–49, 61, 73
Adenyl cyclase, 107
Adipose tissue, 54
ADP. *See* Adenosine diphosphate
Adrenal glands, 105, 108
Adrenaline, 105
Adrenocorticotropic hormone, 105
Aedes mosquito, 147
Aerial roots, 67
Aerobic bacteria, 134
Aerobic respiration, 49, 66
Aflatoxin, 151
African sleeping sickness, 139, 147
Agglutinins, 90
Agglutinogens, 90
Aggregates, 210
Aging, 172–173
AIDS, 146, 153
Air pollution, 237
Alcohol, 129–130
    fermentation of, 50, 66
Alcoholics Anonymous, 130
Alcoholism, 130
Algae, 253, 257–258, 259–260
Alimentary canal:
    description of, 80
    hormones of, 103
Allantois, 169
Alleles, 181, 185
Allergies, 153
Alpha waves, 116
Altitude, 232
Alveolus, 94
Alzheimer's disease, 172–173
Amber, 208
Ambulocetus, 213
Ameba:
    behavior of, 112
    description of, 37, 39
    reproduction of, 159
Amino acids, 46, 64, 74, 191
Ammonia, 98
Amniocentesis, 192
Amnion, 169, 170
Amphetamines, 131
Amphibia, 255, 266
Amylase, 69
Amylopsin, 82
Amyotrophic lateral sclerosis, 191
Anaerobic bacteria, 134
Anaerobic respiration, 49–50, 66
Anaphase, 177

Anemia, 75, 89, 152
Angina pectoris, 150
Angiospermae, 256, 270
Animal(s):
    breeding to improve, 204–205
    conservation measures for, 249–250
    extinction of, 242–243
    inheritance in, 183–184
Animal Kingdom, 254–269
Annelida, 254, 264
Annual rings, 56
Annuals, 270
Anopheles mosquito, 139, 146–147
Anoxia, 96
Anther, 162
Anthrax, 134, 144, 153
Anthropology, 223
Antibiotics, 141–142
Antibodies, 140, 152, 153
Anticodon, 197
Antigen, 152, 153, 190
Antiseptic, 141
Antiseptic surgery, 137
Anus, 83, 100
Aorta, 87, 91
Apes, 223
Appendicitis, 83, 89
Appendix, 83
Aquatic biomes, 232
Aqueous humor, 118
Arachnida, 254, 265–266
Archaea, 211, 252
Archaeopteryx, 208–209, 213
Archaeozoic era, 208
Arsenic compounds, 141
Arteries, 87–88
Arterioles, 88
Arteriosclerosis, 54, 87
Arthritis, 54, 105
Arthropoda, 254–255, 265
Artificial insemination, 170, 205
Artificial pollination, 163
Artificial respiration, 95
Asbestos, 150
Ascorbic acid, 76, 78
Aseptic surgery, 137
Asexual reproduction, 159
Assimilation, 37
Aster, 176
Asthma, 96
Atherosclerosis, 149
Athlete's foot, 148
Atoms, 42
ATP. *See* Adenosine triphosphate
Atria, 86
Atrial natriuretic factor, 106–107
Auditory nerve, 118
Auriculin, 87
Australopithecus afarensis, 223
Autoclave, 134
Autonomic nervous system, 117, 120
Autosomes, 190
Autotrophs, 60, 210
Auxins, 70, 107, 112, 163
Axon, 53, 114
Azidothymidine (AZT), 146
Azoic era, 208

B cells, 140
Bacilli, 133, 137, 154, 237
Bacillus Calmette-Guérin, 143
Bacteria:
    aerobic, 134
    anaerobic, 134
    characteristics of, 133–134, 253, 257
    colonies of, 134
    denitrifying, 137
    discovery of, 133
    diseases caused by, 137, 143–144
    flagella, 134
    growing of, 134–135
    harmful types of, 136–137
    helpful types of, 135–136
    microscopic study of, 135
    nitrifying, 136
    nitrogen-fixing, 69, 136, 229
    reproduction of, 134
    size of, 34, 133
    soil, 135
    spore formation, 134
    staining of, 135
    sterilization of, 137
    types of, 133
Bacteriologists, 137
Bacteriophage, 134
Bacteriophages, 194–195
Barbiturates, 131
Bark, 56–57
Basal metabolism, 73
Basal metabolism test, 104
Base, 44
Behavior:
    intelligent, 125–127
    learned, 123–125
    types of, 126–127
    unlearned, 122–123
Bengal tiger, 243
Benign tumor, 150
Beriberi, 76, 152
Beta waves, 116
Bicuspids, 80
Biennials, 270
Bile, 82
Bile duct, 82
Binomial classification, 252
Biochemistry, 41
Biogeography, 231–232
Biological clock, 126
Biology, 252
Biomass, 230
Biome, 231–232
Biomediation, 238
Biosphere, 227
Bioterrorism, 144
Biotin, 76
Birds:
    conservation laws, 250
    reproduction in, 169
Birth, 171–172
Black plague, 139, 153
Bladderwort, 69, 112
Blade, 62
Bleeding, 152
Blood:
    plasma, 89
    platelets, 89–90
    red blood cells, 89

Notes

Notes

THERE'S ONLY ONE PLACE TO TURN FOR TOP SCORES...

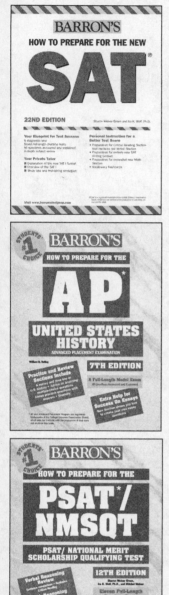

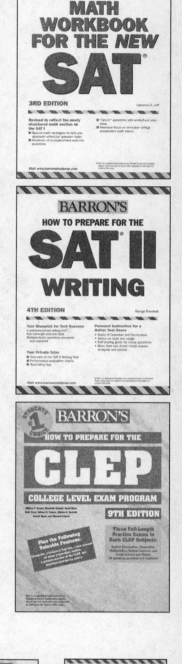

SAT I
How to Prepare for the New SAT, 22nd Ed., *$16.95, Canada $24.50*
Math Workbook for the New SAT, 3rd Ed., *$14.95, Canada $21.95*
Verbal Workbook for the New SAT, 10th Ed., *$12.95, Canada $17.95*
Pass Key to the New SAT, 4th Ed., *$9.95, Canada $13.95*

Hot Words for SAT
$9.95, Canada $14.50

Advanced Placement Examinations in:
Biology, 6th Ed., *$16.95, Canada $24.50*
Calculus, 7th Ed., *$16.95, Canada $23.95*
Calculus, 7th Ed., (Book & CD), *$29.95, Canada $43.50*
Chemistry, 3rd Ed., *$16.95, Canada $23.95*
Computer Science, 2nd Ed., *$16.95, Canada $24.50*
English Language and Composition, *$16.95, Canada $24.50*
English Literature and Composition, *$16.95, Canada $24.50*
Environmental Science Exam, *$16.95, Canada $23.95*
European History, 3rd Ed., *$16.95 Canada $23.95*
French, (Book & 3 CDs), *$24.95, Canada $35.95*
Human Geography, *$16.95, Canada $24.50*
Macroeconomics/Microeconomics, *$15.95, Canada $22.50*
Physics B, 3rd Ed., *$16.95, Canada $24.50*
Physics C, *$16.95, Canada $23.95*
Psychology, *$16.95, Canada $24.50*
Spanish, 3rd, (Book & CDs), *$26.95, Canada $37.95*
Statistics, *$16.95, Canada $24.50*
United States Government & Politics, 3rd, *$16.95, Canada $23.95*
United States History, 7th Ed., *$16.95, Canada $24.50*
United States History, 7th Ed., (Book & CD), *$29.95, Canada $43.50*
World History, *$16.95, Canada $23.95*

PSAT/NMSQT (Preliminary Scholastic Aptitude Test/ National Merit Scholarship Qualifying Test)
How to Prepare for the PSAT/NMSQT, 12th Ed., *$14.95, Canada $21.95*
Pass Key to the PSAT/NMSQT, 3rd Ed., *$8.95, Canada $12.95*

ACT (American College Testing Program Assessment)
How to Prepare for the ACT, 13th Ed., *$14.95, Canada $21.95*
Pass Key to the ACT, 5th Ed., *$8.95, Canada $12.95*

SAT II (Subject Tests) in:
Biology, 13th Ed., *$14.95, Canada $21.00*
Chemistry, 7th Ed., *$14.95, Canada $21.00*
French, (Book & CD), *$19.95, Canada $27.95*
Literature, 2nd Ed., *$14.95, Canada $21.95*
Mathematics IC, 8th Ed., *$13.95, Canada $19.50*
Mathematics IIC, 7th Ed., *$13.95, Canada $19.50*
Physics, 8th Ed., *$14.95, Canada $19.95*
Spanish, 9th Ed., (Book & CD), *$19.95, Canada $27.95*
United States History, 11th Ed., *$14.95, Canada $21.00*
World History, 2nd Ed., *$14.95, Canada $21.00*
Writing, 4th Ed., *$14.95, Canada $21.95*

CLEP (College Level Exam Programs)
How to Prepare for the College Level Exam Program, 9th Ed., (CLEP), *$14.95, Canada $21.00*

COMPUTER STUDY PROGRAMS AVAILABLE!

- SAT I Book with Safari CD-ROM, **$29.95, Canada $41.95**
- ACT Book w/CD-ROM, **$29.95, Canada $43.50**

BARRON'S EDUCATIONAL SERIES, INC.
250 Wireless Boulevard
Hauppauge, NY 11788
In Canada: Georgetown Book Warehouse
34 Armstrong Avenue
Georgetown, Ontario L7G 4R9
www.barronseduc.com

Prices subject to change without notice. Books may be purchased at your local bookstore, or by mail from Barron's. Enclose check or money order for total amount plus sales tax where applicable and 18% for postage and handling (minimum charge $5.95). NY State, Michigan, and California residents add sales tax. All books are paperback editions.

(#4) R 11/04

College-bound students can rely on Barron's for the best in SAT II test preparation...

Every Barron's SAT II test preparation manual contains a diagnostic test and model SAT II tests with answers and explanations. Model tests are similar to the actual SAT II tests in length, format, and degree of difficulty. Manuals also present extensive subject review sections, study tips, and general information on the SAT II. Manuals for foreign language tests, priced slightly higher than the others, come with audiocassettes or CDs that present listening comprehension test sections and additional practice material.
All books are paperback.

SAT II: Biology, 13th Ed.
Maurice Bleifeld
ISBN 0-7641-1788-2, $14.95, *Can$21.00*

SAT II: Chemistry, 7th Ed.
Joseph A. Mascetta
ISBN 0-7641-1666-5, $14.95, *Can$21.00*

SAT II: French
Reneé White
w/CD
ISBN 0-7641-7621-8, $19.95, *Can$27.95*

SAT II: Literature, 2nd Ed.
Christina Myers-Schaffer, M.Ed.
ISBN 0-7641-0769-0, $14.95, *Can$21.95*

SAT II: Math IC, 8th Ed.
James Rizzuto
ISBN 0-7641-0770-4, $13.95, *Can$19.50*

SAT II: Math IIC, 7th Ed.
Howard Dodge
ISBN 0-7641-2019-0, $13.95, *Can$19.50*

SAT II: Physics, 8th Ed.
Herman Gewirtz and Jonathan S. Wolf
ISBN 0-7641-2363-7, $14.95, *Can$21.95*

SAT II: Spanish, 9th Ed.
José Diaz
ISBN 0-7641-7460-6, $19.95, *Can$27.50*

SAT II: United States History, 11th Ed.
David A. Midgley and Philip Lefton
ISBN 0-7641-2023-9, $14.95, *Can$21.00*

SAT II: World History
Marilynn Hitchens and Heidi Roupp
ISBN 0-7641-1385-2, $14.95, *Can$21.00*

SAT II: Writing, 4th Ed.
George Ehrenhaft
ISBN 0-7641-2346-7, $14.95, *Can$21.95*

The publisher's list price in U.S. and Canadian dollars is subject to change without notice. Available at your local bookstore, or order direct, adding 18% for shipping and handling (minimum $5.95). N.Y. State, Michigan, and California residents add sales tax.

Barron's Educational Series, Inc.
250 Wireless Blvd., Hauppauge, NY 11788
Order toll free: 1-800-645-3476
Order by fax: (631) 434-3217

In Canada: Georgetown Book Warehouse
34 Armstrong Ave., Georgetown, Ont. L7G 4R9
Order toll free: 1-800-247-7160
Order by fax toll free: 1-800-887-1594

Visit us on our web site at: www.barronseduc.com

(#87) R11/04

It's Your Personal Online Test Preparation Tutor for the SAT*I, New SAT,* PSAT,* and ACT

BARRON'S TestPrep.com

Log in at:
www.barronstestprep.com

**And discover the most comprehensive
SAT I, New SAT, PSAT, and ACT test preparation available anywhere—**

- Full-length sample SAT I, New SAT, PSAT, and ACT practice exams with automatic online grading.
- Your choice of two test-taking modes:

    Test Mode: Replicates actual test-taking conditions with questions divided into six sections (timed or untimed). Students must complete one section before they can advance to the next.

    Practice Mode: User chooses the particular question type for practice and review— for instance, sentence completion, reading comprehension, quantitative comparisons, etc.—and chooses the number of questions to answer.

- Answers and explanations provided for all test questions.
- Performance evaluation given after each exam taken, and on the total exams taken by any one user. Students can get a comparative ranking against all other web site users.
- Extensive vocabulary list for practice and review.

Plus these great extra features—
- Test-taking advice in all areas covered by the exam • Optional onscreen timer
- Up-to-date SAT I, New SAT, PSAT, and ACT exam schedule
- A personal exam library to evaluate progress
- Latest information on new test formats for 2004 and 2005

tart now with a click of your computer mouse
Just $10.95 per subscription gives you complete preparation in all test areas
Group Subscriptions Available

ll Subscription Information
site at www.barronstestprep.com

BARRON'S

#104 R11/04

ered trademark of the College Entrance Examination Board, which was not involved in
and does not endorse this product.